Topics in Applied Physics
Volume 94

Available online at
SpringerLink.com

Topics in Applied Physics is part of the SpringerLink service. For all customers with standing orders for Topics in Applied Physics we offer the full text in electronic form via SpringerLink free of charge. Please contact your librarian who can receive a password for free access to the full articles by registration at:

springerlink.com → Orders

If you do not have a standing order you can nevertheless browse through the table of contents of the volumes and the abstracts of each article at:

springerlink.com → Browse Publications

There you will also find more information about the series.

Springer
Berlin
Heidelberg
New York
Hong Kong
London
Milan
Paris
Tokyo

Topics in Applied Physics

Topics in Applied Physics is a well-established series of review books, each of which presents a comprehensive survey of a selected topic within the broad area of applied physics. Edited and written by leading research scientists in the field concerned, each volume contains review contributions covering the various aspects of the topic. Together these provide an overview of the state of the art in the respective field, extending from an introduction to the subject right up to the frontiers of contemporary research.

Topics in Applied Physics is addressed to all scientists at universities and in industry who wish to obtain an overview and to keep abreast of advances in applied physics. The series also provides easy but comprehensive access to the fields for newcomers starting research.

Contributions are specially commissioned. The Managing Editors are open to any suggestions for topics coming from the community of applied physicists no matter what the field and encourage prospective editors to approach them with ideas.

Managing Editors

Dr. Claus E. Ascheron

Springer-Verlag Heidelberg
Topics in Applied Physics
Tiergartenstr. 17
69121 Heidelberg
Germany
Email: ascheron@springer.de

Dr. Hans J. Koelsch

Springer-Verlag New York, Inc.
Topics in Applied Physics
175 Fifth Avenue
New York, NY 10010-7858
USA
Email: hkoelsch@springer-ny.com

Assistant Editor

Dr. Werner Skolaut

Springer-Verlag Heidelberg
Topics in Applied Physics
Tiergartenstr. 17
69121 Heidelberg
Germany
Email: skolaut@springer.de

Lorenzo Pavesi
David J. Lockwood (Eds.)

Silicon Photonics

With 262 Figures

Springer

Professor Lorenzo Pavesi
Dipartimento di Fisica
Università di Trento
via Sommarive 14
38050 Povo, Italy
pavesi@science.unitn.it

Professor David J. Lockwood
Institute for Microstructural Sciences
National Research Council
Ottawa, ON K1A 0R6, Canada
david.lockwood@nrc.ca

Library of Congress Cataloging in Publication Data applied for.
Bibliographic information published by Die Deutsche Bibliothek. Die Deutsche Bibliothek lists this publication in the Deutsche Nationalbibliografie; detailed bibliographic data is available in the Internet at http://dnb.ddb.de.

Physics and Astronomy Classification Scheme (PACS):
42.70.-a; 84.40.Ua; 85.60.-q; 42.82.-m; 85.85.+j

ISSN print edition: 0303-4216
ISSN electronic edition: 1437-0859
ISBN 3-540-21022-9 Springer-Verlag Berlin Heidelberg New York

Springer-Verlag is a part of Springer Science+Business Media

springeronline.com

Typesetting: DA-TEX · Gerd Blumenstein · www.da-tex.de
Cover design: *design & production* GmbH, Heidelberg

Printed on acid-free paper 57/3141/tr 5 4 3 2 1 0

To our families:
Anna and Maria Chiara, Matteo, Michele and Tommaso,
Eugenia and Alisa, Ilana, and Lilia.

Preface

Quo vadis?
Henryk Sienkiewicz
(Wola Okrzejska 1846 – Vevey 1916)

Silicon is a material where *the extraordinary is made ordinary.*[1] This statement has been prompted by the rapid (exponential) development of the industry based on manufacturing silicon devices, which has seen enormous performance improvements and cost reductions over the last fifty years. Coming after the stone, iron, and industrial ages, the electronic age commenced with the invention of the transistor in 1947 by J. Bardeen and W. H. Brattain and then became a reality when the first electronic integrated circuit was developed by J. S. Kilby in 1957. The first microprocessor chip (model 4004 with 2300 transistors) was realized by Intel in 1971. During those early days, Gordon Moore coined his famous empirical law: the number of transistors on an integrated circuit would double every 12 months (lately corrected to 18 months).[2] Today we are at the 130 nm feature-size node with hundreds of millions of transistors on a single silicon chip, and the semiconductor industry is on track to have the 65 nm node in production during 2005 by using 300 mm diameter silicon wafers.

In 1999, the semiconductor industry was the largest value-added industry in manufacturing in the USA, almost five times the size of the iron and steel sector that year. The semiconductor industry, based on silicon and the ubiquitous CMOS technology, had a growth rate of about 20% per year over more than three decades until 1996 when there was a significant slow down. It is predicted that the semiconductor industry sales will grow from $163 billion in 2003 to $219 billion in 2006. Communications related sales accounted for almost 23% of the 2003 total, while optoelectronic related devices (lasers, image sensors, etc.) accounted for about 5.5%.[3] The new keywords describing the

[1] H. J. Leamy, J. H. Wernick: *Semiconductor Silicon: The Extraordinary Made Ordinary*, MRS Bull., **22** (1997), pp. 47–55

[2] G. E. Moore: *Cramming more components onto integrated circuits*, Electronics, **38** (1965)

[3] Semiconductor Industry Association, November 2003 forecast

semiconductor industry are cost, performance, and flexibility. Semiconductor industry progress is rooted in its underlying technical structure. Improvements have resulted from parallel advances in processor architecture, design methods and tools, fabrication technology (both new methods and manufacturing technology), and industry infrastructure. A wealth of new materials, processing, and architecture has been introduced in the manufacturing of devices to keep pace with the steady evolution dictated by Moore's law and the increase in chip development costs. Copper interconnects, low-k dielectric materials, Si-Ge alloys, SOI wafers, and strained silicon are all examples of new technologies exploited recently in silicon devices.

Nowadays, we are in the mass communications/information age enabled through the development of the internet. The evolution of the internet and of pervasive computing requires new solutions to enhance their technological foundations. Photonics and electronics are playing a fundament role in this effort. Hence this is an age of golden opportunity to introduce and marry photonics with electronics. Indeed, most of the main semiconductor companies are aggressively addressing the telecommunications, data-communications and optoelectronics markets. The integration of photonics with electronics has two principal driving forces that seemed unimaginable even just a decade ago:[4]

- the requirements for smart optical networks, greater bandwidths, and lower cost are pushing the integration of electronics into photonics
- the interconnect bottleneck for CMOS circuits operating above 10 GHz is pushing the integration of photonics into electronics for timing and possibly signal channels.

To implement these powerful technology drivers, a common platform is required that integrates both optical and electronic functionalities. This then is the next challenge: to turn silicon into *the* photonic material where the convergence between electronics and photonics can be realized, i.e., to make the *extraordinary* happen again in silicon.

This book is aimed at presenting the fascinating picture of the state-of-the-art in silicon photonics and providing perspectives on what can be expected in the near future. In the Chapter by *D. J. Lockwood* and *L. Pavesi* an in-depth introduction to the optical properties of silicon with an emphasis on how it can be turned into a light emitting material is reported. Recent progress towards a silicon laser is also reviewed. The Chapter by *M. Paniccia*, *M. Morse* and *M. Salib* details the view of one major company on the use of silicon in integrated optics. An in-depth comparison between microelectronics and silicon photonics is made. Silicon is weighed against other materials. Possible schemes of convergence between microelectronics and photonics are reported. A coherent monolithical approach is presented in the Chapter by *L. C. Kimerling* et al., where the drivers, barriers, and current solutions for

[4] James S. Harris, Stanford University

the development and deployment of a monolithic silicon microphotonics technology are presented. One of the most promising applications of silicon photonics is to solve the problem related to the interconnection bottleneck. This is detailed in the Chapter by *Z. Gaburro*, where novel technical and architectural solutions based on photonics are reviewed. All the various building blocks necessary to develop silicon photonics are presented. The Chapter by *P. M. Fauchet* discusses one of the most critical issues still to be solved: the lack of a reliable light source, either a light emitting diode or a laser, made entirely from silicon. Recent progress shows that the solution is at hand. The application of stimulated Raman scattering to form in-line optical amplifiers and wavelength converters is reported in the Chapter by *B. Jalali*, *R. Claps*, *D. Dimitropoulos* and *V. Raghunathan*. The Chapter by *H. Zimmermann* is an up-to-date review of state-of-the-art photo-receivers developed with silicon. Light guiding in the silicon chip that can be achieved through the use of silicon-based waveguides is the topic of the Chapter by *R. J. Bozeat* et al. A new approach to light manipulation (guiding, switching, multiplexing, etc.) is achievable by the use of photonic band-gap materials. These are introduced and discussed in the Chapter by *T. Geppert*, *J. Schilling*, *R. B. Wehrspohn* and *U. Gösele*. Photonic crystals will allow the miniaturization of optical devices in silicon on extremely short length scales. Wavelength division multiplexing with silicon waveguides is reported in the Chapter by *S. Janz* and light modulation with silicon devices is presented in the Chapter by *A. Iracc*, *G. Breglio*, *M. Iodice* and *A. Cutolo*.

We foresee numerous future advancements in the field of silicon photonics in the coming years, many of which are already introduced in concept in this book. We thank all of our authors, who not only responded enthusiastically to our invitation but also produced extremely interesting state-of-the-art chapters. The list of colleagues and friends with whom we have shared our enthusiasm over silicon photonics is too great to give all their names here, but we gratefully acknowledge their contributions.

Trento, Ottawa, 6 December 2003

Lorenzo Pavesi
David J. Lockwood

Contents

Silicon Fundamentals for Photonics Applications

David J. Lockwood[1] and Lorenzo Pavesi[2]

[1] Institute for Microstructural Sciences, National Research Council of Canada
Ottawa, ON, Canada K1A 0R6
David.Lockwood@nrc-cnrc.gc.ca

[2] Dipartimento di Fisica, Università di Trento
Via Sommarive 14, 38050-Povo Trento, Italy
pavesi@science.unitn.it

Abstract. The many and diverse approaches to materials science problems have greatly enhanced our ability to engineer the physical properties of semiconductors. Silicon, of all semiconductors, underpins nearly all microelectronics today and will continue to do so for some time to come. However, in optoelectronics and photonics, the severe disadvantage of an indirect band-gap and of a negligible electro-optic coefficient has limited the application of elemental silicon. This chapter provides introductory material on the physical properties of silicon and outlines a number of diverse approaches to engineering efficient photonic components from silicon. The common paradigm is material compatibility with microelectronic components.

1 Introduction

The ubiquitous silicon microelectronic "chip" is taken for granted in modern society. There has been much research involved in producing these high technology marvels and such research continues unabated at a faster and faster pace. Continued developments in Si and, more recently, $Si_{1-x}Ge_x$ alloy and strained silicon technology [1, 2] continue to advance the frontiers of device integration, complexity, and speed. This continued advance has been driven by application requirements in switching technology (e.g., computers) and high-speed electronics (e.g., wireless telecommunications). Today the thrust is towards pervasive computing: the end-user will use all the power of a computer without noticing that the device he is using has a computer inside. Other compound semiconductor materials, such as GaAs or InP or III–V alloys, have, however, maintained a significant role in the construction of optoelectronic and purely photonic devices [3]. Photonics is the technology associated with signal generation, processing, transmission and detection where the signal is carried by photons (i.e., light); principal photonic devices are lasers, waveguides, modulators, detectors, and optical fibers [4].

If one compares today's photonics industry (worth almost 9 b$) with the microelectronics one (worth almost 160 b$) there are many differences:

L. Pavesi, D. J. Lockwood (Eds.): Silicon Photonics, Topics Appl. Phys. **94**, 1–50 (2004)

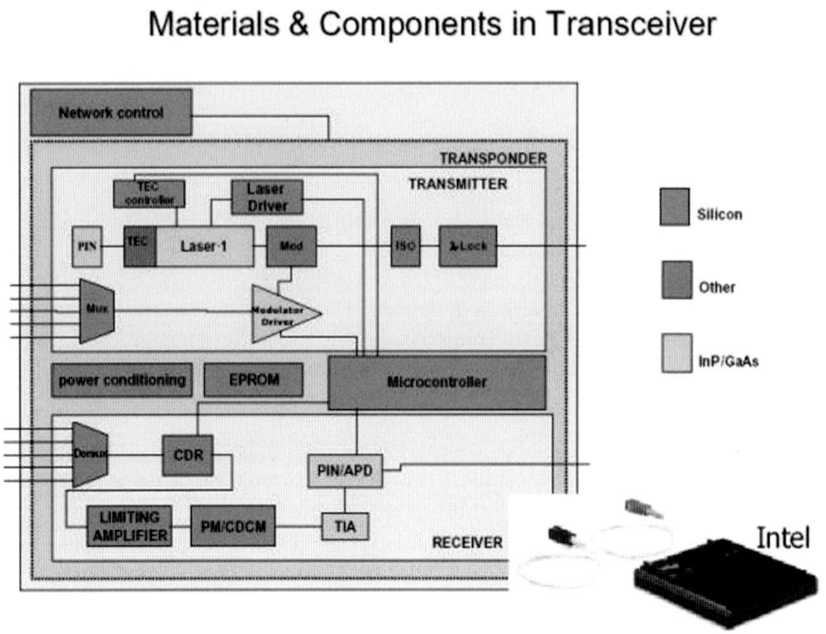

Fig. 1. Materials and components in an optoelectronic transceiver. (Source: Intel ©)

- A variety of different materials is used: e.g., the InP substrate for source development, silica as the material for fibers, lithium niobate for modulators, other materials for DWDM components and EDFA amplifiers, and so on (see Fig. 1).
- No single material or single technology is leading the market. Some convergence is appearing towards the use of InP as the substrate material to integrate different optical functions. However, other materials are potential candidates for overcoming the limitations of InP.
- The industry is characterized by many different small companies that are specialized in producing specific devices: e.g., lasers, modulators, etc. No large companies are dominating this field at present.
- The production technology is still very primitive. Chip scale integration of optical components, which enables low cost and high reproducibility, has not yet been achieved. Neither the standardization of processes nor of the packaging of optical components, which is inherent for mass production and repeatability, are present.
- Roadmaps to dictate and forecast the evolution of photonics are only now being elaborated. It is commonly accepted that the industrial model of microelectronics, if applied to photonics, will give a substantial boost to the development and implementation of photonics. All the big manufactur-

ers of microelectronics have aggressive programs to develop microphotonic devices, mostly based on silicon.

The merging of Si-based electronics with photonics has largely required the pursuit of hybrid technologies for light emitters and modulators (see, for example, [5]), which are often both expensive and complicated to produce. Given that Si light emitters and, in particular, lasers, are not yet commercially available for on-chip optoelectronic applications, considerable effort is being placed on marrying dissimilar materials to fashion hybrid devices utilizing Si microelectronics [5, 6]. For example, monolithically integrated GaAs efficient light emitting diode (LED) arrays with Si driver circuits have been created [7], and integrated photonic circuits where active III-V lasers and diodes are merged with passive silicon based optical components are commercially available [8]. However, the main interest is in combining III–V semiconductor laser diodes with Si integrated circuits for optical fiber communications or optical interconnects. The main issue to be faced is the potential enormous market of fiber-to-home applications. This requires the growth of III–V materials such as GaAs or InP on Si followed by processing, or the direct bonding of III–V laser devices detached from their substrates via an epitaxial lift-off process [9, 10]. Both methods have their disadvantages.

The lattice mismatch of 4% between GaAs and Si, the different thermal expansion coefficients and the fact that Si is a fast diffusing impurity for III–V semiconductors create severe difficulties in maintaining the required low defect density in GaAs for laser production. dislocations produced by relaxation of the GaAs epitaxial layer are detrimental to device performance and life. Various methods are being tried to alleviate this problem [11] including Ge, superlattice, or graded buffer layers between Si and GaAs, but other difficulties arise from the high GaAs growth temperatures. Despite all these difficulties, a few industrial companies have claimed recently the invention of heteroepitaxial procedures that could potentially boost the development of hybrid optoelectronic integration. Nitronex has pioneered a proprietary technology called "Pendeoepitaxy" that grows GaN crystals over SiC substrates or over Si substrates after the growth of a SiC buffer layer. A reduction in defect density of many orders of magnitude relative to other growth techniques can be obtained. This technology enables the integration of high quality GaN with Si [12]. *Motorola Labs* have developed a technique based on a Strontium titanate (STO) buffer layer. By growing thick STO layers, oxygen atoms diffuse through STO and locate at the STO/Si interface, forming an amorphous interface between crystalline Si and crystalline STO. The STO layer relaxes and allows the successive epitaxial growth of GaAs layers [13]. These new techniques have just been developed and thus it is impossible to tell yet how Si optoelectronics will be affected by them.

The epitaxial lift-off technique is more straightforward, involving wet chemical etching of a release layer, floating off the III–V heterostructure, and transferring it to a planar Si substrate, where it bonds via the van der

Waals force [9, 10]. It is essential for this bonding technique to work that the substrate surface be chemically clean and free of particles.

Wafer fusion involves bringing the two materials of high quality into intimate contact under hydrogen ambient at temperatures of around 450 °C. Subject to a uniform directly-applied pressure, the substrates form robust chemical bonds over times of the order of 30 min. Selective chemical etching allows for the removal of one of the substrates, resulting in a thin film of one material bonded to a substrate of another, and the interface between them is quite free of threading dislocations. Wafer fusion has been demonstrated in many materials systems, although from the point of view of devices the technique is still in its infancy. The usual approach comprises bonding composite semiconductor films onto thicker Si substrates. For example, optoelectronic elements such as InGaAs/GaAs strained quantum well lasers [14] and InP lasers [15] have been bonded to Si. GaP/Si wafer-bonded heterojunctions have recently been demonstrated and suggested for optoelectronics [16].

A different approach has been suggested by *Fonstad* et al. [17]. They point out that optoelectronic devices are intrinsically very thick devices (typically at least a micrometer thick), and thus, if they are bonded on a Si substrate in the usual approach, they are very sensitive to stress. On the other hand, Si MOS transistors need only be a few tens of nanometers thick and are much less stress sensitive. Therefore, they suggest retaining a GaAs substrate so that the intrinsically thick, inherently strain- and defect-sensitive optoelectronic devices see their optimum substrate (i.e., GaAs). The SOI technique would be a possibility to produce Si CMOS electronics on GaAs substrates without sacrificing any of the performance of the CMOS. They term this technology silicon-on-gallium arsenide (SonG). The SonG process combines Si and GaAs substrates by wafer bonding.

Other problems are associated with the change in crystal structure from non-polar Si to polar GaAs. A non-planar Si growth surface can lead to stacking faults in the GaAs and there is charge build-up at the Si-GaAs interface [10]. The 8% lattice mismatch for InP on Si is even worse than for GaAs leading to even more severe heteroepitaxy growth problems. Even so, by use of a thick GaAs buffer layer, a 1.54 μm wavelength InGaAs/InGaAsP multiple quantum well laser operating continuously at room temperature has been produced on a Si substrate [18]. Recently InAs quantum dot lasers have been directly grown on Si [19]. In addition the use of Si as a substrate for III–N blue lasers is becoming more and more important [20]. Further significant progress in many of these areas can be anticipated within the next few years, and commercial devices will follow once the scaling up of these techniques to the mass production level has been achieved.

The most satisfactory solution would be optoelectronic and photonic devices created entirely from Si-based materials, where the extensive experience in Si fabrication and processing could be put to best use. The major deficiency in Si-based optoelectronic devices remains the lack of suitable light

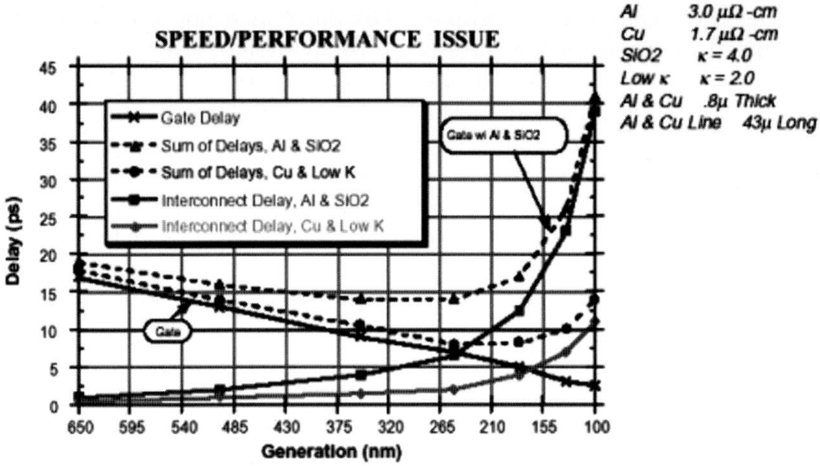

Fig. 2. Calculated gate delay and wire delay as a function of the minimum feature size (device generation). From SIA Roadmap 1997 [27]

emitters and especially lasers. Research in this field is rapidly progressing and amongst the most recent results are the observation of net optical gain in silicon nanocrystals [21], and the demonstration of an LED with a power efficiency in excess of 1% in porous silicon (PS) [22] or more than 20% in Er-doped silicon rich oxide. Some companies are claiming the availability of high-efficiency infrared Si diodes [23], the first release of a commercial device based on an all-silicon opto-coupler [24], and the measurement of net optical gain in semiconductor grade silicon [25].

The general requirements in Si-based light sources are LEDs, lasers, and optical amplifiers for use in electronic devices, optical devices and displays. Operating wavelengths in the range 0.4 μm–1.6 μm are needed to cover both full-color displays and the transparency windows of optical fibers at the operating wavelengths of 1.3 μm and 1.55 μm. Specific applications for such sources include fiber-optic transmitters, optical interconnects within and between computer chips and boards, optical controllers for phased-array microwave antennas, information display screens, printing elements in xerography, and writing and readout of optical compact disc information.

Especially important are applications of optical interconnects related to the limitations of the operating speed of microelectronic devices created by the interconnect bottleneck [26]. Figure 2 shows the signal delay as a function of the transistor generation [27]. For a gate length shorter than 200 nm, a situation is reached where the delay is no longer dictated by the gate switching time but by the wiring delay. In addition, the length of interconnects on a single chip are getting longer and longer. Nowadays chips have a total interconnection length per unit area of the chip of some 5 km/cm^2 with a chip area of 450 mm^2 while in ten years from now these lengths will become

20 km/cm^2 for a chip area of 800 mm^2. The problem is not only related to the length of interconnect but also to the complexity of their architecture. Nowadays, there are 6 levels of metal layers, while in ten years from now they will become greater than 12. All these facts introduce problems related to the delay in signal propagation caused by resistance-capacitive coupling, signal latency, signal cross-talk, and resistance-inductive delays due to the reduction in dimensions and increase in density of the metal lines. Finally, there is the problem of power consumption and heat dissipation, which causes the thermal power generated within the chip to reach a density in excess of a conventional hot-plate and within few years to reach the level typical of nuclear reactors. A possible solution to these problems is looked for in optical interconnects [28]. Nowadays optical interconnects utilizing optical fibers and III–V laser sources are already used to connect different computers or peripherals. It is predicted that optical interconnects will be used to connect computer boards in five years, while optical interconnects within the chip will be used in ten to fifteen years from now [29]. The need for optical interconnects is one of the main motivations to develop silicon photonics.

To alleviate the miserable light emission in bulk Si ($\sim 10^{-6}$ quantum efficiency at 300 K in standard electronic grade silicon, where the quantum efficiency is defined as the ratio of the number of photons generated over the number of excited electron–hole pairs), many quite different approaches have been proposed and are actively being explored [30]. Some, such as $Si_{1-x}Ge_x$ quantum well or Si/Ge superlattice structures, rely on band structure engineering, while others rely on quantum confinement effects in low dimensional structures, as typified by quantum dots or PS. Still another approach is impurity-mediated luminescence from, for example, isoelectronic substitution or by the addition of rare earth ions. A brief overview of results obtained with these and other methods is given below. However, in order to understand more fully the reasons why such different approaches are necessary, it is appropriate to review first what creates the optical emission problem in crystalline Si (c-Si).

2 Physical Properties of Silicon

Silicon crystallizes in the diamond structure [31], which consists of two interpenetrating face-centered cubic lattices displaced from each other by one quarter of the body diagonal. In zinc-blende semiconductors such as GaAs, the Ga and As atoms lie on separate sublattices, and thus the inversion symmetry of Si is lost in polar III–V binary compounds. The difference in the crystal structures underlies the disparate electronic and optical properties of Si and GaAs and results in the slow optical response of Si together with its small electro-optic coefficient. The energy band structure in semiconductors is derived from the relationship between the energy and momentum of a charge carrier, which depends not only on the crystal structure but also on the bonding between atoms, the respective bond lengths, and the chemical species. The band structure is often quite complex and can only be calculated empirically. The results of such calculations [32] for Si and GaAs are shown in Fig. 3. The figure shows the dispersion relations for the energy $E(\boldsymbol{k})$ of an electron (positive energy) or hole (negative energy) for wave vectors $\boldsymbol{k}$ within the first Brillouin zone.

The valence band structure is generally similar for many semiconductors and exhibits a maximum at the Brillouin zone center or Γ point (i.e., at $\boldsymbol{k} = 0$). The notable difference between Si and GaAs is that the degeneracy in the $\Gamma_{25'}$ band maximum at $\boldsymbol{k} = 0$ is removed in the case of GaAs, because of the spin-orbit interaction, into Γ_7 and Γ_8 sub-bands. In general, $E(\boldsymbol{k})$ has maxima or minima at zone center and zone boundary symmetry points, but additional extrema may occur at other points in the Brillouin zone (see Fig. 3). In the case of Si, the lowest point in the conduction band occurs away from high symmetry points near the X point at the Brillouin zone boundary (along $\langle 001 \rangle$), whereas in GaAs it occurs at the Γ point. The energy gap in a semiconductor is defined as the separation between this absolute conduction band minimum and the valence band maximum at the Γ point. For GaAs, the energy gap is classified as direct, because the maximum and minimum occur at the same $\boldsymbol{k}$ and an electronic transition can occur directly at $\boldsymbol{k} = 0$ between initial and final states having the same wave vector. Correspondingly, Si is termed an indirect gap semiconductor, because the initial and final states have different wave vectors.

Optical transitions conserve both energy and momentum. In direct gap GaAs, an excited electron at the bottom of the conduction band can relax spontaneously back into a hole in the valence band by emitting a photon at the band-gap energy with momentum conservation (the photon momentum is negligible with respect to the electronic one). This electron–hole radiative recombination process can only occur in Si with the assistance of a further process to conserve the momentum. This, in pure Si, occurs via the transfer of the electron momentum to a phonon that is created with equal and opposite wave vector to that of the initial state in the conduction band. Such a three-body process is quite inefficient compared with direct gap recombination [33, 34]. Thus, the probability of spontaneous emission is very low for Si and high for GaAs. Alternatively, by relating the recombination probability to a characteristic recombination lifetime, the spontaneous emission or radiative lifetime in silicon is very long (millisecond range) while in direct gap III–V semiconductors it is short (nanosecond range). In general, when excited carriers relax, non-radiative recombination pathways compete with radiative recombination. This competition is strongly thrown off balance for normal Si, because of the long radiative lifetime, and hence non-radiative recombinations are dominant which in turn cause the very low luminescence efficiency. However, when high purity and surface passivated silicon is used,

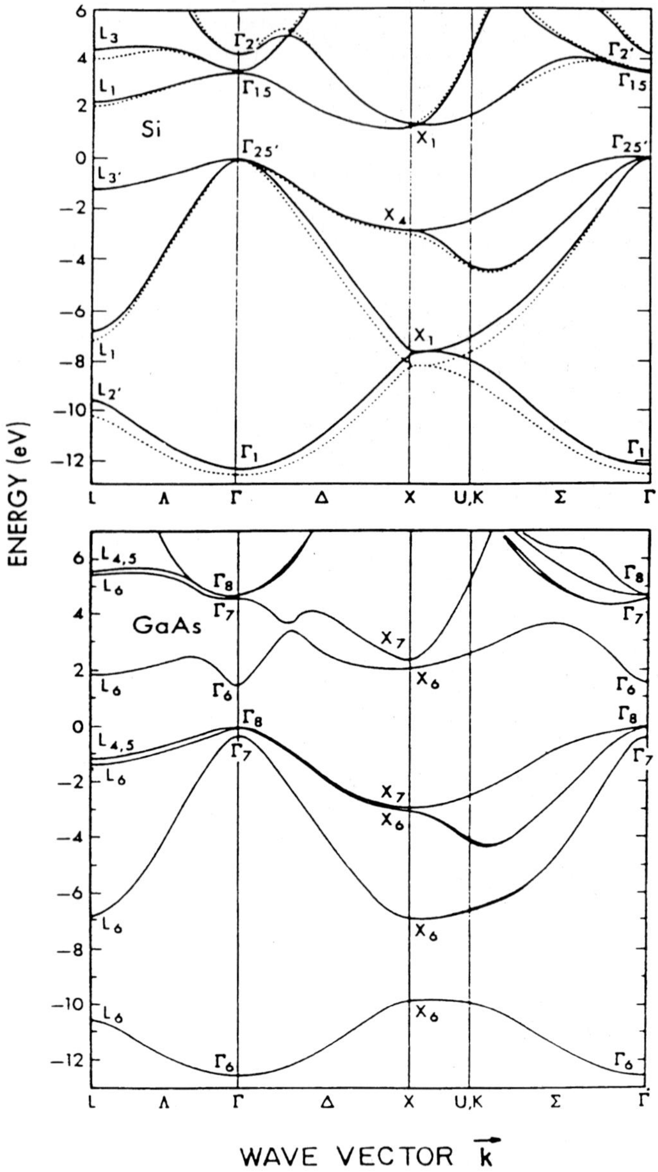

Fig. 3. Theoretical band structures of Si and GaAs. In the case of Si, results are shown for nonlocal (*solid line*) and local (*dashed line*) pseudopotential calculations. After [32]

high luminescence efficiency can be achieved although still with a very long recombination time [35].

The indirect band-gap of Si is 1.12 eV at room temperature (see Fig. 3). The weak band-to-band emission at this energy in the near infrared was first observed by *Haynes* and *Briggs* [36] using visible light excitation or by forward-biasing Si diodes. A review of the work on intrinsic and extrinsic radiative recombination in Si has been given by *Dean* [37] and more recently by *Davies* [38].

Electron–hole pairs created either optically or electrically in Si may bind to each other to form excitons, which can be either free or tied to impurities or defects [33,34]. Exciton recombination dominates the optical emission process at low temperatures and is characterized by very narrow emission lines. At high temperatures, however, excitons are thermally dissociated and emission is due to direct band-to-band recombination. Excitonic emission is under active investigation in quantum well, wire, and dot structures corresponding to carrier confinement in one, two and three dimensions, respectively [39].

Materials engineering, a relatively new phenomenon in materials science, is now being actively applied to Si in an attempt to overcome the indirect band-gap limitations in light emission from Si. In these various attempts, the aims are:

- to increase the efficiency of the luminescence by increasing the overlap of the electron and hole wavefunctions via, for example, quantum confinement and band structure engineering or by spatially confining the exciton in a defect-free region so that nonradiative recombination lifetimes become very long,
- to tune the wavelength of the emission by forming alloys, molecules, and clusters,
- to induce recombination at impurity centers, or
- to use intraband transitions (e.g., within the valence band) where the indirect band-gap of Si does not play any further role.

Such attempts can often involve several of these factors. Each of these methods is outlined below.

3 Band-Gap Engineering

3.1 Brillouin Zone Folding in Atomic Layer Superlattices

Some 25 years ago *Gnutzmann* and *Clausecker* [40] conjectured that Brillouin zone folding in thin layer superlattices where the layer thicknesses were of the order of the unit cell dimensions could result in a direct (or quasi-direct, as it is now termed) band-gap structure. The growth in the 1980s of high quality $(Si_mGe_n)_p$ atomic layer superlattices (m and n are the number of monolayers of Si and Ge in each period and p is the number of periods) by molecular

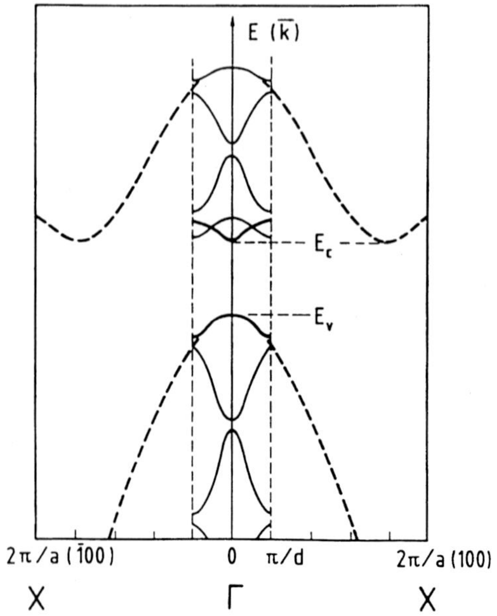

Fig. 4. Schematic representation of the Brillouin zone folding concept in Si resulting from the new superlattice periodicity in the growth direction. Here, the conduction band minimum along the X direction is folded back into the Γ point when the superlattice period is about 10 monolayers of Si. From [1]

beam epitaxy [1] led impetus to this concept, which was revisited by *Jackson* and *People* [41] in 1986 and, subsequently, by a number of other theoreticians. The essence of the idea is conveyed in Fig. 4. The new superlattice periodicity d along the growth direction results in a smaller Brillouin zone of size $\pm\pi/d$ compared with that of the original lattice ($\pm 2\pi/a$, where a is the lattice constant). The electronic band structure is then folded back into this new reduced Brillouin zone. For this simple model, it is apparent that the minimum in the conduction band in bulk Si is folded into the Brillouin zone center for $d \approx 5a/2$, which corresponds to 10 monolayers of Si, and a direct gap is evident. It is relevant that the superlattice zone folding occurs only along the growth direction and, in practice, strains within a Si_mGe_n superlattice together with the band offsets at the heterointerfaces compromise this naive picture (see, for example [1] and [42,43,44,45]). Theory has shown that for certain superlattice periods and when the Si layers are strained a direct energy gap is expected in Si_mGe_n superlattices, but the transition probability is still several orders of magnitude below that of GaAs because the electronic properties of the band-edge states are not sensibly modified by the zone-folding.

The first experimental evidence of modifications to the Si and Ge band structures in such superlattices was obtained from electroreflectance measurements of Si_4Ge_4 superlattices grown on (001) Si [46]. However, it was not until later on when strain-symmetrized Si_mGe_n superlattices were grown on strain-relaxed thick $Si_{1-x}Ge_x$ alloy buffer layers on Si that first indications of the predicted photoluminescence (PL) intensity enhancement and reduced energy gap were obtained [1,47]. Improvements in crystal growth conditions subsequently led to a positive identification of these new features [48]. Further details on the work on band-gaps and light emission in Si/Ge atomic layer structures are given in the review by *Abstreiter* [49].

Although infrared emission can readily be obtained at low temperature from such Si_mGe_n structures at energies useful for fiber-optic transmission work, the PL and electroluminescence (EL) is essentially quenched at room temperature due to exciton dissociation [50, 51]. Some promising steps towards room temperature EL structures have been reported [52, 53] and the development of undulating or wavy Si/Ge superlattice structures has allowed considerable improvements in PL tunability in the wavelength range 1.3 μm–1.55 μm combined with greater mechanical stability [49,54]. Nevertheless, the infrared emission is still weak and more recent results for Si/Ge wavy superlattice p–i–n diodes report an internal quantum efficiency of only $\sim 10^{-5}$ at 300 K [55]. Thus, from a device point of view, unless there are further major improvements in material quality it is most likely that these atomic layer superlattices will find eventual use as infrared detectors rather than as emitters [56, 57].

Quantum cascade and Terahertz (THz) emission structures both utilize Si/Ge superlattices, but they employ intraband transitions rather than the interband transitions of interest here and will be discussed later in Sect. 7.

3.2 Band Structure Engineering via Alloying

Alloying of Ge or C with Si allows engineering of the electronic band structure, where the energy gap may be varied with alloy composition and strain [57,58]. This is shown, for example, in Fig. 5 for strained $Si_{1-x}Ge_x$ on Si, where the tunability range is appropriate for fiber optic communications. Unfortunately, because of heterostructure stability limitations, the $Si_{1-x}Ge_x$ layer thickness must be kept below the critical thickness, which is defined as the maximum thickness for defect-free strained growth of the layer. Larger thicknesses imply relaxation of the layer due to the formation of dislocations and other lattice defects. The $Si_{1-x}Ge_x$ critical thickness decreases rapidly with increasing x (see Fig. 5). Thus the absorbing/emitting regions in infrared detectors/emitters are necessarily thin, although wavy $Si_{1-x}Ge_x$/Si superlattice structures may alleviate this restriction [54]. Also, and most significantly, the band-gap remains indirect. Despite these severe limitations much research has been carried out on the optical properties of Si/$Si_{1-x}Ge_x$ heterostructures [57], which exhibit type I band alignment [58,59], and, to a lesser extent, on Si/$Si_{1-x}C_x$ or even Si/$Si_{1-x-y}Ge_xC_y$ [60, 61, 62, 63].

Electroluminescence and PL have been observed from $Si_{1-x}Ge_x$ in both single layer and superlattice form with increased intensity compared with Si. The luminescence energy tracks the alloy composition dependence shown in Fig. 5, but at a lower energy [64, 65, 66, 67]. The recombination mechanism varies depending on the alloy layer thickness and crystalline quality, resulting

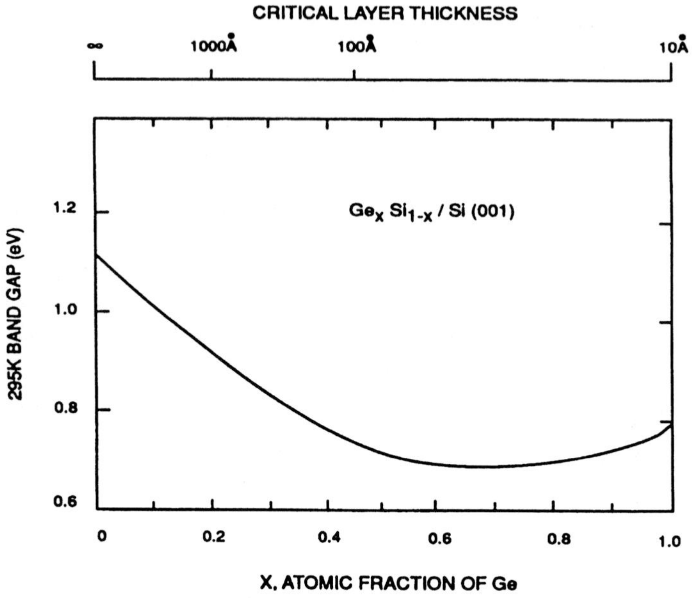

Fig. 5. The band-gap at room temperature of strained $Si_{1-x}Ge_x$ on Si. Also shown is the critical layer thickness as a function of x. From [57]

in near band edge and/or excitonic luminescence [68, 69, 70]. In the earliest work, the EL from $Si_{1-x}Ge_x$/Si p–n diodes was quenched by increasing the temperature above 80 K [64], but EL was soon reported at temperatures up to 220 K in p–i–n diode structures [71]. Progress in materials quality and device design has continued to improve EL device performance (see, for example, [72, 73, 74, 75, 76]) such that room temperature EL has now been reported at wavelengths near 1.3 μm [72, 76]. The major problem with such devices for practical purposes at present is their low efficiency at room temperature, which is again due to exciton thermal dissociation [72, 76]. For example, $Si_{1-x}Ge_x$/Si p–i–n diodes, which were grown on patterned substrates to optimize the unrelaxed alloy layer thickness [77], resulted in an internal quantum efficiency of just $\sim 10^{-4}$ for the EL at 300 K [55].

By alloying C with $Si_{1-x}Ge_x$ it was hoped that the compressive strain in $Si_{1-x}Ge_x$ layers on Si would be offset by the tensile strain exhibited by $Si_{1-x}C_x$ layers on Si and thereby permit relatively thicker strained alloy layers on Si with an improved conduction band offset [63]. Indeed, such strain compensation does occur in $Si_{1-x-y}Ge_xC_y$ layers on Si, but the range of compensation is limited by the low solubility of C in Si and also the small size of the C atom introduces severe distortions in the Si lattice near C sites. In addition, the alloy retains an indirect band-gap and only weak PL is obtained [60]. Much more work remains to be done on this ternary alloy system before its advantages over $Si_{1-x}Ge_x$ can be exploited.

4 Quantum Confinement

Research on the quantum confinement of carriers in silicon-based nanostructures including porous silicon (PS), nanoclusters, and quantum wells, wires, and dots forms a large part of the work on light emission in Si. Much of this work was stimulated by the discovery of bright visible light emission at room temperature in PS reported in 1990 [78]. The interest in nanostructures of Si stems from the effects of confinement on carrier wave functions when the crystallite diameter is less than the size of the free exciton Bohr radius of 4.3 nm in bulk c-Si [39]. The quantum confinement increases the electron–hole wave function overlap, resulting in increased light emission efficiency, and shifts the emission peak to higher energy [33, 79]. In addition, the probability of finding a non-radiative recombination center in these Si quantum dots decreases rapidly with decreasing dot size. When a non-radiative recombination center is present in the dot, exciton recombine non-radiatively and the dot is dark with respect to luminescence. When there is no non-radiative recombination center, the dot is bright and the excited carriers recombine radiatively (i.e., the single dot has almost a 100% quantum efficiency). It is the statistical distribution of dark and bright dots that then determines the overall luminescence efficiency. Last but not least, the fact that the Si dots are usually formed in a dielectric matrix having a lower refractive index than that of c-Si increases the extraction efficiency of the light generated in the active material itself.

4.1 Porous Silicon

Porous silicon was discovered over 35 years ago by *Uhlir* [80]. The porous material is created by a partial electrochemical dissolution of Si wafers in HF-based electrolytes. Hydrofluoric acid, on its own, etches single-crystal Si extremely slowly, at a rate of only nanometers per hour. However, passing an electric current between the acid electrolyte and the Si sample speeds up the process considerably, leaving an array of deep narrow pores. Pores with diameters from nanometers up to micrometers and micrometers deep have been achieved both in disordered or ordered arrangements under specific etching conditions (see Fig. 6).

In July 1989, *Canham* conceived the idea of fabricating Si quantum wires in PS by reverting to the much slower chemical HF etch after electrochemically etching c-Si. In this way Canham proposed to join up the pores, leaving behind an irregular array of undulating freestanding pillars of c-Si only a few nanometers wide. In 1990, *Canham* [78] observed intense visible PL at room temperature from PS that had been etched under carefully controlled conditions. Visible luminescence ranging from green to red in color was soon reported for other PS samples and ascribed to quantum size effects in wires of width ~ 3 nm [78, 81]. Independently, *Lehmann* and *Gösele* [82] reported on the optical absorption properties of PS. They observed a shift

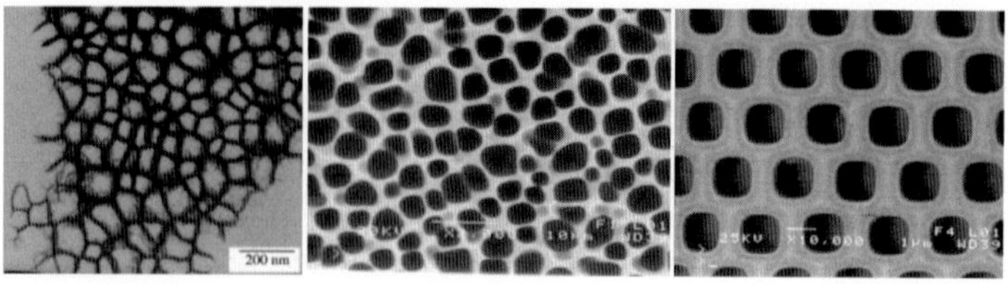

Fig. 6. Various images of PS samples obtained under different etching conditions. Pore sizes (from *left* to *right*) vary from some 100 nm to several micrometers and to 1 μm when arranged in an ordered array

in the bulk Si absorption edge to values as high as 1.76 eV that they also attributed to quantum wire formation. Tremendous activity on research into the physical and associated chemical characteristics of PS has ensued from these early reports. It is impossible to mention all of this work here and interested readers are directed to a number of reviews and books about PS [83, 84, 85, 86, 87, 88, 89, 90, 91, 92, 93, 94, 95, 96] for further aspects of this work.

The most widely studied PL occurs in the far-red to orange-yellow wavelength region, which we shall denote simply as the "red" PL. Interest in red PL stems from the fact that it is the only one that can be electrically excited. This PL shifts to shorter wavelength with increasing chemical dissolution time or increasing anodization current density. It was soon found that much smaller immersion times were required to produce noticeable blue shifts when the chemical dissolution was carried out in the presence of light. The spectra also show a blue shift with increasing anodization current density. As the porosity (defined as the void density) of PS increases with increasing anodization current density, the behavior of the red PL spectra qualitatively reflects the differences in sample porosity and hence in the dimensions of Si nanocrystals within PS (a larger void density is associated with a smaller c-Si skeleton size). The blue shift of the PL and optical absorption with increasing porosity provided the first important evidence that quantum confinement plays a significant role in PS, which has since been confirmed by the observation of the splitting of the luminescent states, the polarization of the PL, and the phonon replicas seen in the PL under resonant excitation [95]. The dynamics of the emission is characterized by a room temperature decay of hundred of microseconds, while lifetimes of some milliseconds were observed at low temperature. The temperature evolution of the lifetimes has been interpreted as due to a thermal balance between dipole allowed recombination from higher energy singlet states and dipole forbidden recombination from lower energy triplet states. The PL peak wavelength and intensity are sensitive to the surface chemistry of PS, particularly with regard to the relative amounts of hydrogen and oxygen on the surface. Thus, besides the quantum confinement mechanism, various surface state and defect models have been invoked to explain the various other results obtained [85, 95, 97].

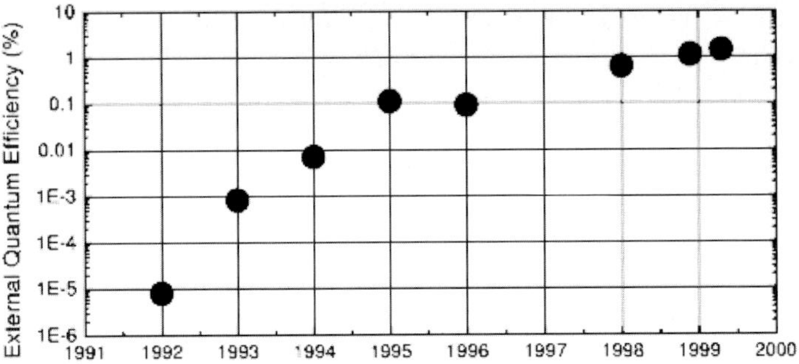

Fig. 7. External quantum efficiency of porous silicon LEDs over the years. The record in efficiency to date is that of [22]

Oxidation of the PS surface has been shown to produce blue PL [85]. The blue PL is quite weak in as-prepared PS, becomes intense only after strong oxidation and has a nanosecond decay time. Models currently under consideration include band-to-band recombination in Si nanocrystals, emission from oxide, and emission due to surface states [96].

Blue PL has been observed also in oxide-free PS simply by decreasing the size of the Si crystallites in accordance with the quantum confinement mechanism [97, 98]. In fact, the PL can be tuned from the c-Si band-gap of 1.1 eV all the way up to $\sim$ 3 eV by a judicious choice of the porosity in unoxidized PS [97, 98, 99].

The near-infrared PL [85] at $\sim$ 0.8 eV (below the bulk Si band-gap) exhibits complex nonexponential dynamics, with a wide distribution of decay times, and has been assigned to deep level transitions associated with dangling bonds on the surface of the Si nanocrystals [100].

From these considerations it is apparent that PL in PS is very sensitive to the chemistry of PS production and treatment. Si wires or nanocrystals and amorphous Si (a-Si) material, or any combination of them, may be formed in a given sample. The PS layers thus formed may be far from uniform, which adds to the difficulties in analyzing their optical properties. Other light emitting species may also be formed on the surfaces of the anodized and otherwise chemically treated Si.

Despite all these disadvantages, the ease of production of PS and the facts that the room temperature PL is very efficient (1%–10% external quantum efficiency [93, 94]) and that it is tunable through blue to near infrared wavelengths have led to impressive efforts to produce practical room-temperature devices. The evolution of PS LED performance over the years is reported in Fig. 5 [22]. The PS approach has, however, a draw-back in the high reactivity of the sponge-like texture, which causes rapid ageing of the LED and uncontrollable variations of the LED performances with time [96]. This major disadvantage has recently been overcome with an appropriate passivation of the PS surface although at the cost of a lower LED efficiency [101].

The latest red LEDs have exhibited external quantum efficiencies greater than 1% and lifetimes of the order of months [22,102]. Such LEDs have been integrated into Si microelectronic circuits to provide an addressable LED display [103]. However, improvements in efficiency and power dissipation are necessary for display applications, while an increased modulation frequency (presently $\sim 1\,\mathrm{MHz}$ [104]) is required for optical interconnects. One way to improve the EL efficiency, to narrow the band width, to improve the directionality, and to increase the long-term stability is to insert the LED into a PS resonant cavity [105]. Even so, the long switching times observed in present PS LEDs may yet prove to be an Achilles' heel in optoelectronic applications.

A large optically induced absorption change has been observed in PS [106]. This phenomenon has been used to form all-optical logic gates in PS and raises the possibility of fabricating all-optical integrated circuits on Si [86, 106, 107].

PS can be also applied to the formation of complex dielectric structures. The facts that during the etching the current density determines the porosity of the etched layer, that the refractive index of a layer is proportional to its porosity, and that the etching process occurs only at the pore tips, allow changing the refractive index of PS simply by changing the current density during the etching. A temporal current profile is reflected in a correlated in-depth refractive index profile. This allows the formation of dielectric multilayers such as Bragg reflectors or microcavities [108, 109], or the study of photon propagation in complex dielectric structures where the analogies between photons and electrons can be investigated [110] and new one-dimensional devices could be designed, e.g., an optical delay line.

If a crystal has uniaxial symmetry, the optical response of the lattice is characterized by two distinct values of the refractive index. A crystal possessing such anisotropic dielectric properties is said to be birefringent. In PS the optical anisotropy is due to the anisotropic geometry of the pores. The anisotropy of PS depends on the crystalline orientation of the silicon substrate. Birefringence has been reported in (100), (111), and (110) oriented PS [111, 112, 113, 114, 115]. In all these cases, the PS layer can be assumed to be uniaxial, and the direction of the optical axis to be normal to the surface for the (100) and (111) cases, and parallel to the surface for the (110) case. The birefringence of (100) oriented PS has been reported to be positive [111, 112, 113] and values of the order of 10% have been demonstrated [116]. This high birefringence has led to the development of polarizers, polarization-sensitive Bragg reflectors, and microcavity resonators based on multilayers of optically nanostructured silicon where their optical response is governed by a three-dimensional (in-plane and in-depth) variation of the refractive index.

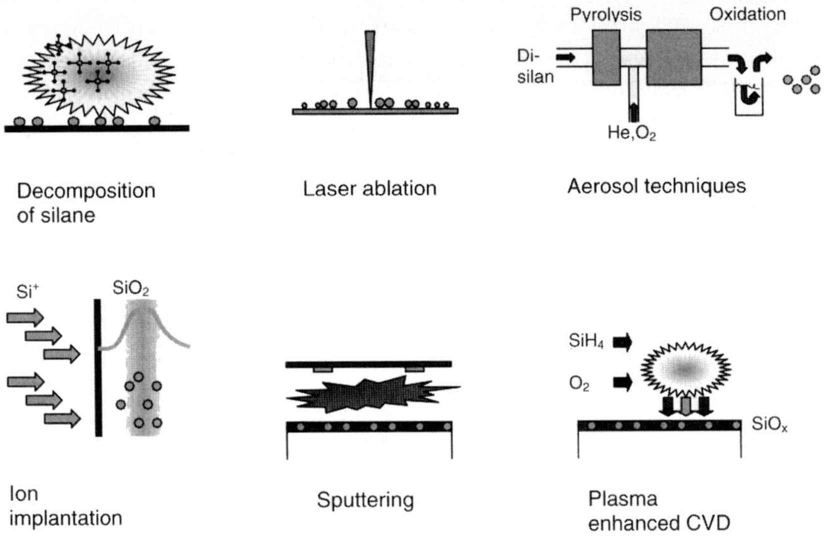

Fig. 8. Silicon nanocrystal fabrication techniques. The three lower techniques produce nanocrystals in a SiO_2 matrix. After [124]

4.2 Silicon Nanocrystals

Rather than produce nanometer-size Si crystals by etching, as in PS, there have been numerous attempts at growing them either directly from a gas phase or indirectly by recrystallization within a matrix (see Fig. 8) [33, 39, 117, 118, 119]. A popular approach is to produce Si nanocrystals (Si-nc) in a silica matrix to exploit the quality and stability of the SiO_2/Si interface and the improved emission properties of quantum confined Si. Many different approaches have been proposed to form the Si-nc [120, 121]. The most widely used are based on the deposition of sub-stoichiometric silica films, with a large excess of Si, followed by a high temperature annealing [122]. The annealing causes a phase separation between the two constituent phases, i.e., Si and SiO_2, with the formation of small Si-nc. The size and density of the Si-nc can be controlled by the deposition and the annealing parameters. Recently, the annealing of amorphous Si/SiO_2 superlattices has been proposed to control the size distribution and an almost monodispersed size distribution has been demonstrated [123].

It is noteworthy that the observation of a nanoparticle-size dependence of the PL energy in small Si-nc passivated with hydrogen [125] predates the similar finding in PS [78]. *Takagi* et al. [125] found that the PL peak energy varied as $1/d^2 (3 \leq d \leq 5\,\mathrm{nm})$, where d is the Si nanocrystal diameter, in accordance with quantum confinement effects predicted by a simple effective mass model. As for PS, however, the emitted light energy generally falls below that expected from calculations of the energy gap for Si spheres [85]. Also, the confinement effect is seen [125, 126, 127] or not seen [128] in emission depending on sample preparation. Interpretation of the nanocrystal PL spectra suffers from the same ambiguities as PS, i.e., nanocrystal size distribution effects and surface chemistry effects [119]. Quite surprisingly, single nanocrystal luminescence experiments show a linewidth of the emission spectra of some hundreds of meV [129], which could suggest either a strong vibronic character of the emission or that recombination is not due to quantum confinement but is related to surface state emission. Indeed, the nanocrystal structure may deviate from the cubic diamond structure for very small Si nanoclusters [33]. Calculations [130, 131] have shown that luminescence in Si-nc can be due to excitons trapped at the surface, which is passivated by hydrogen or silicon oxide, while the optical absorption is characteristic of quantum confinement effects. In addition, localized radiative levels can be formed at the Si-nc interface due to the presence of the Si=O (silanone) double bond. First principle calculations show that multiple Si=O defect centers could form on the Si-nc surface and that the emission energy does not shift appreciably as the number of defects increases [132]. The red PL quantum efficiency and lifetime is similar to that found for PS [133], indicating a similar light emission mechanism involving quantum-confined nanocrystal states. The observation of phonon assisted optical transitions in oxidized Si nanocrystals has also been reported by *Kovalev* et al. [134] and they found that for confinement energies above 0.7 eV the radiative transitions are governed by no-phonon quasidirect processes. Direct measurements of the conduction and valence band edges of Si nanocrystals with diameters from 1 to 5 nm using X-ray absorption and photoemission spectra have confirmed the quantum size effect in the band structure [135], although the increase of the band-gap with decreasing nanocrystal size is lower overall than that predicted by theory. The role of the surface is, however, a key factor in the interpretation of the emission mechanism [136]. Thus, the vagaries and complexities of the nanocrystal-interface-surface system are still proving difficult to unravel.

Nanocrystals of Si embedded in dielectrics form an attractive system for device fabrication when compared with PS, because of the increased surface stability and material rigidity. Visible electroluminescence (EL) has been observed, for example, from Si-nc embedded in films of a-Si:H [137] and from an electrochemically-formed nanocrystalline Si thin film deposited on SnO_2 [138]. In the latter case the p–i–n LED at room temperature emitted orange-red light (1.8 eV) that was readily visible to the eye. The light emission is ascribed variously to near surface states [137] and the quantum size effect [138]. Also, infrared emission near 1.1 eV has been obtained from a room-temperature EL device comprised of Si-nc embedded within a Si-rich SiO_{2-x} matrix [139]. The EL from this structure has an external quantum efficiency of 10^{-3} [140].

The refractive index of Si-nc is larger than the one of SiO_2, thus high quality waveguides can be prepared [141]. In such systems optical gain has

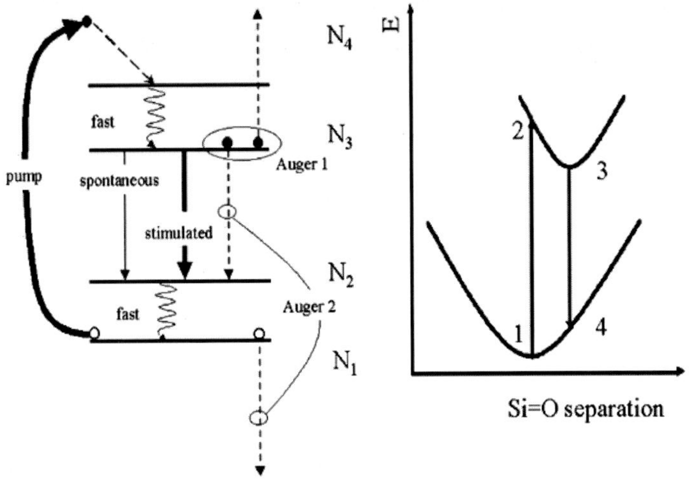

Fig. 9. (*Left*): An effective four level system that has been introduced to model qualitatively the recombination dynamics under gain conditions. From Level 3 the excited carriers can recombine by spontaneous, stimulated, or Auger recombinations. (*Right*): Schematic of the energy configuration diagram of the silicon-nc in an oxygen rich matrix. Localized radiative states are formed inside the nanocrystal band-gap by the interface oxygen atoms. The excited nanocrystal state can occur at a different lattice coordinate than that of the ground state. From [149]

been observed [121,142,143,144,145,146,147,148,149] by using various complementary techniques:

- a superlinear increase of the luminescence intensity as a function of the pumping rate [121, 146],
- amplified spontaneous emission in a waveguide geometry [142, 143, 144, 147, 148, 149],
- probe amplification in transmission experiments under high pumping excitation [142],
- collimated and speckled patterned emissions, which show the coherent properties of the emitted light [146], and
- an increase of the degree of polarization as a function of the pumping rate [150].

Some concerns have been raised about the methods used to measure the gain [151], and other studies have failed to observe light amplification in pump and probe experiments [152]. Here it seems that the main problem is related to the measurement techniques, which although easy in principle have many subtleties that should be considered [153]. Almost all of the authors agree on the fact that the gain is due to localized state recombinations either in the form of silicon dimers or in the form of silanone Si=O bonds created at the interface between the Si-nc and the oxide or within the oxide matrix. The

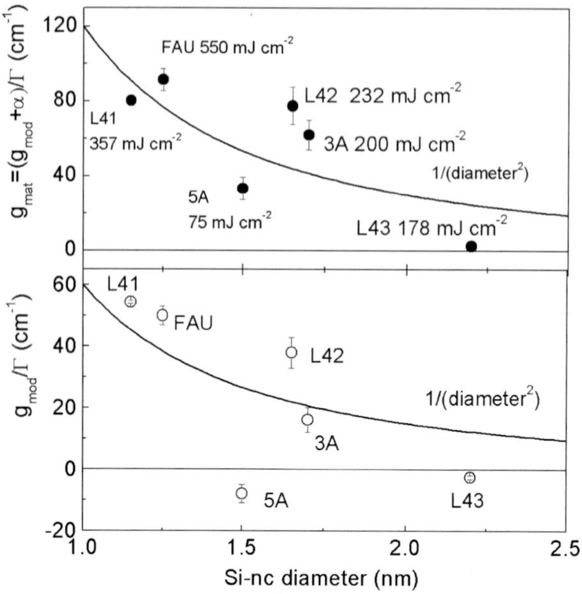

Fig. 10. Modal gain and material gain measured under pulsed conditions (on each data point the fluency is reported under pump conditions of 355 nm, 6 ns, and 10 Hz) for various Si-nc waveguides as a function of the Si-nc diameter. Here, L41, L42, and L43 refer to samples grown at MPI-Halle (Germany) by a size-controlled deposition technique [123], 5A and 3A to samples grown by PECVD in Catania (Italy) [154], and FAU to samples grown in Rochester (USA) [155]. The bottom plot is the net gain obtained by the measured gain coefficient g_{mod} divided by the optical mode overlap with the Si-nc rich layer (optical mode confinement factor Γ). The upper plot refers to the estimated material gain g_{mat} which takes into account the propagation losses α measured at the same wavelength as the gain measurements under CW measurements

suggested scheme to explain population inversion and, hence gain, is a four level model where a large lattice relaxation of the photoexcited localized center gives rise to the four levels (see Fig. 9) [143, 149].

From Fig. 9 it is clear that stimulated emission is competing with other non-radiative recombination mechanisms such as Auger or excited state absorption. In particular, time resolved experiments on the amplified spontaneous emission from Si-nc in a waveguide geometry [143, 144, 149] have shown an extremely fast component (nanosecond lifetime) due to stimulated emission. A delicate interplay between Auger recombination and stimulated emission governs the possibility of observing optical gain in Si-nc. As discussed in [149], the Si-nc density should be large enough to yield a significant optical gain.

Figure 10 shows a summary of various experimental results obtained in Trento University on various samples prepared using different methods. It is

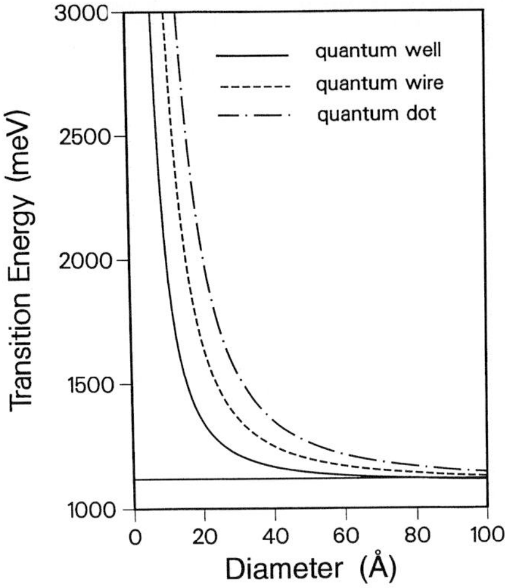

Fig. 11. Optical gap in Si quantum wells, wires, and dots versus system diameter. The transition energy is calculated for the lowest electron and heavy hole energies for infinite confining potentials. From [161]

interesting to note that, in addition to the packing density, another parameter governs the possibility and strength of material gain, i.e., the Si-nc size. As the Si-nc size decreases the material gain increases, and maximum gain values of 100 cm^{-1} under pulsed excitation are reported in the figure.

The problem of gain in Si nc still has some unanswered queries: 1. what is the role played by the Si-nc and by the embedding medium, 2. what are the key parameters that determine the presence of gain in the Si-nc, 3. how is the nanocrystal interaction influencing the gain, and 4. what is the precise nature of the four levels in the model and, in particular, the location and role of Si-O bonds. Substantial progress in the development of such nanocrystalline Si structures exhibiting optical gain can be expected over the next few years.

The Si-nc system is a very promising one in which to achieve a laser. Indeed, all the other key ingredients for a laser have been demonstrated. Vertical optical microcavities based on a Fabry–Pérot structure with mirrors constituted by distributed Bragg reflectors (DBR) and where the central layer is formed by Si-nc dispersed in SiO_2 have been already fabricated [156]. The presence of a thick SiO_2 layer needed to form the DBR can be a problem for electrical injection when current has to flow through the DBR. Lateral injection schemes can avoid these problems. On the other hand, electrical injection into the Si-nc is a delicate task in itself [157]. Bipolar injection is extremely difficult to achieve. Despite some claims, most of the reported Si-nc LEDs are impact ionization devices: electron–hole pairs are generated by impact ionization via the energetic free carriers injected through the electrode. By exploiting impact ionization, Si-nc LEDs have been demonstrated with EL spectra overlapping PL spectra, an onset-voltage as low as 5 V, and efficiencies in excess of 0.1% [158]. Some unconfirmed claims of near-laser action of Si-nc LEDs have appeared in the literature [159, 160].

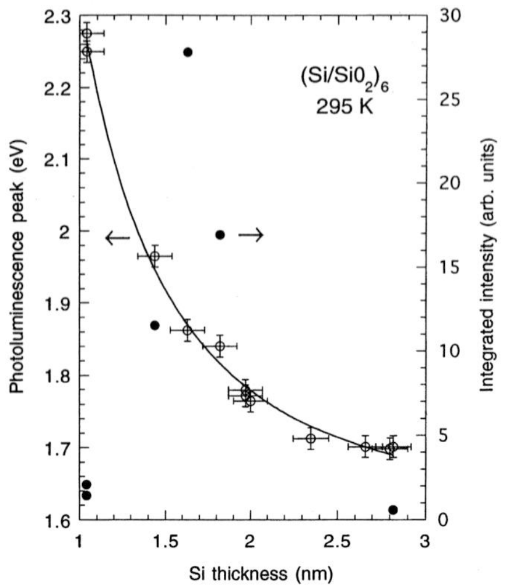

Fig. 12. The PL peak energy (*open circles*) and integrated intensity (*filled circles*) at room temperature in $(Si/SiO_2)_6$ superlattices as a function of Si layer thickness. The *solid line* is a fit with effective mass theory. From [167]

4.3 Quantum Wells, Wires and Dots

One of the major problems involved in PS and Si-nanocluster research and development work is the inhomogeneity of the material. Such inhomogeneous broadening effects in the PL and EL can be minimized by preparing *uniform* Si structures in the form of quantum wells, wires, or dots. Such structures can readily be produced directly by modern epitaxial growth techniques such as planar epitaxy, quantum wire formation along wafer steps, and dot self assembly, or indirectly by etching appropriate planar structures in the case of wires and dots. The predicted Si transition energies [161] due to the different degrees of quantum confinement are shown in Fig. 11 (more sophisticated pseudopotential calculations [162] give qualitatively similar results), where it can be seen that appreciable confinement effects are seen only for diameters less than 3 nm. Etched structures of this size have been difficult to produce in Si until recently.

4.3.1 Wells

The simplest approach is to grow thin quantum wells of Si separated by wide band-gap barriers. Suitable barrier candidates are SiO_2, CaF_2, and Al_2O_3 [163], and SiO_2 has the additional advantage of being an excellent passivator of Si [164]. Although a number of Si/barrier superlattices have been produced in the past [165] none has produced convincing evidence for quantum confinement induced emission until recently. In 1995, *Lu* et al. [166] reported visible light emission at room temperature from ultra thin-layer Si/SiO_2 superlattices grown by molecular beam epitaxy (MBE) that exhibited a clear quantum confinement shift with Si layer thickness, as shown in Fig. 12. The indications of direct band-to-band recombination were confirmed by measurements via X-ray techniques of the conduction and valence band shifts with layer thickness [166, 168]. The integrated intensity at first rises sharply with decreasing Si thickness until $d \approx 1.5\,\mathrm{nm}$ and then decreases again, which is consistent with quantum well exciton emission [169, 170]. The PL intensity is enhanced by factors of up to 100 on annealing and is also selectively enhanced and band-width narrowed by incorporation into an optical microcavity [171, 172]. Such superlattices have now been produced in an industrial environment [173]. The Si quantum wells in these superlattices have a disordered, but nearly crystalline, structure. Recently, the confinement effects have been reproduced in single ultra-thin c-Si wells formed from industrially produced Si-on-insulator wafers [174].

Theoretical calculations of the optical properties of Si/SiO_2 superlattices based on an empirical pseudopotential homojunction model [175] revealed that for Si layers in the (100) direction the energy gap is pseudodirect, i.e., the optical transition matrix element is smaller than that of a direct transition as in bulk GaAs. Nevertheless, the energy increases with decreasing Si quantum-well thickness and the transition matrix element increases steeply up to $\sim 2\,\mathrm{nm}$ thickness and then decreases again in general agreement with the experimental results given in Fig. 12. Tight binding band structure calculations for $Si_m/(SiO_2)_n$ crystalline superlattices with the number of unit cells $m, n \leq 5$ [176] show a striking new feature. Besides the expected increase in the band-gap due to quantum confinement, the confined bands along the Γ–Z symmetry direction are essentially dispersionless and exhibit a strongly-nested direct band-gap character at the minimum energy gap of the system. Thus, these superlattices produce a high radiative efficiency for a 1 nm–3 nm ($m = 2$–5) Si layer thickness [176], as found experimentally.

The bright PL obtained from as-grown and annealed Si/SiO_2 superlattices offers interesting prospects for the fabrication of a Si-based light emitter that can be tuned from 500 to beyond 800 nm by varying the Si layer thickness and/or the annealing conditions, all using available vacuum deposition technology and standard Si wafer processing techniques. The next important step is to develop LEDs based on such superlattices. Several prototype devices have been constructed in Si/SiO_2 [177, 178, 179, 180, 181] and all report visible EL, although as yet there is no strong evidence that the emission originates from confined states in the Si quantum wells. The EL from the Si/SiO_2 superlattices is notably stable [177, 178]. An optically active microcavity with a Si/SiO_2 superlattice inserted in the cavity has also been realized [182].

4.3.2 Wires

Quantum wires obtained by etching $Si/Si_{1-x}Ge_x$ heterostructures have been investigated by several groups (see, for example, [183] and [184]). In PL measurements, wires defined by electron beam lithography and reactive ion etching have shown small blue shifts of up to 30 meV in the $Si_{1-x}Ge_x$ alloy peak at $\sim 1.1\,\mathrm{eV}$ due to a combination of strain and confinement [183, 184].

Alternatively, $Si_{1-x}Ge_x$ wires have been grown on V-groove patterned Si substrates [185]: The infrared emission (PL and EL) in this case exhibits a large optical anisotropy [186]. No significant intensity enhancements compared with PL from quantum well transitions have been realized in these wire structures.

Until recently, it has not been possible to produce thin enough freestanding wires of c-Si by etching techniques to observe quantum confinement effects, although room temperature PL at wavelengths from 400 nm–850 nm is found for pillars with diameters $\sim$ 10 nm (see, for example, [187, 188, 189] and references therein). Recently, an EL device based on such Si nanopillars has been produced: The device emitted red light that was visible to the naked eye [190]. New etching techniques have now been developed for fabricating Si quantum wires with dimensions in the sub-10-nm region that after thermal oxidation should allow investigation of the optical properties of sub-2-nm wires in conventional wire [191] and free-standing pillar [192] form. Two-dimensional lattices of Si nanopillars have been shown to exhibit photonic band-gap properties that can be extended to submicrometer wavelengths [193]. Owing to the huge free space within these photonic crystals, luminescent materials may be readily incorporated to produce a photonic-band-edge laser.

In alternative production methods, Si nanowires have also been synthesized directly [194, 195]. Although the wire diameters obtained so far ($\sim$ 10 nm–30 nm) are too large for the observation of carrier confinement [196], these nanowires [197] and also Si nanoparticle chains [198] exhibit interesting electron field emission properties. Confinement effects could be induced in these wires by using thermal oxidation to reduce the Si nanowire diameter [195].

Theoretical calculations of the electronic structure and optical properties of H-passivated Si quantum wires have been reported by a number of research groups (see, for example [199] and references therein). First principles calculations show the same band nesting phenomenon and near-flat dispersion along the Γ–Z symmetry (wire) direction, as described above for Si quantum wells, and the occurrence of direct gaps [199, 200].

4.3.3 Dots

Attention has now turned mainly to the production of Si and $Si_{1-x}Ge_x$ quantum dots, in addition to the Si/SiO_2 system. Quantum dots fabricated by etching $Si/Si_{1-x}Ge_x$ superlattices have produced 4 K PL at 0.97 eV that is 200 times brighter in 60 nm dots compared with the unetched superlattice PL [201]. Similar studies of $Si_{1-x}Ge_x$ dots fabricated by self-assembling island growth on Si have shown an increased luminescence efficiency due to the localization of excitons in the dots [202]. In the latter case, the dots were buried in Si, which has the advantage of minimizing surface defect recombination. In both cases, EL has been observed from diode structures at low

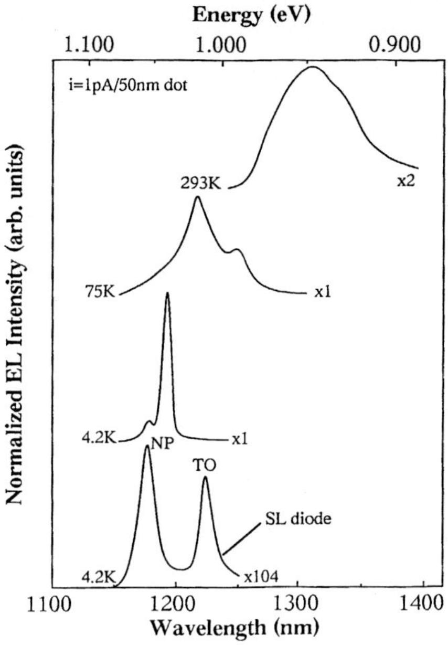

Fig. 13. Temperature dependence of EL spectra of a 50 nm $Si/Si_{0.7}Ge_{0.3}$ quantum dot diode under reverse bias of 0.5 V and an injection current of 1 pA/dot. A reference spectrum from a superlattice (SL) diode is also shown. After [203]

temperatures [202, 203] and at room temperature [203], as shown in Fig. 13. The infrared EL at 4.2 K in the dot is two orders of magnitude higher in intensity than in the as-grown superlattice (see Fig. 13). At room temperature, the dot EL at 1.3 µm is only 50% less efficient, with a threshold injection current of ∼ 0.1 pA/dot and an electrical-input to optical-output power conversion efficiency of 0.14% [203].

In subsequent work on Si quantum dots, ∼ 10 nm in size, formed in a-Si/SiO_2 superlattices by controlled Si crystallization, *Tsybeskov* et al. [204] have shown that the PL emitted near 1.1 eV is remarkably efficient (∼ 0.2% external quantum efficiency at 300 K) compared with c-Si. The high efficiency is due to spatial confinement of carriers in well passivated pure Si dots. A shift of the PL peak toward higher energies was found with smaller-size (< 10 nm) Si dots. Optical gain has since been measured in these Si dots [155, 205]. The PL in such dots can be tuned up in energy by confinement and down in energy by alloying of Ge with Si. The latter case has been investigated in MBE-grown $Si_{0.5}Ge_{0.5}$ dots in coherent-wave superlattice structures. Here, strong no-phonon PL was obtained at ∼ 0.8 eV (1.55 µm) that could easily be varied in wavelength by changing the Ge content [54].

5 Impurities

5.1 Luminescence via Impurity Centers in Silicon

Another approach to increasing the EL efficiency of an indirect band-gap semiconductor is to introduce an impurity that localizes the electron and hole, as pioneered in GaP [206]. This has been done in Si EL diodes by using, for example, rare earth impurities [207], carbon complexes [208], and sulfur/oxygen complexes [209] as localization centers for electron–hole recombination. Extrinsic luminescence in Si can arise from a variety of sources [37,38]. Here, we concentrate on isoelectronic and rare-earth extrinsic centers, as these are presently the most promising for optoelectronic device applications.

Isoelectronic centers are created by doping Si with electrically neutral impurities such as the isovalent elements C, Ge, and Sn or a multiple-atom complex with no dangling bonds. Isoelectronic impurities bind free excitons in Si, which can increase the probability of electron–hole recombination due to spatial confinement of the particles. The resultant recombination energy may appear as light or disappear through phonon generation and other nonradiative decay channels [33,210].

The optical properties of a variety of such isoelectronic impurity centers including In, Al–N, Be, S, and Se have been studied both in Si and $Si_{1-x}Ge_x$ alloys [33,38,210]. A luminescence external quantum efficiency of 5% and a lifetime greater than 1 ms have been reported for the S complex emission at 1.32 μm in Si at low temperatures [211], but the PL intensity and lifetime decrease sharply with increasing temperature. This variation with increasing temperature is due to exciton dissociation and competing nonradiative recombination processes. The low bound-exciton emission intensity at room temperature militates against isoelectronic-impurity based EL devices at present.

The optical properties of rare earth ions in solids have been investigated in great detail and are generally well understood [212]. The optical emission of the Er^{3+} ion is of particular interest for semiconductor device applications, because it occurs near 1.5 μm. The Er^{3+} ion emits photons at 1.54 μm in Si by intracenter transitions between Er^{3+}-ion discrete states ($I_{13/2} \rightarrow I_{15/2}$ transition within the 4f electron shell) [213]. The excitation of the Er^{3+} ions contained in Si is a complicated process [33,214,215] involving first electron–hole carrier generation in Si, then exciton formation, and finally Er excitation by an intracenter Auger process, with a number of competing pathways in the excitation process. High internal quantum efficiencies are observed when Er is surrounded by a strong electronegative ligand field, and room temperature EL devices with improved performance through the use of an oxygen codopant are emerging (see, for example [216,217,218,219,220]). The room temperature EL from Si coimplanted with Er and O is shown in Fig. 14. From such LED under reverse bias, an internal quantum efficiency of 0.05%, an emitted power

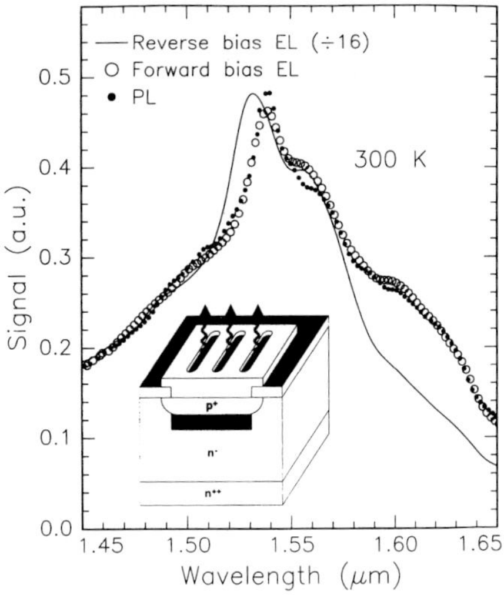

Fig. 14. Room-temperature EL spectra of Si:Er,O under reverse (*solid line*) and forward bias (o) conditions for a diode current density of 2.5 A/cm^2 compared with the PL (•). The *inset* gives a schematic of the LED structure. From [215]

of 0.2 mW, and a modulation frequency of 10 MHz have been achieved [215, 220].

5.2 Erbium Coupled with Silicon Nanocrystals

What it is more interesting for light amplification studies is the experimental finding of a strong enhancement of the Er luminescence when Er is implanted or deposited in a SiO_2 matrix where Si-nc have been formed, i.e., Si-nc acts as sensitizers for erbium ions [221, 222, 223]. Nonradiative de-excitation processes are reduced by widening the Si band-gap and thus avoiding one of the most detrimental sources of Er luminescence quenching. Indeed, the thermally activated back-transfer of excitation from Er^{3+} ions to Si-nc becomes less efficient than in bulk Si since the energy mismatch for the process becomes larger. Widening of the band-gap also produces a reduction in the free carrier concentration thus limiting the Auger processes. As demonstrated in Fig. 15, a strong luminescence comes from Er^{3+} ions that are pumped through an electron–hole mediated process in which photo-excited excitons from Si-nc transfer their energy to Er^{3+} ions [221, 222]. The number (between 1 to 10) of Si-nc coupled to a single Er^{3+} ion is still a debated issue [222, 224]. Regarding where Er is placed, from high resolution luminescence it is clear that most of the Er is in the SiO_2 matrix, which is an ideal situation for reproducing the environment that is found in an Er-doped fiber amplifier. Hence, Er coupled Si-nc has the advantages of both Si (efficient excitation) and SiO_2 (weak non-radiative processes, i.e., negligible temperature quenching of the luminescence), while it avoids their disadvantages (low excitation efficiency in SiO_2 and strong non-radiative processes in bulk Si). Indeed, MOS light-emitting devices operating at room temperature have been made with this system,

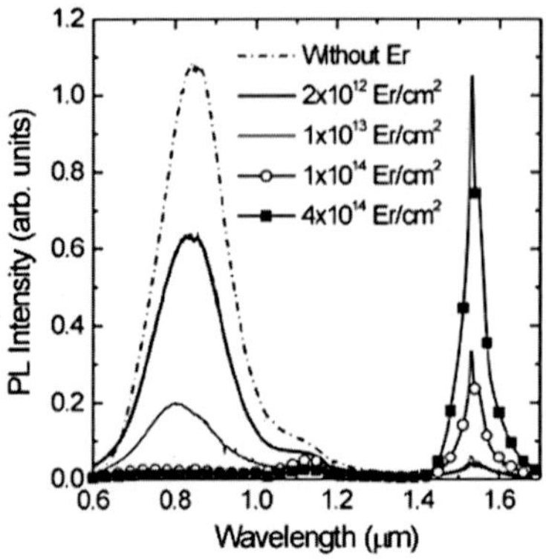

Fig. 15. Room temperature PL spectra of Er-implanted Si nanocrystals at different Er doses. The pump power of the laser beam was 50 mW. After [222]

and a quantum efficiency greater than 1% has been demonstrated [225]. Even higher efficiencies (10%) are reported for Er in silicon rich oxide films; however, in this system reliability is still an issue [226].

The layer co-doped with Si-nc and Er^{3+} ions has a refractive index that is larger that SiO_2, i.e., waveguides can be formed with a core containing Er^{3+} ions coupled to Si-nc. Experiments have shown that the luminescence intensity increases [227] or there is even evidence of signal enhancement [228] in these waveguides. Even though no net optical gain was measured, an enhancement in the probe transmission at 1.535 μm was observed as the pump power was increased. A fit to the experimental data yields an increased Er^{3+} ion emission cross section compared to Er^{3+} ions in silica or in Si [229]. This is a quite unexpected result, which conflicts with the observation of a millisecond lifetime from Er coupled Si-nc. The reason for the emission cross section increase is still unclear; one can speculate about the role of the dielectric environment, which is modified by the presence of the Si-nc [224]. What makes this finding interesting is the possibility to significantly reduce the cavity length in an amplifier or laser than the one usually employed in the Er-doped fiber systems. Sizeable gain can be further obtained by low Er doping concentrations. To summarize the very interesting properties of the Er^{3+} ion coupled Si-nc system, Table 1 compares the observed cross sections of Er^{3+} ions in silica, in Si, and coupled with Si-nc.

The system of Er^{3+} ions coupled with Si-nc is very promising for laser applications, because the active material (Er^{3+} ions in SiO_2) has already shown lasing properties. In addition, the technology to produce the material is very compatible with CMOS processing. Microcavities with excellent luminescence properties have been also demonstrated [156], which allows the design of both edge emitting or vertical emitting laser structures. The issue related to electrical pumping of the active material, which was believed to be a major draw-back of this approach, can be worked out since extremely high efficiency LEDs have been demonstrated [225, 226]. A still open issue is the ability to engineer waveguide losses in order to be able to measure net opti-

Table 1. Summary of the various cross sections related to Er^{3+} ions in various materials

	Er in SiO_2 (cm^2)	Er in Si (cm^2)	Er in Si-nc (cm^2)	Reference for Er in Si-nc
Effective excitation cross section of luminescence at a pumping energy of 488 nm	$(1–8) \times 10^{-21}$	3×10^{-15}	$(1.1–0.7) \times 10^{-16}$	[230, 231]
Effective excitation cross section of electroluminescence		4×10^{-14}	1×10^{-14} by impact ionization	[225]
Emission cross section at 1.535 μm	6×10^{-21}		2×10^{-19}	[229]
Absorption cross section at 1.535 μm	4×10^{-21}	2×10^{-20}	8×10^{-20}	[232]

cal gain and not only signal enhancement in pump and probe experiments. Further work should then be spent on optimizing the gain with respect to the waveguide parameters and developing a suitable optical cavity that can be electrically injected.

6 Photonic Components

It was predicted in the early 1990s that Si based optoelectronics will be a reality before the end of the century [9, 120]. Indeed, except for a Si based laser, all the basic components have been already demonstrated [233], as outlined below and described in greater detail in the following chapters.

6.1 Silicon Based Waveguides

The first essential component in Si microphotonics is the medium through which light propagates: the waveguide. This has to be Si compatible and should withstand normal microelectronics processing. Critical parameters are the refractive index n of the core material, its electro-optical effects, the optical losses, and the transparency region. To realize low-loss optical waveguides, various approaches have been followed (see Fig. 16) [234]: low dielectric mismatch structures (e.g., doped silica [235], silicon nitride [236], silicon oxynitrides on oxide [237], or differently doped Si [238]) or high dielectric mismatch structures (e.g., Si on oxide [239]). Low loss silica waveguides are characterized by large dimensions, typically 50 μm thick, due to the low refractive index mismatch (Δn = 0.1%–0.75%). Silica waveguides have a large mode

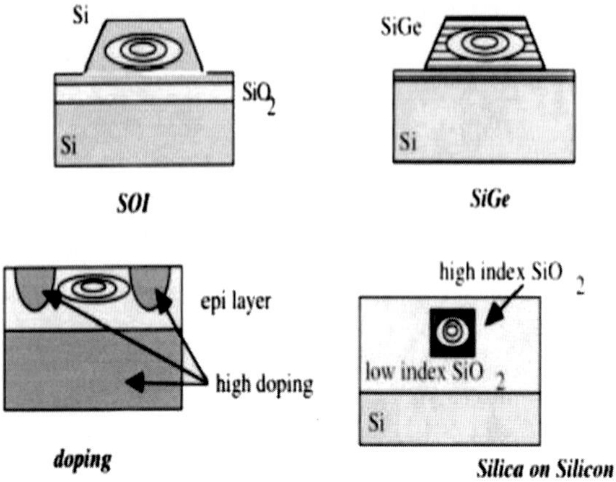

Fig. 16. Different approaches used for the fabrication of optical waveguides on Si substrates. From [233]

spatial extent and, thus, are interesting for coupling with optical fibers but not for integrating into/within electronic circuits because of their significant difference in sizes. The large waveguide size also prevents the integration of a large number of optical components in a single chip. Similar problems exist for Si on Si waveguides where the index difference is obtained by varying the doping density [238]. Si on Si waveguides are very effective for realizing free-carrier injection active devices (e.g., modulators) as well as fast thermo-optic switches owing to the high thermal conductivity of Si. A major problem with these waveguides is the large free carrier absorption, which causes optical losses of several dB/cm for single-mode waveguides at 1.55 μm. Silicon nitride based waveguides [236] and silicon oxynitrides waveguides [237] show losses lower than 0.5 dB/cm at 633 nm and bending radii of less than 200 μm. The nitride based waveguides are extremely flexible with respect to the wavelength of the signal light: both visible and IR.

At the other extreme, Si on insulator (SOI) or polysilicon-based waveguides allow for a large refractive index mismatch and, hence, for small size waveguides in the sub-micrometer range. This allows a large number of optical components to be integrated within a small area. Optical losses as low as 0.1 dB/cm at 1.55 μm have been reported for channel waveguides in SOI (optical mode cross-section $0.2\times4\ \mu m^2$) [240]. Ideal for on-chip transmission, SOI waveguides have coupling problems with silica optical fibers due to both the large size difference and the different optical impedance of the two systems. Various techniques have been proposed to solve these problems, among which are adiabatic tapers, V-grooves, and grating couplers [241,242]. Large single-mode stripe-loaded waveguides on SOI can be achieved provided that the stripe and slab are both made of Si [243]. This SOI system provides low loss waveguides ($<$ 0.2 dB/cm) with single mode operation, large rib structures

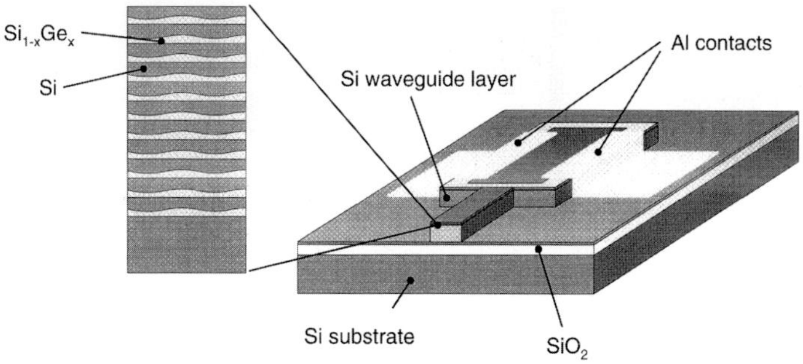

Fig. 17. Schematic representation of a $Si/Si_{0.5}Ge_{0.5}/Si$ waveguide photodetector based on a coherent wave $Si/Si_{0.5}Ge_{0.5}$ superlattice. The photocurrent quantum efficiency for the 240-mm-long waveguide detector was 0.1 A/W at 1.55 μm. (Source: S. Janz, National Research Council of Canada)

(optical mode cross-section $4.5 \times 4\,\mu m^2$), and low birefringence ($< 10^{-3}$). An appropriate geometry, with the use of an asymmetric waveguide, allows bend radii as short as 0.1 mm [244]. A number of photonic components in SOI have been demonstrated [243] and commercialized [244]: directional couplers, dense WDM arrayed waveguide gratings, Mach–Zehnder filters, Star couplers, etc.

6.2 Detectors

The optical signal is converted into an electrical signal by using Si based photodetectors. Detectors for Si photonics are based on three different approaches [245]: Si photoreceivers for wavelengths less than 1.1 μm, hybrid systems (mostly III–V on Si), and heterostructure based systems. High speed (up to 8 Gb/s) monolithically integrated Si photoreceivers at 850 nm have been fabricated by using 130-nm CMOS technology on a SOI wafer [246]. Other recent results confirm the ability of Si integrated photoreceivers to detect signals with a high responsivity of 0.46 A/W at 3.3 V for 845 nm light and a 2.5 Gb/s data rate [247]. The heterostructure approach is mainly based on the heterogrowth of Ge rich SiGe alloys: Ge-on-Si photodetectors have been reported with a responsivity of 0.89 A/W at 1.3 μm and a 50 ps response time [248]. A 1% quantum efficiency at 1.55 μm in a metal-semiconductor-metal (MSM) detector based on a Si/SiGe superlattice shows that promising developments are possible [249]. Similarly, waveguide photodetectors with Ge/Si self-assembled islands show responsivities of 0.25 mW at 1.55 μm with zero bias [250]. An example of a $Si/Si_{1-x}Ge_x$ waveguide photodetector is given in Fig. 17. The MSM detector is based on a $Si_{0.5}Ge_{0.5}/Si$ strained undulated superlattice and the waveguiding is provided by the SOI structure with SiO_2 and air cladding [251].

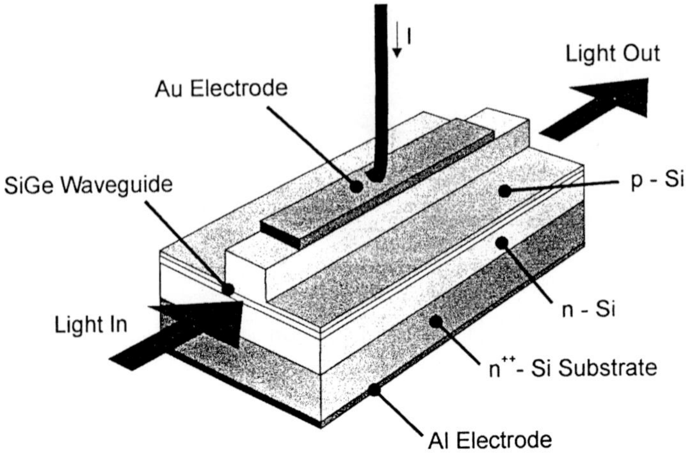

Fig. 18. Schematic representation of a Si/$Si_{0.85}Ge_{0.15}$/Si p–i–n waveguide modulator. A maximum modulation depth of 66% was obtained at 1.3 μm for a 2-mm-long waveguide using a peak pulse current density of 2700 A/cm^2. After [253]

6.3 Other Photonic Components

Almost all the other required photonic components have been demonstrated in Si microphotonics [120, 245]. Optical modulators, optical routers, and optical switching systems have all been integrated into Si waveguides [252]. A discussion of a series of photonic component realized with SOI waveguides is reported in [243], including plasma-dispersion-effect based active gratings, evanescent waveguide coupled Si-Ge based photodetectors, and Bragg cavity resonant photodetectors. An example of a Si/$Si_{1-x}Ge_x$ photonic device is the rib-waveguide optical modulator shown in Fig. 18, where carrier injection in a $Si_{1-x}Ge_x$ p–i–n heterostructure allows optical intensity modulation in Si at megahertz frequencies [253]. Such monolithic modulators operating at much higher (gigahertz) frequencies are required in practice.

6.4 Silicon Photonic Integrated Circuits

Based on the technologies reported in the previous sections, various demonstrator models of photonic integrated circuits based on Si have been reported. Here we present some examples.

Hybrid integration of active components and silica-based planar ligthwave circuits provide a full scheme for photonic component integration within a chip [254]. Passive components are realized by using silica waveguides while active components are hybridized within the silica. Active components (laser diodes, semiconductor optical amplifiers, and photodiodes) are flip-chip bonded on Si terraces where also the optical waveguides are formed. By using this approach various photonic components have been integrated such as multi wavelength light sources, optical wavelength selectors, wavelength converters, all optical time-division multiplexers, etc. [254]. Foreseen applications of this technology include WDM transceiver modules for fiber-to-the-home use.

A fully integrated optical system based on silicon oxynitrides waveguides, Si photodetectors, and CMOS transimpedance amplifiers has been realized [237]. Coupling of visible radiation to a Si photodetector can be achieved by using mirrors at the end of the waveguide. These are obtained by etching the end of the waveguide with an angle so that the light is reflected at almost 90 degrees into the underlying photodetector.

Commercial systems for the access network telecommunications market have been realized by using SOI waveguides and the Si optical bench (SOB) approach to interface the waveguides with both III–V laser sources and III–V photodetectors. The SOB is a technology where the Si wafer is used as a substrate (optical bench) where the various optical components are inserted by micromachining a suitable lodging. In [244], lasers and photodetectors are inserted into etched holes in Si and bump soldered in place. The system operates at 1.55 μm with a typical bit rate of 155 Mb/s [244]. A further advantage of using a large optical mode waveguide is the ease of interfacing to a single-mode optical fiber. In the approach of [244], these are located in V-grooves etched into Si.

A fully integrated system working at 1.55 μm has been demonstrated based on Si waveguides of very small optical mode size (cross-section $0.5 \times 0.2\,\mu m^2$), which allows extremely small turn radii (1 μm) [239]. In this way a large number of optical components can be integrated on a small surface ($\geq 10\,000$ components per cm^2). Detectors are integrated within Si by using Ge hetero-growth on Si itself. a responsivity of 250 mA/W at 1.55 μm and response times shorter than 0.8 ns have been achieved [248]. Here the laser source is external to the chip and acts as a *photon battery* similarly to usual batteries for electrons.

A realistic bidirectional optical bus architecture for clock distribution on a Cray T-90 supercomputer board based on polyamide waveguides (loss of 0.21 dB/cm at 850 nm), GaAs VCSEL, and Si MSM photodetectors has been developed [255]. By using 45° total-internal-reflection mirror coupling, efficiencies as high as 100% amongst the sources or the detectors and the waveguides have been demonstrated.

7 New Directions for Silicon Lasers

As mentioned earlier, the main limitation in achieving monolithically integrated Si microphotonics is the lack of any practical Si-based light sources: either efficient LEDs or Si lasers. A laser is preferred, as incoherent emission is likely not sufficient for dense high-speed interconnects. To produce a Si laser three key ingredients are required: 1. an active material that should

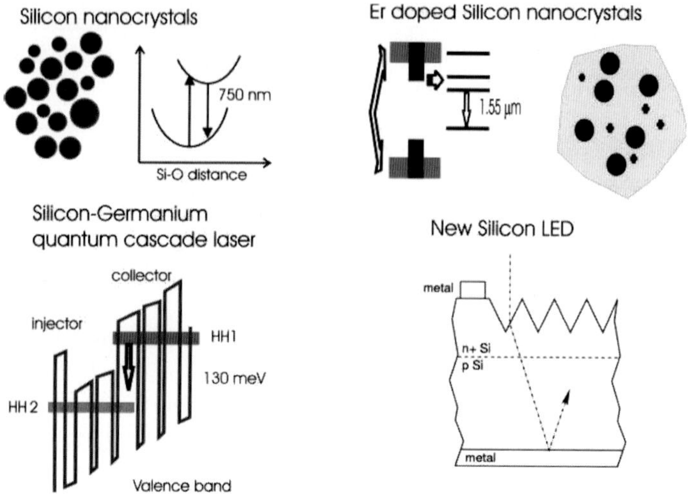

Fig. 19. Various approaches proposed to realize a silicon laser

luminesce in the region of interest and that should also be able to amplify light, 2. an optical cavity in which the active material is placed to provide the positive optical feedback, and 3. a suitable and efficient pumping scheme to achieve and sustain the laser action, preferably via electrical injection.

In the following sections, we outline some new approaches to producing Si lasers that are summarized in Fig. 19.

7.1 Bulk Silicon

Two different approaches to developing Si LEDs have been taken [256, 257]. The first approach is based on the results achieved in high efficiency solar cells and on the consideration that, within thermodynamic arguments, absorption and emission are two reciprocal processes [256]. As the first step, the non-radiative recombination rates are reduced by using high-quality intrinsic Si substrates, passivation of surfaces by high-quality thermal oxide, small metal areas, and limiting the high doping regions to contact areas. The parasitic absorption of photons, once they have been generated, is then reduced to a minimum. Finally, the extraction efficiency of light from bulk Si can be enhanced by suitably texturizing the Si surface. [256] reports the highest power efficiency to date for a Si based LED, approaching 1%. Electroluminescence spectra of these devices are typical for band-to-band recombination in Si. In addition a fully integrated opto-coupler device (LED coupled to a photodetector) was also demonstrated on the basis of this technology [258]. The main drawbacks of this approach for an integrated laser or LED are: 1. the need for both high purity (low doping concentration) and surface texturing, which renders the device processing incompatible with standard CMOS processing;

2. the strong and fast free carrier absorption typical of bulk Si, which can prevent the attainment of population inversion, is not addressed [259]; 3. appropriate integration of the active bulk Si into an optical cavity to achieve the required optical feedback to sustain laser action can be a problem; and 4. appropriate integration of the active bulk Si into an optical cavity to achieve the required optical feedback to sustain laser action can be a problem; and 5. the modulation speed of the device, which can be limited by the long lifetime of the excited carriers (milliseconds) and the need for a large optical cavity. Theoretical evidence concerning the possibility of achieving net optical gain in bulk silicon has been published recently [260].

A somewhat different approach towards achieving a reduction in the non-radiative channels in bulk Si involves exploiting the strain produced by localized dislocation loops [257, 261, 262]. Dislocations form potential pockets (i.e., energy barriers) close to the LED junction, which block the carriers and enhance radiative decay by localizing them in defect-free regions. The size of the dislocation loops was $\sim 100\,\mathrm{nm}$, i.e., not small enough to cause quantum confinement of the carriers, and the loop distances were $\sim 20\,\mathrm{nm}$. Free carriers injected through the top electrode are not able to diffuse away and are then constrained to recombine in the vicinity of the junction region. Onset of the EL at the band edge was observed when the diode was turned on under forward bias. No EL was observed under reverse bias. An ultimate external quantum efficiency of about 1% is claimed for these LEDs. The EL spectrum does not present significant differences in lineshape and peak position compared to that of bulk Si. A remarkable feature of this device is the high injection efficiency into the confined regions, due to the lack of quantum effects. In fact, since the density of states in the active zone is large (comparable to the bulk value), it is not a limiting factor for free carrier injection, contrary to quantum confined structures. The injection is also smooth, because there is no wide band-gap material acting as a confining barrier. This device has the additional and unexplained interesting feature of increasing efficiency with increasing temperature. The main problem of this approach for a Si laser is that it does not remove the two main problems of Auger recombination and free carrier absorption [259].

Finally, there is also a problem related to the bulk-Si LED emission wavelength, which is resonant with the Si band-gap. It is very difficult to control the region where the light is channeled in Si if these LEDs are used as a source for optical interconnects. Light will propagate throughout the wafer and will be absorbed in unwanted places.

7.2 Si/Ge Quantum Cascade Structures

One promising route that bypasses the fundamental limitation of the indirect band-gap in Si is to avoid using interband transitions. Indeed, if one exploits only intraband transitions, e.g., intravalence band transitions, no fundamental problems exist to impede lasing action in Si [263]. This is the approach taken in the quantum cascade (QC) Si/Ge system. With Si/Ge mid-infrared QC lasers, the concept that has already proved successful for III–V semiconductors is applied [264]. The idea behind the device is shown in Fig. 19. The QC scheme can be implemented in the conduction or valence band. To achieve a conduction band discontinuity, the growth of a Si/Ge superlattice on a relaxed SiGe buffer layer is necessary. For pseudomorphic growth on a Si substrate most of the band offset occurs in the valence band. Hence, the cascading scheme is usually designed in the valence band [265]. This differs from QC lasers based on III–V semiconductors that employ electron cascade structures. In Fig. 19, the valence band diagram of a cascading stage of a hole-injected p-i-p valence band device is shown. Injected holes make a vertical transition between sub-bands, and they then cascade down the electrically biased staircase. In order to assist population inversion, the lower laser level is rapidly depopulated by relaxation within the miniband. In practice, there are two identical active regions connected by an injector. Electroluminescence from a Si/Ge QC structure grown on Si has recently been demonstrated [265,266]. The quantum efficiency estimate is about 10^{-5} for EL [266,267,268,269]. Temperature-dependent measurements show nearly identical spectra between 20 and 90 K and a broadening and vanishing of the peak at about 160 K. It is possible to improve these results by controlling the large accumulation of strain imposed by the use of a Si substrate [266].

Although the QC concept works well for III–V semiconductors, the Si/Ge system has a fundamental limit posed on the number of periods of successive cascades that is given by the critical thickness for the formation of misfit dislocations. Hence, even though these devices show interesting EL properties for the development of a Si-based laser, highly evolved cascade structures must still be realized. As the gain per single element is low, due to the nature of the intraband transition, a large number of cascading structures will be needed to accumulate a macroscopic gain. In fact, no stimulated emission in SiGe QC structures has been reported to date. In addition, these structures have to be integrated within a waveguide cavity and the emission wavelength is different from those commonly used for optical interconnects. Waveguiding at these wavelengths can be realized by using SOI substrates or thick relaxed SiGe graded buffer layers. Other needed photonic components have still to be developed to achieve a photonic integrated system. Despite this, some authors propose the use of the Si/Ge QC laser for free-air optical interconnects, but such a laser will be of little use for Si photonics if the other required elements are not developed.

7.3 Terahertz Emission

A gap in the frequency spectrum of electromagnetic waves opens up in the Terahertz (THz) region, where there were no semiconductor sources until recently. At lower frequencies, conventional sources are made via electronic oscillators (high speed transistors) while at higher frequencies the sources are injection lasers. Recently a III–V semiconductor THz laser has been demonstrated [270]. Using the many advantages of the SiGe system over the III–V system for THz frequencies, a research effort is underway to implement the QC concept and make a laser covering this frequency region [271, 272, 273]. A typical structure employs p-type heterostructures to emit radiation from light hole–heavy hole transitions. In this way both edge emission and surface-normal THz emission might be obtained. Growth of p-Si/SiGe QC structures comprising up to 100 periods has been demonstrated using low pressure CVD via a strain-balanced approach on virtual substrates. Intersubband THz EL from a range of Si/SiGe QC structures has been observed in both edge and surface emission geometries. The light to heavy hole intersubband lifetime was measured to be ∼20 ps, which is over an order of magnitude longer than high temperature values in III–V heterostructures, implying that a Si/SiGe THz QC laser may be capable of much higher operating temperatures than corresponding III–V devices. Emission power levels comparable to those reported for III–V devices before laser processing have been measured, which indicate that there are good prospects for the realization of a THz Si/SiGe QC laser [273].

Another approach to THz laser emission in Si has also been developed [274, 275, 276, 277]. The idea is to make a THz laser using ultra-shallow donor optical transitions in Si. Very narrow spectral emission and a light intensity threshold versus pumping power were reported. These results purport that lasing has been achieved in this system. However, some points need to be clarified, such as the optical mode pattern in the simple cavity structure employed, the evolution from spontaneous to stimulated emission, and the coherent property of the light. Other concerns are related to the dilute doping of the system in order to avoid impurity–impurity interaction, which will prevent population inversion and complicate schemes for electrical injection.

7.4 Low Dimensional Silicon Structures

Light emission from quantum well and dot structures may yet hold the most promise for producing lasers at wavelengths across the visible into the infrared. The Si/SiO_2 multiple quantum well structures or silicon nanocrystals [165] are well suited for visible wavelength lasers at room temperature. We already discussed the observation of optical gain in such a system. Their optical absorption characteristics are ideal for optical pumping in a planar microcavity, but it is not yet certain if their electrical characteristics are amenable to injection laser design. Quantum dot LEDs made from $Si/Si_{1-x}Ge_x$ [203] show considerable potential for laser applications at 1.3 and possibly 1.55 μm [54]. However, much more research and development work on these structures is required before this potential can be realized.

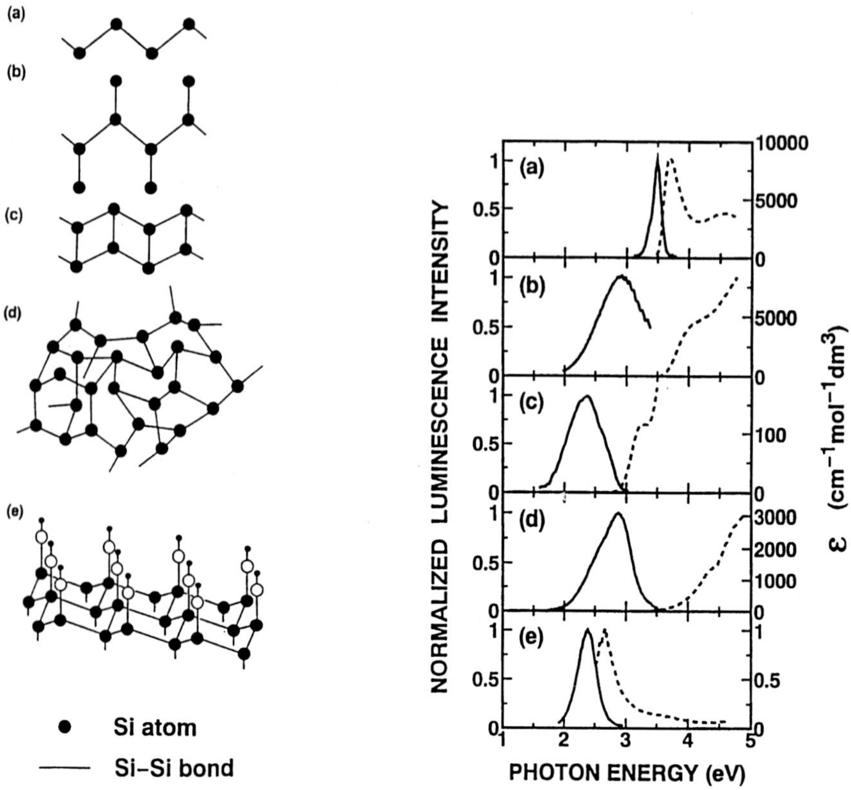

Fig. 20. Schematics of Si polymers with different backbones and their corresponding optical absorption and normalized PL spectra: (a) chain, (b) branch, (c) ladder, (d) network, and (e) planar Siloxene. After [88]

7.5 Polymers and Molecules Containing Silicon

Bright visible luminescence has been found in a number of Si polymer and molecular compounds [88, 117, 278]. The most prominent of these is Siloxene, $Si_6O_3H_6$, (see, for example [279]) and polysilane and its polymer derivatives (see, for example [280]).

Siloxene produces bright red PL with characteristics similar to that of the red PL in PS [281]. Its optical properties have been investigated with the hope of producing Siloxene-based devices. Siloxene has a direct band-gap [282] and the PL can be wavelength tuned across the visible region [283], but its chemical instability at higher temperatures limits its practical usefulness in devices requiring thermal processing.

Polysilane compounds are polymers based on a Si backbone with H atom termination of Si dangling bonds [284]. Derivatives of polysilane are obtained by modification of the Si skeleton structure and by the bonding of various atomic and molecular species to the backbone. Many such derivatives are possible, including Siloxene [280, 284]. The electronic structure and optical properties of many of the simpler compounds are now understood [117]. Linear trans polysilane, $(SiH_2)_n$, comprises a zigzag one-dimensional backbone of Si atoms with each Si atom bonded to two other Si atoms and to two H atoms above and below the Si atom. As such, it can be considered the one-dimensional limit of c-Si. This form of polysilane exhibits a direct band-gap of 3.9 eV and efficiently emits ultraviolet light [88, 280]. The PL characteristics of a variety of Si polymers are shown in Fig. 20. In the case of the chain, sharp ultraviolet emission and absorption peaks are observed, which are attributed to one-dimensional excitons delocalized on the backbone chain. In the branch and ladder structures, broad PL occurs with low quantum efficiency (10^{-3}–10^{-5}) at visible wavelengths and the excitons are strongly localized. The electronic properties of two-dimensional Si backbone polymers are intermediate between those of $(SiH_2)_n$ and three-dimensional bulk Si: Either a direct or an indirect band-gap can be obtained depending on the particular configuration [280]. The PL properties of broad-band visible wavelength emission and relatively long lifetime ($\sim$ 1 ns) possessed by the branch, ladder, and network Si polymers resemble those of amorphous Si, because real polymers exhibit structural disorders. Although chain-like Si polymers possess the desired characteristics of an ultraviolet light emitter, much more development work is required before they can be considered for ultraviolet devices based on Si.

8 Conclusions

Throughout this introductory overview, we have emphasized the present status of Si microphotonics and the recent advances that have caused people to be optimistic about the realization of a practical Si light source. Indeed, there have been many claims from many of the researchers involved in this field that a Si based laser will be realized within a short period of time [285]. If this objective can be realized, all the major building blocks for monolithic Si microphotonics will then be available. The final vision is to have Si microphotonics participating in every global application of the photonics industry including communications, computing, information displays, optical and infrared imaging, medicine, optical printing, optical command and control, optical sensing of physical chemical and biological inputs, optical signal processing, optical storage, and optical control of microwave devices or systems [286]. Indeed, we propose Si as the unifying material wherein the next generation of photonic devices will be realized.

References

1. E. Kasper, F. Schäffler: Group IV Compounds. In: T. P. Pearsall: *Strained-Layer Superlattices: Materials Science and Technology*. Semiconductors and Semimetals **33** (Academic Press, Boston 1991), p. 223

2. G. Abstreiter: Phys. World **5** (3), 36 (1992)
3. P. Bhattacharya: *Semiconductor Optoelectronic Devices*, Second Edition (Prentice Hall, New York 1996)
4. B. A. Saleh, M. C. Teich: *Fundamentals of Photonics* (Wiley, New York 1991)
5. R. Soref: MRS Bull. **23** (4), 20 (1998)
6. A. Mills: III–Vs Review **15** (4), 30 (2002)
7. B. D. Dingle, M. B. Spitzer, R. W. McClelland, J. C. C. Fan, P. M. Zavracky: Appl. Phys. Lett. **62**, 2760 (1993)
8. http://www.bookham.com
9. R. A. Soref: Proc. IEEE **81**, 1687 (1993)
10. S. S. Iyer, Y.-H. Xie: Science **260**, 40 (1993)
11. E. A. Fitzgerald, L. C. Kimmerling: MRS Bull. **23** (4), 39 (1998)
12. K. J. Linthicum, T. Gehrke, R. F. Davis, D. B. Thomson, K. M. Tracy: U. S. Patent 6,255,198 (2001)
13. R. Droopad, Z. Yu, W. J. Ooms, J. Ramdani: U. S. Patent Application 20010013313 (2001)
14. Y. H. Lo, R. Bhat, D. M. Hwang, C. Chua, C. H. Lin: Appl. Phys. Lett. **62**, 1038 (1993)
15. K. Mori, K. Tokutome, S. Sugou: Electron. Lett. **31**, 284 (1995)
16. A. Soeno, D. Kajita, J. Suda, H. Matsunami: Jpn. J. Appl. Phys. **39**, L905 (2000)
17. J. M. London, A. H. Loomis, J. F. Ahadian, C. G. Fonstad: Photon. Tech. Lett. **11**, 958 (1999)
18. M. Sugo, H. Mori, Y. Sakai, Y. Itoh: Appl. Phys. Lett. **60**, 472 (1992)
19. R. Leon: *Silicon Monolithic Integrated Circuits in RF Systems* (Digest of Papers IEE 2001), pp. 79–87
20. A. Hoffmann, A. Rizzi (Ed.): Papers presented at the International Workshop on Nitride Semiconductors (IWN 2002), Special issue Phys. Stat. Sol. (a) **194**, (2) (2003), pp. 359–590
21. L. Pavesi, L. Dal Negro, C. Mazzoleni, G. Franzò, F. Priolo: Nature **408**, 440 (2000)
22. B. Gelloz, N. Koshida: J. Appl. Phys. **88**, 4319 (2000)
23. http://www.eetimes.com/at/news/OEG20031104S0004
24. http://us.st.com/stonline/press/news/year2003/t1354h.htm
25. http://www.lightreading.com/document.asp?doc_id=43400
26. T. N. Theis: IBM J. Res. Develop. **44**, 379 (2000)
27. ftp://download.intel.com/labs/eml/download/EML_opportunity.pdf
28. D. A. B. Miller: Proc. IEEE **88**, 728 (2000)
29. S. K. Moore: IEEE Spectrum (2002)
30. S. Ossicini, L. Pavesi, F. Priolo: *Light Emitting Silicon for Microphotonics* (Springer, Berlin, Heidelberg 2003)
31. *Properties of Silicon* (INSPEC, London 1988)
32. J. R. Chelikowsky, M. L. Cohen: Phys. Rev. B **14**, 556 (1976)
33. L. C. Kimerling, K. D. Kolenbrander, J. Michel, J. Palm: Solid State Phys. **50**, 333 (1997)
34. J. I. Pankove: *Optical Processes in Semiconductors* (Dover, New York 1971)
35. T. Trupke, J. Zhao, A. Wang, R. Corkish, M. A. Green: Appl. Phys. Lett. **82**, 2996 (2003)
36. J. R. Haynes, H. B. Briggs: Phys. Rev. **86**, 647 (1952)

37. P. J. Dean: *Luminescence of Inorganic Solids* (Academic Press, New York 1966), chap. 3
38. G. Davies: Physics Reports **176**, 83 (1989)
39. A. D. Yoffe: Advan. Phys. **42**, 173 (1993)
40. U. Gnutzman, K. Clausecker: Appl. Phys. **3**, 9 (1974)
41. S. A. Jackson, R. People: Mat. Res. Soc. Symp. Proc. **56**, 365 (1986)
42. R. People, S. A. Jackson: Phys. Rev. B **36**, 1310 (1987)
43. L. Brey, C. Tejedor: Phys. Rev. Lett. **59**, 1022 (1987)
44. S. Froyen, D. M. Wood, A. Zunger: Phys. Rev. B **36**, 4547 (1987)
45. M. S. Hybertsen, M. Schlüter: Phys. Rev. B **36**, 9683 (1987)
46. T. P. Pearsall, J. Bevk, L. C. Feldman, J. M. Bonar, J. P. Mannaerts, A. Ourmarzd: Phys. Rev. Lett. **58**, 729 (1987)
47. R. Zachai, K. Eberl, G. Abstreiter, H. Kasper, H. Kibbel: Phys. Rev. Lett. **64**, 1055 (1990)
48. U. Menczigar, G. Abstreiter, J. Olajos, H. G. Grimmeiss, H. Kibbel, H. Presting, E. Kasper: Phys. Rev. B **47**, 4099 (1993)
49. G. Abstreiter: *Light Emission in Silicon*, ed. by D. J. Lockwood (Academic Press, San Diego 1997), chap. 2
50. H. Presting, U. Menczigar, G. Abstreiter, H. Kibbel, E. Kasper: Mat. Res. Soc. Symp. Proc. **256**, 83 (1992)
51. U. Menczigar, J. Brunner, E. Freiss, M. Gail, G. Abstreiter, H. Kibbel, H. Presting, E. Kasper: Thin Solid Films **222**, 227 (1992)
52. J. Engvall, J. Olajos, H. G. Grimmeiss, H. Presting, H. Kibbel, E. Kasper: Appl. Phys. Lett. **63**, 491 (1993)
53. J. Engvall, J. Olajos, H. G. Grimmeiss, H. Kibbel, H. Presting: Phys. Rev. B **51**, 2001 (1995)
54. J. M. Baribeau, A. Delâge, S. Janz, H. Lafontaine, D. J. Lockwood, J. P. McCaffrey, S. Moisa, N. L. Rowell, D. X. Xu: *Advanced Luminescent Materials, Quantum Confinement*, ed. by M. Cahay, S. Bandyopadhyay, N. Koshida, J. P. Leburton, D. J. Lockwood, M. Meyyappan, T. Sakamoto (Electrochemical Society, Pennington, 1999), p. 45
55. L. Vescan, T. Stoica: J. Lumin. **80**, 485 (1998)
56. H. Presting, H. Kibbel, M. Jaros, R. M. Turton, U. Menczigar, G. Abstreiter, H. G. Grimmeiss: Semicond. Sci. Technol. **7**, 1127 (1992)
57. T. P. Pearsall: Prog. Quantum Optics **18**, 97 (1994)
58. R. People, S. A. Jackson: *Strained Layer Superlattices: Physics*, ed. by T. P. Pearsall (Academic, Boston 1990), p. 119
59. D. C. Houghton, G. C. Aers, S.-R. E. Yang, E. Wang, N. L. Rowell: Phys. Rev. Lett. **75**, 866 (1995)
60. A. St. Amour, C. W. Liu, J. C. Sturm, Y. Lacroix, M. L. W. Thewalt: Appl. Phys. Lett. **67**, 3915 (1995)
61. B. A. Orner, J. Olowolafe, K. Roe, J. Kolodzey, T. Laursen, J. W. Mayer, J. Spear: Appl. Phys. Lett. **69**, 2557 (1996)
62. R. A. Soref, Z. Atzman, F. Shaapur, M. Robinson, R. Westhoff: Opt. Lett. **21**, 345 (1996)
63. J. C. Sturm: MRS Bull. **23** (4), 60 (1998)
64. N. L. Rowell, J.-P. Noël, D. C. Houghton, M. Buchanan: Appl. Phys. Lett. **58**, 957 (1990)
65. J.-P. Noël, N. L. Rowell, D. C. Houghton, D. D. Perovic: Appl. Phys. Lett. **57**, 1037 (1990)

66. J. C. Sturm, H. Manohoran, L. C. Lenchyshyn, M. L. W. Thewalt, N. L. Rowell, J.-P. Noël, D. C. Houghton: Phys. Rev. Lett. **66**, 1362 (1991)
67. L. C. Lenchyshyn, M. L. W. Thewalt, J. C. Sturm, P. V. Schwartz, E. J. Prince, N. L. Rowell, J.-P. Noël, D. C. Houghton: Appl. Phys. Lett. **60**, 3174 (1992)
68. J.-P. Noël, N. L. Rowell, D. C. Houghton, A. Wang, D. D. Perovic: Appl. Phys. Lett. **61**, 690 (1992)
69. L. C. Lenchyshyn, M. L. W. Thewalt, D. C. Houghton, J.-P. Noël, N. L. Rowell, J. C. Sturm, X. Xiao: Phys. Rev. B **47**, 16 655 (1993)
70. N. L. Rowell, J.-P. Noël, D. C. Houghton, A. Wang, L. C. Lenchyshyn, M. L.W. Thewalt, D. D. Perovic: J. Appl. Phys. **74**, 2790 (1993)
71. D. J. Robbins, P. Calcott, W. Y. Leong: Appl. Phys. Lett. **59**, 1350 (1991)
72. Q. Mi, X. Xiao, J. C. Sturm, L. C. Lenchyshyn, M. L. W. Thewalt: Appl. Phys. Lett. **60**, 3177 (1992)
73. S. Fukatsu, N. Usami, T. Chinzei, Y. Shiraki, A. Nishida, K. Nakagawa: Jpn. J. Appl. Phys. **31**, L1015 (1992)
74. Y. Kato, S. Fukatsu, Y. Shiraki: J. Vac. Sci. Technol. B **13**, 111 (1995)
75. M. Förster, U. Mantz, S. Ramminger, K. Thonke, R. Sauer, H. Kibbel, F. Schäffler, H.-J. Herzog: J. Appl. Phys. **80**, 3017 (1996)
76. H. Presting, T. Zinke, A. Splett, H. Kibbel, M. Jaros: Appl. Phys. Lett. **69**, 2376 (1996)
77. T. Stoica, L. Vescan, M. Goryll: J. Appl. Phys. **83**, 3367 (1998)
78. L. T. Canham: Appl. Phys. Lett. **57**, 1046 (1990)
79. L. Brus: Appl. Phys. A **53**, 465 (1991)
80. A. Uhlir, Jr.: Bell Syst. Tech. J. **35**, 333 (1956)
81. A. G. Cullis, L. T. Canham: Nature **353**, 335 (1991)
82. V. Lehmann, U. Gösele: Appl. Phys. Lett. **58**, 856 (1991)
83. D. C. Bensahel, L. T. Canham, S. Ossicini: *Optical Properties of Low Dimensional Structures* (Kluwer, Dordrecht 1993)
84. P. M. Fauchet, C. C. Tsai, L. T. Canham, I. Shimizu, Y. Aoyagi: *Microcrystalline Semiconductors: Materials Science and Devices* (Materials Research Society, Pittsburgh 1993)
85. D. J. Lockwood: Solid State Commun. **92**, 101 (1994)
86. Z. C. Feng, R. Tsu: *Porous Silicon* (World Scientific, Singapore 1994)
87. B. Hamilton: Semicond. Sci. Technol. **10**, 1187 (1995)
88. Y. Kanemitsu: Phys. Rep. **263**, 1 (1995)
89. J. C. Vial, J. Derrien: *Porous Silicon Science and Technology* (Springer, Berlin, Heidelberg 1995)
90. R. Hérino, W. Lang: *Porous Silicon and Related Materials* (Elsevier, Amsterdam 1995)
91. D. J. Lockwood, P. M. Fauchet, N. Koshida, S. R. J. Brueck: *Advanced Luminescent Materials* (Electrochemical Society, Pennington 1996)
92. P. M. Fauchet: J. Lumin. **70**, 294 (1996)
93. P. M. Fauchet: In: *Light Emission in Silicon*, ed. by D. J. Lockwood (Academic, San Diego 1998), chap. 6
94. A. G. Cullis, L. T. Canham, P. D. J. Calcott: J. Appl. Phys. **82**, 909 (1997)
95. P. D. J. Calcott: Mat. Sci. Eng. B **51**, 132 (1998)
96. O. Bisi, S. Ossicini, L. Pavesi: Surf. Sci. Rep. **264**, 1 (2000)
97. M. V. Wolkin, J. Jorne, P. M. Fauchet, G. Allan, C. Delarue: Phys. Rev. Lett. **82**, 197 (1999)

98. H. Mizuno, H. Koyama, N. Koshida: Appl. Phys. Lett. **69**, 3779 (1996)
99. G. Polisski, H. Heckler, D. Kovalev, M. Schwartzkoff, F. Koch: Appl. Phys. Lett. **73**, 1107 (1998)
100. F. Koch: Mater. Res. Soc. Symp. Proc. **298**, 319 (1993)
101. B. Gelloz, H. Sano, R. Boukherroub, D. D. M. Wayner, D. J. Lockwood, N. Koshida: Appl. Phys. Lett. **83**, 2342 (2003)
102. B. Gelloz, T. Nakagawa, N. Koshida: Appl. Phys. Lett. **73**, 2021 (1998)
103. K. D. Hirschman, L. Tsybeskov, S. P. Duttagupta, P. M. Fauchet: Nature **384**, 338 (1996)
104. L. Tsybeskov, S. P. Duttagupta, K. D. Hirschman, P. M. Fauchet: Appl. Phys. Lett. **68**, 2058 (1996)
105. L. Pavesi, R. Guardini, C. Mazzoleni: Solid State Commun. **97**, 1051 (1996)
106. T. Matsumoto, N. Hasegawa, T. Tamaki, K. Ueda, T. Futagi, H. Mimura, Y. Kanemitsu: Jpn. J. Appl. Phys. **33**, L35 (1994)
107. T. Matsumoto, M. Daimon, H. Mimura, Y. Kanemitsu, N. Koshida: J. Electrochem. Soc. **142**, 3528 (1995)
108. M. Ghulinyan, C. J. Oton, Z. Gaburro, P. Bettotti, L. Pavesi: Appl. Phys. Lett. **82**, 1550 (2003)
109. M. Ghulinyan, C. J. Oton, G. Bonetti, Z. Gaburro, L. Pavesi: J. Appl. Phys. **93**, 9724 (2003)
110. R. Sapienza, P. Costantino, D. Wiersma, M. Ghulinyan, C. J. Oton, L. Pavesi: Phys. Rev. Lett. **91**, 263902 (2003)
111. F. Ferrieu, A. Halimaoui, D. Bensahel: Solid State Commun. **84**, 293 (1992)
112. P. Basmaji, G. Surdutovich, R. Vitlina, J. Kolenda, V. S. Bagnato, H. Mohajerimoghaddam, N. Peyghambarian: Solid State Commun. **91**, 91 (1994)
113. I. Mihalcescu, G. Lerondel, R. Romestain: Thin Solid Films **297**, 245 (1997)
114. H. Kryzanowska, M. Kulik, J. Zuk: J. Lumin. **80**, 183 (1999)
115. D. Kovalev, G. Polisski, J. Diener, H. Heckler, N. Künzner, V. Y. Timoshenko, F. Koch: Appl. Phys. Lett. **78**, 916 (2001)
116. C. J. Oton, Z. Gaburro, M. Ghulinyan, L. Pancheri, P. Bettotti, L. Dal Negro, L. Pavesi: Appl. Phys. Lett. **81**, 4919 (2002)
117. H. Kamimura: *Light Emission from Novel Silicon Materials*, Supplement B to J. Phys. Soc. Japan **63** (Phys. Soc. Jpn., Tokyo 1994)
118. T. Ogawa, Y. Kanemitsu: *Optical Properties of Low-Dimensional Materials* (World Scientific, Singapore 1995)
119. Y. Kanemitsu: In: *Light Emission in Silicon*, ed. by D. J. Lockwood (Academic, San Diego 1998), chap. 5
120. O. Bisi, S. U. Campisano, L. Pavesi, F. Priolo: *Silicon Based Microphotonics: from Basics to Applications* (IOS Press, Amsterdam 1999)
121. L. C. Lenchyshyn, M. L. W. Thewalt, J. C. Sturm, P. V. Schwartz, E. J. Prince, N. L. Rowell, J.-P. Noël, D. C. Houghton: Appl. Phys. Lett. **60**, 3174 (1992)
122. J.-P. Noël, N. L. Rowell, D. C. Houghton, A. Wang, D. D. Perovic: Appl. Phys. Lett. **61**, 690 (1992)
123. L. C. Lenchyshyn, M. L. W. Thewalt, D. C. Houghton, J.-P. Noël, N. L. Rowell, J. C. Sturm, X. Xiao: Phys. Rev. B **47**, 16655 (1993)
124. J. Linnros: *Silicon Based Microphotonics: from Basics to Applications*, ed. by O. Bisi, S. U. Campisano, L. Pavesi, F. Priolo (IOS Press, Amsterdam 1999), pp. 47–86
125. H. Takagi, H. Ogawa, Y. Yamazaki, A. Ishizaki, T. Nakagiri: Appl. Phys. Lett. **56**, 2379 (1990)

126. S. Schuppler, S. L. Friedman, M. A. Marcus, D. L. Adler, Y.-H. Xie, F. M. Ross, T. D. Harris, W. L. Brown, Y. J. Chabal, L. E. Brus, P. H. Citrin: Phys. Rev. Lett. **72**, 2648 (1994)
127. Y. Kanzawa, T. Kageyama, S. Takeoka, M. Fujii, S. Hayashi, K. Yamamoto: Solid State Commun. **102**, 533 (1997)
128. Y. Kanemitsu, T. Ogawa, K. Shiraishi, K. Takeda: Phys. Rev. B **48**, 4883 (1993)
129. J. Valenta, R. Juhasz, J. Linnros: Appl. Phys. Lett. **80**, 1070 (2002)
130. G. Allan, C. Delerue, M. Lannoo: Phys. Rev. Lett. **76**, 2961 (1996)
131. C. Delerue, G. Allan, M. Lannoo: *Light Emission in Silicon*, ed. by D. J. Lockwood (Academic, San Diego 1998), chap. 7
132. M. Luppi, S. Ossicini: J. Appl. Phys. **94**, 2130 (2003)
133. L. E. Brus, P. F. Szajowski, W. L. Wilson, T. D. Harris, S. Schuppler, P. H. Citrin: J. Amer. Chem. Soc. **117**, 2915 (1995)
134. D. Kovalev, H. Heckler, M. Ben-Chorin, G. Polisski, M. Schwartzkopff, F. Koch: Phys. Rev. Lett. **81**, 2803 (1998)
135. T. van Buuren, L. N. Dinh, L. L. Chase, W. J. Siekhaus, L. J. Terminello: Phys. Rev. Lett. **80**, 3803 (1998)
136. N. Daldosso, M. Luppi, S. Ossicini, E. Degoli, R. Magri, G. Dalba, P. Fornasini, R. Grisenti, F. Rocca, L. Pavesi, S. Boninelli, F. Priolo, C. Bongiorno, F. Iacona: Phys. Rev. B **68**, 085327 (2003)
137. S. Tong, X.-N. Liu, L.-C. Wang, F. Yan, X.-M. Bao: Appl. Phys. Lett. **69**, 596 (1996)
138. T. Toyama, T. Matsui, Y. Kurokawa, H. Okamoto, Y. Hamakawa: Appl. Phys. Lett. **69**, 1261 (1996)
139. L. Tsybeskov, K. L. Moore, D. G. Hall, P. M. Fauchet: Phys. Rev. B **54**, R8361 (1996)
140. G. Franzo, A. Irreira, E. C. Moreira, M. Miritello, F. Iacona, D. Sanfilippo, G. F. Di Stefano, F. Fallica, F. Priolo: Appl. Phys. A **74**, 1 (2002)
141. R. T. Neal, M. D. C. Charlton, G. J. Parker, C. E. Finlayson, M. C. Netti, J. J. Baumberg: Appl. Phys. Lett. **83**, 4598 (2003)
142. L. Pavesi, L. Dal Negro, C. Mazzoleni, G. Franzò, F. Priolo: Nature **408**, 440 (2000)
143. L. Dal Negro, M. Cazzanelli, N. Daldosso, Z. Gaburro, L. Pavesi, F. Priolo, D. Pacifici, G. Franzò, F. Iacona: Physica E **16**, 297 (2003)
144. L. Khriachtchev, M. Rasanen, S. Novikov, J. Sinkkonen: Appl. Phys. Lett. **79**, 1249 (2001)
145. M. H. Nayfeh, S. Rao, N. Barry: Appl. Phys. Lett. **80**, 121 (2002)
146. K. Luterova, I. Pelant, I. Mikulskas, R. Tomasiunaset, D. Muller, J.-J. Grob, J.-L. Rehspringer, B. Honerlage: J. Appl. Phys. **91**, 2896 (2002)
147. P. M. Fauchet, J. Ruan: *Towards the First Silicon Laser*, ed. by L. Pavesi, S. Gaponenko, L. Dal Negro (Kluwer/Academic, New York 2003), p. 197
148. M. Ivanda, U. V. Densica, C. W. White, W. Kiefer: *Towards the First Silicon Laser*, ed. by L. Pavesi, S. Gaponenko, L. Dal Negro (Kluwer/Academic, New York 2003), p. 191
149. L. Dal Negro, M. Cazzanelli, Z. Gaburro, P. Bettotti, L. Pavesi, F. Priolo, G. Franzò, D. Pacifici, F. Iacona: *Towards the First Silicon Laser*, ed. by L. Pavesi, S. Gaponenko, L. Dal Negro (Kluwer/Academic, New York 2003), p. 145

150. M. Cazzanelli, Z. Gaburro, L. Dal Negro, L. Pavesi, D. Kovalev et al.: in press
151. J. Valenta, I. Pelant, J. Linnros: Appl. Phys. Lett. **81**, 1396 (2002)
152. R. G. Elliman, M. J. Lederer, N. Smith, B. Luther-Davies: Nucl. Instr. Meth. B **206**, 427 (2003)
153. L. Dal Negro, P. Bettotti, M. Cazzanelli, L. Pavesi, D. Pacifici: Optics Comm. **229**, 337 (2003)
154. L. Dal Negro, M. Cazzanelli, L. Pavesi, S. Ossicini, D. Pacifici, G. Franzò, F. Priolo, F. Iacona: Appl. Phys. Lett. **82**, 4636 (2003)
155. J. Ruan, H. Chen, P. M. Fauchet: Mat. Res. Soc. Symp. Proc. **737**, 407 (2003)
156. F. Iacona, G. Franzò, E. C. Moreira, F. Priolo: J. Appl. Phys. **89**, 8354 (2001)
157. Z. Gaburro, L. Pavesi: *Handbook of luminescence, Display Materials, and Nanocomposites*, ed. by H. S. Nalwa, L. S. Rohwer (American Scientific, Stevenson Ranch 2003)
158. G. Franzò, A. Irreira, E. C. Moreira, M. Miritello, F. Iacona, D. Sanfilippo, G. F. Di Stefano, F. Fallica, F. Priolo: Appl. Phys. A **74**, 1 (2002)
159. C. F. Lin, P. F. Chung, M. J. Chen, W. F. Su: Opt. Lett. **27**, 713 (2002)
160. L. Heikkila, T. T. Kuusela, H. P. Hedman: Superlattice. Microst. **26**, 157 (1999)
161. D. J. Lockwood, G. C. Aers, L. B. Allard, B. Bryskiewicz, S. Charbonneau, D. C. Houghton, J. P. McCaffrey, A. Wang: Can. J. Phys. **70**, 1184 (1992)
162. A. Zunger, L.-W. Wang: Appl. Surf. Sci. **102**, 350 (1996)
163. R. Tsu: Nature **364**, 19 (1993)
164. C. J. Frosch, L. Derick: J. Electrochem. Soc. **104**, 547 (1957)
165. D. J. Lockwood: Phase Transit. **68**, 151 (1999)
166. Z. H. Lu, D. J. Lockwood, J.-M. Baribeau: Nature **378**, 258 (1995)
167. D. J. Lockwood, Z. H. Lu, J.-M. Baribeau: Phys. Rev. Lett. **76**, 539 (1996)
168. D. J. Lockwood, J.-M. Baribeau, Z. H. Lu: *Advanced Luminescent Materials*, ed. by D. J. Lockwood, P. M. Fauchet, N. Koshida, S. R. J. Brueck (Electrochemical Society, Pennington 1996), p. 339
169. J. A. Brum, G. Bastard: J. Phys. C. Solid State Phys. **18**, L789 (1985)
170. Z. H. Lu, J.-M. Baribeau, D. J. Lockwood, M. Buchanan, N. Tit, C. Dharmawardana, G. C. Aers: SPIE Proc. **3491**, 457 (1998)
171. B. T. Sullivan, D. J. Lockwood, H. J. Labbé, Z. H. Lu: Appl. Phys. Lett. **69**, 3149 (1996)
172. D. J. Lockwood, B. T. Sullivan, H. J. Labbé: J. Lumin. **80**, 75 (1999)
173. G. Pucker, P. Bellutti, C. Spinella, K. Gatterer, M. Cazzanelli, L. Pavesi: J. Appl. Phys. **88**, 6044 (2000)
174. D. J. Lockwood, M. W. C. Dharmawardana, Z. H. Lu, D. H. Grozea, P. Carrier, L. J. Lewis: Mat. Res. Soc. Symp. Proc. **737**, F1.1.1 (2003)
175. J.-B. Xia, K. W. Cheah: Phys. Rev. B **56**, 14925 (1997)
176. N. Tit, M. W. C. Dharmawardana: Solid State Commun. **106**, 121 (1998)
177. R. Tsu, Q. Zhang, A. Filios: SPIE Proc. **3290**, 246 (1997)
178. A. G. Nassiopoulou, V. Ioannou-Sougleridis, P. Photopoulos, A. Travlos, V. Tsakiri, D. Papadimitriou: Phys. Stat. Sol. (a) **165**, 79 (1998)
179. G. G. Qin, S. Y. Ma, Z. C. Ma, W. H. Zong, L. P. You: Solid State Commun. **106**, 329 (1998)
180. L. Heikkilä, T. Kuusela, H.-P. Hedman, H. Ihantola: Appl. Surf. Sci. **133**, 84 (1998)
181. Z. Gaburro, G. Pucker, P. Bellutti, L. Pavesi: Solid State Commun. **114**, 33 (2000)

182. G. Pucker, P. Bellutti, L. Pavesi: Spectrochim. Acta A **57**, 2019 (2001)
183. Y. S. Tang, C. D. W. Wilkinson, C. M. Sotomayor Torres, D. W. Smith, T. E. Whall, E. H. C. Parker: Solid State Commun. **85**, 199 (1993)
184. J. Lee, S. H. Li, J. Singh, P. K. Bhattacharaya: J. Electron. Mat. **23**, 831 (1994)
185. N. Usami, T. Mine, S. Fukatsu, Y. Shiraki: Appl. Phys. Lett. **63**, 2789 (1993)
186. N. Usami, T. Mine, S. Fukatsu, Y. Shiraki: Appl. Phys. Lett. **64**, 1126 (1994)
187. H. I. Liu, N. I. Maluf, R. F. W. Pease, D. K. Biegelsen, N. M. Johnson, F. A. Ponce: J. Vac. Sci. Technol. B **10**, 2846 (1992)
188. A. G. Nassiopoulos, S. Grigoropoulos, D. Papadimitriou: *Advanced Luminescent Materials*, ed. by D. J. Lockwood, P. M. Fauchet, N. Koshida, S. R.J. Brueck (Electrochemical Society, Pennington 1996), p. 296
189. S. H. Zaidi, A.-S. Chu, S. R. J. Brueck: In: *Advanced Luminescent Materials*, ed. by D. J. Lockwood, P. M. Fauchet, N. Koshida, S. R. J. Brueck (Electrochemical Society, Pennington 1996), p. 307
190. A. G. Nassiopoulos, S. Grigoropoulos, D. Papadimitriou: Appl. Phys. Lett. **69**, 2267 (1996)
191. H. Namatsu, K. Kurihara, M. Nagase, T. Makino: Appl. Phys. Lett. **70**, 619 (1997)
192. H. I. Liu, D. K. Biegelsen, F. A. Ponce, N. M. Johnson, R. F. W. Pease: Appl. Phys. Lett. **64**, 1363 (1994)
193. V. V. Poborchii, T. Tada, T. Kanayama: J. Appl. Phys. **91**, 3299 (2002)
194. Y. F. Zhang, Y. H. Tang, N. Wang, D. P. Yu, C. S. Lee, I. Bello, S. T. Lee: Appl. Phys. Lett. **72**, 1835 (1998)
195. D. P. Yu, Z. G. Bai, J. J. Wang, Y. H. Zou, W. Qian, J. S. Fu, H. Z. Zhang, Y. Ding, G. C. Xiong, L. P. You, J. Xu, S. Q. Feng: Phys. Rev. B **59**, R2498 (1999)
196. X.-H. Sun, Y.-H. Tang, P. Zhang, S. J. Naftel, R. Sammynaiken, T. K. Sham, H. Y. Peng, Y.-F. Zhang, N. B. Wong, S. T. Lee: J. Appl. Phys. **89**, 6396 (2001)
197. F. C.K. Au, K. W. Wong, Y. H. Tang, Y. F. Zang, I. Bello, S. T. Lee: Appl. Phys. Lett. **75**, 1700 (1999)
198. Y. H. Tang, X. H. Sun, F. C. K. Au, L. S. Liao, H. Y. Peng, C. S. Lee, S. T. Lee, T. K. Sham: Appl. Phys. Lett. **79**, 1673 (2001)
199. F. Buda, J. Kohanoff: Prog. Quant. Electr. **18**, 201 (1994)
200. S. Ossicini, C. M. Bertoni, M. Biagini, A. Lugli, G. Roma, O. Bisi: Thin Solid Films **297**, 154 (1997)
201. Y. S. Tang, C. M. Sotomayor Torres, R. A. Kubiak, D. A. Smith, T. E. Whall, E. H. C. Parker, H. Presting, H. Kibbel: *The Physics of Semiconductors*, vol. 2, ed. by D. J. Lockwood (World Scientific, Singapore 1995), p. 1735
202. R. Apetz, L. Vescan, A. Hartmann, C. Dieker, H. Lüth: Appl. Phys. Lett. **66**, 445 (1995)
203. Y. S. Tang, W.-X. Ni, C. M. Sotomayor Torres, G. V. Hansson: Electron. Lett. **31**, 1385 (1995)
204. L. Tsybeskov, K. D. Hirschman, S. P. Duttagupta, M. Zacharias, P. M. Fauchet, J. P. McCaffrey, D. J. Lockwood: Appl. Phys. Lett. **72**, 43 (1998)
205. J. Ruan, P. M. Fauchet, L. Dal Negro, M. Cazzanelli, L. Pavesi: Appl. Phys. Lett. **83**, 5479 (2003)
206. D. G. Thomas, M. Gershenzon, J. J. Hopfield: Phys. Rev. **131**, 2397 (1963)
207. H. Ennen, G. Pomrenke, A. Axmann, K. Eisele, W. Haydl, J. Schneider: Appl. Phys. Lett. **46**, 381 (1985)

208. L. T. Canham, K. G. Barraclough, D. J. Robbins: Appl. Phys. Lett. **51**, 1509 (1987)
209. P. L. Bradfield, T. G. Brown, D. G. Hall: Appl. Phys. Lett. **55**, 100 (1989)
210. T. G. Brown, D. G. Hall: *Light Emission in Silicon*, ed. by D. J. Lockwood (Academic Press, San Diego 1998), chap. 3
211. T. G. Brown, D. G. Hall: Appl. Phys. Lett. **49**, 245 (1986)
212. G. H. Dieke: *Spectra and Energy Levels of Rare Earth Ions in Crystals* (Wiley, New York 1968)
213. H. Ennen, J. Schneider, G. Pomrenke, A. Axmann: Appl. Phys. Lett. **43**, 943 (1983)
214. J. Michel, L. V. C. Assali, M. T. Morse, L. C. Kimmerling: *Light Emission in Silicon*, ed. by D. J. Lockwood (Academic, San Diego 1998), chap. 4
215. S. Coffa, G. Franzò, F. Priolo: MRS Bulletin **23** (4), 25 (1998)
216. J. Michel, B. Zheng, J. Palm, E. Ouellette, F. Gan, L. C. Kimerling: Mat. Res. Soc. Symp. Proc. **422**, 317 (1996)
217. J. Stimmer, A. Reittinger, G. Abstreiter, H. Holzbrecher, Ch. Buchal: Mat. Res. Soc. Symp. Proc. **422**, 15 (1996)
218. S. Coffa, G. Franzò, F. Priolo: Appl. Phys. Lett. **69**, 2077 (1996)
219. L. Tsybeskov, S. P. Duttagupta, K. D. Hirschman, P. M. Fauchet, K. L. Moore, D. G. Hall: Appl. Phys. Lett. **70**, 1790 (1997)
220. G. Franzò, S. Coffa, F. Priolo, C. Spinella: J. Appl. Phys. **81**, 2784 (1997)
221. G. Franzò, V. Vinciguerra, F. Priolo: Appl. Phys. a **69**, 3 (1999)
222. G. Franzò, D. Pacifici, V. Vinciguerra, F. Priolo, F. Iacona: Appl. Phys. Lett. **76**, 2167 (2000)
223. M. Zacharias, M. S. J. Heitmann, P. Streitenberger: Physica E **11**, 245 (2001)
224. P. G. Kik, A. Polman: *Towards the First Silicon Laser*, ed. by L. Pavesi, S. Gaponenko, L. Dal Negro (Kluwer/Academic, New York 2003), p. 383
225. F. Iacona, D. Pacifici, A. Irrera, M. Miritello, G. Franzó, F. Priolo, D. Sanfilippo, G. Di Stefano, P. G. Fallica: Appl. Phys. Lett. **81**, 3242 (2002)
226. M. E. Castagna, S. Coffa, L. Caristia, A. Messina, C. Bongiorno: Proceedings of ESSDDERC 2002 (2002), p. 439
227. X. Zhao, S. Komuro, H. Isshiki, Y. Aoyagi, T. Sugano: Appl. Phys. Lett. **74**, 120 (1999)
228. H.-S. Han, S.-Y. Seo, J. H. Shin: Appl. Phys. Lett. **79**, 4568 (2001)
229. H.-S. Han, S.-Y. Seo, J. H. Shin, N. Park: Appl. Phys. Lett. **81**, 3720 (2002)
230. F. Priolo, G. Franzò, D. Pacifici, V. Vinciguerra, F. Iacona, A. Irrera: J. Appl. Phys. **89**, 264 (2001)
231. A. J. Kenyon, C. E. Chryssou, C. W. Pitt, T. Shimizu-Iwayama, D. E. Hole, N. Sharma, C. J. Humphreys: J. Appl. Phys. **91**, 367 (2002)
232. P. G. Kik, A. Polman: J. Appl. Phys. **91**, 534 (2002)
233. G. Masini, L. Colace, G. Assanto: Mat. Sci. Eng. B **89**, 2 (2002)
234. B. P. Pal: *Progress in Optics XXXII*, ed. by E. Wolf (Elsevier, Amsterdam 1993), p. 1
235. T. Miya: IEEE J. Sel. Top. Quantum Electron. **6**, 38 (2000)
236. D. A. P. Bulla, B. V. Borges, M. A. Romero, N. I. Morimoto, L. G. Neto, A. L. Cortes: IMOC'99 IEEE Proc. (1999), p. 454
237. U. Hilleringmann, K. Goser: IEEE Trans. Electron. Dev. **42**, 841 (1995)
238. G. Cocorullo, F. G. Della Corte, M. Iodice, I. Rendina, P. M. Sarro: IEEE J. Sel. Top. Quantum Electron. **4**, 983 (1998)

239. L. C. Kimerling: Appl. Surf. Science **159–160**, 8 (2000)
240. K. K. Lee, D. R. Lim, H.-C. Luan, A. Agarwal, J. Foresi, L. C. Kimerling: Appl. Phys. Lett. **77**, 1617 (2000)
241. M. A. Rosa, N. Q. Ngo, D. Sweatman, S. Dimitrijev, H. B. Harrison: IEEE J. Sel. Top. Quantum Electron. **5**, 1249 (1999)
242. T. W. Ang, G. T. Reed, A. Vonsovici, A. G. R. Evans, P. R. Routley, M. R. Josey: Electron. Lett. **35**, 977 (1999)
243. B. Jalali, S. Yegnanarayanan, T. Yoon, T. Yoshimoto, I. Rendina, F. Coppinger: IEEE J. Sel. Top. Quantum Electron. **4**, 938 (1998)
244. T. Bestwick: Proc. 48th IEEE Electronic Components and Technology Conference, 25–28 May 1998 (1988), pp. 566–571
245. H. Zimmermann: *Integrated Silicon Optoelectronics* (Springer, New York 2000)
246. S. M. Csutak, J. D. Schaub, W. E. Wu, R. Shimer, J. C. Campbell: J. Lightwave Technol. **20**, 1724 (2002)
247. M. Yang, K. Rim, D. L. Rogers, J. D. Schaub, J. J. Wleser, D. M. Kuchta, D. C. Boyd, F. Rodier, P. A. Rabidoux, J. T. Marsh, A. D. Icknor, Q. Yang, A. Upham, S. C. Ramac: IEEE Elect. Dev. Lett. **23**, 395 (2002)
248. G. Masini, L. Colace, G. Assanto, K. Wada, L. C. Kimerling: Electron. Lett. **35**, 1467 (1999)
L. Colace, G. Masini, G. Assanto, H. C. Luan, K. Wada, L. C. Kimerling: Appl. Phys. Lett. **76**, 1231 (2000)
249. S. Winnerl, D. Buca, S. Lenk, Ch. Buchal, S. Mantl, D. X. Xu: Mat. Sci. Eng. B **89**, 73 (2002)
250. M. El kurdi, P. Boucaud, S. Sauvage, G. Fishman, O. Kermarrec, Y. Campidelli, D. Bensahel, G. Saint-Girons, I. Sagnes, G. Patriarche: J. Appl. Phys. **92**, 1858 (2002)
251. D.-X. Xu, S. Janz, H. Lafontaine, M. R. T. Pearson: SPIE Proc. **3630**, 50 (1999)
252. G. Coppola, A. Irace, G. Breglio, M. Iodice, L. Zeni, A. Cutulo, P. M. Sarro: Optics Laser Eng. **39**, 317 (2003)
253. C. Fernando, S. Janz, N. G. Tarr, R. Normandin, J.-P. Noël, J. S. Wight: SPIE Proc. **2402**, 131 (1995)
254. K. Kato, Y. Tohmori: IEEE J. Sel. Top. Quantum Electron. **6**, 4 (2000)
255. R. T. Chen, L. Lin, C. Choi, Y. J. Liu, B. Bihari, L. Wu, S. Tang, R. Wickman, B. Picor, M. K. Hibbs-Brenner, J. Bristow, Y. S. Liu: IEEE Proc. **88**, 780 (2000)
256. M. A. Green, J. Zhao, A. Wang, P. J. Reece, M. Gal: Nature **412**, 805 (2001)
257. W. L. Ng, M. A. Lourenço, R. M. Gwilliam, S. Ledain, G. Shao, K. P. Homewood: Nature **410**, 192 (2001)
258. J. Zhao, M. A. Green, A. Wang: J. Appl. Phys. **92**, 2977 (2002)
259. W. P. Dumke: Phys. Rev. **127**, 1559 (1962)
260. T. Trupke, M. Green: J. Appl. Phys. **93**, 9058 (2003)
261. C. Delerue, M. Lannoo, G. Allan, E. Martin, I. Mihalcescu, J. C. Vial, R. Romestain, F. Muller, A. Bsiesy: Phys. Rev. Lett. **75**, 2229 (1995)
262. P. Jonsson, H. Bleichner, M. Isberg, E. Nordlander: J. Appl. Phys. **81**, 2256 (1997)
263. R. A. Soref: Thin Solid Films **294**, 325 (1997)
264. C. Gmachl, F. Capasso, D. L. Sivco, A. Y. Cho: Rep. Prog. Phys. **64**, 1533 (2001)

265. G. Dehlinger, L. Diehl, U. Gennser, H. Sigg, J. Faist, K. Ensslin, D. Grützmacher: Science **290**, 2277 (2000)
266. L. Diehl, S. Mentese, H. Sigg, E. Müller, D. Grützmacher, U. Gennser, I. Sagnes, T. Fromherz, J. Stangl, T. Roch, G. Bauer, Y. Campidelli, O. Kermarrec, D. Bensahel, J. Faist: Appl. Phys. Lett. **81**, 4700 (2002)
267. G. Dehlinger, L. Diehl, U. Gennser, H. Sigg, E. Muller, S. Stutz, J. Faist, J. Stangl, T. Roch, G. Bauer, D. Grützmacher: Mat. Sci. Eng. B **89**, 30 (2002)
268. L. Diehl, S. Mentese, H. Sigg, E. Müller, D.Grützmacher, U. Gennser, I. Sagnes, T. Fromherz, J. Stangl, T. Roch, G. Bauer, Y. Campidelli, O. Kermarrec, D. Bensahel, J. Faist: *Towards the First Silicon Laser*, ed. by L. Pavesi, S. Gaponenko, L. Dal Negro (Kluwer/Academic, New York 2003), p. 325
269. I. Bormann, K. Brunner, S. Hackenbuchner, G. Zandler, G. Abstreiter, S. Schmult, W. Wegscheider: Appl. Phys. Lett. **80**, 2260 (2002)
270. R. Kohler, A. Tredicucci, F. Beltram, H. Beere, E. Linfield, G. Davies, D. Ritchie, R. C. Iotti, F. Rossi: Nature **417**, 156 (2002)
271. S. A. Lynch, R. Bates, D. J. Paul, D. J. Norris, A. G. Cullis, Z. Ikonic, R. W. Kelsall, P. Harrison, D. D. Arnone, C. R. Pidgeon: Appl. Phys. Lett. **81**, 1543 (2002)
272. S. A. Lynch, S. S. Dhillon, R. Bates, D. J. Paul, D. D. Arnone, D. J. Robbins, Z. Ikonic, R. W. Kelsall, P. Harrison, D. J. Norris, A. G. Cullis, C. R. Pidgeon, P. Murzyn, A. Loudon: Mat. Sci. Eng. B **89**, 10 (2002)
273. R. W. Kelsall, Z. Ikonic, P. Harrison, S. A. Lynch, R. Bates, D. J. Paul, D. J. Norris, S. L. Liew, A. G. Cullis, D. J. Robbins, P. Murzyn, C. R. Pidgeon, D. D. Arnone, R. A. Soref: *Towards the First Silicon Laser*, ed. by L. Pavesi, S. Gaponenko, L. Dal Negro (Kluwer/Academic, New York, 2003), p. 367
274. S. G. Pavlov, H.-W. Hübers, M. H. Rümmeli, J. N. Hovenier, T. O. Klaasen, R. Kh. Zhukavon, A. V. Muravjov, V. N. Shastin: *Towards the First Silicon Laser*, ed. by L. Pavesi, S. Gaponenko, L. Dal Negro (Kluwer/Academic, New York, 2003), p. 331
V. N. Shastin, E. E. Orlova, R.Kh.Zhukavin, S. G. Pavlov, H.-W. Hübers, H. Riemann: *Towards the First Silicon Laser*, ed. by L. Pavesi, S. Gaponenko, L. Dal Negro (Kluwer/Academic, New York 2003), p. 341
275. N. Shastin, R. Kh. Zhukavin, E. E. Orlova, S. G. Pavlov, M. H. Rummeli, H.-W. Hubers, J. N. Hovenier, T. O. Klaassen, H. Riemann, I. V. Bradley, A. F. G. van der Meer: Appl. Phys. Lett. **80**, 3512 (2002)
276. A. Blom, M. A. Odnoblyudov, H. H. Cheng, I. N. Yassievich, K. A. Chao: Appl. Phys. Lett. **79**, 713 (2001)
277. S. G. Pavlov, R. Kh. Zhukavin, E. E. Orlova, V. N. Shastin, A. V. Kirsanov, H.-W. Hubers, K. Auen, H. Riemann: Phys. Rev. Lett. **84**, 5220 (2000); S. G. Pavlov, H.-W. Hubers, H. Riemann, R. Kh. Zhukavin, E. E. Orlova, V. N. Shastin: J. Appl. Phys. **92**, 5632 (2002)
278. L. E. Brus: *Light Emission in Silicon*, ed. by D. J. Lockwood (Academic Press, San Diego 1998), chap. 8
279. M. Stutzmann, M. S. Brandt, H. D. Fuchs, M. Rosenbauer, M. K. Kelly, P. Deak, J. Weber, S. Finkbeiner: *Optical Phenomena in Semiconductor Structures of Reduced Dimensions*, ed. by D. J. Lockwood, A. Pinczuk (Kluwer, Dordrecht 1993), p. 427
280. K. Takeda: *Light Emission from Novel Silicon Materials*, ed. by H. Kamimura (Physical Society of Japan, Tokyo 1994), p. 1

281. M. S. Brandt, H. D. Fuchs, M. Stutzmann, J. Weber, M. Cardona: Solid State Commun. **81**, 307 (1992)
282. P. Deák, M. Rosenbauer, M. Stutzmann, J. Weber, M. S. Brandt: Phys. Rev. Lett. **69**, 2531 (1992)
283. M. Stutzmann, M. S. Brandt, M. Rosenbauer, H. D. Fuchs, S. Finkbeiner, J. Weber, P. Deak: J. Lumin. **57**, 321 (1993)
284. R. D. Miller, J. Michl: Chem. Rev. **89**, 1359 (1989)
285. N. Ascombe: Seeking a Silicon Laser, Photonics Spectra (February 2003), p. 62
286. R. A. Soref: *Silicon Based Microphotonics: from Basics to Applications*, ed. by O. Bisi, S. U. Campisano, L. Pavesi, F. Priolo (IOS Press, Amsterdam 1999), p. 1

Integrated Photonics

Mario Paniccia, Mike Morse, and Michael Salib

Intel Corporation, Santa Clara CA 95054, USA
{mario.paniccia,mike.morse,michael.s.salib}@intel.com

Abstract. This chapter attempts to bring an industrial perspective to the topic of silicon photonics integration. Comparisons between microelectronics and silicon photonics are made. Monolithic and hybrid integration are discussed throughout the chapter with attention to cost, manufacturability, and performance. Potential applications, design issues, and packaging issues are addressed. Special attention is given to processing problems common in silicon-based optical systems.

1 Introduction

The need to transmit signals optically is becoming increasingly apparent as data rates exceed 1 Gb/s. Such high rates are straining the capabilities of electronics signaling technology. High-speed electronic transmission over copper is currently limited to distances around 100 m, and this distance will certainly shrink as data rates rise. If an integrated silicon photonic chip could be developed, it could bring significant, even revolutionary changes, to the next generation communications industry, by radically altering the price, power, and size for photonic components.

This chapter will draw parallels between the manufacturing of photonic devices and transistors in silicon. From the first integrated circuit in 1958 to today's high-end microprocessors that contain over 500 million transistors on a single chip, silicon remains the material of choice for electronics. Decades of industrial research and investment in silicon manufacturing have brought significant economic and technological advances to the semiconductor world, largely driven by the ability to reduce transistor size. These advances enable the creation of devices with increased performance at significantly lower costs. Figure 1 plots the number of integrated transistors per chip as a function of time since 1970. At this rate, there will be over one billion transistors on a single die before 2007. Figure 2 shows the resulting increase in processor performance as measured in Million Instructions Per Second (MIPS) and the corresponding decrease in cost per MIPS since 1985. If silicon photonics were to make similar advances as those made in microelectronics over the past thirty years, then this would create a radically new communications landscape.

Optical devices are limited in size by the wavelength of light routed through them. Due to the high refractive index of silicon one can make significantly smaller devices as compared to those made in a low index medium

L. Pavesi, D. J. Lockwood (Eds.): Silicon Photonics, Topics Appl. Phys. **94**, 51–88 (2004)

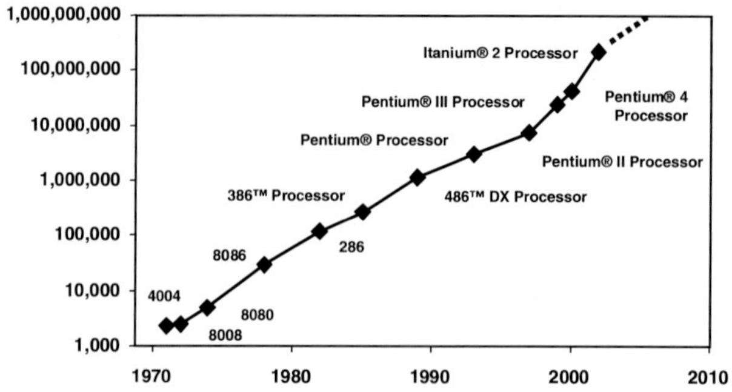

Fig. 1. Number of transistors on a die as a function of time since 1970

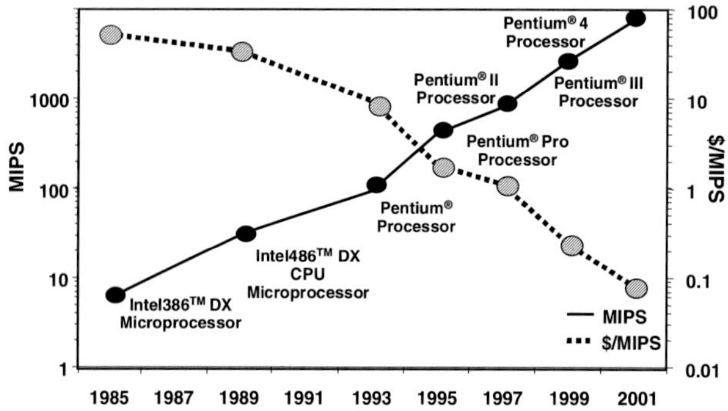

Fig. 2. Plot depicting the net result of Moore's law scaling as measured in $/MIPS (Million Instructions Per Second) as a function of time

(e.g. silica). All of this, however, comes with challenges such as increased scattering losses due to sidewall roughness, increased sensitivity to Polarization Dependant Loss (PDL), and increased coupling losses [1, 2].

Many reviews have been written since the early 1990's describing silicon based optoelectronics devices [3, 4, 5, 6, 7, 8, 9], and although much progress has been made, many technical challenges remain. Most experts see silicon as lacking in many properties needed for high quality optical performance. Three key properties are light emission, fast modulation for data encoding and low coupling loss between silicon waveguides and optical fibers.

Silicon is not the only material in which optical integration is proving difficult. To date, there has not been much commercial success with monolithic integration, even with preferred III–V optical materials. This is because integration for most devices has not demonstrated improved performance or a reduced cost. The reasons for this include complicated epitaxial regrowth steps needed for III–V materials that significantly reduce yield and issues with facet formation. Weak market demand for integrated optics has also been a factor. If, however, silicon based photonic devices could be developed using processes similar to baseline CMOS processes, one could achieve much higher yielding optical devices at lower cost.

Developing cost effective integrated optical devices in silicon will require an understanding of the issues related to processing, testing, and packaging. The ability to take advantage of existing fabrication infrastructure and knowledge greatly improves the chance that silicon-based optical devices could be produced in high volume at a low cost. This chapter will present an industrial perspective of creating integrated silicon photonic components.

2 Applications for Integrated Silicon Photonics

To overcome the limitations imposed by electronic communication, optics has been introduced at longer distances. Currently, it is used for almost all high-speed (> 1 Gb/s) interconnects running longer than 100 m, and is competing in 1–100 m commercial applications. Figure 3 shows a diagram of the applications for which photonics already exist, or is being considered. At the longest distances, dominated by telecommunications at 1550 nm, photonic components are being used for lasing, amplifying, multiplexing/demultiplexing (MUX/DEMUX), and attenuating functions among others. Ethernet and Fibre Channel protocols are typically used for shorter distances including enterprise and storage area networks (SAN). In general, these systems are less complex than their long-haul telecommunication counterparts, but some are implemented in dense wavelength division multiplexing (DWDM) systems. At intermediate distances of 1–100 m, both copper and fiber solutions coexist for applications such as rack-to-rack connections. Optical solutions here are relatively simple with no amplification or multiplexing involved. At even shorter distances, electrical signaling dominates for technical and economic reasons. Photonic components will have to be tightly integrated with the pre-existing infrastructure and have very low costs to be accepted in this space.

A discussion about the merits of optical interconnects is beyond the scope of this chapter, but others have written on this matter [10, 11, 12]. Instead, the insertion point of optics will be summarized. Figure 4 shows an estimate of the transition from electrical to optical communication over different distances and data rates. There is clearly some uncertainty in the timing because the electrical performance of silicon is constantly improving. Electrical data rates of greater than 8 Gb/s have already been demonstrated with limited equalization [13, 14, 15, 16] and modeling has predicted 10 Gb/s data rates for up to half a meter [17]. Companies like Intel are even more optimistic about extending the performance of copper to at least 20 GHz [18]. Since chip-to-chip interconnects are not predicted to reach such bandwidth until the next decade, optical signaling will not be implemented soon for this application. However, board-to board and rack-to-rack applications will need optical interconnect sooner. Since the requirements for silicon photonics

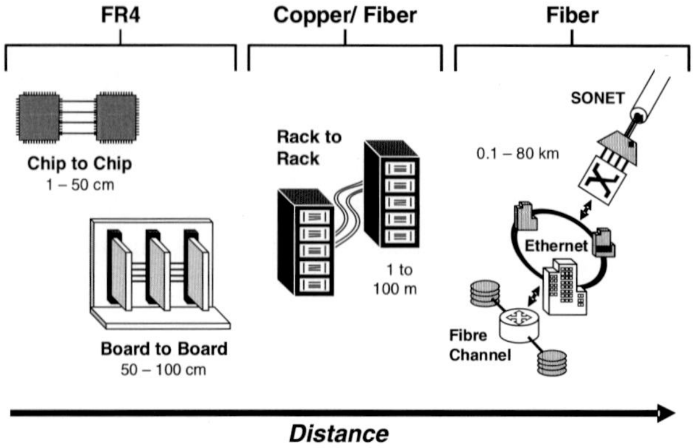

Fig. 3. Applications for communications. Optical technologies are currently only being used for longer links

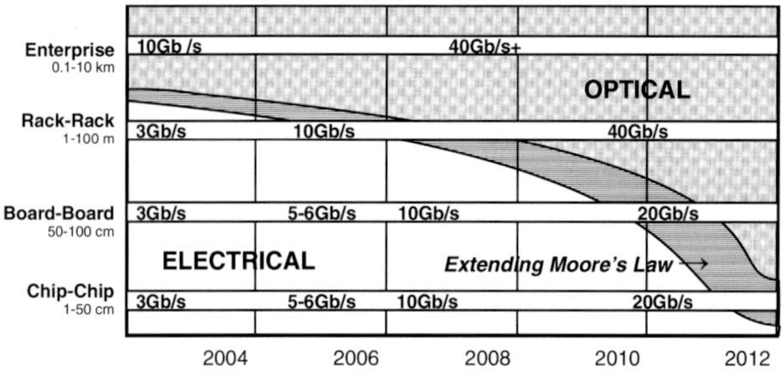

Fig. 4. Timeline for the transition between electrical and optical communication for various distances

are very different for the chip-to-chip application as compared to all other applications, separate sections of this chapter will cover the possible integration routes for each. The majority of the attention will be devoted to the chip-to-chip communications since that application offers the largest volume opportunity for the integration of photonics and microelectronics.

2.1 Chip-to-Chip Applications

Several reviews have discussed the prospects of optical interconnects for chip-to-chip communication [17,19,20,21,22] in depth, and the Chapter by *Gaburro* in this book will more thoroughly cover the material. It is important however to understand the current system design and performance as a starting point to a discussion of system constraints.

Figure 5 shows a typical architecture of an Intel CPU and chipset. The most demanding interface is the front side bus between the microprocessor and the memory controller hub (MCH), a chip that serves as the link to the graphics and memory functions. Below this is the input/output (I/O)

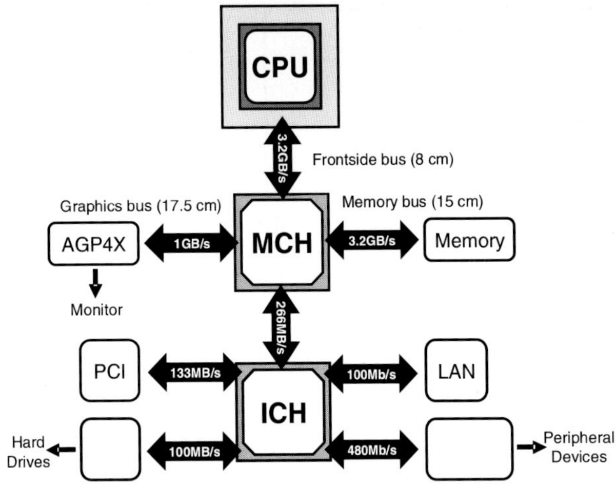

Fig. 5. System architecture of a desktop PC, showing the connectivity of the microprocessor and chipsets with the aggregate data rate and length

controller hub (ICH) that provides a connection to external ports. The speed of the link between the processor and off-chip memory is important because the processor will have to interrupt execution to wait for needed data to be fetched from memory. While microprocessor speeds have been growing exponentially for over twenty years, the front side bus speed has not kept up (Fig. 6). This means that the computer's performance is not scaling with the speed of the microprocessor. Similar problems also exist in multiple processor systems that share a front side bus. To minimize the off-chip latency, cache memory has been put directly on the microprocessor. This trend will continue, with even more cache memory being added, but architectural changes such as moving from a bus to point-to-point architectures will also help. Some of the first results of these concentrated efforts to improve the front side bus speed are seen in Fig. 6, where the most recent processors show a significant jump in performance.

Current proposals for chip-to-chip optical interconnects can be grouped into three camps: monolithic integration of emitters and detectors into silicon [23], hybrid integration of an off-chip light source with an integrated photodetector [24], and a totally hybrid approach where a III–V emitter and detector are placed on top of the silicon die [25]. In order to be competitive with electrical interconnects, any optical solution must have similar bandwidth, cost, power, and latency as the copper solution, while being capable of very high volume manufacture. The implications are that each optical channel must have a data rate of at least 20 Gb/s by 2010, and the entire cost of the link should be less than $10.

Monolithic integration is the most challenging approach due to the limitations of silicon-based light emission. Extremely low optical power output

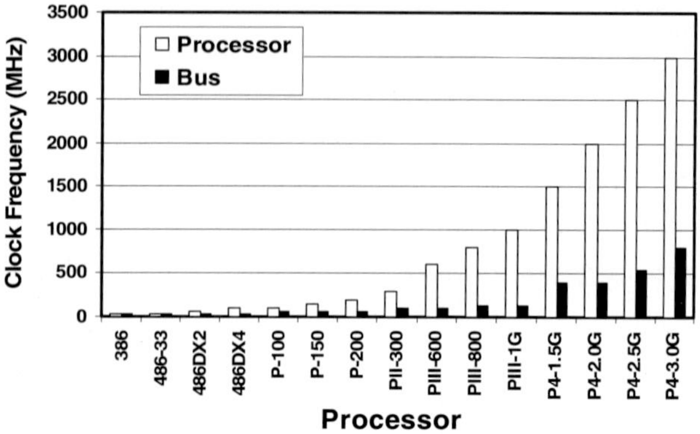

Fig. 6. Frequency gap between the microprocessor and the front side bus over the last 15 years

has plagued silicon devices, such as erbium-doped light emitting diodes, for more than a decade [26, 27]. This is discussed in more detail in the Chapter by *Fauchet* in this book. Another issue is that none of the current research in silicon light emitters promises modulation bandwidths greater than about 1 Mb/s, because of long carrier lifetimes in either silicon or excited states in the rare-earth centers. There is hope that this can be overcome by using a fast silicon-based modulator to encode data on a continuous wave light source. The vast majority of research on these modulators has been focused on waveguide-based devices, which tend to be fabricated either on thick Silicon-on-Insulator (SOI) wafers or on specialty epitaxial wafers. Neither of these two substrate solutions is particularly amenable to integration with leading edge microelectronics, which needs a substrate optimized for electrical performance. It is critical, therefore, to design a compatible solution before significant time and effort are expended on development of the optical devices. The fastest modulator data published to date is based on current injection in silicon waveguides, with a reported bandwidth of 20 MHz [4, 5], so there is a significant technology gap with the requirements of chip-to-chip communications. However, several groups around the world are addressing this issue, and current work is advancing the performance rapidly.

In the case that total monolithic integration of silicon optical and electrical components fails, or indeed as an alternative solution, III–V materials could be used to add optical functionality to silicon microelectronics. GaAs or InP-based lasers could either be bonded to the chip or packaged beside the silicon. Alternatively, an on-chip modulator could be used in conjunction with a III–V laser [20]. These modulators could be silicon, III–V, or polymer based. The integration challenges of all of these approaches will be discussed later.

The selection of the wavelength for optical communication is critical in determining candidate components for integration onto silicon. For wavelengths below 1100 nm, silicon is opaque. This means that silicon waveguides and planar modulators are very lossy, but that silicon photodetection is efficient. Above 1100 nm the converse happens, with the absorption coefficient of silicon falling to insignificant levels. SiGe alloys could be integrated on silicon to potentially stretch the absorption edge to 1550 nm, but no commercially available, integrated SiGe photodetector has yet been made.

If the goal is to achieve true monolithic integration on silicon it makes much more sense to operate at a wavelength that is more easily detected (1310 nm or less). The design of any silicon-based light emitter needs to take this into account. If planar silicon modulators are also required, the wavelength of operation for the emitter is similarly restricted due to waveguide absorption, resulting in an approximate usable range of 1000–1310 nm.

2.1.1 Temperature Issues

One of the practical constraints for integrated optics on silicon is the performance at elevated temperatures. The worst-case thermal environment occurs when the optical devices are located directly on the microprocessor. Figure 7 shows the modeled junction temperature of a proposed server die with four microprocessor cores and a large memory cache, under two conditions. In Fig. 7a, the top two microprocessor cores are active while the two bottom cores are off. This causes a hot spot of 105 °C to form over the active cores, with the remaining surface temperature ranging from 55–95 °C. In Fig. 7b, all four cores are active, but due to the system constraints on the processor, each is running at lower power. In this case, the hot spot has been reduced to 85 °C. It is expected that devices directly bonded to the top surface would actually experience a slightly hotter temperature by some 5–10 °C. However it should not be assumed that the microprocessor acts as a heat sink for the optical devices. It is much more accurate to describe the microprocessor as a heating plate for the optoelectronics. It is very unlikely that III V light emitters could operate for long times if they were subjected to temperatures above 80 °C. Future silicon-based light emitters might have fewer catastrophic device failures due to resistance to dark line defects, but the output power might fall so much at elevated temperatures that the optical loss budget is exceeded. In addition to surviving at the maximum temperature, the optical device has to operate over a range of temperatures as the device heats and cools under normal operating conditions for a lifetime of the order of ten years. Many of the proposed optical solutions located at the processor should be re-evaluated in light of this thermal environment [28]. Receivers and modulators are not expected to be as problematic as the light emitters since their failure mechanisms are less temperature sensitive, but their performance may be seriously compromised at high temperatures due to variations in current, refractive index and efficiency.

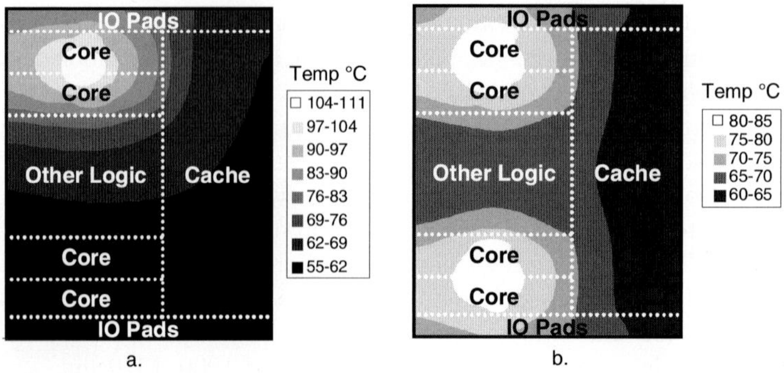

Fig. 7. Modeled thermal maps of the junction temperature for a potential server microprocessor. The die consists of four logic cores and a large cache. In (**a**), the top two cores are active and the bottom two inactive. This is the worst-case scenario for the part. (**b**) shows the case when all four cores are active, but with reduced power

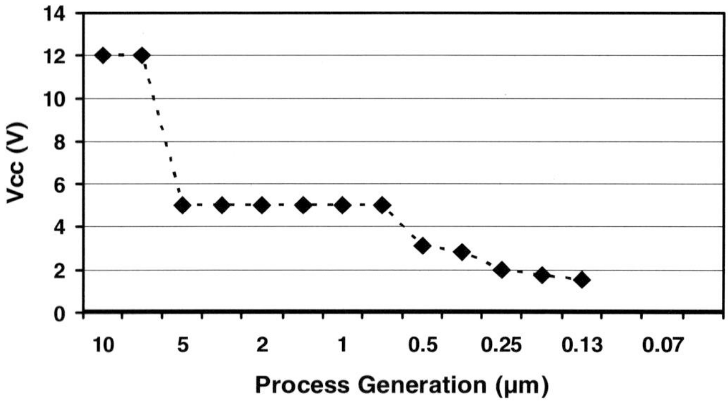

Fig. 8. Voltage scaling of microprocessors

2.1.2 Voltage Issues

The relentless pursuit of Moore's Law necessitates the reduction of voltage for every new CMOS generation. Current microprocessors are running at 1.3 V. Special circuits on the chip, commonly called charge pumps, can generate higher voltages with proper design, some possibly reaching five times the supply voltage, but they are the exception and not the rule. A higher DC source on the circuit board, for example 3.3 V, would be required at the starting point for the charge pumps in order to reach higher voltages. Figure 8 shows the voltage scaling that has been achieved, as function device critical dimension. The key point is that optical devices that operate at voltages lower than 10 V have a less of a barrier to entry into CMOS designs.

2.2 Longer Reach Applications

As mentioned earlier, several applications areas beyond chip-to-chip are viable for integrated silicon photonics, including telecommunications and data communications. The recent collapse of the telecommunications market has dramatically slowed the development and rollout of integrated optical components. Nevertheless, the convergence with data communication at 10 Gb/s is putting significant downward pressure on cost and will force operators to adopt lower cost equipment. A high level of performance must be maintained while power consumption and size are decreased, and ease of management is increased.

The application (and market size) determines which devices should be integrated. In serial integration, consecutive components of a single channel are integrated together to form a larger part of a photonic integrated circuit. An example would be the Externally Modulated Laser (EML) where a laser and modulator are integrated. Parallel integration occurs when multiple copies of a specific device exist across different channels. Banks of photodetectors or attenuators for different wavelengths in a wavelength division multiplexed (WDM) system would be good examples. The reader is referred to the Chapter by *Janz* on silicon DWDM in this book for more details. Components using parallel integration are usually much larger in order to accommodate the physical spacing between channels. Their cost can therefore be prohibitive unless low cost materials or reduced pitch devices are used. Since no semiconductor can compete with silicon in cost and device size, silicon will have an advantage for parallel integration applications that do not require light emission.

Silicon photonics will have a better chance to gain a foothold into data communications than telecommunications for three reasons; some performance can be sacrificed because the distances are smaller than in telecommunications, the market is extremely cost sensitive, and a large part of the data communication is anchored to servers and desktop computers where silicon in ubiquitous. For the shorter distances, fiber loss is not significant so higher losses can be tolerated at the interfaces between components. The large potential volume of the market and the competition with copper will necessitate inexpensive optical solutions.

Figure 9 shows a schematic of an integrated transceiver that could be implemented for data communications, combining an InP laser with silicon-based photodetection and waveguiding. SiGe photodetectors would be necessary if InP is used due to the low absorption coefficient of silicon at wavelengths above 1100 nm. Transceivers like this could be used for board-to-board and rack-to-rack communications as depicted in Fig. 10, as well as for enterprise applications. Additional channels could be integrated together to increase the aggregate data rate if needed (Fig. 11).

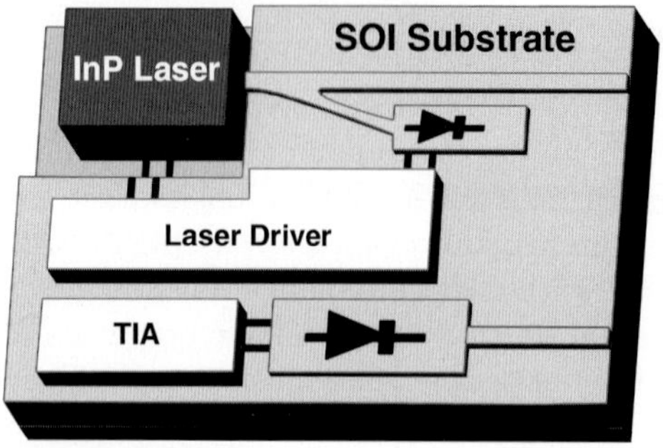

Fig. 9. Example of the integration of silicon and III–V to make a transceiver. Two silicon-based photodetectors are included, one for power monitoring and another for receiving data

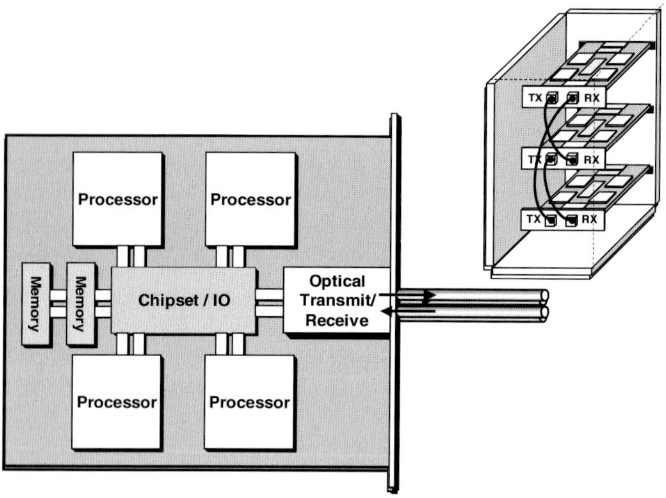

Fig. 10. Example of a potential board-to-board optical interconnet. Aggregated electrical data from several processors is converted to optical data at the edge of the board before being sent to other boards

The environment for these applications is much less harsh than the chip-to-chip case. Expected operating temperatures for the components are less than 60 °C, with available power supplies of greater than 10 V.

3 Integration

There are two primary benefits from integration: increasing performance and decreasing cost. Unlike the case of electronic integration that results in a greater number of identical, higher performance, transistors on a chip,

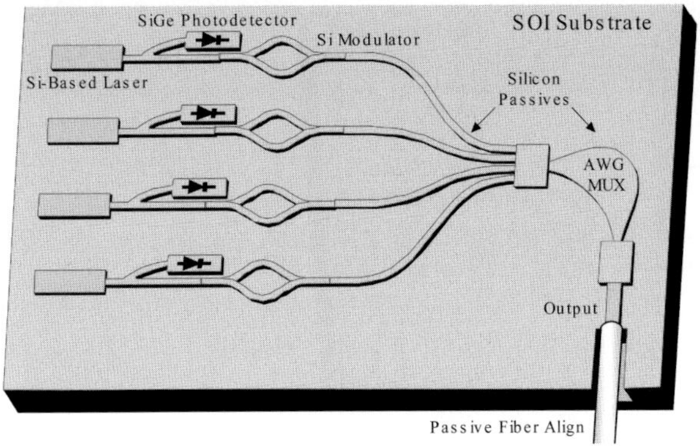

Fig. 11. Example of monolithic integration on silicon

optical integration involves assembling components that either perform different functions, or that are made from dissimilar materials. This has profound implications for the gains in performance and cost. Since the method of integration is critical in determining the quality of the component, it is necessary to first define the different integration approaches before discussing their potential benefits.

3.1 Types of Integration

The most common definition of monolithic integration refers to a process by which all of the components (e.g. electronic circuits, light sources, photodetectors, modulators, waveguides, and multiplexers) are manufactured on the same piece of semiconductor substrate. The example shown in Fig. 11 assumes a Si-based light source exists. A more liberal definition of monolithic integration would allow the inclusion of III–V compounds or polymers (either grown or deposited) processed on a silicon platform. This latter definition is convenient because it allows for different materials to be combined during processing. Such a mixed-material monolithic process provides an opportunity for greater flexibility in the choice of devices to be integrated. This would be an option for integrating a non-silicon-based light source on an otherwise all-silicon chip. To fully realize mixed-material integration, compromises in processing may need to be made which could severely affect both performance and yield. Despite these concerns, researchers continue to work on both single-material and mixed-material monolithically integrated planar optical chips [28, 31, 32, 33, 34].

Hybrid integration is accomplished by assembling disparate parts onto one common platform; the example depicted in Fig. 12 uses SOI as the base material. Common hybrid optical components are those in which III–V compound light sources and detectors are attached onto SOI, silica, or polymer platforms. As can be seen from Fig. 12, the power of hybrid integration lies

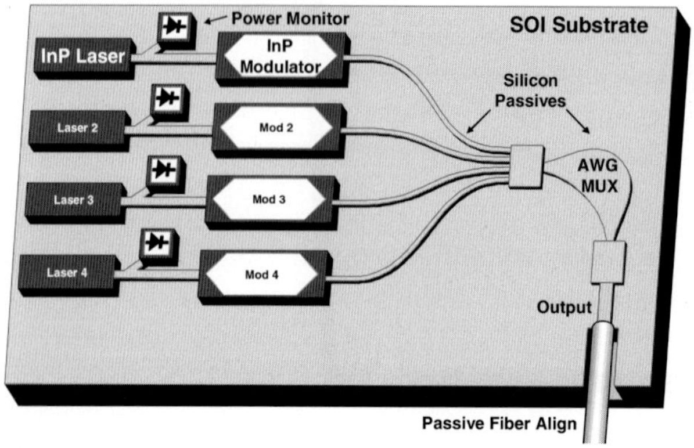

Fig. 12. Example of Hybrid Integration on silicon

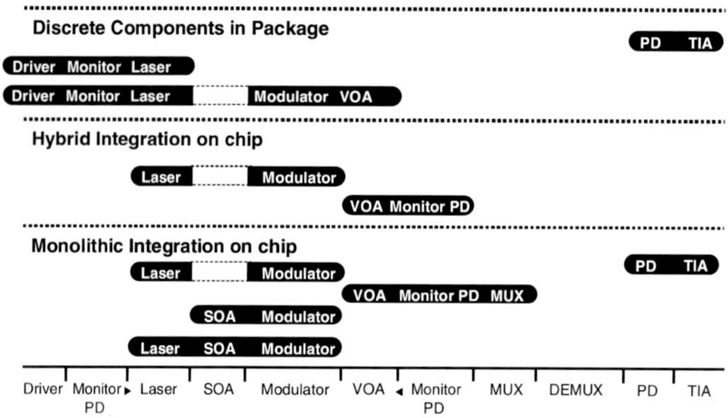

Fig. 13. Level of device integration in commercially available products

in the ability to combine the best performing devices onto one common chip, despite the different technologies used to manufacture each individual device.

Optical components using both types of integration are commercially available. Figure 13 shows which devices have been integrated together on a single chip in comparison to the standard packaging of discrete devices. This latter category is typical for components used in telecommunication applications. The order of devices on the axis follows the order that is found in an optical link. Examples of monolithic integration are almost exclusively made in InP, with the only exception being the integration of a normal incidence photodetector and amplifier in silicon. The integration of optical and electrical devices is the exception rather than the rule however. Combining optical devices is more common because it offers clear performance advantages while maintaining production yields. This will be discussed in depth in the next section.

3.2 Cost

Although integrated optical components had their beginnings over thirty years ago, they have not evolved with the same success, neither in complexity nor in functionality, as integrated electronics [29]. Most of the credit for the technological successes of integrated electronics can be attributed to the advancements in silicon processing. With the opportunity to use these mature manufacturing processes, semiconductor electronics manufacturers are in an exceptional position to drive optoelectronic research, development, and product implementation. An additional benefit of silicon photonics is the opportunity to utilize older technology process tools. Using depreciated factories dramatically reduces the costs of research and development, making investment and production of silicon optical devices more attractive.

Other factors in the cost of integrated optics are the lack of convergence to one material and the absence of device standards, as in the case for volume manufacturing of electronics. With the successful development of additional optical components in silicon, a standardization of market specifications, and focus on a common material platform, silicon-based integrated optics may realize the economy-of-scale benefits that integrated electronics has enjoyed for so long. However, convergence alone will not drive down costs.

In order for silicon to be a viable material for monolithic optical integration, it will need to compete against the high yields associated with hybrid integration. This will require tapping into available processing knowledge and infrastructure to create a stable environment for research and development of new optical components. Monolithic integration will also need to manage production costs, which rise as the complexity and number of processing operations increase. Each process technology is more expensive to implement than the one before it. Modern lithography reticle costs are rising with every technology generation (about $1 M for a 0.13 μm CMOS reticle set), resulting in a need for high volume production over which to amortize such large initial costs. Fortunately many of today's optical designs can be made using older generation processing equipment.

With the cost of new CMOS manufacturing fabrication facilities in the several billions of dollars, competition for volume silicon processing is becoming more and more cost prohibitive. Though the same number of processing steps are not needed in optical processing, large costs of operation are promoting more 'fabless' start-up companies as the number of optical manufacturing companies dwindles. Figure 14 shows the costs of several optical semiconductor materials ($\$/\mathrm{in}^2$) as well as the costs typically charged by foundries to manufacture some common devices. It can be seen that InP area costs can be almost an order of magnitude larger than either GaAs or SOI. In addition, although GaAs and SOI wafer area costs are almost equal, it is expected that SOI costs will drop as the demand for volume production in the electronics industry increases. Similarly, as the electronics industry matures on 12 in silicon wafers, a reduction in the initial wafer area cost is also expected.

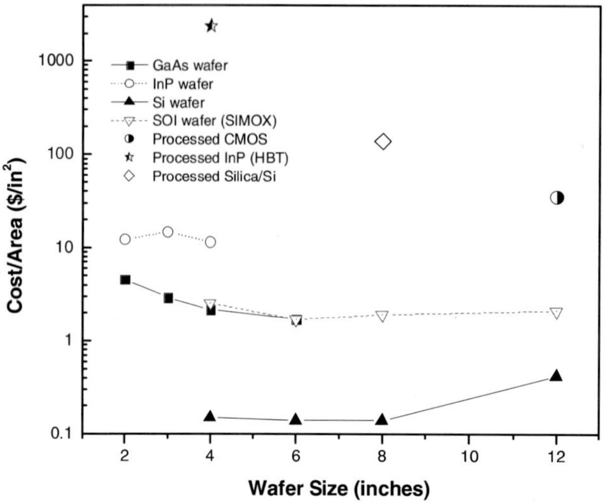

Fig. 14. Start wafer costs and wafer fabrication costs per unit area for several semiconductor materials

Wafer area costs alone do not provide enough insight into the total cost structure of optical devices. Device size and manufacturability are also large factors in determining whether or not there exists an economic incentive to pursue a particular material. As a point of reference, the foundry cost to process a 5-metal layer silicon electronic device (about 25 total mask layers) on a 300 mm wafer using 0.13 μm process technology can be \$5000. The price per square-inch is around \$40, as shown in Fig. 14. The foundry cost of an InP Distributed Feedback (DFB) laser can be over $\$2000/\text{in}^2$ while that of a silica-on-silicon based device is about $\$150/\text{in}^2$. These costs should be considered when designing optical components that require large areas (e.g. waveguides, AWGs).

Although monolithic components in silicon may be simpler to manufacture than those in III–V materials, the potential to add electronics to active silicon-based optical components is an attractive feature to corporations. Unfortunately, the challenges associated with marrying silicon-based planar optical devices and conventional electronics on the same die have not been overcome. Instead, several research groups have combined normal incident silicon photodetectors with receiver circuitry [30]. These efforts have been successful because almost no new processing steps were required to produce the photodetector. The manufacturing cost and yield of the integrated part is similar to the cost of a single discrete part. This is not always the case when two devices are integrated in silicon. If a SiGe process is used for the photodetector, it can increase the number of mask layers and impose thermal constraints required for film stability. In addition to decreasing the total yield, there could be a negative impact on the throughput as thick buffer layers or slower growth rates are required. To better anticipate some of the potential issues that await silicon-based optoelectronic integration, it may be helpful to briefly review, as a case study, the history of optical integration on InP.

InP is the quintessential optoelectronic material because high performance electrical and optical devices can be made on it. Groups in universities and companies have been doing research on the integration of either all optical devices, or optical and electric devices [31,32]. Examples include the complete integration of WDM optical devices from the lasers through a MUX/DEMUX to photodetectors [33]. Integration of a photodiode and the amplifier has also been shown [34]. However, many companies have given up on commercializing these devices, either due to lack of demand or technical difficulty.

The reason that integration on InP has largely been relegated to research is that it has failed to decrease the cost of production or improve performance. Putting different types of devices on InP usually entails epitaxial regrowth steps that often *significantly* decrease the yield. The number of processing steps such as lithography, etching, and cleaning are also increased, further reducing yield. With these increased number of processing steps come increased wafer handling, which, since InP is such a brittle material, results in higher incidence of wafer breakage. Some devices such as AWGs consume a large fraction of the incredibly valuable substrate surface, even though no device specification is reliant on the materials properties of the III–V material. Added to these cost disincentives is the R & D budget required to make the process functional at all. In order to recoup these costs, there has to be a very large market for the components, and the recent telecommunication collapse has prevented that from happening for all but the EML, which is currently the only commercially successful component.

Since there are so many differences between InP and Si integration, there is reason to hope for a better outcome in silicon. Most of the optical devices that are being considered for silicon are based on either intrinsic silicon or silicon doped with another element, so the need to resort to yield-killing epitaxial regrowth steps can be avoided. If the optical functionality can be added while keeping the process similar to the baseline process for electronic devices, it is reasonable to expect very high yields as compared to yields of 5–15% for InP EML. It is still an open question as to how close the optical fabrication process flow will be to a normal CMOS process. It is also likely that fewer device compromises will be required in silicon than in InP if everything is nominally made from the same material.

Examining only manufacturing costs of devices is not sufficient since the costs of testing and packaging PLCs can be as much as 60–80% of the total cost [35]. Therefore, unless significant improvements are made in these areas, simply reducing substrate manufacturing costs will not provide the necessary economic incentive to pursue highly integrated devices. Although both monolithic and hybrid integrated devices require testing and packaging, the time and cost is significantly reduced in the monolithic case. Testing and assembly times of hybrid integrated devices increases with the number of parts on the substrate. In contrast, the time to test and assemble monolithic devices can be significantly reduced since there are fewer mechanically aligned interfaces to test, because many of the interfaces are lithographically defined in the substrate. As a result, the costs associated with assembly and testing can be shared by many devices on one chip. Another benefit of monolithic integration is the potential to increase the number of devices with little or no impact to production time. This is not true for hybrid integration since an increase in the number of devices necessitates a proportional increase in the amount of processing time.

3.3 Performance

One of the most contentious issues in integrated optics is determining the impact of integration on the component performance. The fundamental question hinges on the trade-offs that must be made in the integration process, and it is different for the two types of integration described previously.

Monolithic integration on silicon has distinct advantages associated with alignment and interfaces, as compared to hybrid integration. Misalignment between devices in the monolithic circuit is defined by the accuracy of the lithography tool. Integration of components entirely based on a silicon waveguide platform, such as the integration of VOAs and AWGs, are defined in a single step and are not subject to misalignment at all. This minimizes or eliminates the losses between the individual devices. Interfacial losses can also be reduced if the mode is always in a constant refractive index material with constant dimensions. Since losses between discrete components can be 0.1 dB/interface, the savings in an integrated system can quickly accumulate.

The problem with monolithic integration is the poor device performance resulting from constraints placed on material selection. This is the most important for devices that actively manipulate the light, such as emitters, modulators, and detectors. Most of these devices operate near the band edge of the material. Since photodetection only becomes efficient when the wavelength is well above the band-gap, a silicon-based detector with a much smaller band-gap than the silicon-based emitter has to be integrated. Unfortunately, band-gap engineering in silicon-based materials is constrained by problems in lattice matching, so compromises in device performance have to be made. The result is that totally monolithic integration will be best suited for less demanding applications.

Hybrid integration offers more degrees of freedom for the design of each optical component. Each individual component – emitter, detector, etc. – is optimized during fabrication and integrated later. The penalties paid for this are the additional interfaces and misalignment between components. When the interfaces are designed correctly with anti-reflection coatings and mode matching, the losses can be on the order of 0.1 dB, but this requires an extensive alignment and packaging effort. Issues associated with this process will be discussed later in the chapter.

3.4 Processing

Over the past decade, there have been many reviews [1, 2, 3, 36] on the opportunities associated with optical devices and the potential integration with silicon electronics. In particular, researchers have presented a vision of optical devices monolithically integrated with modern CMOS devices [36, 37]. Though quite realizable, the details of such a vision must be completely compatible with the physical constraints associated with silicon-based processing. Optical devices in silicon have been, to date, made without the monolithic integration of silicon electronic devices. Albeit a promising step towards compatibility, it is an altogether different problem to integrate these research devices with silicon electronics. In this section, we hope to inspire more serious considerations with regard to the feasibility of creating high-volume silicon-compatible optical devices.

Although there are many silicon electronic processes, we will use CMOS processing as a working example for the remainder of this section. Throughout the section, we will try to point out several instances in which devices can utilize existing infrastructure and processing techniques yet remain incompatible with the technology to which they are being integrated, and vice versa. In order to successfully manufacture a monolithic circuit, one must carefully choose the order of each process step to avoid significant physical conflicts. Conflicts such as chemical incompatibility, thermal instabilities, or challenges to lithography and etching processes due to underlying topography, can present stubborn processing roadblocks to successful integration. Realizing CMOS compatible monolithic integration of optical devices will require special attention to processing details. In this section we will discuss several processing challenges.

3.4.1 Thermal Stability

Thermal budget limits are critical when integrating different devices into the same process. In the case of CMOS devices, inserting additional processing steps must be done while observing the original processing limits. Failing to account for thermal budgets could result in the interdiffusion of dopant species, weakening of metal layers, and the introduction of stress due to differing coefficients of thermal expansion (see Table 1).

CMOS processes are carefully tuned to optimize transistor performance. In order to avoid compromising electrical performance, additional steps should either conform to existing thermal budgets or be completed before electrical processing begins. The same holds true for ensuring optical performance is maintained; a Si modulator may operate at slower switching speeds

Table 1. List of refractive index and linear coefficient of thermal expansion for common optical materials at $T = 300\,\mathrm{K}$

Material	n (1550 nm)	CTE ($10^{-6}\,\mathrm{K}^{-1}$)
Si	3.45	2.56
SiO_2	1.5	0.55
Si_3N_4	2.0	2.50
$Si_{1-x}Ge_x$	3.45–4	2.56–5.7
GaAs	3.38	6
InP	3.18	5.5
Polymers	1.6–1.7	10–100

or suffer an increase in optical loss if dopants diffuse from their original implanted regions.

Some of the most common CMOS thermal processing steps are listed below in Table 2.

Table 2. List of common front end thermal steps used in a typical CMOS process

Process Step	Approx. Operating Temp (°C)
TEOS anneal	1000
Gate anneal	1050
Source/drain anneal	1050
Silicide formation	650–1100
Tungsten deposition	500–650
Cobalt RTA	600
Gate oxidation	1050
Activation RTA	950

Table 2 ignores the total thermal budget, that is, the temperature *and* time of heating; we are also not addressing any complications associated with thermal cycling or ramp rates for a particular step. Spike anneals and fast thermal cycles are required to reduce transient enhanced diffusion and dopant deactivation. The effects of transient diffusion are the main reasons why the electronics industry moved from slow furnace batch annealing to single-wafer Rapid Thermal Annealing (RTA) techniques such as spike and flash anneals. One difficulty for optical integration would be the incorporation of an optical process step that requires growing a thick SiO_2, layer. For such a large thermal budget, usually many hours at temperatures above 900 °C, this step would need to be done prior to any backend (metal) or critical ion implantation steps. Thermal budgets for both aluminum and copper back-end processes with oxide-based dielectrics are limited by the Interlayer Dielectric (ILD). Although some low-k dielectrics ($k < 3.9$) are being used (e.g. Carbon-Doped Oxide (CDO)), the thermal stresses and the weaker mechanical properties of CDO will not be able to tolerate temperatures in excess of 450 °C; low-

k solutions such as Spin-on-Dielectrics will drive the thermal budgets even lower. In the case of electro-optical (EO) polymers, the thermal budget can decrease dramatically since typical glass transition temperatures are in the range of 150 °C–250 °C. For all practical purposes, these polymers can only be integrated as the last steps of the process.

In addition to the thermal issues of decomposition and softening mentioned above, the problem of strain relaxation must be considered for processes that use heteroepitaxy. While this could be the case for III–V growth on silicon, it is far more likely to be seen if germanium is introduced to create photodetector production. Germanium concentrations in excess of 50% are desired to achieve acceptable responsivity from the SiGe photodetector at 1310 nm and 1550 nm. At these concentrations, it is very easy to exceed the thermodynamic critical thickness [38] and grow a thermally metastable film. If latter processing is sufficiently hot to allow the formation of misfit dislocations to relax the film, the optical properties of the device can be degraded. This could restrict the processing of thermal oxides and RTA steps.

Another factor in the successful processing of monolithic optical devices is the choice of materials. Silica-on-silicon technology typically requires that thick layers (several microns) of SiO_2 be processed in contact with crystalline silicon. After cooling from the high-temperature (approximately 1000 °C) oxidation step, a great deal of strain is created between the glass and crystalline layers. Interlayer strain can cause undesirable effects including strain in the optical layer, de-lamination of the oxide layer from the silicon substrate, or cracking of the glass layer to name just a few.

The thermal budgets of hybrid integration mainly revolve around the alignment stability of the packaged part and the degree of thermal isolation between devices. Each of the optical components may be secured in place with a different method (e.g. gold eutectic bonding, laser welding, epoxy, etc.), which will impose different thermal constraints on the operating temperatures of the device. Thermal crosstalk can be damaging to the alignment between two optical devices. Several options exist to ensure protection of the hybrid component from thermally-induced misalignment. Some options include keeping operating and environment temperatures at the lowest value common to all sub-components, ensuring thermal isolation between the individual components, or actively controlling individual component temperature to compensate for thermal crosstalk.

3.4.2 Chemical Compatibility

During integration, chemical compatibility between different processes is a cumbersome problem to deal with. Most conflicts can be addressed by incorporating extra de-contamination steps or adding dedicated tools for the problematic operations. As one would expect, these solutions can be extremely expensive. Depressed yields due to chemical contamination have been brought under control in standard silicon processing, but would need to be addressed if the process demanded the inclusion of elements and compounds not used in standard processing. Integration problems related to chemical contamination are likely to pose practical as opposed to physical limitations.

The term 'contamination' can encompass problems related to incompatible chemical processes in addition to particle defects. Contamination of the wafer surface or of thin-film materials can lead to adhesion and delamination issues. Certain techniques commonly used in university research processing are avoided in silicon fabrication due to their high likelihood of contaminating the production line. An example of this is the case of fast anisotropic wet-etching of silicon with KOH to form V-grooves for fiber alignment. In a commercial environment the risk of alkali contamination in silicon is too high. As a result, defining *v*-grooves must be done after all electronics processing is complete. Similarly, gold metallization (typical for III–V devices) is avoided in the silicon process until the final step, to prevent diffusion of metal into silicon. Even after its introduction over three years ago, the semiconductor industry is still trying to address issues associated with copper (Cu) contamination. The problem of Cu contamination extends beyond the factory and to the vendors and suppliers who perform cleaning services and other handling steps. Copper protocols can dramatically impact cost by requiring a special cleanroom layout and air flow as well as a duplication of all process equipment, tools used to service semiconductor equipment, wafer carriers, and even the gowns worn by the fab personnel. There are still no universally acceptable levels of Cu contamination, and many questions still remain about how far beyond the fabrication tools and chambers one must de-contaminate. Similar logistical problems may be waiting for the monolithic integration of optical devices.

3.4.3 Lithography

Minimum dimensions of photonic devices are limited by the wavelength of light in the medium which for which they are designed; ($\lambda = \lambda_o/n$), where λ_o is the free-space wavelength and n is the index of refraction of the medium. In the case of silicon ($n_{\mathrm{Si}} = 3.45$), devices designed for a free-space wavelength of 1550 nm have minimum widths of approximately 450 nm for an operating wavelength of 1310 nm, widths are approximately 350 nm. In optical systems, unlike electronic circuits, bends in transmission lines with small radii of curvature can cause unacceptably high radiation losses and transition losses. The large refractive index contrast of SOI ($\Delta n = 2$) provides better confinement than lower index contrast systems, allowing for relatively compact designs. Silica-on-silicon based Arrayed Waveguide Gratings (AWGs), with an index contrast of $\Delta n < 0.01$, have typical design sizes of approximately 12.5 cm^2 for a 1×4 channel multiplexer/demultiplexer. The same device can be designed in SOI at 0.5 cm^2, providing a reduction in device size of $25\times$.

The ability to faithfully reproduce designed features on a wafer is important to the success of optical devices. Critical Dimension (CD) control and

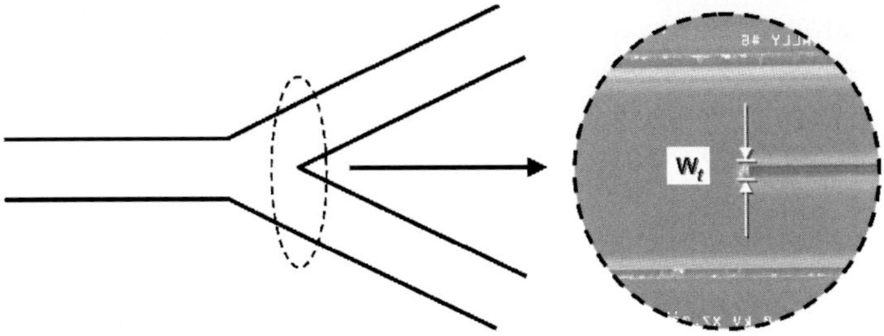

Fig. 15. Picture of a Y-splitter in SOI. Scattering occurs primarily at the junction tip labeled w_t

smooth device surfaces are two of the most important parameters that must be controlled during processing. Producing CDs of approximately 350 nm is hardly a challenge for modern optical lithography processes, whose minimum features have already been demonstrated below 60 nm. This does not, however, reflect the minimum CD in the design. In devices such as Y-splitters, AWGs, etc., the reduction of the splitter region to a point will always benefit from state-of-the-art patterning technology. In the case of a Y-splitter (Fig. 15), rounding and increased tip size in the splitter region increases the scattering loss.

The ideal situation would be one that reduces the tip width (w_t) to zero, resulting in zero scattered light. Since such a design cannot be patterned reproducibly, the minimum size of the splitting region is limited to the minimum resolvable feature size available by photolithography. Thus, the better the lithography and etching capability, the lower the scattering losses associated with the splitting region.

Another example of the benefits high-resolution patterning brings to planar optical circuits can be demonstrated by a Bragg grating. These gratings can be used as wavelength selective filters or laser cavity mirrors. To define Bragg gratings for a specific wavelength λ_b, the relation $\lambda_b = 2n_{\text{eff}}\Lambda$ must be satisfied, where Λ is the grating pitch and n_{eff} is the effective index. To create a first-order Bragg grating in silicon for $\lambda_b = 1550$ nm and $n_{\text{eff}} = 3.45$, a pitch of 225 nm is required. Thus, for a line/space duty cycle of 50%, lithographic patterning must be capable of 110 nm lines or spaces. Such features are common for the 90 nm lithography node. Although it is possible to create higher-order Bragg gratings with lower resolution lithography, the resulting loss is greater.

State-of-the-art lithography resolution may not be required for most conventional waveguide designs, but ring resonators and photonic crystals might require CD values on the order of 100 nm. Resolution and CD control are both important to the success of the devices. In such cases, CD control that accompanies modern patterning techniques (about 1 nm over an 8 in wafer) can be a benefit. In the case of Si-based ring resonators, one of the critical parameters to control is the gap (about 100 nm) between the ring and bus. Since the device operates through evanescent coupling, the coupling is exponentially dependent on the size of the separating gap. Thus, in order to reliably process high-Q devices, control of a few nm demands CD control readily achieved by modern 130 nm or 90 nm optical lithographic techniques [39, 40, 41]. This level of CD control would also be required for other small coupler designs.

More recently, SOI-based photonic crystals have been receiving considerable attention. Though we leave a more detailed discussion for another chapter, we would like to point out the most dramatic benefits of a high-index contrast system like SOI for novel physical systems such as photonic crystals. Among the many advantages to be discussed later, two are of special note: compatibility with modern lithographic resolution and reduced device form factor. For structures designed to operate at wavelengths of 1550 nm and 1310 nm, photonic crystal lattice structures (holes) are of the same dimensions/design rules as those for 130 nm and 90 nm CMOS transistor designs. This is of particular interest since, to date, almost all research efforts, especially photonic crystals, have been restricted to the use of e-beam lithography for pattern definition.

Although small CDs for optical designs (about 100 nm) can be patterned using modern 130 nm lithography tools, it is not a viable research process. Furthermore, since optical lithography minimum resolution is still many years away from that of today's standard research e-beam writers, e-beam tools are preferred for research and small commercial manufacturers. A few of the greatest advantages of e-beam lithography are the relatively low cost, the ability to pattern < 20 nm features, and their relative ease of use for processing pieces of wafers. In contrast, today's 130 nm optical lithography scanners have a resolution of approximately 100 nm, can process only whole wafers, and are prohibitively expensive. Owing to factors such as long patterning times, relatively large field-stitching errors, and susceptibility to particle contamination, e-beam lithography is a long way from being a competitive commercial alternative to conventional optical microlithography. A state-of-the-art research e-beam system can have patterning times of 30 minutes or more for a 10 μm^2 field with a 50% pattern density of 100 nm minimum feature sizes. Should minimum resolution be desired (20 nm), the patterning time can increase. With exposure times of less than 1 s for a standard scan field, optical lithography offers an incredible advantage in processing time if an optical design requires a larger area than that provided by one optical lithography scan field (approximately 25 mm $\times$ 35 mm), the lithographic stepping error due to stitching two fields together is < 25 nm 3σ. If however the design requires an area less than or equal to one scanning field, modern optical lithography is a better tool since the stitching error of the e-beam field (60 nm 3σ) can result in additional optical losses due to the many discontinuous sections that compose the device. Overall, standard 130 nm optical lithography can deliver equal or better results as compared to most research e-beam systems for today's planar optical designs.

High resolution lithography requires flat surfaces on which to pattern features. Height variations in wafer topography must be less than the lithography tool's usable depth of focus (DOF). Since typical waveguide heights exceed the DOF, process steps must include extra layers that can be mechanically polished. Even the shortest SOI single-mode waveguides must be covered in a planarized layer (e.g. SiO_2, SiON) before the next layer is patterned.

3.4.4 Etching

In addition to lithography is the subsequent etching of the patterned area. There are two common methods used in semiconductor processing: wet chemical etching and reactive ion etching (RIE) or 'dry etching'. Wet etching is a purely chemical etch and can be isotropic or anisotropic, depending on the chemicals used and the materials being etched. In some cases, researchers have exploited the fact that certain crystal planes etch faster than others to achieve anisotropic etch rates, but the anisotropy and precise CD control have not matched those found in dry etching. RIE is a technique that uses a plasma consisting of ions, free electrons, free radicals, and neutral molecules. By controlling the electrical potential of the wafer, positive ions can be guided to the wafer surface in a directional manner. This combination of chemical etching (free radicals reacting with the wafer) and physical etching (ion bombardment) provides a highly anisotropic etch; the vertical etch rate is much greater than the horizontal etch rate. Most optical and electrical designs demand vertical sidewalls, and therefore require the use of highly anisotropic etch techniques. In addition, unlike purely wet chemical etching, RIE provides excellent critical dimension control and the ability to etch high aspect-ratio patterns.

Most critical etching is accomplished by controlling the time of the process. Until recently, researchers could not rely on RIE control for accurate patterning of critical dimensions. In the case of ring resonators, researchers have used thin film deposition to manage the separation distance of resonantly coupled features (the space between the bus and the ring). Such a technique requires additional mask layers to accomplish what can now be done with standard lithography and RIE.

Since RIE relies in part on physically etching the wafer surface, the etch selectivity (selectivity = etch rate of material to be etched/etch rate of patterned protective mask layer) must be relatively high in order to achieve high aspect ratio etching. If the selectivity is low, the masking layer (e.g. photoresist) thickness is increased. Unfortunately, such a situation cannot be supported by modern lithography since a consequence of reduced DOF is the need for a reduction in photoresist thickness. The DOF puts a limit on the photoresist thickness, which is rapidly approaching $< 100\,\text{nm}$. This is not an issue for CMOS processing since typical etch depths are on the order of 0.1–0.3 μm for critical layers requiring minimum feature sizes. Typical etch depths for optical designs are between 0.1–6 μm.

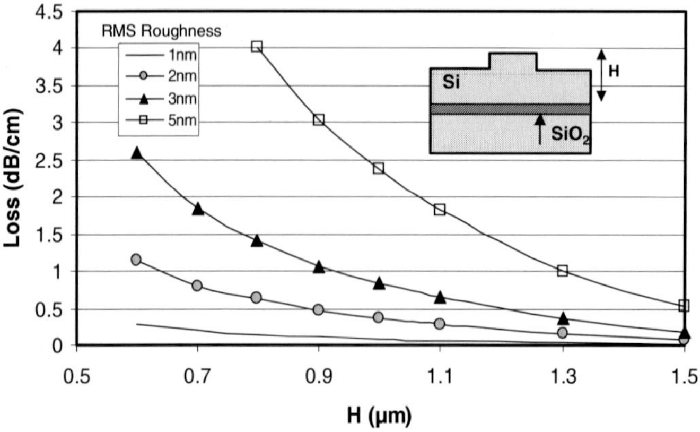

Fig. 16. Modeled scattering loss for SOI rib waveguides. The dimensions of the etched rib height were chosen to satisfy the single-mode condition for a particular total WG height (H)

RIE processing results in roughened sidewall surfaces, which, if significant, can result in additional optical loss. Contributions to waveguide propagation loss are scattering and absorption. If optical transmission losses in waveguide structures are reduced to negligible levels, the scattering associated with sidewall/interface roughness plays the dominant role in the overall loss of undoped crystalline materials. Unlike the scattering associated with defects in the waveguide medium, interface scattering is a direct result of lithography and etching performance. *Tien* derived an approximation to the loss related to interfacial scattering [42, 43, 44]. Today's conventional 180 nm lithography/etch processes can reliably produce sidewall roughness figures on the order of several nm rms. Unfortunately, since the scattering losses vary rapidly with the rms interface roughness as waveguide dimensions are reduced, several nm may not suffice for certain applications. An example of this dependence can be seen in Fig. 16, which shows modeled results of the scattering loss of SOI rib waveguides. The modeling was performed by using a commercial software package called BeamPROP [45], which incorporates advanced finite-difference beam propagation techniques for the simulation. The plot demonstrates the effect of increased roughness and decreased waveguide dimensions on optical loss. In addition to optimizing etch parameters to reduce silicon sidewall roughness, post-etch oxidation of the silicon surface has been used to reduce the roughness and hence the loss [46, 56] relying on the fact that convex silicon features oxidize faster than concave areas near the surface.

3.5 Yield

Today's electronic devices are highly complex and continually shrinking, requiring fabrication methodology to include parametric testing for process control. Electronic devices are inherently easy to probe. Even before they are completely processed, they can be accessed across an entire wafer. It is for this reason that today's typical CMOS fabrication testing plan can include thousands of parameters to be monitored in the hope of identifying potential threats to device performance and reliability.

Over several decades of CMOS development, the automated testing of devices before completion has been instrumental in reducing per-wafer manufacturing costs. This so-called 'in-line' testing provides performance and defect density information essential to predicting and rapidly correcting for process errors and depressed yields. Thus, metrology collected after only a fraction of the total processing is complete can provide excellent insight into the final performance of the device under test.

Optical devices, however, do not readily lend themselves to convenient in-line testing. There are at present no convenient, fast, automated, and nondestructive probing techniques that can be well correlated to final, packaged optical device performance. Much like the electronic test cells used for collecting parametric data, optical device test structures can be designed and dedicated to the collection of relevant inline information. Some efforts have been made to use top down probing of optical devices but none of these techniques have been introduced into the manufacturing process.

One such technique for top-down optical probing of a waveguide is through the use of a prism [42,47]. In Fig. 17a are depicted two prism couplers, held above the surface of the waveguide at a distance of approximately $\lambda_o/4$. After light enters the prism it is totally reflected at the base. This reflection creates a standing wave in the prism whose field extends below the base and into the waveguide below (evanescent tail). The evanescent tail excites a light wave in the waveguide; coupling light to the waveguide through the air gap is known as optical tunneling. Prisms can be used as both input and output couplers and, since they are not a physical part of the device under test, the distance between them can be adjusted to collect loss measurements as well as transmission data. In addition, it is possible to test the product directly without the need for dedicated test structures that monopolize expensive field area. However, positioning the prisms within one-quarter of a wavelength to the waveguide surface can easily lead to surface damage. This, coupled with the need for prism material with a higher index than the waveguide material, makes prism couplers a difficult practical solution.

Another successful technique is to couple the incident light perpendicular or nearly perpendicular to the plane of the wafer via a grating [48, 49, 50] (Fig. 17b). By appropriately designing the grating period, efficient coupling can be achieved. Modeled efficiencies up to 95% have been reported using this technique, with measured efficiencies of 70% for rectangular gratings [51] and

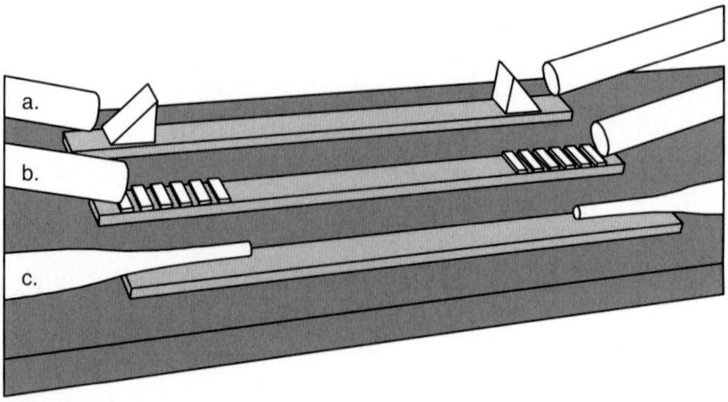

Fig. 17. Waveguide coupling via (**a**) prisms (**b**) gratings, and (**c**) evanescent tapers

84% for blazed gratings [52]. Since a grating coupler must be monolithically integrated into the waveguide under test, a special test structure would need to be created specifically for in-line optical testing. Fortunately, most planar waveguides are only a few microns wide, allowing for test structures to be placed in noncritical scribe area, similar to metal test structures in electrical test chips. The practical difficulty of implementing this technique will be in accurately positioning the waveguide probes since the grating coupler efficiency is highly sensitive to the angle of incidence.

Yet another in-situ test technique relies on evanescent coupling between a tapered optical fiber and a planar waveguide. As the optical fiber is brought into close proximity with the waveguide (Fig. 17c), resonant coupling of the evanescent field transfers the optical power into the waveguide. Modeled coupling efficiencies of approximately 98% have been reported for photonic crystal waveguides [53]. Like the coupling via a prism, this technique can also be used to couple directly to waveguides without the need for dedicated test structures.

In addition to increasing complexity in parametric characterization, reductions in the size of electronic devices demands tight processing control over unwanted defects. Some examples of defects are metal shorts, particle contamination, bumps in thin films (low-k deposition), etch residues left after etch cleaning, post-contact etch cracking due to Inter-Layer Dielectric (ILD) stress, and adhesion loss causing blistering of bond pads, to name a few. These defects can be monitored during processing, and appropriate measures can be taken to correct for them. However, the nature and effect of these defects are understood and relatively easy to detect. Most problems occur at the surface of the wafer after a given process step, allowing for surface inspection (e.g. optical microscope, laser, or pattern recognition/comparison).

Defect detection and characterization are not mature sciences for the production of optical devices. For a defect to impact the operation of an optical device, it need not be detectable from the surface of the wafer. One of the most common material issues in photonic devices is a variation in the refractive index of the guiding/cladding layers. These variations can be due to stress, defects, or nonuniformities in thickness and composition. Effective in-line techniques to inspect optical devices for these variations have yet to be developed.

3.6 Coupling and Packaging

Coupling light into and out of a silicon chip is very challenging. Particularly difficult is the coupling of light from a standard optical fiber to a silicon waveguide due to the large modal mismatch. Furthermore, lacking efficient silicon-based emitters, external light sources must be directly coupled into silicon devices. These challenges require the development of additional processes and structures beyond those of the core device. This section will survey some of the commonly used practices for the assembly of commercial products.

3.6.1 Tapers

Coupling light into and out of a silicon chip is very challenging. Particularly difficult is the coupling of light from a standard optical fiber to a silicon waveguide due to the large modal mismatch. A single mode fiber core ($n = 1.5$) usually has a diameter of 8 µm with a symmetric mode while a silicon waveguide ($n = 3.45$) is typically only a few microns in width with an asymmetric mode. One way to overcome these large differences in effective index, core size, and symmetry is with a waveguide taper. This allows for a reduction in coupling loss through a modal transformation. A taper can also be used to increase the alignment tolerance to other optical devices, such as III–V lasers.

Several methods have been proposed and have demonstrated efficient coupling from a relatively large silicon waveguide into an optical fiber. These include pseudo-vertical tapering and using gray-scale lithography for more gradual horizontal and vertical modal tapering [2]. Figure 18 is a schematic depicting these two mode transform designs. Results using a pseudo-vertical taper (Fig. 18a), tapering from a 12 µm by 12 µm input down to waveguides on order of 4–5 µm have given coupling losses as low as 0.5 dB/facet [54].

There are a few key processing parameters that must be controlled in order for successful processing of either of these mode-transfer devices. Specifically these are the lithography and etching processes.

For pseudo-vertical tapering, a horizontally tapered waveguide is patterned on top of another waveguide and the optical mode is gradually squeezed from the top taper to the smaller, lower waveguide. The key parameters for this transition are the length of the taper (the longer the length the more slowly one can transform the mode resulting in lower loss) and the taper tip width (see Fig. 18a). In order to reduce optical losses associated with the finite size of the tip width, the tip should be designed such that the

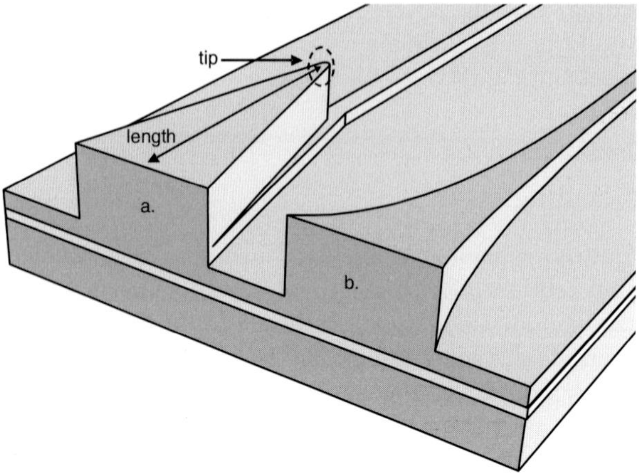

Fig. 18. Two examples of 3-D modal transformers used for coupling light from a fiber into a silicon waveguide: (**a**) pseudo-vertical taper, (**b**) grayscale taper

minimum width is substantially smaller than the wavelength of light transmitted in the waveguide. Optimally, the tip width should be so small that there is almost no optical mode in the tip region. For wavelengths of interest around $\lambda_o \approx 1550\,\text{nm}$ this corresponds to a wavelength in silicon of around $\lambda \approx 450\,\text{nm}$. This final tip dependency is very critical and the effect of tip width variation on taper loss is shown in Fig. 19. The plot shows results of simulation of a pseudo-vertical taper design starting from 10 μm by 10 μm and tapering down to 2.5 μm by 2.3 μm. The modeling was performed by using a commercial software package called BeamPROP [45] which incorporates advanced finite-difference beam propagation techniques for the simulation. In the simulation, the loss due to sidewall roughness was assumed to be zero. The taper length was 1 mm and free space wavelength of 1550 nm was used. One can see from Fig. 19 that for a fixed sidewall angle of 80° the change in the tip point from 2μm down to 0.5μm improves the total taper loss by more than 20 dB. The higher loss for wider tip point is due to the fact that there is a more significant proportion of optical field existing in the tip region that can not propagate along the final waveguide so that it radiates out of the waveguide.

Unfortunately, to fabricatE such a taper, having the lithographic patterning and photo resist capability alone is not enough. In order to complete the process, etching these taper devices invariably leads to some rounding of all sharp edges, thus never allowing one to really achieve the desired sharp point.

In the case of grayscale techniques, appropriate resists and etch recipes must be created and controlled to provide smooth surfaces to slowly vary the height of the taper from the input down to the final waveguide dimension. The technique of grayscale lithography has been known for several years, and

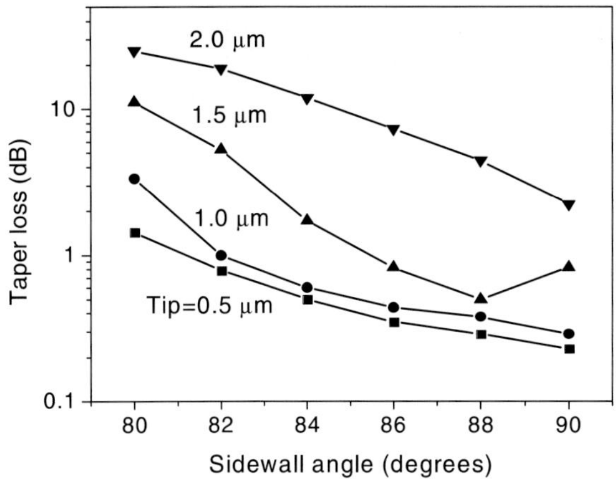

Fig. 19. Plot showing dependency of coupling loss for a silicon taper on both final tip dimension and sidewall angle. The sidewall roughness contribution to loss was assumed to be zero

is widely used in optical Micro Electrical Mechanical Systems (MEMS) to create lenses, prisms, and various other structures [55]. However, grayscale based lithography adds some additional complexity and cost. A grayscale reticle is typically five times more expensive than a standard lithography reticle and process refinement to develop and pattern the resist is not trivial. Although fundamentally tapers based on gray scale should result in much lower loss, these types of tapers have not been as prominent as top tapers.

Another key parameter for processing tapers is that of sidewall roughness and sidewall angle as a result of the silicon etch. Creating taper devices usually relies on significant etching away of silicon material in order to produce the final taper design. Although one can use wet-chemistry for etching, it is often difficult to control and it is not practical, especially if one would like to obtain vertical sidewalls or more exotic profiles, such as a parabolic design. Thus the most common approach is RIE.

RIE also has its drawbacks as it produces surface roughness on the exposed sidewall. The effect of sidewall roughness as a function of waveguide height was discussed previously (Fig. 16). As the optical loss due to the surface roughness is strongly dependent on the waveguide size, the loss due to roughness should be minimal at the beginning of the taper since the taper dimension is large. At the taper end, however, the waveguide becomes smaller and the loss contribution from the roughness becomes larger. Therefore, it is critical to control the surface roughness at the smaller end of the taper. Techniques to reduce loss have been proposed and demonstrated, which involve smoothing waveguides by use of repeated oxidation and stripping steps [46, 56]. However, oxide smoothing also results in rounding of sharp points which could add additional loss due to tip width rounding.

Another approach to reduce the effect due to surface roughness would be to pattern the taper into a thick oxide layer and then grow the taper using an epitaxial growth step. This may result in lower loss due to the reduction of the silicon etch step but adds complexity into the process which may jeopardize the entire process flow as discussed previously.

In addition to surface roughness one must monitor the resulting sidewall angle of the pseudo-vertical taper after silicon etching is complete. Figure 19 shows the effect of sidewall angle on the taper loss. It can been seen that a sidewall angle variation from 90° to 80° can increase the loss by up to 20 dB for a given tip width.

Top tapers have been successful for coupling to waveguides with cross sectional dimensions of 4–5 μm. Modeling has shown that these designs will not work well when the waveguides are smaller than 2 μm. Assuming no loss due to roughness and perfectly vertical sidewalls, modeling [45] shows that tapering from a 13 μm by 13 μm input to a 1 μm by 1 μm waveguide still gives a loss of 2.3 dB/facet. This is true even with a tip width as small as 0.05 μm and a tapering length of 5.2 mm. Thus a different approach must be used for coupling to very small waveguides (on the order of 1 μm or less) to standard optical fibers. Alternative approaches have recently been proposed for coupling to sub-micron waveguides, including the use of grating assisted couplers [1, 2].

3.6.2 Laser Die Alignment

Barring a breakthrough in the power output of silicon light emitters, III–V lasers will be required as on-chip light sources. These laser die are attached either by an active or passive alignment procedure. Active alignment refers to a process by which the laser die is turned on during placement, and the die position is optimized through real-time feedback. The required electrical connections for this process make the alignment difficult. An addition complication is the size of the sub-mount for these electrical connections. The sub-mount must be large enough to be probed electrically and yet small enough to be easily aligned to the waveguide and bonded to the substrate [57].

In contrast, passive alignment of a light emitting die is performed without electrically connecting or probing the III–V device. This removes the requirement for a sub-mount, so the die placement can be performed using a vacuum tip chuck to lift and position the III–V die onto the silicon substrate. In order for the final integrated module to be cost effective, the passive approach should drive as much cost out of the die placement process as possible. The silicon substrate could be used to increase placement accuracy by using mechanical etch stops. Also, the need to place the III–V die with high accuracy could be reduced by adding tapers or other loss reducing devices, to the silicon waveguide (Fig. 20).

The mechanical challenges of the die attachment process center around the issue of alignment accuracy. Depending upon the relative size and shape of the

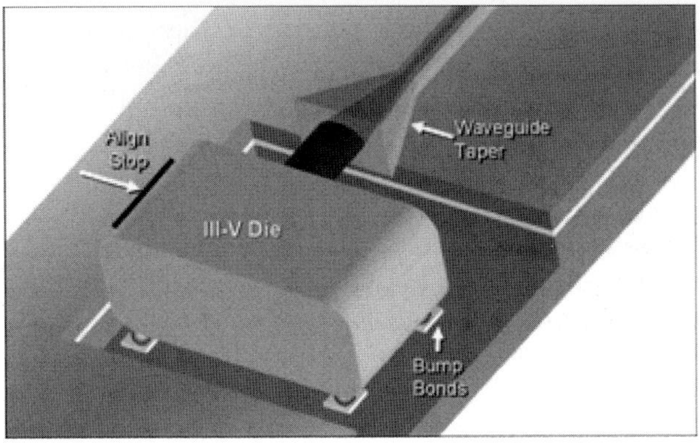

Fig. 20. Simple schematic showing a III–V die placed onto a silicon on insulator platform. The diagram depicts key features that could be used to enable passive alignment

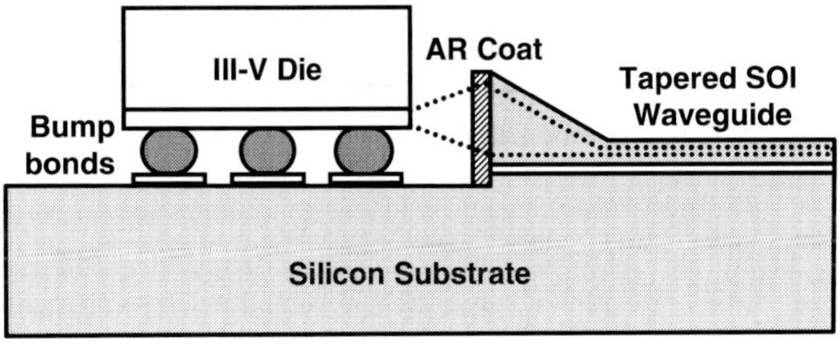

Fig. 21. Simple schematic showing a cross-sectional view of a III–V die placed active side down onto a silicon platform and aligned to silicon waveguide. Shown in figure are key process related activities that needed to be developed

waveguide mode, the mode of the III–V die, and the tolerable system loss, the required placement accuracy could be tighter than 1 μm, with angular tolerances of several degrees. This required placement accuracy imposes tight requirements on the die attachment technique, especially under both broad temperature and long lifetime requirements. In addition very accurate control of the etch depth is required to ensure that the vertical alignment can also meet the comparable alignment tolerance.

Many different types of die attach and alignment methods are possible for a passive alignment scheme. High precision die placement is commonly accomplished using accurate pick-and-place machines outfitted with vision recognition systems [58, 59]. Self-aligned solder reflow techniques with and without mechanical stops to control die placement have been published [60, 61]. Commercial laser die typically have x-y dimensional tolerances on the order of ±5 μm due to the limitations of the cleaving method used to separate individual die. In order to overcome these errors, alignment techniques often use pillars on the die surface. The position and size of these pillars are formed through photolithography and etching, so the distance between the edge of the pillars and the center of the laser is strictly controlled. High quality pick-and-place tools equipped with machine vision align lithographically defined marks on the laser die to alignment marks on the silicon. However the greater the required alignment accuracy the longer the cycle time needed for die placement and the more expensive the pick-and-place machine. Typical die, pick and place machines with accuracies of around 3 μm are five times lower in cost than a similar die placement machine that has < 1 μm accuracy. In order to reduce the overall cost of assembly both the tool cost and cycle time must be accounted for. A system designed with large tolerances for the manufacturing process (> 1 μm) will drive down the cost of the tool and decrease the cycle time resulting in net lower cost of assembly.

Typical die attachment techniques involve using gold as the bonding metal with either thermo compression or soldering as the die attachment mechanism [62, 63]. As mentioned previously, gold is well known in the electronic industry as an electronic trap and thus deposition and patterning of gold cannot be performed inside a CMOS fabrication facility. As a result all gold processing must be done at the back end of device fabrication and isolated from the remainder of the CMOS fabrication facility. This could present significant challenges with the die attachment and integration process. Increasing the challenge, the metal (see Fig. 21), must be patterned at the bottom of the trench into which the III–V die is placed. The trench is typically 5–15 μm in depth and thus control over the size, height and potentially the volume uniformity of the metal pattern will be affected by how well the lithography process at the bottom of such a trench can be patterned. Additionally solders are often deposited by an electroplating process in order to achieve the correct stoichiometry and the control over the deposited thickness can be challenging to control to better than 5%.

Low loss passive alignment requires that the mode between the light emitting die and the SOI waveguide be well matched. The similar index between silicon and most III–V materials improves the situation as compared to mode matching a III–V light emitting die to say for example a silica (SiO_2) waveguide ($n \approx 1.5$). However mode matching still requires careful design. Microlenses have commonly been used to shape and focus the laser output mode before coupling into an optical fiber, but are not low cost as compared to integrated solutions.

The mode of a typical edge emitting laser diode is elliptical, horizontal in the near field and vertical in the far field and this mode diverges rapidly (within a few wavelengths of the edge of the laser facet). On the silicon side a typical SOI single mode rib waveguide mode is also asymmetric. The amount of this asymmetry is dependant on the actual size and dimensions of the silicon waveguide [2]. In order to achieve better mode overlap between the laser die and the SOI rib waveguide one or both of the optical modes must be transformed. This can be achieved by either integrating a mode converter into the light emitting die or into the SOI waveguide. Mode shaping at either the III–V dies or waveguide input has the additional benefit of allowing the mode to be expanded, decreasing the required die placement accuracy. Mode transfer devices have not yet been integrated into commercial III–V laser die, requiring tapers to be made on the silicon waveguide.

Another challenge for successful die placement is the formation of high quality etched waveguide facet (typically rms roughness of 10 nm or less is required) in order to reduce the scattering loss at the air-waveguide interface. Due to the large index difference between air and silicon ($\Delta n \approx 2.5$) minimization of back reflections from the waveguide facet requires the deposition of an antireflection coating (ARC) on the etched facet. This ARC needs to be processed at the wafer level and could be combined with an angled facet in either the waveguide or light emitting die to produce net back reflection of $< 10^{-3}$. Integration of an ARC with the process of gold bump patterning while maintaining smooth waveguide facets can be accomplished with significant process development.

3.6.3 Fiber Attachment

The final steps of a processed optical chip include testing and packaging. These steps often demand the alignment and attachment of fibers, which can add significantly to the final cost of the device. Active alignment is inherently more precise due to the real-time feedback in positioning, but can be significantly more expensive due to the time required for alignment and the specialized equipment. A lower cost, passive alignment approach is possible in silicon through the use of conventional lithography. The drawback to this technique is a potential reduction in coupling efficiency due to variations in the in the fiber geometry and in the silicon processing. A photograph of passive alignment using a U-groove is shown in Fig. 22. In this approach, trenches are etched into the SOI substrates which are used to hold and self-align the fiber to the silicon waveguide.

The alignment accuracy of a passively aligned fiber in a U-groove depends not only on the etching tolerances of the groove, but also on the manufacturing tolerances of the fiber. Standard single mode fiber cladding specifications are typically 125 µm ± 1 µm. Newer fibers with tighter specifications are now available, but come at a higher price. The lateral alignment tolerance of an 8 µm single-mode fiber core to a 13 µm × 13 µm square planar waveguide input facet was modeled using FIMMPROP [64]. The modeled results, based on film mode matching and local mode expansion methods, show that the lateral alignment tolerance is ±2 µm for 1 dB excess loss. By increasing the input taper to a sufficiently larger size (e.g. 15 µm by 15 µm), it could be possible to relax the alignment tolerance somewhat and help overcome the manufacturing variations of the fiber geometry. Producing such large tapers is extremely challenging. Due to the large index difference between silicon

Fig. 22. SEM photograph of passive fiber alignment using a U-groove

and glass fiber, an ARC will also need to be deposited onto the etched facet of the waveguide.

4 Summary

Microelectronics manufacturers, through their years of processing experience, are uniquely positioned to advance integrated silicon photonics. With increased investment in silicon photonics, manufacturers will begin to overcome some of the issues associated with optical and electrical process integration. The fabrication requirements for most optical devices do not currently require state of the art processing capability. By using depreciated infrastructure, companies should anticipate developing low-cost devices for applications that demand high-volume. Unfortunately, cost-effective processing alone will not guarantee success; coupling and packaging will remain as two of the most stubborn obstacles to achieving low-cost optical solutions.

Several things must happen before integrated silicon photonics becomes commercially sustainable. First and foremost, there must be an application for integrated photonics where either an electrical or discrete optical solution will not work. Electrical signaling of 20 Gb/s over meter distances will occur in the near future, so photonics will not be used for chip-to-chip communication until the next decade. Longer distance applications are suitable candidates for silicon photonics over the next 10 years, but there must be a large enough market to justify the cost of integration. Standardization of component designs will go a long way to create demand. The development of a silicon-based light emitter, modulated directly or indirectly at data rates > 100 Mb/s, would also greatly expand the areas for which silicon photonics could have large impact. The monolithic integration of other silicon photonic components such as detectors, couplers, and multiplexers will also be useful if processing can be kept similar to that of baseline semiconductor electronics, thereby reducing cost.

The integration of photonics and electronics involves many constraints that need to be considered. There will always be compromises between transistor performance and optical performance. High speed electronics require expensive, cutting-edge processing. Monolithic integration of electronics and optical devices, which do not require such a level of processing sophistication, could be cost prohibitive. Therefore, the choice between hybrid and monolithic integration will be largely driven by economic forces.

The electronics industry has seen revolutionary advances in the last thirty years. Increased research in silicon photonics is just beginning, and one or two major advances in this area could spark a revolution in photonics. Something similar occurred fifty years ago during the transition from vacuum tubes to transistors. In discussions at Bell Laboratories then about the value of transistors, an observation was made that "the significance of the transistor is not that it can replace the tube but that it can do things that the vacuum tube could never do" [65]. The same may hold true for silicon photonics.

Acknowledgements

The authors would like to acknowledge Sean Koehl his help with figures and editing, Ansheng Liu for his assistance with simulations and plots, Drew Alduino for discussions on passive alignment. In addition the authors would like to thank David Ayer, Jerry Bautista, Shekar Bokhar, Randy Mooney, and Valluri Rao for many useful discussions. Finally, the authors would like to thank Prof. Graham Reed of the University of Surrey, UK, for advice on the content and presentation of this text.

Legal Notice

References

1. G. T. Reed, C. E. Png, G. Masanovic, S. T. Lim, S. P. Chan, W. Headley, T. W. Ang, A. Liu, M. Paniccia: *Silicon nanophotonic circuits* (Invited Paper, XLVII ETRAN Conf. 813, Herceg Novi, Serbia and Montenegro 2003)
2. G. T. Reed, A. P Knights: *Silicon Photonics: An Introduction* (Wiley Interscience, New York 2004)
3. R. A. Soref: Proc. IEEE **81**, 1687–1706 (1993)
4. C. K. Tang, G. T. Reed: Electron. Lett. **31**, 451–452 (1995)

5. P. Dainesi, A. Kung, M. Chabloz, A. Lagos, P. Fluckiger, A. Ionescu, P. Fazan, M. Declerq, P. Renaud, P. Robert: IEE Photon. Technol. Lett. **12**, 660–662 (2000)
6. G. Franzò, S. Coffa, F. Priolo, C. Spinella: J. Appl. Phys. **81**, 2784–2793 (1997)
7. L. T. Canham: Appl. Phys. Lett. **57**, 1046–1048 (1990)
8. L. Pavesi, L. Dal Negro, G. Mazzoleni, G. Franzò, F. Priolo: Nature **408**, 440–444 (2000)
9. W. L. Ng, M. A. Lorenço, R. M. Gwilliam, S. Ledain, G. Shao, K. P. Homewood: Nature **410**, 192–194, (2001)
10. J. Trezza, H. Hamster, J. Iamartino, H. Bagheri, C. D. Cusatis: IEEE OCS **1**, 1 (February 2003), S36–S42, suppl. to IEEE Comm. Mag. **41** 2 (2003)
11. Q. Tan, G. De Pestel, G. Willems, J. Cannaerts, W. Rehm, N. Kaiser, K. Vandeputte, A. Van Hove, J. Van Koetsem: Proc. 47th Electron. Comp. Techn. Conf. (ECTC), San Jose, USA, May 1997, pp. 234–239
12. E. Griese: IEEE Trans. Adv. Packaging **24**, 375–383 (2001)
13. M.-J. Lee, W. J. Dally, P. Chiang: IEEE J. Solid-State Circ. **35**, 1591–1599, (2000)
14. R. Mooney: *Scaling methods for electrical interconnects to meet the performance requirements of microprocessor platforms*, Workshop Notes, 14th Annual Workshop on Interconnections Within High Speed Digital Systems, Santa Fe, NM, USA, May 2003
15. J. Jaussi, G. Balamurugan, D. Johnson, B. Casper, A. Martin, J. Kennedy, R. Mooney, N. Shanbhag: *An 8 Gb/s source-synchronous I/O link with adaptive receiver equalization, offset cancellation, and clock deskew*, to be presented at 2004 IEEE Int. Solid-State Circuits Conf. (ISSCC), San Francisco, CA, USA, February 2004
16. B. Casper, A. Martin, J. Jaussi, J. Kennedy, R. Mooney: *8 Gb/s differential simultaneous bidirectional link with 4 mv 9 ps waveform capture diagnostic capability*, Tech. Dig. 2003 IEEE Int. Solid-State Circuits Conf. (ISSCC), San Francisco, CA, USA, 2003, pp. 78–79
17. C. Svensson: IEEE Trans. VLSI **10**, 777–788 (2002)
18. N. Savage: IEEE Spectrum, 32–36 (2002)
19. D. A. B. Miller, H. Ozaktas: J. Parallel Dist. Comp. **41**, 42–52 (1997)
20. A. Krishnamoorthy, D. A. B. Miller: IEEE J. Sel. Top. Quant. **2**, 55 (1996)
21. P. Wijetunga, A. F. J. Levi: Proc. Electrochem. Soc. **2002-4**, 381–397 (2002)
22. V. Rao: Proc. UCSB Solid State Technol. Rev. (UCSB SSTR '02), University of California, Santa Barbara, 2002 `http://www.ece.ucsb.edu/OTC/SSTR02/`
23. R. Soref: Proc. IEEE **81**, 1687–1706 (1993)
24. B. Schuppert, J. Schmidtchen, A. Splett, U. Fischer, T. Zinke, R. Moosburger, K. Petermann: J. Lightwave Tech. **14**, 2311 (1996)
25. Y. Liu: IEE Trans. Adv. Packaging **25**, 43–49 (2002)
26. M. Markmann, A. Sticht, F. Bobe, G. Zandler, K. Brunner, G. Abstreiter, E. Muller: J. Appl. Phys. **91**, 9764–9771 (2002)
27. M. Castagna, S. Coffa, M. Monaco, L. Caristia, A. Messina, R. Mangano, C. Bongiorno: Physica E **16**, 547–553 (2003)
28. D. Fehly, A. Schlachetzhi, A. Bakin, A. Guttzeit, G. Wehmann: IEEE J. Quant. Electron **37**, 1246–1252 (2001)
29. H. Kogelnik: IEEE Trans. Microwave Theory Tech. **MTT-23**, 2–16 (1975)
30. S. M. Csutak, J. D. Schaub, S. Wang, J. Mogab, J. C. Campbell: IEEE Proc. Optoelectron. **150**, 3, 235–238 (2003)

31. M. K. Smit: Advanced Components for WDM, Proc. ECOC99, Nice, France, Sept. 1999, pp. 98–99
32. M. Masanovic, L. Val, J. S. Burton, E. J. Skogen, L. A. Coldren, D. J. Blumenthal: IEEE Photo. Tech. Lett. **15**, 1117–1119 (2003)
33. White paper from III–V Photonics, See www.35ph.com for details
34. G. Rondeau, S. Biblemont, J. Decobert, G. Post: Compound Semicond. **6**, 83–83 (2000)
35. J. Mills: Photonic Integration Combats Network Complexity, fibers.org (2002)
36. H. Zimmermann: *Integrated Silicon Optoelectronics* (Springer, Berlin, Heidelberg 2000)
37. K. W. Goossen: Integration of GaAs modulators to silicon IC's by monolithic and hybrid techniques, Sarnoff Symposium 1994 (IEEE Princeton/Central Jersey 1994), 100–107
38. D. C. Houghton: J. Appl. Phys. **70**, 2136–1327 (1991)
39. B. E. Little et al.: IEEE Photonics Tech. Lett. **10**, 549–551 (1998)
40. B. E. Little et al.: J. Lightwave Tech. **15**, 998–1005 (1997)
41. D. V. Tishinin et al.: IEEE Photonics Tech. Lett. **11**, 1003–1005 (1999)
42. P. K. Tien: Appl. Opt. **10**, 2395–2419 (1971)
43. D. Marcuse: *Light Transmission Optics*, (Van Nostrand Reinhold, New York 1972), Chap. 9
44. F. Ladouceur, J. D. Love, T. J. Senden: IEEE Proc. Optoelectron. **141**, 242–248 (1994)
45. www.rsoftdesign.com
46. K. Lee et al.: Opt. Lett. **26**, 1888–1890 (2001)
47. W. A. Pasmooij et al.: J. Lightwave Tech. **7**, 175–180 (1989)
48. T. Tamir, S. T. Peng: Appl. Phys. **14**, 235 (1977)
49. T. W. Ang, G. T. Reed: A. Vonsovici: Appl. Phys. Lett. **77**, 4214 (2000)
50. D. Taillaert et al.: J. Quant. Electron **38**, 949–955 (2002)
51. T. W. Ang et al.: Proc. SPIE **3620**, 79–86 (1999)
52. T. W. Ang et al.: Appl. Phys. Lett. **77**, 4214–4216 (2006)
53. P.E. Barclay et al.: Conf. Lasers Electro-Optics (CLEO), Baltimore, MD, USA, May 2003
54. I. Day et al.: Tapered silicon waveguides for low insertion loss highly efficient high speed electronic variable attenuators, IEEE Opt. Fiber Comm. Conf. (OFC 2003), Atlanta, GA, USA, March 2003
55. M. B. Frish et al.: Proc. 2002 IEEE Int. SOI Conf. (IEEE, 2002), pp. 165–166
56. L. Lai, E. A. Irene: J. Appl. Phys. **86**, 1729-1735 (1999)
57. W. Kowbel et al.: Aerospace Conf. Proc. **5**, 2423–2428 (2002)
58. G. Lecarpentier, J. Mottet, J. Dumas, K. Cooper: High accuracy machine automated assembly for opto electronics, 50th Electron. Comp. Technol. Conf. (ECTC), Las Vegas, USA, May 2000, pp. 1–3
59. T. Hashimoto, Y. Nakasuga, Y. Yamada, H. Terui, M. Yanagisawa, K. Moriwaki, y. Suzaki, Y. Tohmori, Y. Sakai, H. Okamoto: IEEE Photon. Technol. Lett. **8**, 1504–1507 (1996)
60. K. P. Jackson, E. B. Flint, M. F. Cina, D. Lacey, Y. Kwark, J. M. Trewhella, T. Caulfield, P. Buchmann, C. Harder, P. Vettiger: J. Lightwave Technol. **12**, 1185–1191 (1994)
61. R. A. Nordin, D. B. Buchholz, R. F. Huisman, N. R. Basavanhally, A. F. J. Levi: IEEE Transactions on Components, Hybrids, and Manufacturing Technology **16**, 783–788 (1993)

62. Q. Tan, Y. C. Lee: Soldering technology for optoelectronic packaging, Proc. 46th Electron. Comp. Technol. Conf. (ECTC), Orlando, FL, USA, May 1996, pp. 26–36
63. A. Ambrosy, H. Richter, J. Hehmann, D. Ferling: IEEE Transancations of Components, Packaging and Manufacturing Technology, Part A **19**, 34–40 (1996)
64. www.photond.com
65. H. Huff: J. Electrochem. Soc. **149**, 35–58 (2002)

Monolithic Silicon Microphotonics

L. C. Kimerling, L. Dal Negro, S. Saini, Y. Yi, D. Ahn, S. Akiyama, D. Cannon, J. Liu, J. G. Sandland, D. Sparacin, J. Michel, K. Wada, and M. R. Watts

Massachusetts Institute of Technology, Cambridge, MA, 02139, USA
lckim@mit.edu

Abstract. The technology evolution of optical interconnection is driven by the distance $\times$ bandwidth product required for a given link. As such optical fiber deployment in telecommunications has become the dominant paradigm since the early 1980s. Today, with excess telecommunications capacity, optical interconnection is driven by data communications needs throughout the interconnection hierarchy, from the network to the chip. Silicon Microphotonics is the only current approach that offers a scalable solution to the anticipated barriers of I/O density, interconnection bandwidth and latency, and electronic/photonic partitioning through monolithic integration. This chapter reviews the drivers, barriers and current solutions for the development and deployment of a monolithic silicon microphotonic technology. Light sources, photodetectors, waveguides, photonic filters, optical amplifiers and photonic crystals are discussed in the contexts of materials design and systems applications.

1 Introduction

1.1 The Silicon Paradigm

The legacy of Silicon Microelectronics has made silicon the most studied material in the history of civilization. As one of the most abundant elements on the earth's crust, and as the basis of a \$250B semiconductor industry, silicon (Si) has become more pervasive than steel in its affect on quality of life. Silicon integrated circuits have provided the computation capacity that created the Information Age. The keys to the success of silicon in Microelectronics have been 1. compatible, multifunctional materials: Si, SiO_2, Al; 2. scalable, planar device interconnection; and 3. integrated vs. single component process yield management. With these tools in place, the variable of technology shrink has produced faster, lower cost components with each new generation. This technology shrink has progressed with the help of a Roadmap [1] that coordinates materials, processing and design for successive generations. The Roadmap (ITRS) projects that the benefits of continued scaling will be limited by the performance and complexity of interconnects when the device density is $> 10^7$ devices/chip.

L. Pavesi, D. J. Lockwood (Eds.): Silicon Photonics, Topics Appl. Phys. **94**, 89–119 (2004)

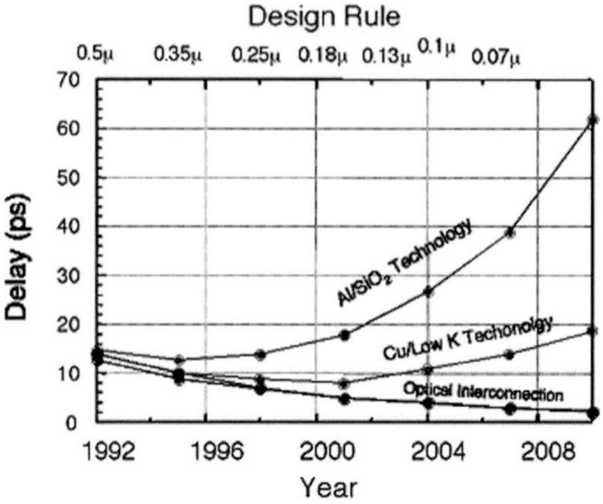

Fig. 1. Trend in interconnect propagation delay (1 cm length) with technology linewidth and time for traditional aluminium metal and SiO2 insulator; projected copper and low-k dielectric; and projected optical waveguide technologies [2]

1.2 The Interconnect Bottleneck

The 'interconnect bottleneck', is an RC circuit delay occurring due to the smaller cross-section and closer spacing of conducting metal lines above the integrated circuit chip. In addition, these metal interconnects account for the majority of chip power dissipation,information latency and cross talk noise. Figure 1 shows that while aluminum-based interconnect technology reached a performance limit at the 0.55 μm lithographic technology node, the introduction of copper conductors to reduce resistance and low-k dielectrics to reduce capacitance has sustained a performance improvement limit up to 0.18 μm [2].

Silicon microphotonics represents a potential solution to the resistance and capacitance limitations of traditional metal interconnects. Using photons as bits of information instead of electrons, enormous transfer rates due to the high frequencies of optical carrier waves can be achieved. In addition, since photons are electrically neutral massless particles, they propagate in transparent media with negligible heat dissipation and no cross talk. A silicon based optical interconnection technology has the appeal of a well understood materials and process technology with the performance advantage of optics.

1.3 A Roadmap for Communication Technology

The distance × data rate product is the performance metric for point-to-point communications. Twisted pair electronic conductors has provided the bandwidth for interpersonal voice communication with enhanced distance. As the number of users increased it became necessary to compress bandwidth/user and multiplex multiple users on a single twisted pair line.

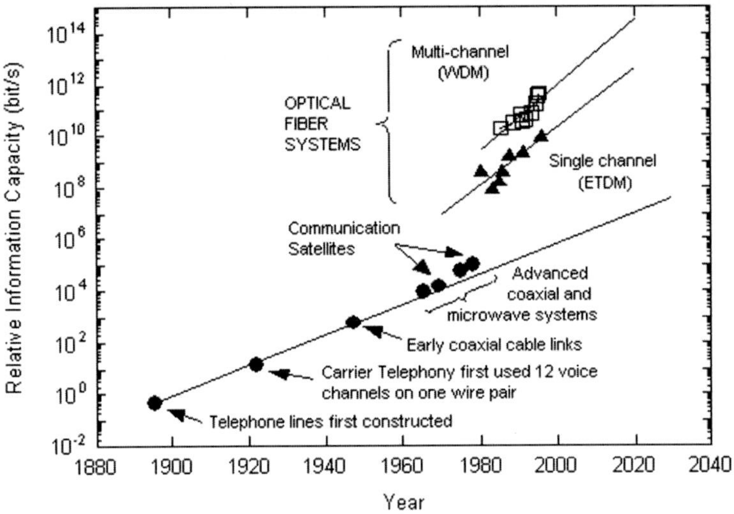

Fig. 2. Growth in information carrying capacity of a single communications line [3]

This capacity growth increased at a remarkable semi-logarithmic rate as coaxial cable, microwave and satellite platforms were adopted. The fiber platform for ligthwave communications represented the first deviation from this trend. The crossover from electronic to optical communication (see Fig. 2) occurred at the distance $\times$ bandwidth product of $\sim$ 10 Mb/s $\times$ km. This performance metric scales to 1 Tb/s $\times$ cm for local interconnection. The characteristics of this cross-over are both a discontinuous increase in performance and change of slope to a sustainable increase in performance/cost of 100 $\times$ /10 years [3].

1.4 Photonic Interconnection

The use of photonic interconnection has revolutionized telecommunication systems with the deployment of low-loss, single mode silica fibers, efficient photodetectors and double heterostructure, single mode injection lasers. Long haul fiber optic systems were ultimately enabled with the development of erbium-doped fiber amplifiers (EDFAs), operating around the 1.55 μm fiber transparency window, dubbed the C-band.

Present optical networks contain several optoelectronic components such as lasers, modulators, detectors, beam splitters and multiplexers all connected by glass fibers. This long haul component paradigm does not translate well to higher complexity and functionality at shorter distances.

A significant reduction in costs and system size can be reached by the development of Microphotonics, in which all the optical components are arranged on a single silicon substrate, resulting in an optoelectronic integrated circuit (OEIC). On an integrated optical chip, light is directed through optical

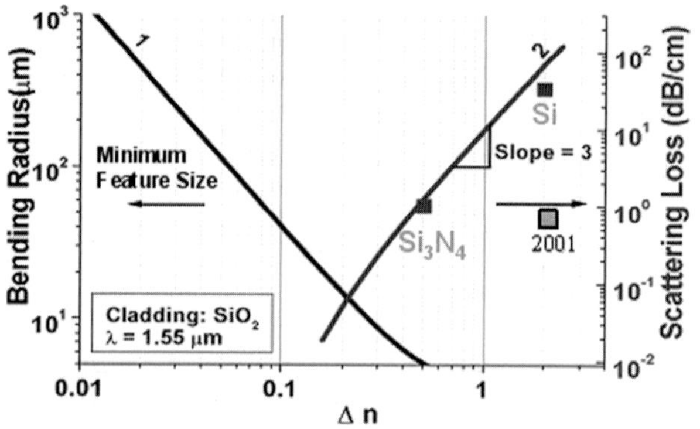

Fig. 3. Dimensional scaling for microphotonic component with index difference Δn and the consequent increase in scattering loss with with current pattern transfer technology [5]

circuits by means of planar waveguides, fabricated on silicon with standard lithographic techniques.

These micron-scale structures present a convergence of solutions for both the interconnect bottleneck problem, and the MAN requirement of low-cost high volume components. At present there are two main approaches to achieve silicon OEICs: hybrid and monolithic integration. Since silicon is a poor light emitter, the hybrid approach tends to combine efficient light emitting III–V semiconductor materials or device components by expensive bonding technologies to Si substrates. In the monolithic approach, III–V and Si based light emitting materials are integrated by sequential deposition, growth and pattern transfer onto Si substrates, producing a complete set of optoelecrtronic functionalities [4].

1.5 Microphotonic Integrated Circuits

It is generally accepted that optical devices will scale to larger dimensions than electrical devices due to the size of the photon relative to that of the electron in a medium. However, this ultimate limitation should not hinder the integration of microphotonic components with current ULSI electronic circuits, and it will likely not affect the ultimate architecture and applications of optoelectronic integrated circuits (OEICs). Current Planar ligthwave Circuits (PLCs) are constructed of lightly doped SiO_2 waveguides with index contrast ($\Delta n = 0.005$) closely matching that of the long haul, optical fiber that feed signals to them. The core dimension is about 10 μm and the cladding is about 20 μm on either side, resulting in a total film thickness approaching 50 μm.

While this design serves the constraint of low loss, it does not meet the critical microphotonics requirement of small bend radius and silicon CMOS compatibility. High Index Contrast (HIC) design is consistent with 1 μm turn radii and total film thicknesses of $< 5\,\mu m$. These results meet both the performance and fabrication targets for a low cost, high functionality platform. Figure 3 gives a representation of the scaling law for silicon compatible HIC design. As Δn approaches 1 the minimum device dimension falls below $10\,\mu m$, and scattering loss from edge roughness becomes a dominant concern. The data for silicon nitride and silicon waveguides are taken post-lithography and match the theoretical (solid line) expectation. The data point labelled 2001 represents the measured result achieved by post-lithography processing (oxidation smoothing) [6]. The challenge is to bring the scattering loss line down by process innovation to realize radiation limited (bending loss) performance. The propagation loss target of optical fiber $\sim 0.1\,dB/km$ scales to a value of $\sim 0.1\,dB/cm$ for microphotonic integrated circuits.

1.6 Optical Interconnects in Computing

Within a computing system, optics can afford superior performance for a variety of communication needs ranging from the board-to-board to the intrachip level. This section introduces two applications of optical interconnects in computing systems: optical clocking and the optical data bus. The figure of merit for electronic/photonic convergence is bandwidth $\times$ power dissipation/footprint.

1.6.1 Optical Clock Signal Distribution

Distributing clock signals throughout the electronic computing system is one of the important functions that must be fulfilled by interconnection. In current prevalent synchronous design of computing architecture, periodic pulse signals travel along the interconnects and deliver to each register a timing reference for the movement of data within a system. As a result, the synchronized computing system will be able to operate by sampling data in the registers at precise time intervals, controlled by a central clock. This clocking synchronization applies to elements within an individual microprocessor, or at a higher hierarchy level, between chips and boards. In conventional clocking, where an oscillator clock generates an electrical signal and transports it through a conducting line, increasing technical difficulties arise at the higher frequencies required for faster operation.

Since clock signal interconnects are often loaded with the greatest fanout, travel the largest distances, and operate at the highest speeds within the system, clocking is the application where electrical interconnect problems arise first: prohibitive power consumption, crosstalk and RC delay make it an increasingly difficult task to lower the skew and jitter of signals at higher clock frequencies [7, 8, 9].

Optical clocking, which transports the clock signal by optical interconnects, is an alternative solution overcoming this barrier. Its fast transfer of signal and relative independence of power consumption on clocking speed and

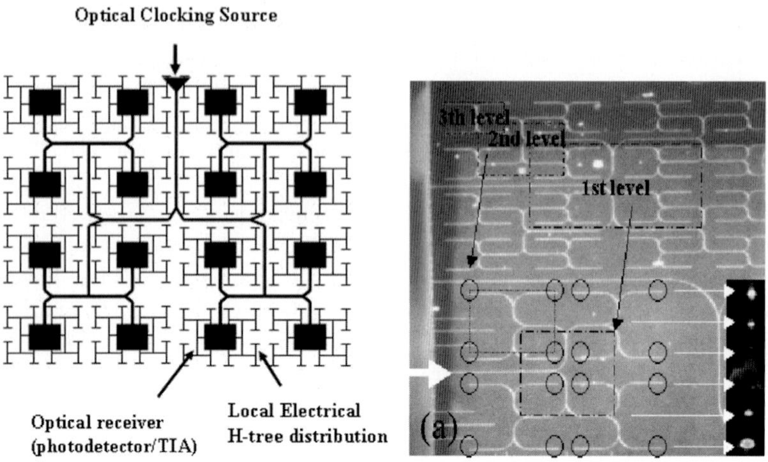

Fig. 4. *Left*: Monolithically integrated optical clocking structure for the Si microprocessor. *Right*: Photographic image of an H-tree optical clock signal distribution network circuitries

propagation length will make optical clocking attractive for new technology platforms.

Figure 4 shows the architectures for a reliable intra-chip optical clocking [10]. A single light source that generates a periodic pulse clock signal can be externally attached to the waveguide and monolithically integrated onto the chip. Light signals propagate along the waveguide in a H-tree network [see Fig. 4 (left)] ensuring that equal path length and fast propagation of photons will minimize skew of the clocking signal. At the end of the optical interconnect, photon signals are converted back to electrical signals by receiver units comprised of a photodetector and transimpedance amplifier. The electric clock signal continues to travel locally via electrical interconnects to the end register. The crossover level from optical interconnect to electrical interconnect is determined by the relative technical advantage between the two technologies at each H-tree hierarchy level.

Key technology elements are monolithically-integrable with Si waveguides, photodetectors and their coupling between components. These components and technologies are discussed in the following sections. optical clocking, unlike an optical data communication bus, provides a simplified technical challenge because there is no need for arrays of on-chip light sources and data modulators.

Since clocking applications see the electrical technology platform quickly approaching its technical limit and offer a relatively simple transition into optical technology, optical clocking posits itself as a significant milestone to the demonstration of the feasibility and applicability of microphotonics.

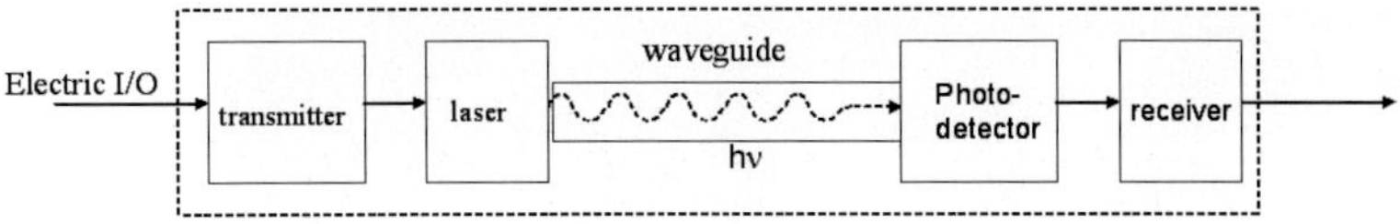

Fig. 5. Schematic block diagram of point-to-point optical data bus

1.6.2 Optical Data Bus

The optical data bus is a device for transporting data in an optical interconnect from one electronic device to another, enabling data transmission at faster rates and with less heat dissipation. The optical data bus is comprised of five basic components: a transmitter, laser, waveguide, photodetector, and receiver, as shown in Fig. 5. The transmitter translates electrical data into drive currents for the laser. The modulated laser light is then injected into the waveguide (optical interconnect) and transmitted to the photodetector. The photodetector converts the photons into an electrical signal, which is then reconditioned by the receiver and outputted to another electrical device.

Implementation of the optical data bus into higher hierarchy level applications (i.e. board to board), where fast data buses are needed is a rapidly developing area of research. One major issue in implementing the optical data bus is the material compatibility of the laser. Most lasers in the telecom window are made up of III–V semiconductors, requiring difficult hybrid integration onto a Si substrate [11,12], increasing fabrication complexity and cost.

2 Photonic Components

Section 2 presents the results of our work in developing passive Photonic device building blocks for the PLC. These waveguide structures guide light by the phenomena of total internal reflection (*Index Guiding*), and are based on Si, silicon nitride (Si_3N_4) and silicon oxynitride (SiON) waveguide cores surrounded by a silicon oxide (SiO_2) cladding, resulting in index guided structures with a core-cladding index difference $\Delta n = 0.1 \div 2.0$. These microphotonic device elements are realized by Si-on-Insulator technologies or deposition on thermal oxidized Si substrates, where the thermal oxide is a $0.5 \div 10$ μm layer optically isolating waveguide modes from the Si substrate.

2.1 Optical Waveguides

As with most technologies, PLC design comes with tradeoffs, specifically a tradeoff between the photonic device size and sensitivity to surface roughness. Both properties are dictated by the refractive index difference Δn between waveguide core and cladding. A high Δn allows for smaller turning

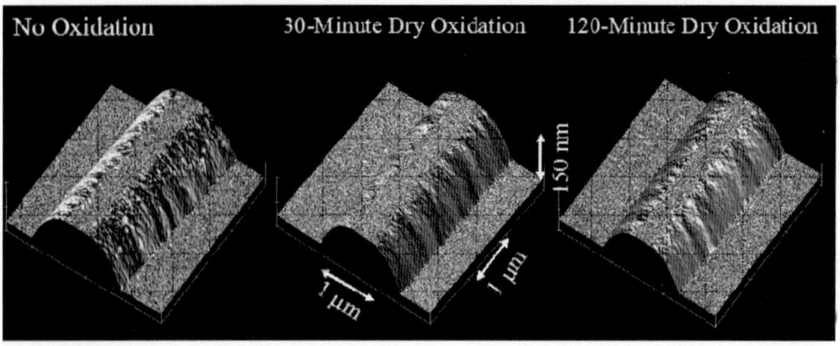

Fig. 6. AFM images of silicon waveguides, the *left image* is as fabricated whereas the *center* and *right* waveguides have undergone 30 and 120 minutes of dry oxidation and subsequent oxide etch respectively

radii, due to increased light confinement, which leads to smaller, densely integrated PLC components. However, as Δn is increased, index guided modes become increasingly sensitive to the roughness of device surfaces and results in appreciable transmission loss. Waveguide roughness running parallel to the direction of the waveguide yields an effective dielectric stack, in terms of varying effective index, resulting in destructive interference and scattering of the optical signal. This effect is augmented in high Δn systems since the strength of a dielectric stack is proportional to Δn^3.

An acceptable scattering loss for PLCs is less than 0.1 dB/cm. This loss value is mainly affected by fabrication related scattering loss. Therefore, there is a strong need to reduce scattering loss through the introduction of additional processing steps. Surface roughness in microphotonics is analogous to the role of impurities in microelectronics, with respect to processing difficulties affecting device performance. There are several ways to reduce surface roughness: oxidation [5], wet etching, and laser annealing [13]. Recently the focus has been on oxidation smoothing, a technique that can be used in any microphotonics system that uses a core material that forms a stable native oxide. Silicon is one such material with the added benefit of having SiO_2 as a compatible, high index contrast, cladding material ($\Delta n = 2$). The reduction of surface roughness is apparent from the AFM images shown in Fig. 6. As oxidation time increases the waveguide surface roughness decreases. From our initial work, we believe the transmission loss in silicon waveguides can be reduced to less than 0.2 dB/cm [14], matching the demands of future PLCs.

2.2 Bends

Current spliced-fiber photonic circuits are limited in complexity by their footprint. Each fiber 'pigtail' must be about 15 cm in length, because prohibitive radiative losses are introduced at fiber bend radii of < 1 cm. The platform elements must, therefore, provide open space for the fiber bends. While this

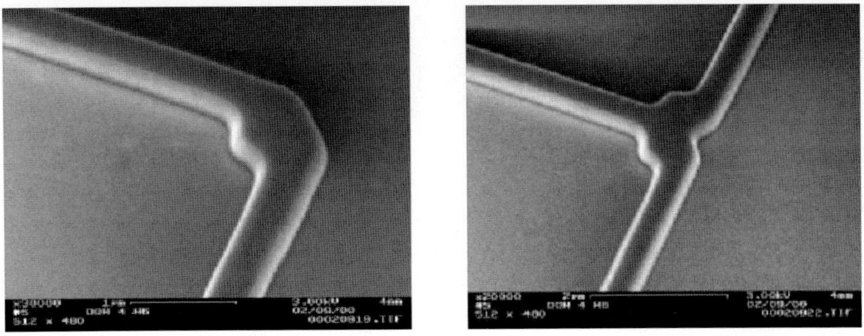

Fig. 7. *Left*: HTC bend device: the device area is $0.5\,\mu m^2$ and the loss are $< 0.3\,dB$/turn for $1\,\mu m$ bend. *Right*: HTC junction device: optical loss are $< 1\,dB$

architecture has been adequate for long haul telecommunications applications where the interconnection density is limited, it clearly does not scale to server buses and chip-to-chip interconnects. This section reviews the principles of design for small bend radii. They are predicated on both higher optical confinement through 1. high index contrast (HIC) materials systems and 2. resonant structures that confine light.

2.2.1 High Transmission Cavity (HTC) Bends

Using High Index Contrast (HIC) material systems such as Si/SiO_2 or SiON/SiO_2, the bending radii can be minimized on the order of $1 \div 10\,\mu m$ at the cost of 0.1 dB/point of bending loss [16]. Due to its strong light confinement, the HIC platform provides excellent dense optical integration [17]. To reduce the footprint further, High Transmission Cavity (HTC) bends were devised and fabricated [18].

Motivated by the optical clock H-tree structure, right angle bends have been prototyped using Si and SiON HIC platforms (Fig. 7). The bending loss was minimized up to 0.3 dB/turn for $1\,\mu m$ bend and 1 dB loss was achieved for T-splitter junctions [18]. The issue for HIC platforms is high propagation loss when sharp bends are realized.

2.2.2 Air Trench Bends for Sharp Bends

The low index contrast (LIC) silica-based platform has been produced using silica optical fiber technology where Δn is less than 0.01. In this LIC platform, propagation loss is not an issue, and several products such as arrayed waveguide gratings (AWGs) are commercially available. The challenge for the LIC platform is to minimize its large footprint applicability to optical clocking. To miniaturize the footprint, the bending radii should be minimized. However, sharp bends of LIC waveguides lead to substantial loss of light power due to poor light confinement.

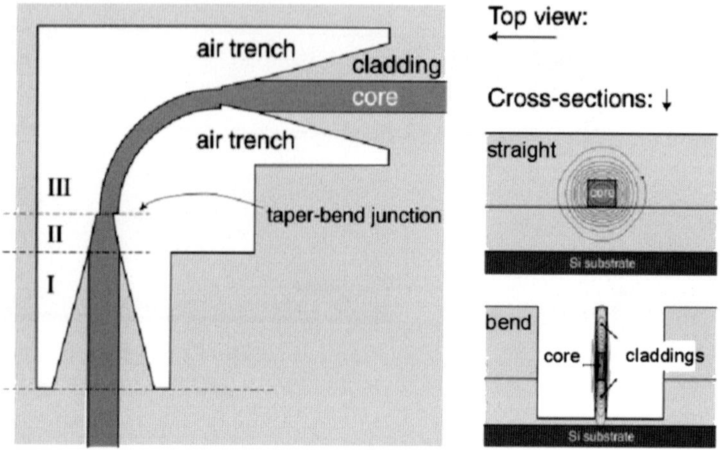

Fig. 8. Schematic diagram of air trench bend

The bending radius is typically 10 mm for $\Delta n = 0.01$, and thus on-chip integration of Si-LSICs (Large Scale Integrated Circuits) is almost impossible using the LIC platform. To minimize bending loss in LIC systems, we introduce air cladding at the bends (in the form of etched air trenches) to locally increase the index difference [19]. The index difference at the bends increases by approximately 0.5, enabling the minimum bending radii to 5–10 µm. The LIC platform with air trench bends becomes a potential candidate for on-chip integration. Figure 8 shows a schematic diagram of the air trench bends, consisting of two tapered parts and a bending part with offset junctions between the taper and the bend. The adiabatic tapering is to avoid abrupt junction-induced mode mismatch and Fresnel reflection. This results in drastic reduction of bending radii while preserving the low-loss performance of the LIC platform. Two-dimensional finite difference time-domain (FDTD) simulation shows how the inclusion of air trenches strongly reduces the bending radius (10–1000 ×) enabling smaller footprints (4–60 ×) for index guided systems with $\Delta n = 0.003$–0.1.

Table 1. Experimentally measured Bend transmission (%) per right turn (TM mode data, ±3% error bars) and FDTD calculation for the structure with $\Delta n = 0.1$

Radius (µm)	2	3	4.5	7	9
Theoretical (%)	92.82	98.12	98.85	99.57	99.67
Experimental (%)	86.87	93.92	96.31	96.54	97.17

For $\Delta n = 0.007$ the bending loss of 0.1 dB/turn at 1.55 µm is obtained with an air trench bend size of 96 µm (trench region bend radius of 9 µm), as compared to a regular bend of 2.5 mm. The fabrication was done based on $\Delta n = 0.1$ and showed good agreement with FDTD simulations (Table 1).

2.3 WDM Functions and Add/Drop Micro-rings

The fundamental element required for bringing WDM communications onto a PLC chip is a compact, high-speed add/drop multiplexerdemultiplexer ("mux/demux"). Coupling of individual channels is the dominant design and cost factor in WDM optical signal distribution. For Microphotonic PLCs utilizing High Index Contrast (HIC) components, resonant coupling provides effective coupling with minimum footprint. A passive device that is resonant with the optical carrier frequency (wavelength) acts similar to an inductive tank circuit in electronics as power is coupled into the device with great efficiency. microcavity resonant structures, designed with free spectral ranges (FSR) larger than the WDM communications window, comfortably satisfy the mux/demux requirement within a compact area, for individual wavelength channels pulse encoded at 1–100 Gb/s.

Micro-ring resonators are travelling wave resonant structures which couple to an incident WDM waveguide bus by evanescent field overlap. Figure 9 shows an SEM of a micro-ring resonator coupled to input and output waveguides for channel drop functionality [20]. The ring becomes a sink for signal power when an integral number of wavelengths match its circumference, producing resonant whispering gallery modes. In Fig. 9 four Si_3N_4 rings, each with a 10% difference in circumference, couple to the same input waveguide. The power loss in each channel is shown in the through port data (upper plot in Fig. 9 on the left). The power distributed to each channel drop port is shown in the drop-port data (lower plot in Fig. 9 on the left). Each micro-ring resonator acts as a filter with Lorentzian line shape. The quality factor, Q, of these filters is ~ 500, and the drop efficiency is nearly 100%.

The value of Q decreases with increased waveguide coupling, increased propagation loss in the micro-ring and radiative loss due to decreasing ring circumference. The micro-rings's FSR, determining add/drop channel selectivity, increases with decreasing circumference, thus correlating optimizing WDM add/drop performance with minimizing device footprint.

In contrast, increasing the Q of a linear resonator however requires increasing the number of Bragg reflector pairs, and this increases the amount of device scattering loss, due to the addition of interface roughness with each additional Bragg reflector pair. Increasing the Q of a micro-ring resonator requires increases the gap spacing between micro-ring and input WDM bus, and this does not increase the device scattering loss. The evanescent coupling strategy of a micro-ring resonator enables the design of ultra-high Q structures: we have demonstrated passive ring resonator structures with $Q = 5000$. The micro-ring architecture is thus ideally suited for long-term dense-WDM (DWDM) PLC designs.

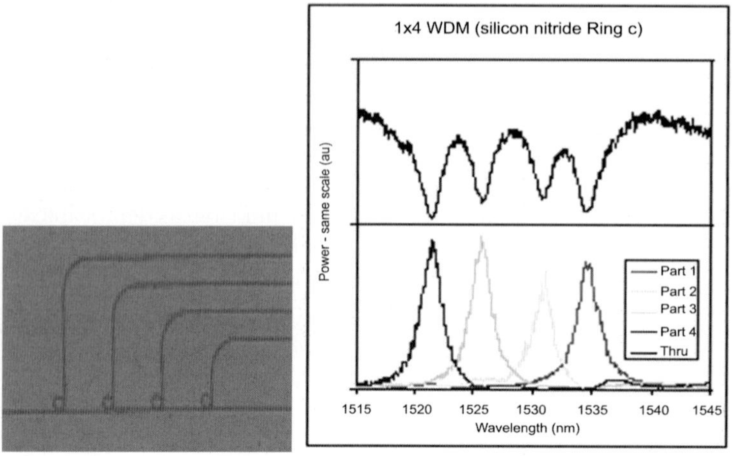

Fig. 9. Four channel add/drop configuration utilizing silicon nitride waveguides and micro-ring resonators with the performance data in [20]

2.4 Polarization Splitters and Rotators

Fiber coupling to a HIC integrated optic circuit is substantially more difficult than coupling to an integrated optic circuit with an index contrast similar to that of the fiber system for the following reasons: 1. The HIC waveguides used on the chip tightly confine the optical field thereby preventing a solid mode match between the chip and fiber modes, 2. The impedance mismatch between the low index fiber and high index chip results in substantial reflections and 3. HIC structures tend to be wavelength sensitive. The mode mismatch may readily be corrected by choosing an appropriate waveguide height and tapering the waveguide to a fine point thereby forcing the waveguide mode into a loosely confined state at the input of the chip. Reflection at the interface may then be mitigated by introducing a quarter-wave antireflection coating on the edge of the chip. However, because the HIC devices on the chip tend to be wavelength sensitive, connecting the input waveguide directly to these components would result in a polarization sensitive component in an otherwise polarization insensitive fiber optic communications link. Rather than attempt to correct the problem at the device level, a polarization diversity scheme may be applied whereby the input polarization is split into orthogonal TE-like and TM-like states, TM-like state is then rotated by 90° and the two identical TE-like chip states are then processed in parallel with identical polarization dependent structures. After passage through the devices on the chip, a second polarization conversion is applied to one of the paths (preferably, the path which it had not been previously applied to) and the two polarizations are recombined [21].

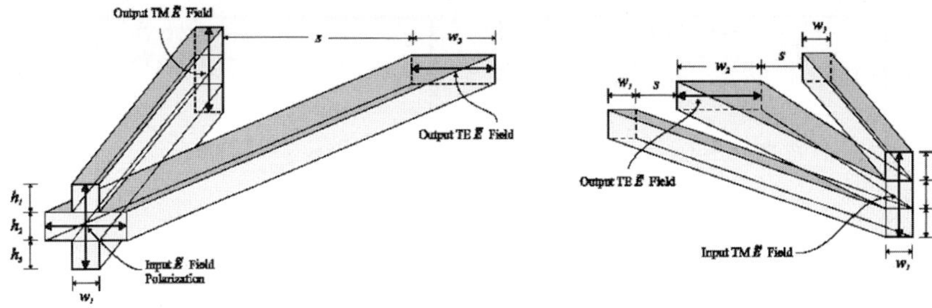

Fig. 10. A polarization splitter (*left*) and a polarization rotator (*right*) based on adiabatic following are shown. Three dimensional FDTD simulations were performed on each structure with core and cladding indices of 2.2 and 1.445 (SiN to SiO_2), layer thicknesses of $h_1 = h_2 = h_3 = 0.25\,\mu m$, and widths $w_1 = 0.25\,\mu m$ and $w_2 = 0.75\,\mu m$. The separations s for the beam splitter and the rotator are $2\,\mu m$ and $0.125\,\mu m$ respectively

In order for this approach to be successful, broadband and low loss polarization splitters and rotators must be integrated on the chip. Several proposals for integrated optic polarization splitters and rotators have been previously published [22,23,24]. However, with the exception of the polarization splitter offered by [25], all such proposals rely on mode coupling to achieve the desired result. As a result of modal dispersion, approaches based on mode coupling tend to be wavelength sensitive. Moreover, in order for the modes to couple efficiently, the modes must be phase-matched and precisely spaced leading to strict fabrication tolerances.

However structures that rely only on adiabatic following [26] do not suffer from these limitations. Essentially, a mode will follow and evolve along a perturbed structure so long as the coupling to other modes introduced by the perturbation is sufficiently small to allow the modes to de-phase before substantial power exchange occurs. The polarization splitter and rotator presented in Fig. 10 (left) and Fig. 10 (right), respectively, are based on this principle. The polarization splitter begins from a cross-shaped waveguide with degenerate polarization states and gradually shifts the two arms of the cross apart. Although the TE-like and TM-like modes of this structure are phase-matched, coupling between them is prevented by mode symmetry and the TE-like and TM-like modes follow the horizontal and vertical waveguides, respectively. The polarization rotator mates up to either of the output arms of the polarization splitter depending on whether a TE or TM on-chip polarization is desired, and is essentially an approximation of a twisted waveguide.

The upper and lower layers are progressively moved into the evanescent field of the mode while core material is added to the middle layer. The results of three dimensional FDTD simulations shown in Fig. 11 (left) and Fig. 11 (right) indicate that near perfect splitting and rotating of polarizations is achieved across the entire $1.45\,\mu m$ to $1.65\,\mu m$ band in structures only

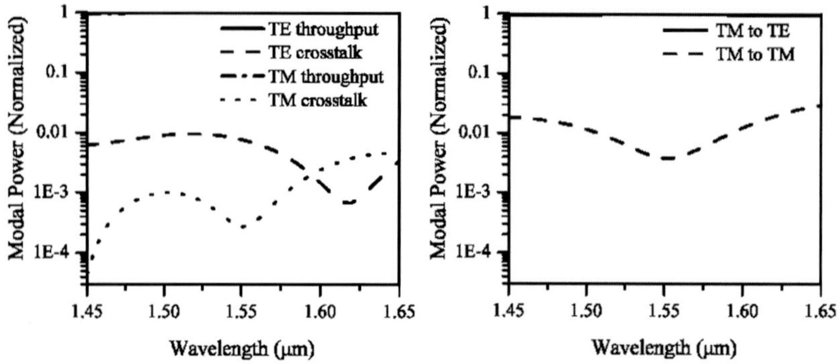

Fig. 11. *Left*: Three-dimensional FDTD simulations results for the polarization splitter with a device length of 50 μm (and separation $s = 2$ μm). *Right*: Results for the polarization rotator with a device length of 50 μm

50 μm long. a simple taper as described in the beginning of this section may be used to couple into the polarization splitter.

2.5 Silicon Light Sources

The scientific and technological breakthrough in the current stage of silicon microphotonics would be the demonstration of an efficient Si light emitter in order to combine Si light emission with all the different functionalities of passive devices described in the preceding sections.

Light emission from silicon is an indirect phonon mediated processes with low probability. In addition, fast nonradiative transitions such as Auger or free carrier absorption [27] severely prevent population inversion at the high pumping rates needed to achieve optical amplification and lasing. Despite of these fundamental limitations, during the 1990s several different strategies have been deeply investigated to increase light emission from different silicon structures [28]. Among the different approaches developed so far, quantum confinement and rare earth doping of silicon have dominated scientific efforts towards the goal of achieving active Si microphotonics. Due to the favorable modification of their optical properties, numerous silicon nanostructures, such as porous silicon [29], silicon nanocrystals embedded in an SiO_2 matrix [30, 31, 32], and Si/SiO_2 superlattices [33, 34] have been widely studied. In particular, silicon nanocrystal research has recently led to exciting results, related to the possibility of fast optical gain in closely-packed silicon nanocrystals, stimulating a rich debate around the feasibility of a future silicon laser device [35,36,37]. In addition, external quantum efficiencies in excess of 1% have been very recently reported in high purity bulk silicon LEDs with surface texturing [38], and the occurrence of optical gain in indirect band-gap materials has been theoretically reassessed [39].

After the discovery of room temperature light emission in Erbium (Er) doped Si [40], Erbium doping has been considered as an ideal candidate to realize the goal of integrated silicon optoelectronics satisfying VLSI requirements. Despite a strong temperature quenching of the luminescence, room temperature eletroluminescence from an Er-doped silicon LED structure with an emission linewidth of 130 Å, was demonstrated for the first time at MIT in 1994 [41]. Despite this breakthrough, it became clear that light emission from Er in silicon is limited by the presence of efficient nonradiative channels like energy back transfer (occurring from the internal first excited state of Er to the external Si lattice) and non-radiative Auger recombination involving Si excitons and Er complexes. A detailed study of the recombination kinetics of Er in silicon is reported in [42]. Using Er as an optical dopant in a Si host material results in a room temperature light emitting device with 10^4 lower quantum efficiency than III–V materials. In addition, parasitic gain limiting effects like free carrier absorption in bulk Si and Er-related up-conversion prevents population inversion of Er in Si, making such a materials system unsuitable for a laser light source or an optical amplifier. Er-doping of glass structures continues to be an ideal approach for the low loss third Telecom window, producing an almost temperature independent emission line at 1.55 μm. Even though Er-doped SiO_2 is already used commercially in optical fiber amplifiers, the application of Er-based structures in Si microphotonics is limited thus far by the small optical cross section of the Er^{3+} transition, and represents a major challenge towards silicon-based microphotonics.

Si CMOS compatible dielectrics, specifically SiO_2, Si_3N_4, and SiON mixtures of the two, form an amorphous insulator host material system for the Er (or other ms-lifetime rare-earth) optical dopant, resulting in 1. efficient room temperature light emission, and 2. negligible Auger, free carrier absorption and energy back-transfer contributions.

Recently, Er doping of Si nanocrystals has been recognized as a hybrid approach combining the promising features of both quantum confinement and impurity doping. It has been demonstrated that Si nanocrystals in the presence of Er act as efficient sensitizers for Er light emission [43], enhancing the Er excitation cross section by more than two orders of magnitude due to a quasi-resonant energy transfer process.

Recent observations of net optical gain at 1.54 μm with enhanced Er emission cross section in Er-doped Si nanocluster sensitized waveguides [44], and the demonstration of efficient room temperature electro-luminescence from Er-silicon nanocrystal devices [45] has opened the route towards the future fabrication of CMOS compatible Er-based devices with efficient electron injection.

2.6 Microphotonic Amplifiers and Lasers

From the perspective of an optical amplifier, the choice of working with ms-lifetime light emitting rare earth atoms, within an amorphous host material,

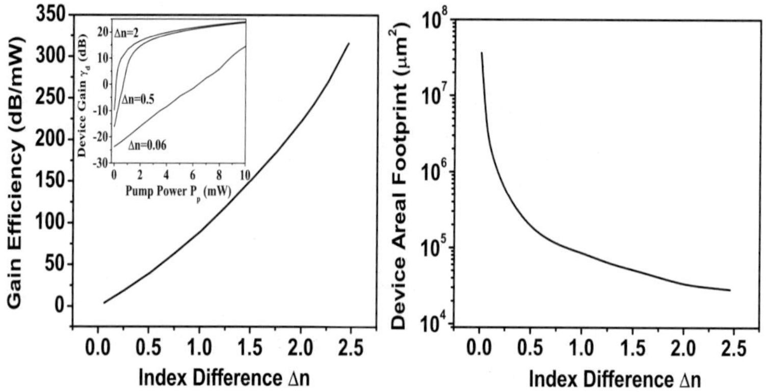

Fig. 12. *Left*: Performance Scaling Advantage [46]: gain efficiency versus Δn for an Er-doped, 1 m long HIC-WOA. *Inset*: plot of Device Gain γ_{eff} vs Pump Power P_{p} shows increasing slope with higher Δn. This data does not include the simulation of scattering transmission loss. *Right*: Size Scaling Advantage: plot of a coiled structure (with a 1 cm device length) areal footprint versus Δn

enables an optimal design by ensuring 1. negligible WDM crosstalk at 1 Gb/s, 2. noise figure values approaching the theoretical limit, and 3. a broad gain bandwidth for full WDM spectrum amplification. From the perspective of a laser light source, the choice of working with light emitting atoms, which are diffused/implanted into waveguide structures, enables the design of in-plane laser structures with zero coupling loss to other passive PLC devices. Using the SiO_2/Si_3N_4 system as a host material, we envision Si microphotonic laser light sources and optical amplifiers as optically pumped waveguide structures with $\Delta n = 0.1$–0.75.

2.6.1 Index Scaling and the Design of Efficient Microphotonic Optical Amplifiers

As microphotonic PLCs scale to higher complexity, the circuit fanout (number of splits) becomes a major source of signal attenuation. WDM long haul transport architecture was enabled by ~ 20 dB gain full spectrum optical amplifiers, utilizing high power III–V pump lasers and > 20 m of Er-doped fiber.

For PLCs a 3 dB signal gain at each Y-split would be sufficient to compensate for fanout. To be practical as a planar device, such 3 dB optical amplifiers must occupy a negligible footprint relative to the circuit size. A reasonable design benchmark for compact amplifier device is a footprint of $< 2\,\text{mm}^2$, 3 dB optical gain, 1.75 dB noise figure, 1 mW pump power source and compatibility with 1 Gb/s data transmission. We have recently derived scaling laws for the design of HIC waveguide optical amplifiers (HIC-WOAs) [46], observing that higher Δn waveguide structures result in 1. performance and 2. size scaling advantage for HIC-WOAs.

The increase in optical confinement with higher Δn results in an increase of signal and pump flux within the waveguide core, creating an increase in amplifier gain efficiency (dB/mW). In a co-propagating pump design, this performance scaling advantage occurs for a constant output signal-to-noise ratio, implying constant noise figure performance, as a function of Δn. We observe (see Fig. 12 (left) a gain efficiency increase of $25\times$ for a microphotonic scale $\Delta n = 1.0$ system, compared to an EDFA scale $\Delta n = 0.06$ system. This means $25\times$ less pump power is required to achieve the same small signal gain coefficient in a HIC-WOA, compared to an EDFA.

The waveguide wound into a two-dimensional coil structure [46] scales dimensionally to a smaller areal footprint with higher Δn due to the smaller turn radii enabled by HIC design. A constant device length [$L = 1\,\mathrm{cm}$, see Fig. 12 (right)] amplifier shows the impact Δn has on its footprint size scaling: a microphotonic scale $\Delta n = 1.0$ system can deliver the same amount of device gain from an area that is $500\times$ smaller than the EDFA scale $\Delta n = 0.06$ system. Gain per unit area on a PLC has thus increased 500 times.

We define a cumulative Figure Of Merit (FOM) for HIC-WOA performance as the device's Areal Gain Efficiency: device gain/(pumppower $\times$ area). This FOM scales dramatically as $(\Delta n)^{2.6}$, summarizing the major benefits due to performance and size scaling for microphotonic HIC-WOA design [46].

2.6.2 Compact Lasers

Micro-ring resonator amplifiers merge both light amplification and WDM filter functions, resulting in flexible devices with high wavelength selectivity and low inversion threshold (very high quality factor Q). In addition, such devices realize the conditions for strong light-matter coupling, yielding a variety of interesting applications for all-optical signal processing related to the enhancement in single channel intensity and nonlinear materials interactions.

In addition, it has been recently demonstrated [47] that Micro-ring laser light sources may potentially realize compact optically pumped lasers for future Si microphotonic platforms.

We show that laser oscillations are possible within these microring structures if fabrication techniques can considerably reduce optical losses. Figure 13 shows the theoretically computed minimum resonator quality factor, Q_{th}, for lasing to occur at a given pump power density, considering an Er-doped Si_3N_4/SiO_2 system.

Several loss coefficients (representing the overall device losses) are shown as horizontal lines related to an equivalent loss quality factor Q_{loss}. For a given value of the total device losses α (yielding a power independent Q_{loss} factor), pump power density must be high enough to ensure the gain clamping relation: $Q_{\mathrm{th}} \leq Q_{\mathrm{loss}}$. Under this condition the total device quality factor given by

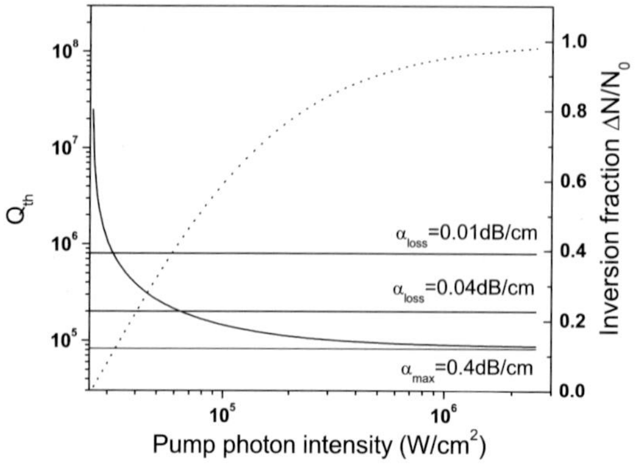

Fig. 13. Threshold quality factor versus pump power density for an Er-doped Si_3N_4/SiO_2 micro-ring resonator. The *dotted curve* shows the fractional population inversion in the device. The *horizontal lines* correspond to the overall device loss reported in the figure

$Q = (Q_{th}^{-1} + Q_{loss}^{-1})^{-1} \approx Q_{th}$, and laser oscillation can be sustained. This analysis makes manifest the critical device processing challenge for realization of optically pumped laser light: the reduction of optical loss (mainly scattering losses) to below 0.4 dB/cm for $\Delta n = 0.1$–0.75 waveguide structures.

3 Photodetectors

3.1 Heteroepitaxial Growth of On-Chip Detectors

Microphotonics is inherently a materials diverse technology. Compound semiconductor sources, dielectric waveguides and ceramic isolators and modulators offer a much greater integration challenge than silicon, silicon dioxide and aluminium did for electronic circuits. The Large Scale Integration (LSI) of microphotonic components to device densities greater than 10^4 devices per chip, encounters two primary barriers: cost and complexity. At device densities of less than 10 devices per chip, hybrid integration by pick-and-place flip-chip, solder-bumps or wafer bonding have the lowest barrier to technology implementation. The advantages of the hybrid approach are 1. best in class optoelectronic devices, and 2. separation of yield issues to device fabrication and packaging [48]. The fact that the packaging of single components dominates cost today is the dominant driver for monolithic integration at even moderate levels of integration. LSI is predicated on the establishment of high yield process technologies and process integration strategies. There is a clear need to develop heteroepitaxial growth and fabrication technologies and concepts for process integration to create a credible vision for LSI microphotonics. The most likely first step for the technology evolution is heteroepitaxy on CMOS or BiCMOS circuits. a particularly important application is the integration of Ge photodetectors with polycrystalline Si waveguides and Si based electronic devices for the distribution and detection of optical signals at wavelengths of 1.3 μm and 1.55 μm on Si. High-quality Ge on Si can also be used for the integration of Ge-based transistors on Si.

The primary difficulty in Ge on Si epitaxy is the 4% lattice mismatch between Si and Ge. This lattice mismatch must be accommodated either through strain or dislocations. In particular, threading dislocations degrade device performance by reducing carrier collection efficiency and increasing leakage current strain may also lead to surface roughness or islanding of films.

One successful approach to reducing the dislocation density is through the use of buffer layers. The first effective linearly graded buffers were grown by *Fitzgerald* et al. [49,50] by MBE at AT & T Bell Labs. In one attempt, pure Ge epilayers were obtained by deposition of amorphous Ge on Si, followed recrystallization by annealing [51]. Despite encouraging results, the problem with direct epitaxy of Ge or SiGe alloys is the high dislocation density.

Due to the lattice mismatch, SiGe alloys with Ge concentrations greater than 30% generally have dislocation densities of 10^{11}–10^{12} cm^{-2} [52]. Post-growth cyclic annealing can reduce the dislocation density to 10^{7} cm^{-2} [53]. This method can produce Ge films suitable for high quality photodetectors.

3.2 Ge Detectors Integrated on Si

Ge is an ideal candidate for low cost, high performance photodetectors because of its compatibility with existing silicon processing, its high carrier mobility, and its high optical absorption at telecommunication wavelengths [54]. CMOS compatible photodiodes were fabricated from films described in the previous section, resulting in the diode structure shown in the inset of Fig. 14. With low temperature oxide (LTO) as an insulation and passivation layer, p–i–n diode structure can be realized in a way that is fully compatible with CMOS technology. At a reverse bias of −2 V and a Ge epitaxial film thickness of 1.3 μm, the responsivity at 1.3 μm and 1.55 μm are 0.64 A/W and 0.40 A/W, respectively.

The experimentally measured responsivity data are shown in Fig. 14 (black curve). As will be explained in the following section, the stress-induced bandgap shrinkage increases the device responsivity with respect to the one calculated for a Ge film without strain, making this approach appealing for achieving broadband detection.

3.3 Broadband High Speed Detectors

The development of Dense Wavelength Division Multiplexing (DWDM) technology and Fiber to the Home (FTTH) technology require broad band, high speed Ge detectors integrated on Si microchips to convert optical signals

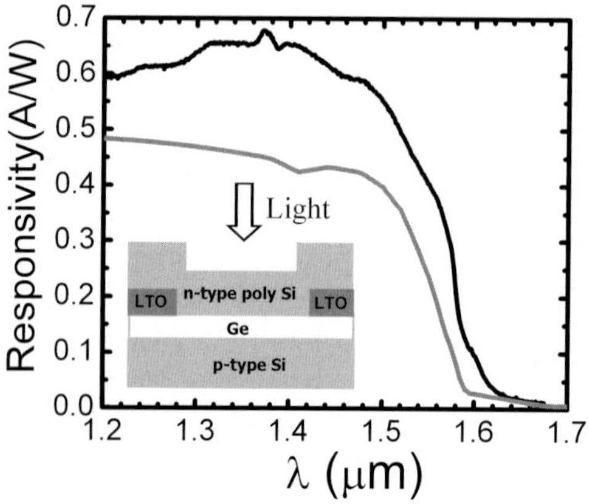

Fig. 14. *Black curve*: Experimentally measured responsivity of the Ge detector device shown in the *inset*. The measured in-plain strain in the Ge film is 0.13%. *Grey curve*: Calculated device responsivity assuming no strain in the Ge film

to electronic signals after demultiplexing by ring resonators. Currently the wavelength range used in telecommunications is expanding from the C-band (1528–1561 nm) to the L-band (1561–1620 nm) [55]. However, the absorption of bulk Ge falls rapidly for energies below the direct band-gap at Γ point (Eg^{Γ}) of 0.8 eV, corresponding to 1550 nm, limiting the application of Ge photodetectors to wavelengths of 1550 nm or shorter. Tensile strained Ge has been found to be a promising candidate for achieving photodetection up to the L-band [56,57,58]. When a thin Ge film of a few atomic layers is pseudomorphically grown on Si, a compressive stress in the film will be induced. As the Ge film grows thicker, misfit dislocations generate at the Ge/Si interface to relax the elastic strain and minimize the free energy of the system [59]. When the sample is cooled down to room temperature, tensile strain dominates the Ge film due to thermal expansion coefficient of Ge being greater than that of Si.

Under tensile biaxial stress the direct band-gap of Ge film shrinks and the valence band becomes non-degenerate, as schematically shown in Fig. 15. The direct band-gap of the tensile stressed Ge film is determined by the gap between the light hole band and the conduction band [$\mathrm{E}_g^{\Gamma} = \mathrm{E}_g^{\Gamma}(\mathrm{lh})$]. The grey shadow areas in Fig. 15 (left) shows $\mathrm{E}_g^{\Gamma}(\mathrm{lh})$ and $\mathrm{E}_g^{\Gamma}(\mathrm{hh})$ as a function of tensile strain. To take into account the error ranges of the deformation potential reported in [60], we have plotted the upper and lower limits of theoretical predictions. Epitaxial Ge films were grown at temperatures from 600 °C to 800 °C to study the direct band-gap versus strain relationship experimentally, and the results are shown by the solid circles in the Fig. 15.

Backside silicidation can be employed to further increase the tensile strain and decrease the direct band-gap of Ge film [58]. The highly conductive silicide layer also forms a good back contact for the devices. With this method, a tensile strain of 0.24% and a direct band-gap of 0.7656 ± 0.0004 eV was achieved, as shown by the hollow circle in Fig. 15. This result indicates

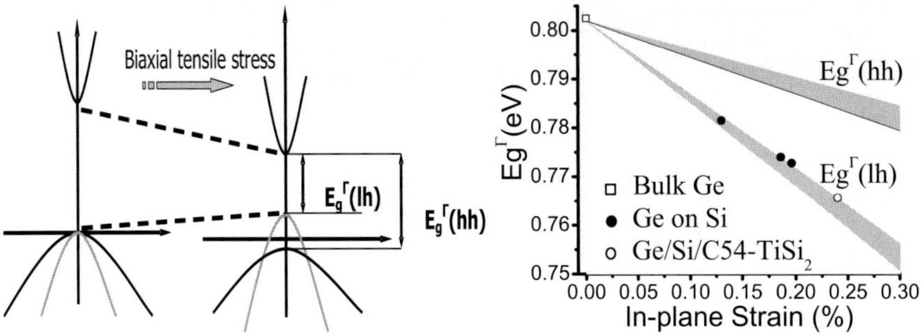

Fig. 15. *Left*: Schematic diagram showing the effect of biaxial stress on the band structure of Ge. *Right*: Theoretical calculation and experimental data of Ge direct band-gap as a function of in-plane strain. The *grey shadow* shows the range of theoretical prediction calculated by the deformation potential theory, using the deformation potential values reported in [60] and taking into account the error bars of these data. The *open square* is the datum of unstrained bulk Ge(100) single crystal. The *solid circles* show the band-gap shrinkage of Ge on Si samples grown at different temperatures, while the *open circle* shows further band-gap shrinkage induced by the backside silicidation

that with backside silicidation efficient light detection up to 1620 nm can be achieved, covering the entire L-band.

The speed of a photodetector is mainly determined by two factors: RC delay and carrier transit time. The RC delay of Ge p–i–n photodiodes can be effectively reduced by decreasing the area so that the capacitance is reduced. For a well designed photodetector, the response time is mainly determined by the carrier transit time. With an electron mobility of $3900\,\mathrm{cm}^2/\mathrm{V}\cdot\mathrm{s}$ and a hole mobility of $1900\,\mathrm{cm}^2/\mathrm{V}\cdot\mathrm{s}$ for bulk Ge material [61], assuming an intrinsic Ge layer thickness of 1 µm and an applied reverse bias of 1 V, the maximum working frequency is calculated to be 25 GHz. High-speed Ge p–i–n detectors with a response time < 200 ps have already been demonstrated, corresponding to a frequency of 2 GHz, and the responsivity was as high as 0.89 A/W and 0.75 A/W at 1.3 µm and 1.55 µm, respectively [62]. p–i–n diodes made of tensile strained Ge grown on Si substrates are expected to achieve a high speed of > 25 GHz with high responsivity over a broad band up to 1620 nm after device optimization (see Fig. 14).

3.4 Lightwave Circuit Integration with Photodetector

Efficient photodetector/waveguide coupling represents an issue of primary importance for the achievement of dense optical integration. Typical schemes for the integration of photodetectors with planar ligthwave circuits can be grouped into two main categories (see Fig. 16): Butt coupling [63, 64, 65] (or End-Fire coupling) and evanescent coupling. In the case of the Butt coupling scheme the photodetector is aligned in series with the waveguide to increase

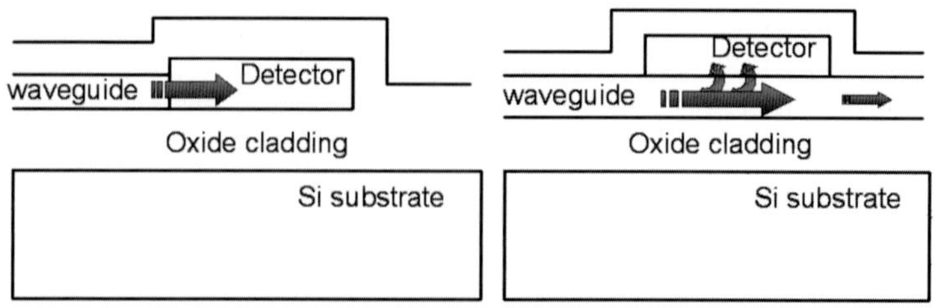

Fig. 16. Coupling structure of waveguide and photodetector integration. *Left*: Butt-coupling structure. *Right*: Evanescent-wave coupling structure

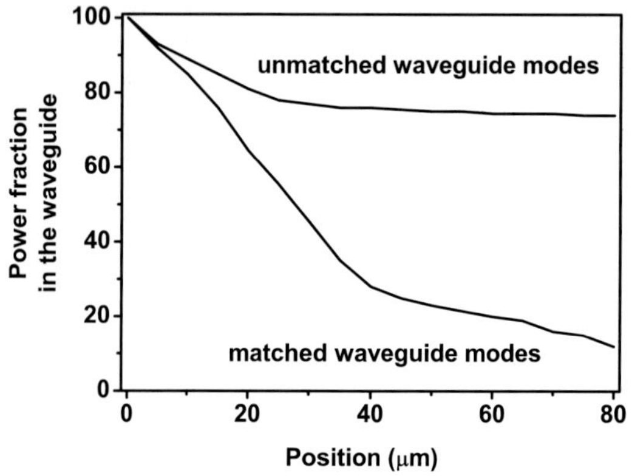

Fig. 17. Two dimensional beam propagation method (BPM) simulation of integrated structure showing power coupling for an out of phase matching condition and for a phase-matched condition

photon absorption rate. However, precise Butt coupling alignment demands complex fabrication capabilities with high geometrical accuracy. In addition, back reflection at the detector/waveguide interface (especially in the case of high Δn materials) has to be suppressed through an anti reflection (AR) coating.

In the case of evanescent coupling structures [66, 67, 68], the detector is positioned on top of the waveguide and light coupling occurs through the evanescent tails of the waveguide modes. When waveguides and detectors are grown by epitaxial methods (Si/Ge, Si/SiGe, or III–V semiconductor materials system), the evanescent coupling approach provides monolithic process control of the waveguide/photodetector interface and eliminates the requirement of precise lateral alignment, yielding more robust coupling structures for reliable integration.

A particularly powerful geometry for weakly absorbing (indirect gap, Ge or Si) detector materials is the 'waveguide configuration': for high index contrast between the waveguide and detector materials, a phase matched design is required for efficient evanescent coupling.

Figure 17 shows the critical dependence of coupling behavior on Si layer thickness for an SiON ($n = 1.5$) waveguide. When calculated index guided mode propagation constants are not matched, photons are not coupled into the absorbing Si layer and almost no coupling occurs. For a calculated Si thickness that matches the Si layer propagation constants to the SiON waveguide, the coupling is strongly enhanced.

4 Photonic Crystal Structures

Photonic crystals provide the ultimate in photon control and device shrink. A photonic crystal is a periodic composite of high and low index materials that, as a medium, influence the propagation of light in a way similar to atomic potentials with the flow of electrons in a semiconductor. The fabrication of photonic crystals is difficult, and HIC index guided structures (see Sect. 2) can achieve similar performance/area for many applications. Nevertheless, photonic crystals provide a dominant design paradigm for microphotonics, including HIC components. Four specialized applications are discussed below: small modal volume for in-line filters and low threshold sources; multichannel tuneable devices for ubiquitous placement in microphotonic circuits; photonic band-gap waveguide amplifiers and complex photonic structures for future active and multi-resonant devices.

4.1 Ultrasmall Devices

An in-line add/drop device can be constructed as a microcavity inserted in a waveguide [69]. One dimensional photonic crystals are inserted as dielectric stack mirrors. As the index contrast in the mirror increases, a larger spectrum of high reflectivity (photonic band-gap) results. In the example of Fig. 18, air sections ($n = 1$) are separated by silicon sections ($n = 3.5$) with the spacings chosen to give a photonic band-gap overlapping the WDM gain spectrum of an Erbium Doped Fiber amplifier (EDFA).

The missing air section in the middle of the device defines the pass band (wavelength of the channel to be dropped). The simulated and measured performance on the right indicates the strong theoretical foundation behind photonic crystal design. The Q of this device is 265 at the resonant wavelength of 1564 nm. Most impressive is the modal volume of $V = 0.055\,\mu\text{m}^3$. Low threshold Er-based light sources require high values of Q/V. High Q can be designed with large numbers of sections, but low V has been the challenge until the advent of photonic crystals.

4.2 Multichannel Design

The design paradigm of photonic crystals creates an analogy between channel drop frequencies of an optical filter and bound defect states in the gap of

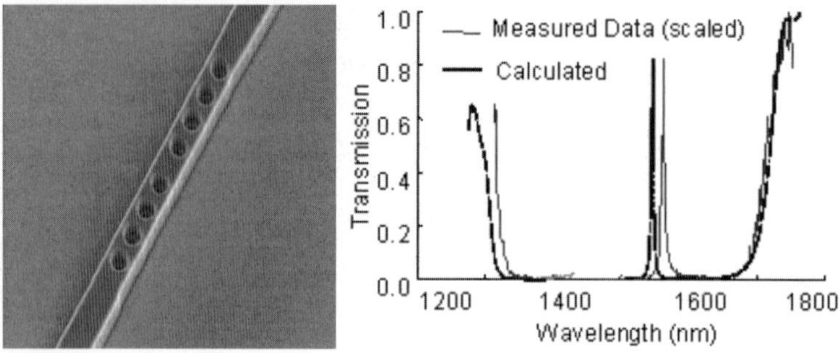

Fig. 18. An inline photonic crystal add/drop filter in a silicon waveguide with measured and calculated performance

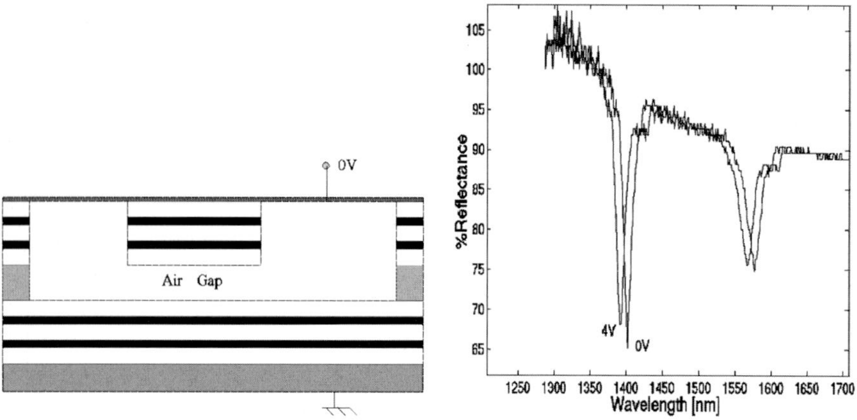

Fig. 19. Device cross-section and performance for a multichannel, tuneable add/drop filter [70]

a semiconductor. As some defects can present more than one state in the gap of a semiconductor, a photonic crystal filter can be designed as a multichannel device. For microphotonic circuits to grow in complexity beyond 10 devices/chip, a ubiquitous device, akin to a transistor, must perform switching and amplification functions. Since photons possess no charge, their interaction is mediated by wavelength or frequency.

A variable frequency (wavelength) filter can act as a switch, add/drop path and multiplexer/demultiplexer for telecommunication channels. If the same device can yield all functions, it can be replicated in a simplified design/process integration scheme. The photonic crystal device in Fig. 19 is such a device with operation in the 1310–1600 nm regime [70]. For this microelectromechanical actuated device a 12 V bias results in 100 nm tuning for each channel.

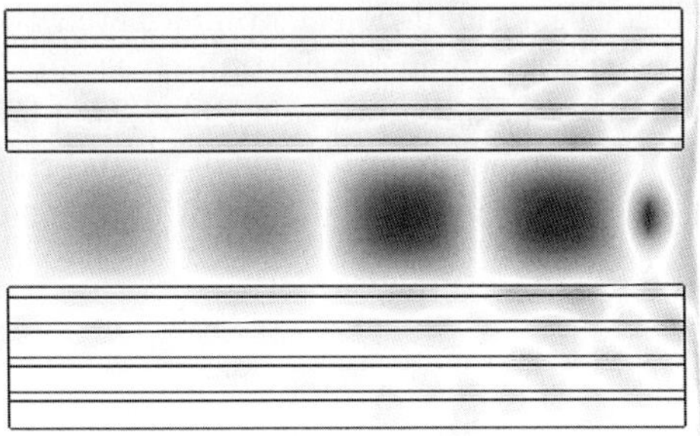

Fig. 20. FDTD simulation of the PBG waveguide, showing light propagation in the low index core material (SiO_2)

4.3 Photonic Band-Gap Waveguide Amplifier

Total internal reflection (TIR) has been traditionally the only mechanism to confine light within the core of optical waveguides whose core refractive index is higher than the index of the cladding. Optical fiber and silica-bench optical waveguides are entirely based on TIR. An interesting departure from this traditional index guiding scheme is based on the realization of photonic band Gaps (PBG), forbidden frequency ranges in periodic dielectric structures capable of full light confinement also in low-index materials or hollow waveguide cores. A new type of silicon waveguide-PBG cladding waveguide can be developed based on the PBG principles. The light in the core of the waveguide is confined by destructive Bragg scattering as opposed to TIR mechanism.

The low refractive index in a PBG cladding waveguide core allows for a large device flexibility, realizing sharp bends with negligible radiation losses due to the PBG confining mechanism. The PBG waveguide can be designed as a slab waveguide, ridge waveguide, and channel waveguide with low index core (SiO_2 or air) surrounded by high index contrast dielectric cladding stacks (Si/SiO_2 or Si/Si_3N_4). The resulting waveguides are fully compatible with CMOS process.

Potential applications include optical amplifiers where these structures are doped with optically active materials (e.g. Er atoms, quantum dots, etc.) in the SiO_2 core. Scattering losses out of the active guiding region can be suppressed in these waveguide devices, allowing for low threshold operation.

The PBG waveguide amplifier can be optically pumped from the top or in a co-propagating geometry. The stop band of the PBG cladding layers can be designed so that the propagation modes are around the 1.54 µm range, while the pumping wavelength can have full transmission. The FDTD simulation of the modes confined in the low index core are shown in Fig. 20.

Near 100% absolute reflectivity of 6 pair Si/SiO_2 PBG cladding layers has been achieved using novel thermal oxidation process.

While photonic crystals and the formation of light defect states can afford unique properties like strong dispersion at the bulk localized band edge states, strong field enhancement effects in a small defect volume, etc., the strict requirement of periodicity gives rise to some fabrication constraints. Recently, new types of photonic structures – random or quasi periodic photonic structures, have been found to be more flexible in terms of engineering of the dispersion properties, affording more freedom in terms of design parameters and defect engineering.

4.4 Complex Photonic Structures

The interest in random dielectrics and complex photonic structures, materials where the refractive index fluctuates over length scales comparable with the wavelength of light, has increased in the last few years. The phenomena related to coherent backscattering of light [71, 72] and Anderson photon localisation [73, 74] have been demonstrated experimentally [75, 76] and the study of disordered dielectrics with gain [77, 78] has demonstrated the feasibility of random lasers [79] – mirrorless lasers where the required optical feedback is achieved through strong multiple scattering inside the inverted disordered medium. The class of photonic complex structures can be enlarged by considering fractal and quasiperiodic structures, characterized by refractive index profiles that fluctuate neither randomly nor periodically, though the structures are deterministically generated according to simple iterative prescriptions (such as the Fibonacci sequence $F_{j+1} = F_j + F_{j-1}$). Despite their attractive physical properties, little attention has been devoted so far to the experimental realization of this intermediate (between random and perfectly ordered structures) class complex dielectrics, namely photonic quasicrystals (PQs) [80].

Photonic quasicrystals, realized by the stacking together different dielectric materials, exhibit many novel phenomena such as a fractal transmission spectrum of zero Lebesgue measure, field quasi-localisation (field states decaying as a power law inside the structure) and sizeable field enhancement, fractal field states and several interesting anomalies both in the electron and photon transport (marginal diffusion, subdiffusive behaviour, etc.) related to the fractal characteristics [81, 82, 83, 84] of the structures.

The rich transmission spectra of fractal photonic structures, the strong band-edge dispersion and the occurrence of quasilocalized modes with field enhancement are attractive characteristics from a device application point of view [85].

In particular, the band-edge zone of quasicrystalline photonic structures can considerably slow down light diffusion because of the combined presence of auto-similar (fractal) localized states and the finite size band-edge resonances with enhanced mode density [86, 87] (see Fig. 22).

The strong light-matter interaction that occurs for high index contrast PQs both at the quasilocalized or at the strongly dispersive band-edge states can lead to gain enhancement [89] in the material with appealing implications for future multifrequency active devices, such as fractal band-edge lasers,

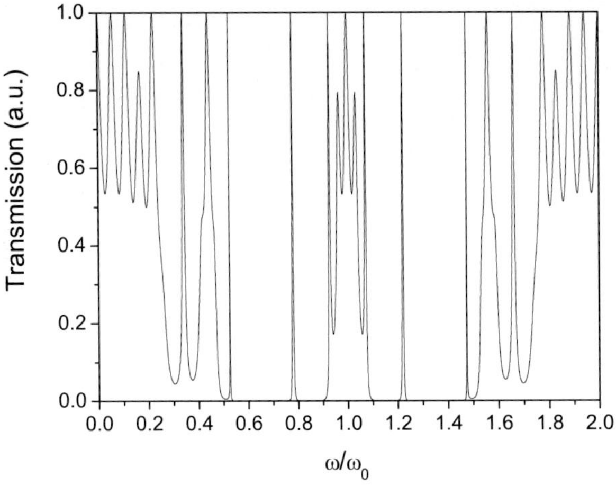

Fig. 21. Calculation of the transmission spectrum versus the normalized frequency for a 32 layers Thue-Morse nonperiodic structure realized with Si and SiO_2 layers. The multilayers structure satisfies the $\lambda/4$ condition around the operation wavelength $\lambda_0 = 1.54\,\mu m$

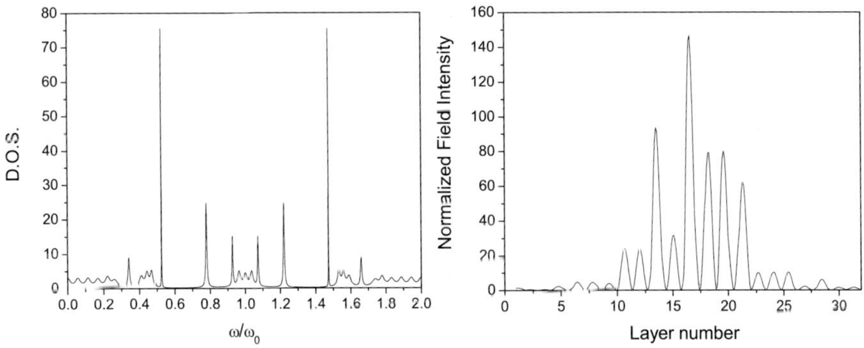

Fig. 22. *Left*: Calculation of the adimensional density of optical modes for the 32 layers Si/SiO_2 Thue-Morse structure [88]. *Right*: Calculation of the electric field intensity at the higher energy band-edge normalized to the input field

multiwavelength light sources and amplifiers, frequency selective filters for DWDM.

The occurrence of strong band edge group velocity reduction and enhanced band edge dispersion (with respect to periodic structures) in quasiperiodic 1D structures have been recently demonstrated experimentally [90, 91].

The possibility to fabricate PQs alternating Si and SiO_2 layers with standard CMOS compatible deposition techniques represents a promising route to demonstrate significant localization features (see Figs. 21 and 22) in a PQ structure with a limited number of dielectric layers, paving the way for the realization of a variety of optical devices based on localized or heavy photons.

5 Conclusion

Monolithic Silicon Microphotonics offers a scalable solution to interconnection density, bandwidth, latency and electronic/photonic partitioning. The use of standard silicon-compatible materials and process tools is highly leveraged with the $250B/year integrated circuit industry supply chain. It is both surprising and gratifying that Silicon Microphotonics can produce the entire range of photonic component functionalities with reasonable performance. One can project a vision for the technology as outlined below. As photonics emerges from its near term niche market character (< $20B/year) to universal deployment in all intelligent systems, silicon will become the dominant platform. With performance/cost as the driving metric, the high levels of parallelism made possible by photonic interconnection will assume control of systems design. The silicon platform will drive standards and cost reduction until photonic circuits are commoditized, and then photonic functionality through repartitioning of the electronic/photonic interface will drive the new value proposition.

References

1. ITRS (2001), International Technology Roadmap for Semiconductors: 2001, `http://public.itrs.net/`
2. K. Wada, H. S. Luan, K. K. Lee, S. Akiyama, J. Michel, L. C. Kimerling, M. Popovic, H. A. Haus: *Silicon and Silica Platform for On-chip Optical Interconnection* (Proc. LEOS Annual Meeting, 2002)
3. C. H. Fine, L. C. Kimerling: Biography of a killer technology: Optoelectronics Drives Industrial Growth with the Speed of Light, in *OIDA Future Vision Program* (Optoelectronics Industry Development Association, Washington, DC 1997), pp. 1–22
4. L. C. Kimerling, D. M. Koker, B. Zheng, F. Y. G. Ren, J. Michel: Erbium-Doped Silicon for Integrated Optical Interconnects, in H. R. Huff, W. Bergholz, K. Su-mino (Eds.): *Semiconductor Silicon* (The Electrochemical Society, Pennington, NJ 1994), 486
5. K. K. Lee, D. R. Lim, H-C Luan, A. Agarwal, J. Foresi, L. C. Kimerling: Appl. Phys. Lett. **77**, 1617 (2000)
6. K. K. Lee, D. R. Lim, L. C. Kimerling: Opt. Lett. **26** (23), 1888 (2001)
7. D. A. B. Miller: Proc. IEEE **88**, 728 (2000)
8. S. K. Tewksbury, L. A. Hornak: J. VLSI Signal Proc. **16** (23), 225 (1997)
9. J-F. Zheng, F. H. Robertson, E. Mohammed, I. Young, D. Ahn, K. Wada, J. Michel, L. C. Kimerling: On-chip optical clock signal distribution, Tech. Dig. OSA Topical Meeting on Optics in Computing, Washington, D.C 2003, Paper OWB3-4
10. E. A. Fitzgerald, L. C. Kimerling: Silicon-based microphotonics and integrated optoelectronics, MRS Bulletin **23**, 39–47 (1998)
11. Y. Liu: Heterogeneous integration of OE arrays with Si electronics and microoptics, IEEE Trans. Adv. Packaging **25** (1), 43–49 (2002)

12. J. J. Liu, B. Riely, P. H. Shen, N. Das, P. Newman, W. Chang, G. J. Simonis: Ultra-low threshold sapphire substrate-bonded top-emitting 850 nm VCSEL array, IEEE Photon. Tech. Lett. **14**, 1234 (2002)
13. D. K. Armani, T. J. Kippenberg, S. M. Spillane, K. J. Vahala: Nature **421**, 925 (2003)
14. D. K. Sparacin, K. Wada, L. C. Kimerling: Oxidation kinetics of waveguide roughness minimization in silicon microphotonics, Proc. OSA Integrated Photonics Research Conference (IPR 2003), Washington D.C. (2003)
15. D. Marcuse: Bell System Tech. J. **50** (8), 2551–2563 (1971)
16. K. K. Lee: Ph.D. Thesis (MIT 2000)
17. D. R. C. Lim: Ph.D. Thesis (MIT 2000)
18. C. Manolatou, S. G. Johnson, S. Fan, P. R. Villeneuve, H. A. Haus, J. D. Joannopoulos: J. Lightwave Technol. **17**, 1682 (1999)
19. M. Popovic, K. Wada, S. Akiyama, H. A. Haus, J. Michel: J. Lightwave Technol. **20**, 1762 (2002)
20. D. R. Lim, B. E. Little, K. K. Lee, M. Morse, H. H. Fujimoto, H. A. Haus, L. C. Kimerling: Proc. SPIE **3847**, 65–71 (1999)
21. C. K. Madsen: Opt. Lett. **15**, 878 (2000)
22. J. Z.Huang, R.Scarmozzino, G.Nagy, M. J.Steel, R. M.Osgood: IEEE Photonics Techn.Lett. **12**, 317 (2000)
23. V. P.Tvolov, M.Fontaine: Opt. Comm. **127** 7 (1996)
24. W. W. Lui, T. Hirono, K. Yokoyama, W. P. Huang: J. Lightwave Technol. **16**, 929 (1998)
25. Y. Shani, C. H. Henry, R. C. Kistler, R. F. Kazarinov, K. J. Orlowsky: Appl. Phys. Lett **56**, 8 (1990)
26. M. R. Watts, H. A. Haus, G. Gorni, M. Cherchi: Proc. OSA Integrated Photonics Research Conference (IPR 2003) (Washington D.C. 2003), p. 26
27. W. P. Dumke: Phys. Rev. **127**, 1559 (1962)
28. O. Bisi, U. Campisano, L. Pavesi, F. Priolo: *Silicon based microphotonics: from basics to applications* (IOS, Amsterdam 1999)
29. L. T. Canham: Appl. Phys. Lett **57**, 1046 (1990)
30. T. Shimizu-Iwayama, K. Fujita, S. Nakao, K. Saitoh, T. Fujita, N. Itoh: J. Appl. Phys. **75**, 7779 (1994)
31. M. L. Brongersma, A. Polman, K. S. Min, E. Boer, T. Tambo, H. A. Atwater: Appl. Phys. Lett. **72**, 2577 (1998)
32. F. Iacona, G. Franzó, C. Spinella: J. Appl. Phys. **87**, 1295 (2000)
33. Z. H. Lu, D. J. Lockwood, J. M. Baribeau: Nature **258**, 378 (1995)
34. D. J. Lockwood, Z. H. Lu, J. M. Baribeau: Phys. Rev. Lett. **258**, 539 (1996)
35. L. Pavesi, L. Dal Negro, C. Mazzoleni, G. Franzó, F. Priolo: Nature **408**, 440 (2000)
36. L. Khriachtchev, M. Rasanen, S. Novikov, J. Sinkkonen: Appl. Phys. Lett. **79**, 1249 (2001)
37. L. Dal Negro, M. Cazzanelli, Z. Gaburro, P. Bettotti, L. Pavesi, F. Priolo, G. Franzó, D. Pacifici, F. Iacona: In: *Towards the First Silicon Laser*, ed. by L. Pavesi, S. Gaponenko, L. Dal Negro, NATO Science Series (Kluwer Academic, Dordrecht 2003)
38. T. Trupke, J. Zhao, A. Wang, R. Corkish, M. Green: Appl. Phys. Lett. **82**, 2996 (2003)
39. T. Trupke, M. Green, P. Würfel: J. Appl. Phys. **93**, 9058 (2003)

40. J. Michel, J. L.Benton, R. F. Ferrante, D. J. Jacobson, D. J. Eaglesham, E. A. Fitzgerald, Y. H. Xie, J. M. Poate, L. C. Kimerling: J. Appl. Phys. **70**, 2672 (1991)
41. B. Zheng, J. Michel, F. Y. G. Ren, L. C. Kimerling, D. C. Jacobson, J. M. Poate: Appl. Phys. Lett. **64**, 2842 (1994)
42. J. Palm, F. Gan, B. Zheng, J. Michel, L. C. Kimerling: Phys. Rev. B **54**, 17603 (1996)
43. G. Franzó, F. Iacona, V. Vinciguerra, F. Priolo: Mat. Sci. Eng. **B 69/70**, 338 (1999)
44. H. S. Han, S. Y. Seo, J. H. Shin: Appl. Phys. Lett. **81**, 3720 (2002)
45. A. Irrera, D. Pacifici, M. Miritello, G. Franzó, F. Priolo, F. Iacona, D. Sanfilippo, G. Di Stefano, P. G. Fallica: Appl. Phys. Lett. **81**, 1866 (2002)
46. S. Saini, J. Michel, L. C. Kimerling: IEEE JLT **21**, 10 (2003) in press
47. L.Yang, D. K. Armani, K. J. Vahala: Appl. Phys. Lett. **83**, 825 (2003)
48. MIT Communications Technology Roadmap, *Interim Report of the Technical Working Group on the Silicon Platform* (2002)
49. E. A. Fitzgerald, Y. H. Xie, M. L. Green, D. Brasen, A. R. Kortan, J. Michel, Y. J. Mii, B. E. Weir: Appl. Phys. Lett. **59**, 811 (1991)
50. Y. H. Xie, E. A. Fitzgerald, P. J. Silverman, A. R. Kortan, B. E. Weir: Mater. Sci. Eng. B **14**, 332 (1992)
51. L. Colace, G. Masini, G. Assanto, H. C. Luan, K. Wada, L. C. Kimerling: Appl. Phys. Lett. **76**, 1231 (2000)
52. F. K. LeGoues, B. S. Meyerson, J. F. Morar, P. D. Kirchner: J. Appl. Phys. **71**, 4230 (1992)
53. H. C. Luan, D. R. Lim, K. K. Lee, K. M. Chen, J. G. Sandland, K. Wada, L. C. Kimerling: Appl. Phys. Lett. **75**, 2909 (1999)
54. L. C. Kimerling: *Photons to the Rescue: Microelectronics Becomes Microphotonics*, The Electrochemical Society Interface, 28 (2000)
55. S. V. Kartalopoulos: IEEE Circuits Device **18**, 8 (2002)
56. Y. Ishikawa, K. Wada, D. D. Cannon, J. Liu, H. Luan, L. C. Kimerling: Appl. Phys. Lett. **82**, 2044 (2003)
57. D. D. Cannon, J. Liu, D. T. Danielson, S. Jongthanmmanurak, H. C. Luan, J. Michael, K. Wada, L. C. Kimerling: Optoelectronics of group-IV-based materials, Mater. Res. Symp. Proc. **770** (San Francisco 2003)
58. J. Liu, D. D. Cannon, K. Wada, Y. Ishikawa, S. Jongthammanurak, D. T. Danielson, J. Michel, L. C. Kimerling: submitted to Appl. Phys. Lett
59. E. A. Fitzgerald: Mater. Sci. Rep. **7**, 87 (1991)
60. F. H. Pollak, M. Cardona: Phys. Rev. **172**, 816 (1968)
61. F. Schaffler: *Properties of advanced semiconductor materials : GaN, AlN, InN, BN, SiC, SiGe*, ed. by M. E. Levinshtein, S. L. Rumyantsev, M. S. Shur, (Wiley, New York 2001), Chap. 6
62. S. Fama, L. Colace, G. Masini, G. Assanto, H.-C. Luan: Appl. Phys. Lett. **81**, 586 (2002)
63. D. Zurhelle, O. Blume, S. Popp, J. Muller: J. Lightwave Technol. **14**, 3 (1996)
64. S. Wunderlich, J. P. Schmidt, J. Muller: Appl. Opt. **31**, 21 (1992)
65. W. Doldissen, F. Fiedler, R. Kaiser, L. Morl: Electon. Lett. **14**, 35 (1989)
66. M. Erman, Ph. Jarry, R. Gamonal, J.-L. Gentner, P. Stephan, C. Guedon: J. Lightwave Technol. **6**, 399 (1988)

67. M. Caldararu, F. Craciunoiu, A. Paraschiv, D. Cristea, F. Caldararu, D. Dascalu, C. Klier, G. Mirea, A. Vasile, M. Nicolae: Solid-State Electron. **39**, 1649 (1996)
68. G. Masini, L. Colace, G. Assanto: Opt. Mat. **317**, 243 (2001)
69. J. S. Foresi, P. R. Villeneuve, J. Ferrera, E. R. Thoen, G. Steinmeyer, S. Fan, J. D. Joannopoulos, L. C. Kimerling, H. I. Smith, E. P. Ippen: Nature **39**, 143 (1997)
70. Y. Yi, P. Bermel, K. Wada, X. Duan, J. D. Joannopoulos, L. C. Kimerling: Appl. Phys. Lett. **81**, 4112 (2002)
71. M. P. van Albada, A. Lagendijk: Phys. Rev. Lett. **55**, 2692 (1985)
72. Y. Kuga, A. Ishimaru: J. Opt. Soc. Am. A1 **831** (1984)
73. P. W. Anderson: Philos. Mag. **52**, 505 (1985)
74. S. John: Phys. Rev. Lett. **53**, 2169 (1984)
75. D. S. Wiersma, P. Bartolini, A. Lagendijk, R. Righini: Nature **390**, 671 (London 1997)
76. D. S. Wiersma, J. Gòmez Rivas, P. Bartolini, A. Lagendijk, R. Righini: Nature **398**, 207 (London 1999)
77. D. S. Wiersma, A. Lagendijk: Phys. Rev. E. **54**, 6118 (1997)
78. H. Cao, Y. G. Zhao, S. T. Ho, E. W. Seelig, Q. H. Wang, R. P. H. Chang: Phys. Rev. Lett. **82**, 278 (1999)
79. V. S. Letokhov: Zh. Eksp. Teor. Fiz. **53**, 1442 (1967)
80. T. Fujiwara, T. Ogawa: *Quasicrystals* (Springer, Berlin, Heidelberg 1990)
81. E. Maciá, F. Domínguez-Adame: Phys. Rew. Lett. **76**, 2957 (1996)
82. W. Gellermann, M. Kohmoto, B. Sutherland, P. Taylor: Phys. Rev. Lett. **72**, 633 (1994)
83. N. Fujita, K. Niizeki: Phys. Rev. B **64**, 144207 (2001)
84. M. Kohmoto, B. Sutherland: Phys. Rev. B **35**, 1020 (1987)
85. E. Maciá: Phys. Rev. B **63**, 205421 (2001)
86. J. M. Benedickson, J. P. Dowling, M. Scalora: Phys. Rev. E **53**, 4107 (1996)
87. C. Sibilia, I. Nefedov, M. Scalora, M. Bertolotti: J. Opt. Soc. Am. B **15**, 1947 (1998)
88. N. Liu: Phys. Rev. B **55**, 55 (1997)
89. S. Nojima: J. Appl. Phys. **90**, 545 (2001)
90. L. Dal Negro, C. J. Oton, Z. Gaburro, L. Pavesi, P. Johnson, Ad. Lagendijk, R. Righini, M. Colocci, D. Wiersma: Phys. Rev. Lett **90**, 055501 (2003)
91. M. Gerken, D. Miller: Appl. Opt. **42**, 1330 (2003)

Optical Interconnect

Zeno Gaburro

INFM and Dipartimento di Fisica
Università di Trento, Italy
gaburro@science.unitn.it

Abstract. The progress of silicon electronic industry is based on scaling down the minimum feature size of integrated circuits. Speed, density and costs of devices improve with scaling, but unfortunately the performance of interconnect worsens, both in terms of speed and power consumption. This issue – the "interconnect bottleneck" – is envisioned as a critical showstopper of electronic industry in the near future. The physical reason behind the interconnect bottleneck is the resistive nature of metals. The introduction of copper in place of aluminum has temporarily improved the interconnect performance, but on the other hand in a few years a more disruptive solution will be required in order to keep the current pace of progress. Optical interconnect is an intriguing alternative to metallic wires, because light can travel in dielectrics, and even in vacuum. At present, optical technology in silicon is not mature for industrial implementation. For this very same reason, however, it is also rich of research opportunities with large potential payoff. The rationale of this chapter has been to trace a snapshot of the current interconnect limitations; to point out the basic differences between the electrical and optical interconnect from different perspectives, ranging from basic physics up to system layouts; and to give a flavor of suggested practical realization of optical solutions. Any of these goals would be too ambitious for a book chapter without a hopefully rich and up-to-date bibliography.

1 Introduction

Since the 60s, the cost-per-function in integrated circuit (IC) technology has kept decreasing exponentially[1] [1,2]. Such growth is completely unparalleled by any other manufacturing industry. The driving force has been scaling down of devices. Smaller devices on larger wafers led to larger yield, cost cut and faster circuits. Minimum feature size (linewidth) of current[2] complementary metal-oxide-semiconductor (CMOS) process has reached 130 nm, i.e. the

[1] The growth rate of the number of components per chip is known as Moore's law, after *Gordon E. Moore*, co-founder of Intel corporation. It is commonly quoted as "the number of components per IC doubles every 18–24 months". *Moore*'s original statement (1965) was "the complexity for minimum component costs has increased at a rate of roughly a factor of two per year", which he expected to last "at least 10 years" [1]. In 1975, *Moore* updated the statement arguing that the slope changed to doubling every two years.

[2] Intel Corp., March 2001, Process Px60, 300 mm wafer diameter, 130 nm lithography limit, 65 nm gate length [3].

L. Pavesi, D. J. Lockwood (Eds.): Silicon Photonics, Topics Appl. Phys. **94**, 121–176 (2004)

"deep submicrometer" (DSM) regime. Unfortunately, no exponential is forever [2]. There is large agreement that DSM effects related to interconnect will be critical "showstoppers" for semiconductor industry [4, 5, 6, 7]. The interconnect problem is related to signal speed, power consumption and signal integrity. Smart implementation of interconnect is required to stick to the scheduled progress.

There are however even deeper long-term issues. Moore's law would predict that DRAM cell size would be less than that of one atom by 2020. This means that IC evolution will rely less and less on device down scaling [8]. Performance improvements will be expected to come from elsewhere. One can devise for example the following two potential paths. The first is a shift towards miniaturized systems, which is motivated by the combined recent achievements of nanotechnology, nanophotonics, MEMS and scanning microscopy. A paradigm of this trend is the "smart dust", as its developers at University of California at Berkeley have dubbed it [9]. Smart dust "motes" are micro opto-electro-mechanical systems of the size of few cubic mm, which are supposed to find applications in sensing, medicine, computing, meteorology and much else [9, 10, 11]. The other path is towards introduction of disruptive solutions, such as photonic circuits, which could become a reality if photonic transistors are invented [12]. In both these cases, micro photonics will play a major role.

2 Electrical Interconnect Showstoppers

2.1 The Electrical Intra-Chip Interconnect and Its Scaling

Technologically, on-chip interconnect is identified as electrical wiring. The International Technology Roadmap for Semiconductors (ITRS) defines interconnect as an "electrical wiring system which distributes clock and other signals, and provides power/ground to and among the various circuits/systems functions on a chip"[3] [5].

A typical interconnect structure is shown in Fig. 1. Interconnects are realized in ICs as a dedicated process. In fact, during fabrication of MOS transistor, devices are realized as stand-alone entities, using doping steps, dielectric gate growth and lithographic techniques. This is referred to as *front-end* process.

The process devoted to metallization and interconnect is called *back-end* process. It involves deposition of metal, inter level dielectric (ILD) and inter metal dielectric (IMD) deposition and etching steps [13]. The characteristic of the back-end process is a sequential stacking of additional layers on top

[3] The inclusion of power wires into the interconnect definition is fully justifiable, not only because they are fabricated simultaneously and with the same technology of data wires, but also because power wires have heavy influence in many aspects of data transmission, including speed, cross-talk and power restriction.

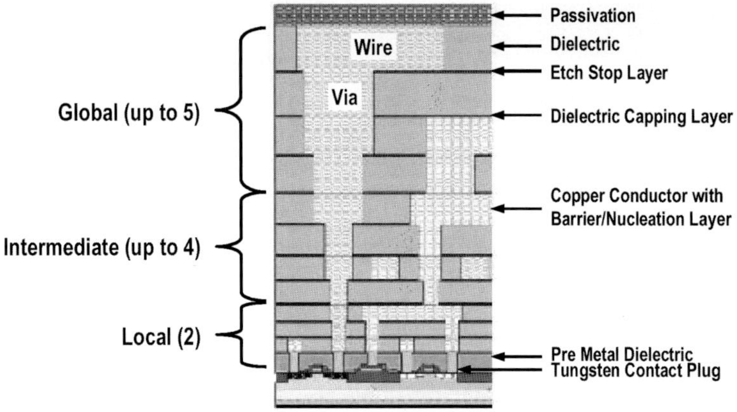

Fig. 1. Schematic view of the back-end structure for electrical interconnect. After [5]

of each other. Before the introduction of copper, the conventional fabrication procedure was by aluminum deposition and dry-etch definition. Due to the poor adhesion of copper to SiO_2, a different fabrication procedure for copper wires has been introduced. In the currently most employed technology – called *damascene* [14] – the interconnect pattern is defined by lithography and etch of dielectric layer. The metal is used to fill the resulting trenches, and finally the excess metal is removed by chemical-mechanical planarization (CMP).

By definition, *global* wires are wires connecting different functional units, and distributing clock signals and power among them. They can be much longer than gate size, and span over the whole chip size, reaching lengths of the order of cm. Thus, the length of global wires scales down with *chip* rather than *gate* size. There are two types of global wires. Semiglobal wires connect functional units within isochronous regions, i.e., regions whose gates can communicate with each other within a cycle of the local clock. Semiglobal wires are not "fat" (their cross section scales with the process). Fully global wires (hereafter, referred to as just "global wires" for simplicity), on the contrary, connect units which leave isochronous regions, and run at slower clock frequency. Global wires must be "fat" [6], as discussed in more detail below (Fig. 8 at page 140). *Local* wires, on the other hand, connect gates, sources and drains of close MOSFETs of the same functional unit, and their length scales with gate size [13].

The metallizations are designed in hierarchical fashion. Hierarchical entities (called *tiers*, Fig. 2) are organized with different wire size, to alleviate interconnect delays.

A tier is defined as a collection of levels that have the same cross-sectional dimensions [15]. Connections between levels are realized by vertical channels called *vias* [5]. Adjacent wire levels are laid out at 90 degrees to each other, a grid scheme that's referred to as *Manhattan* chip design. Local interconnect

Fig. 2. A *tier* is the collection of metallic levels that have the same cross section. The scaling of CMOS chips led to reverse scaling strategy (cross sections increase as opposed to gates' decreasing dimensions) for global wires. Reverse scaling is required to reduce global wire resistance

of polycrystalline Si can also serve as gate electrodes. The length is normally 10 μm to 500 μm. Local wires occupy the lower levels of interconnect (Fig. 1) and often can afford somewhat significant resistivity if they are very short, but must withstand higher process temperature than global interconnect. Materials used for local interconnect include polysilicon, silicided gates, TiN and W. It is paramount that the resistivity of global wires is as low as possible, thus only metals (Al and Cu) are employed for global interconnect.

2.2 The Interconnect Speed Bottleneck

Contrary to transistors, downscaling of interconnect does not enhance speed performance. Scaling effects for transistors, local and global wires are summarized, respectively, in Tables 1, 2 and 3. In 1980, a MOSFET's delay[4] was about 20 ps, whereas the latency[4] associated to a typical aluminum electrical interconnect 1.0 mm long, surrounded by SiO_2, was 6.6 ps. At that time, the interconnect delay was propagation limited, i.e. the delay bottleneck was the

[4] The *delay* or *latency* of a 2-port device is the time that elapses between the availability of the information at the first port (sender's port, in a network) and its *complete* availability at the output port (receiver's end, in a network). The term *latency* is commonly used for networks, and for sequences of connected processing units and devices. The term *delay* is preferred for single units (e.g., a transistor's delay) or specific effects (e.g., the *RC* delay). In the above definitions, *complete* means that all the devices involved in the transmission process are to be included. For instance, optical interconnects have a total latency which is the sum of the delay of the driver circuit, the additional delay of the light source or modulator, the *time-of-flight* and the delay of the receiving diode and its electrical amplifier. The *time-of-flight* is the *delay* effect associated with the transmission at light speed. An electromagnetic signal travelling at light speed is referred to as *propagation* limited.

time-of-flight[4] of electromagnetic waves associated to SiO_2, rather than the RC time constant, which was around 1 ps [16]. According to predictions for the 35-nm technology generation (expected to appear in 2014), the estimated interconnect response time of a 1.0 mm copper line with a low-k dielectric[5] ($k \simeq 2$) is $\tau = 250\,\text{ps}$ [5, 16, 18]. Such value is essentially the RC time constant, which has grown by more than 2 orders of magnitude since the 80s. The global RC delay grows because of scaling, according to Table 3. In comparison, the switching delay of a minimum-geometry 35-nm generation MOSFET is $\tau = 2.5\,\text{ps}$, 100 times faster than interconnect. This simple numerical example emphasizes how the speed performance bottleneck has shifted over the years from gates to interconnect. For 10-nm technology, a 1000 ratio between interconnect latency and gate delay is expected [18].

Table 1. MOSFET scaling. K is the scaling factor ($K > 1$ implies that devices are shrinking). Data of first three columns are partially taken from [19, 20]. α scaling rules have been suggested by Sakurai and Newton, on the grounds that transistor drain current is $I_\text{d} \propto (V_\text{gs} - V_\text{th})^{1.3}/X_\text{gox}$ [21]. k_OX is the relative dielectric constant of the dielectric. ϵ_0 is the dielectric constant of vacuum. See Fig. 3 for the meaning of labels. C_out includes both gate's (C_G) and wires' (C_OX, C_I, see Table 2) capacitance. Wires to be considered for the capacitance are usually local. In any case the wires' capacitance is assumed here to scale as $1/K$

MOSFET parameter	Full Scaling	General Scaling	Fixed Voltage Scaling	α Scaling
Dimension: $X_\text{gox}, L_c, W_c, X_j$	$1/K$	$1/K$	$1/K$	$1/K$
Supply voltage: V	$1/K$	$1/U$	1	$1/K$
Supply current: I	$1/K$	$1/U$	1	$1/K^{0.3}$
Substrate doping: N_a	K	K^2/U	K^2	K
Gate Capacitance: $C_G = k_\text{OX}\epsilon_0 W_c L_c / X_\text{gox}$	$1/K$	$1/K$	$1/K$	$1/K$
Gate delay: $\tau_\text{d} = C_\text{tot}V/I$	$1/K$	$1/K$	$1/K$	$1/K^{1.7}$
Dynamic power dissipation				
at unscaled clock: $C_\text{tot}V^2\alpha_\text{sw}f_\text{clk}$	$1/K^3$	$1/U^2K$	$1/K$	
at fastest switching: $C_\text{tot}V^2/\tau_\text{d}$	$1/K^2$	$1/U^2$	1	$1/K^{1.3}$
Dynamic power dissipation density				
at unscaled clock: $C_\text{tot}V^2\alpha_\text{sw}f_\text{clk}$	$1/K$	K/U^2	K	
at fastest switching: $C_\text{tot}V^2/\tau_\text{d}$	1	K^2/U^2	K^2	$K^{0.7}$

[5] The reader is directed to [17] for an excellent review on low-k materials.

Table 2. Local interconnect scaling. K is the scaling factor ($K > 1$ implies that devices are shrinking). ϵ_0 is k_{OX} is the relative dielectric constant of the dielectric. ρ is the resistivity of the conductor. See Fig. 3 for the meaning of labels

Interconnection parameter	Scaling factor $K > 1$
Interconnect dimensions: L_L, H, W, L_S, X_{OX}	$1/K$
Line resistance: $R_L = \rho L_L/WH$	K
Line capacitance: $C_{\mathrm{OX}} = k_{\mathrm{OX}}\epsilon_0 L_L W/X_{\mathrm{ox}}$	$1/K$
Interelectrode capacitance: $C_I = k_{\mathrm{OX}}\epsilon_0 LH/L_S$	$1/K$
Line response time: $R_L C$	1
Line voltage drop: IR_L	1
Line current density	K

Table 3. Global interconnect scaling. K is the scaling factor ($K > 1$ implies that devices are shrinking). K_{c} is the *chip* scaling factor. ϵ_0 is the dielectric constant of vacuum. k_{OX} is the relative dielectric constant of the dielectric. ρ is the resistivity of the conductor. See Fig. 3 for the meaning of labels

Interconnection parameter	Scaling factor $K > 1$
Interconnect dimensions: H, W, L_S, X_{OX}	$1/K$
Interconnect length L_{max}	$1/K_{\mathrm{c}}$
Line resistance: $R_{\mathrm{G}} = \rho L_{\mathrm{max}}/WH$	K^2/K_{c}
Line capacitance: $C_{\mathrm{OX}} = k_{\mathrm{OX}}\epsilon_0 L_{\mathrm{max}} W/X_{\mathrm{OX}}$	$1/K_{\mathrm{c}}$
Interelectrode capacitance: $C_I = k_{\mathrm{OX}}\epsilon_0 L_{\mathrm{max}} H/L_S$	$1/K_{\mathrm{c}}$
Line response time: $R_{\mathrm{G}} C$	K^2/K_{c}^2
Line voltage drop: IR_{G}	K/K_{c}
Line current density	K

2.3 The Power Consumption, Signal Integrity and Trade-offs

The second issue with electrical interconnect is power consumption. The consumed power density in ICs is rising exponentially along scaling [5]. This can be quantitatively understood as follows. A scaling factor $K > 1$ means that linear dimensions of gates are multiplied by $1/K$. The capacitance at gate output C scales as $1/K$ ($C \propto 1/K$). This applies to all capacitive elements because capacitor's area scales as $1/K^2$ and capacitor's thickness scales as $1/K$.

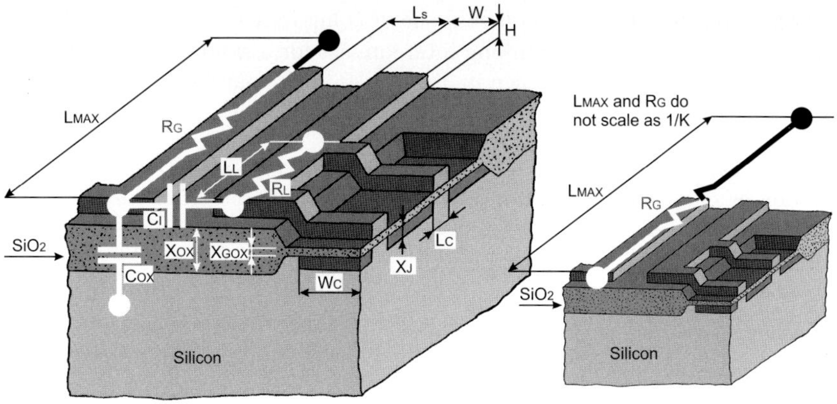

Fig. 3. Global lines (*left* metal) do not scale their length $L_{\max}$, whereas local lines (right metal, length L_L) and MOSFETs do. The purpose of the figure is not to represent functioning devices or connections. Scaling rules are shown in Tables 1, 2 and 3

Local wires – connecting one gate's output to a following gate – are the dominant contribution to this capacitive load C. If DC supply voltage V_{dd} does not change, the dynamically moving charge per gate is $Q = CV \propto 1/K$, and energy per cycle (a cycle is one V_{dd}-to-0 and one 0-to-V_{dd} transition) per gate is

$$E = CV_{dd}^2 \propto 1/K\,. \tag{1}$$

The energy in (1) gets all dissipated. The dissipated power *per device*[6] P is E times the gate switching frequency f, which is proportional to clock frequency f_{clk}. The proportionality factor α_{sw} is called gate switching activity ($f = \alpha_{\mathrm{sw}} \times f_{\mathrm{clk}}$). For example, reasonable assumptions can be that $\alpha_{\mathrm{sw}} \simeq 1/4$ for non-return to zero[7] transmission [22]. Therefore, given the number of gates per unit area ($N \propto K^2$), the power density (power *per unit area*) $\mathcal{P}$ is given by

$$\mathcal{P} = NCV_{dd}^2\alpha_{\mathrm{sw}}f_{\mathrm{clk}} \propto KV_{dd}f_{\mathrm{clk}}\,. \tag{2}$$

Equation (2) tells that dynamic power density is inversely proportional to gate length, that is advantageous for power dissipation to lower DC supply voltage, and that increasing the clock frequency f_{clk} exacerbates the

[6] Equation (1) is very simplified. The total dissipated energy by the MOSFET should also include the short-circuit dissipation [22,23]. For on-chip interconnect, short-circuit component can be 25% of total. For off-chip interconnect, the short-circuit component of the energy is about 20% of the other components for short wires, and reduces to about 10% for longer wires [22]. Thus, (1) only refers to the largest power contribution.

[7] Non-return to zero coding is discussed at page 131.

power density problem. *Meindl* et al. have compared the energy dissipated by a MOSFET and by a 1 mm long interconnect for a single switch, for three technology generations, as shown in Table 4 [16]. The numbers of Table 4 emphasize that, in earlier technologies, the power dissipated by the benchmark interconnect was comparable to MOSFET's dissipation, whereas the projection for the 35 nm generation reveals a ratio of about 30 between interconnect and MOSFET dissipation [16].

Table 4. Projected comparison between the dissipated energy in a single switching operation by one MOSFET and by a 1 mm long interconnect. After [16]

Technology Generation	1.0 μm	100 nm	35 nm
MOSFET switching energy (fJ)	$\simeq 300$	$\simeq 2$	$\simeq 0.1$
Interconnect switching energy (fJ)	$\simeq 400$	$\simeq 10$	$\simeq 3$
Clock Frequency f_{clk}	$\simeq 30$ MHz	$\simeq 1$–3.5 GHz	$\simeq 3.6$–13.5 GHz
Supply Voltage V_{dd} (V)	5	1	0.5

Scaling rules which maintain invariant electric field[8] and power density have been initially suggested by Dennard in 1974 [19]. According to such rules, voltages and electrical currents have to be scaled as the linear dimension (see Table 1). In fact, DC supply voltages have indeed been lowered over the years, but ideal scaling is prevented by other major effects, such as a decrease of MOSFET gain at lower V_{dd} [20], a decrease of signal-to-noise ratio and an increase of gate leakage power [6].

Scaling is also worsening the IC's heating by Joule effect in interconnect. Although power dissipation by Joule effect is lower than the dynamic contribution introduced in (2), interconnects lay inside a thermally insulating dielectric matrix. Logic gates, on the contrary, are in contact with the Si substrate, which has much lower thermal resistivity. Thus, temperature in gates does not rise as much as in metal wires. The further apart is the interconnect wire from the Si substrate, the higher is its temperature. High temperature is undesirable in wires because of electromigration, which eventually leads to wire failure according to Black's equation [25]

$$\mathrm{TTF} = Aj^{-n} \exp\left(\frac{E_q}{k_{\mathrm{B}} T_{\mathrm{m}}}\right) , \tag{3}$$

where TTF stands for *time-to-fail*, A is a constant, j is the average DC current density, n is typically 2, E_q is the activation energy for grain boundary diffusion, k_{B} is Boltzmann constant and T_{m} is the metal temperature [26]. High temperature also leads to open circuits failures [26]. All the temperature

[8] Low fields are needed for reliability. For instance, gate oxides tend to break down if exposed to electric fields in excess of 5 MV/cm to 6 MV/cm [24, 6].

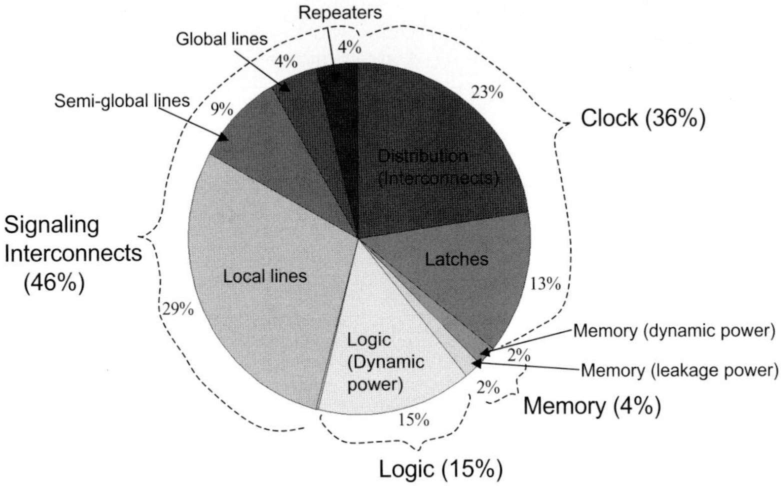

Fig. 4. Breakout of power dissipation in 180 nm IC technology. Clock distribution is heavily power consuming because it is constantly switching at the highest rate, whereas signal wires have a switching activity of about 0.1–0.15 compared to clock. After [27]

problems are further exacerbated by the introduction of low-k dielectrics[5] due to their lower thermal conductivity compared to SiO_2 [26].

Power dissipated by wires is the dominant power dissipation mechanism in modern IC's technology. The power dissipated by digital data lines and clock distribution constitutes 82% of the total dissipation in 180 nm technology (see Fig. 4) [27].

Finally, there is signal integrity. Some integrity issues originated by scaling, connected to physical wire failures, have already been mentioned (V_{dd} lowering, electromigration and open circuit failure). There are also nondestructive integrity scaling problems. The most important is only mentioned here, and is wire-to-wire cross-talk (scaling leads to closer wires) [6, 27].

All the mentioned problems are often complicated by trade-off situations, as already seen for V_{dd} scaling (V_{dd} lowering improves power performance but worsens signal integrity) and for low-k dielectrics (low-k leads to lower capacitive loads but larger power dissipation)[5].

Despite all such critical problems, there is a common positive attitude in looking at this situation as at an opportunity of for ingenious research [16, 28, 29].

2.4 Interconnect as a Global Design Issue

The design of new solutions for interconnects requires global approaches which go well beyond the physical implementation of the circuit level. Research is active in all aspects of interconnect design. Several papers have

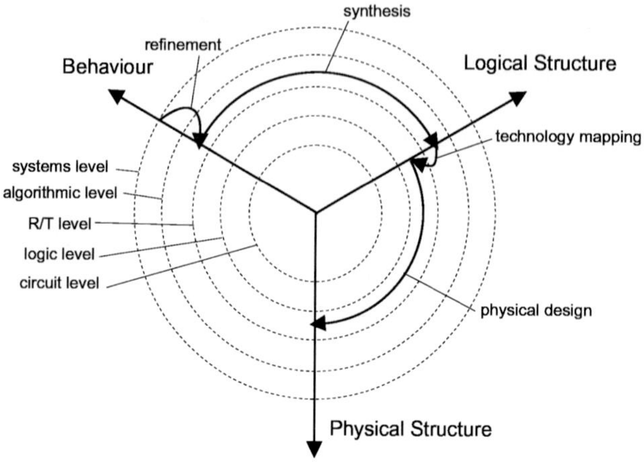

Fig. 5. Gajski–Kuhn diagram [31]

addressed specific issues in great detail under diverse points of view, so that a general and comprehensive overview is a hard task [30].

Interplay of physics and engineering in any system design and, in particular, in VSLI, has been know for long time [31]. The target of a design is the realization of a system with a desired behavior. One must look for a logical structure towards such goal. Typically, a logical structure can be visualized as a block diagram. Finally, the design is translated to a physical structure. These three phases are not independent, as emphasized by the Y-diagram of Fig. 5.

Approaches toward a solution of interconnect limitations can be worked out at several design stages. At the logical synthesis stage, for example, potentiality might emerge in a different partitioning of the logic circuits between ICs, printed boards, busses and wiring, perhaps including disruptive layout strategies – such as for example 3D interconnect or even exterconnect[9]. Diverse interconnect designing approaches have been proposed, which address the interconnect problem including higher-level strategies, such as simultaneous interconnect and packaging design [33] and on-chip interconnection routing networks substituting global wiring [34].

In this work, some aspects of optical interconnect are compared with electrical counterparts. It is reasonable to assume that, in the future, interconnect solutions will be always based on electromagnetic signals[10]. Radio transmission, which is an interesting as well as promising option, is outside the topic

[9] The word *exterconnect* designates an electrical connection which connects two points of an IC using a nonintegrated wire technology, such as a board-level metal path [32].

[10] We exclude from our consideration other types of signals, primarily because of speed limitations. Only electromagnetic signals can approach the fundamental limit of the speed of light in vacuum $c_0 \simeq 3 \times 10^8$ m/s.

of this chapter, although it has already been proposed as an interconnect technology. For example, clock radio distribution using integrated antennas has been demonstrated [35]. As good starting points, the reader is directed to the introductory discussion on emerging optical and radio networking by *El-Sayed* et al. [36], to the overview on wireless networking by *Jordan* et al. [37], and to the overview of Bluetooth Wireless Technology™ by *Davies* [38].

3 Properties of Electromagnetic Interconnect

3.1 Definitions, Figures of Merit and Characterizations

Signals considered in this work are coded and processed as binary digital signals. Transmitter and receiver must agree on encoding scheme and transmission timing. For example, transistors can use different voltage levels to represent 0s and 1s. Binary encoding can be performed by representing a logical 0 as a low voltage signal, and a 1 as a high voltage signal. This encoding is known as On-Off Keying (OOK). In a Non-Return to Zero (NRZ) encoding, the signal is maintained at 1 or 0 levels for the whole duration of a clock cycle. There is no level switching between subsequent bits if they are equal[11]. Digital signals encoded as OOK-NRZ are used for eye diagrams, as discussed below.

In digital transmissions, signals should ideally switch between levels with steep, step-like fronts. In practice, squared pulses cannot be obtained, because signals are bandwidth limited [39]. Finite bandwidth implies finite rise and fall times of clock and digital signals. Assuming – in place of ideal step transitions – single-exponential transitions with time constant τ, the bandwidth f_{max}, defined as the 3 dB frequency of the Fourier spectrum, is $f_{\mathrm{max}} = 1/2\pi\tau$, and the rise time t_{r}, defined as the 10%–90% transition interval, is $t_{\mathrm{r}} = \tau \log(9) \simeq 2.2\tau$, leading to the known approximation

$$f_{\mathrm{max}} = 0.35/t_{\mathrm{r}} \,. \tag{4}$$

Equation (4) refers to *analog* bandwidth. For digital signals, one can also define a *digital* bandwidth f_{dig} as the number of bits per second. The two definitions are essentially equivalent. For example, f_{dig} of a digital channel would be the maximum number of bits per second which can be transmitted over that channel. Assuming $2 \times t_{\mathrm{r}}$ as the minimum pulses width to allow error-free transmission, it is

$$f_{\mathrm{dig}} = 0.5/t_{\mathrm{r}} \,, \tag{5}$$

[11] This might be undesirable, because the clocking signal does not get automatically encoded in a NRZ transmission. If it is desired to keep track of clock, a parallel, dedicated clock line is necessary, or a different coding scheme, such as the Manchester code, in which there is at least a signal transition at any single clock period.

which is the same as (4) except for a constant of the order of 1.

The time delay t_{d}, defined as the time to 50% transition, is also used as figure of merit. Both the signal's finite speed and the line's bandwidth limitation contribute to t_{d}. A line where only the former contribution is significant is a pure delay element. It must have infinite bandwidth[12]. The delay in this case is simply the distance from the source divided by the signal speed. For electromagnetic signals, this delay is the time-of-flight $t_{\mathrm{d(tof)}}$. On the other hand, bandwidth limitations imply a nonzero delay time $t_{\mathrm{d(bw)}}$ even for infinite speed signals. For instance, in exponential transitions,

$$t_{\mathrm{d(bw)}} = \tau \log(2) \simeq 0.7\tau \,. \tag{6}$$

The 50% time delay t_{d} of a step signal carried by an electromagnetic wave across a bandwidth-limited line is the result of both effects. Its evaluation is a mathematical complex problem. A detailed treatment can be found in [15]. The delay of the *wavefront* is always the time-of-flight. However, in a bandwidth-limited (infinite) line, the *level* of the signal at the wavefront decreases exponentially with the distance from the source. If the line is finite, there is the additional complication of the signal bouncing back and forth between the source and the receiver. Thus, the effects arising from the bandwidth limit add up non trivially with the pure time-of-flight delay $t_{\mathrm{d(tof)}}$. Although (4) to (6) are derived for the case of exponential functions, they are widely used as quite general tools for integrated circuits [4, 15, 40, 41, 42].

Eye diagrams (Fig. 6) give a quick way to evaluate the performance of processing units with respect to signal transitions.

They are used for both checking the system operation and evaluation system performance in research and development work. An eye diagram consists of many synchronized, overlaid traces of small sections of a signal, using OOK with NRZ. A few symbols – i.e., definite series of bits – are used. Symbols are randomly and independently arranged, to ensure that any possible significant combination is represented in the diagram. Eye diagram can be measured by oscilloscope or by computer simulations. The wider the vertical opening, the greater the noisy immunity. Intersymbol interference (ISI) can easily be seen in the eye diagram, as it reduces the vertical opening. The ideal sampling instant is at the point of maximum vertical eye opening. The smaller the horizontal opening, the greater the sensitivity to errors in timing phase (jitter)[13].

---

[12] Pure delay lines have 1 amplitude and all frequencies and linear phase characteristic [39].

[13] The *skew* is defined as the difference in arrival time of bits transmitted at the same time over two different transmission lines. Skew is a static phase error, and is defined for *different* lines. The *jitter* is a dynamic phase fluctuation on a *single* line, which can be originated by time-variable conditions, for instance cross-talk phenomena.

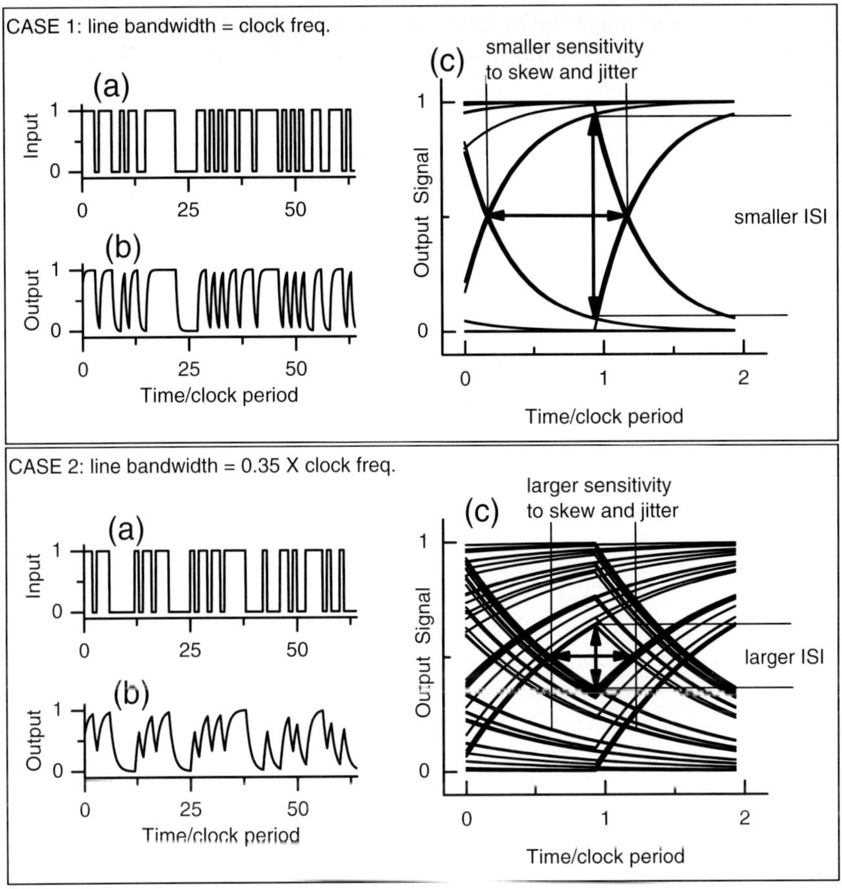

Fig. 6. Simulations of transmitted OOK/NRZ signals (a) over finite-bandwidth lines. Case 1 *top*: transmission line bandwidth is equal to clock frequency. Case 2 *bottom*: bandwidth is 0.35 times the clock frequency. Time scale is normalized to clock period. Eye diagrams (c) are directly extracted from shown output plots (b). Intersymbol interference (ISI) narrows the eye height, whereas jitter (not introduced in this simulation) narrows eye width. Thus, large width means robustness to jitter

3.2 Physical Appeal of Optical Interconnect

Both optical and electrical signals are electromagnetic waves. The difference is in their oscillation frequency f – approximatively $10^{14} \leq f_{\mathrm{opt}} \leq 10^{15}$ and $f_{\mathrm{el}} \leq 10^{11}$ Hz, respectively, for optical and electrical waves. The bandwidth of typical signals Δf is limited by the performance of electronic interconnects and gates. An aggressive estimate, using (4) and very short transition (rise) times t_{r}, say few ps, leads to $\Delta f \leq 10^{11}$ Hz. Thus, a typical optical frequency is at least 3 orders of magnitude larger than the signal bandwidth. This means that a modulated optical wave has an enormous capability of handling the signal bandwidth. Moreover, it is reasonable to expect that the properties of the transmission medium will be fairly homogeneous over the signal bandwidth, leading to small distortion and homogeneous attenuation. The large ratio $f_{\text{opt}}/\Delta f$ allows to accommodate even several channels in reasonably small frequency windows, and still observe little frequency dependence. This technique, known as wavelength division multiplexing (WDM), is routinely used in fiber optics for long-haul connections. Several modulated electromagnetic carriers are sent over the same medium. The frequency of each carrier is displaced by about Δf from adjacent carriers. In traditional electrical connections, this is obviously not possible, since the frequency f_{el} of electrical signals is by our definition included within the band Δf.

The large ratio $f_{\text{opt}}/\Delta f$ also implies that devices based on resonance effects, designed for the frequency f_{opt}, can also work nicely over the whole bandwidth Δf. This option is best viewed in wavelength. It is also $\lambda_{\text{opt}}/\Delta \lambda \geq 10^3$, where $\lambda_{\text{opt}} = c/f_{\text{opt}}$ and $\Delta \lambda \simeq (c/f_{\text{opt}}^2)\Delta f$. Considering a WDM transmission, a carrier with $\lambda_{\text{opt}} \simeq 1.5\,\mu\text{m}$ might be separated from adjacent channels – for example – with Fabry–Perot resonators. Cheap Si based Fabry–Perot resonators have been shown, in whose transmission spectrum the full width half maximum (FWHM) of resonance was of the order of 1 nm [43]. On the other hand, quarter-wavelength dielectric stacks also can work as anti-reflection coating with flat spectra over several nm (several THz, expressed in frequency). The possibility of non-dissipative impedance matching over such large band contrasts with the scarce flexibility of the electrical alternatives, where only dissipative resistive matching is achieved [44].

The large bandwidth capability makes the optical devices robust with respect to technological development. In fact, improvements (i.e., increases) in signal bandwidth will have a little impact on laid out optical transmission systems. The reason is that the optical systems and devices are handling waves oscillating at $f_{\text{opt}} \sim 10^{14}\,\text{Hz}$. It does not matter is the signal band changes some $10^{11}\,\text{Hz}$ around f_{opt}. No upgrade costs are a valuable advantage, especially for long-haul networks.

The large optical frequency gives the capability of generating short pulses. This is a consequence of the well known relationship between the width Δt of a time pulse and the width $\Delta f = K/\Delta t$ of its Fourier transform, where K is a constant of the order of unit, which depends on how the widths are exactly defined. A ps time pulse is obtained by coherently grouping optical waves spanning a frequency interval $10^{12}\,\text{Hz}$. This corresponds, in the optical range, to a wave packet just a few nm wide. In order to achieve such short pulse electrically, all the frequency range from CW regime up to 1 THz would be required. The handling of such pulse would be practically impossible: any transmitting medium will have spectrally non-uniform attenuation and distortion over such span. The ability of short pulses is useful in eliminating clock skew, as discussed in [44].

The photonic granularity in energy ($E = hf_{\mathrm{opt}} \simeq 1\,\mathrm{eV}$, where h is the Planck constant) is coarse with respect to thermal granularity ($k_{\mathrm{B}}T \simeq 25\,\mathrm{meV}$, where k_{B} is the Boltzmann constant and T is the absolute temperature). Coarse granularity implies robustness to thermal noise [45]. It has also undesirable implications as a source of quantum noise, as discussed in Sect. 4.1, but the quantum effects are only observable at very low power.

The large optical frequency f implies small wavelength $\lambda = c_0/f$, where c_0 is the velocity of light. If light is emitted by a spacial coherent source of finite size, the beam has a finite divergence

$$\theta = \beta\lambda/D\,, \tag{7}$$

where $\beta \simeq 1$ is a coefficient which depends on the aperture shape, and D is the aperture diameter. Light sources with aperture as small as few μm can emit with small divergence, allowing unguided beam propagation through homogeneous dielectrics (e.g. free-space propagation) and providing the capability to use imaging optics [44]. Free space interconnects are appealing because they do not require material connectors and are easily reconfigurable.

At large wavelengths, such as in the microwave range and below, interconnect guiding can only be obtained using large index contrasts, which in practice forces the use of conductors – which have "infinite" dielectric constants [44]. Metals provide free carriers which can force electromagnetic fields to be transversal, quasi-plane waves: metals "steer" the Poynting vector in the surrounding dielectrics towards the receiver stage. The wave travels in the dielectric. Unfortunately, microstrip and coplanar waveguides circuits are open waveguides, meaning that the dielectric is in contact of basically all the IC's metallic wires. This is the physical origin of cross-talk problems in electrical connections [46,45]. In contrast, small optical wavelength permit wave guiding using total internal reflection in dielectric materials. The wave is effectively confined and smart cladding design will allow dense packing with no major cross-talk problem. Avoidance of metals also permits galvanic isolation of circuits [45].

Finally, there is the speed of the transmission. In principle, optical connections are not faster than electrical, if one can ensure that the electrical connections are limited by the electromagnetic propagation across the dielectric. Unfortunately, metals are lossy. Finite resistivity often gives rise to dominating wire resistance and to RC time-limited lines. Although resistance can be lowered by increasing the wire section, such solution is often unsatisfactory for several reasons. It is undesirable because it conflicts with downscaling trends. Moreover, efforts towards larger sections are somewhat wasted by skin effect, i.e., the confinement of the current to outer conductor surfaces at high frequency. The skin effect diminishes the effective conductor cross section and increases its resistance at high frequency.

4 Interconnect Limits

4.1 Fundamental Limits

The limits of interconnect are classified following the original scheme of *Meindl* [47, 48] as fundamental, material, device, circuit and system limits. The electrical interconnect are only partially discussed here; the reader is redirected to [47, 48] for complete analysis.

A designer of a communication system should always understand what is the physical fundamental limit of the communication system that is being designed and try to approach that limit as closely as possible in the design. If the design is far away from fundamental limits, there is a chance of significant pay-offs on looking how to improve system performance [49]. We consider in this section the physical laws that set unsurpassable bounds, no matter which material or architecture is used for the interconnect.

The fundamental speed limit is determined – both for optical and electrical transmission – by the speed of light in vacuum. The fundamental speed of electrical interconnect is not the speed of electron motion [29, 44]. The finite electromagnetic velocity implies a non-zero travelling time (time-of-flight)[4]. The latency τ of an interconnect is therefore always larger that the time-of-flight in vacuum, i.e. [48]

$$\tau \geq \frac{L}{c_0}, \tag{8}$$

where L is the interconnect length, c_0 is the velocity of light in vacuum.

The second fundamental limit is about the minimum required signal power for reliable transmission in noisy environments. One wants to minimize power consumption, to save energy and to limit the generated heat that must to be removed. Assuming gaussian white noise, the channel capacity CC (in b/s) is given by *Shannon*'s theorem [50, 51]

$$\mathrm{CC} = \Delta f \log_2 \left(1 + \frac{P_\mathrm{S}}{P_\mathrm{N}}\right) . \tag{9}$$

CC depends on the channel bandwidth Δf and on the signal-to-noise ratio $P_\mathrm{S}/P_\mathrm{N}$, where P_S and P_N are the power, respectively, of signal and noise. Strictly, (9) is not correct for electromagnetic signals. In fact, electromagnetic waves carry signals as photons. Photon states must be read out by detectors. Detection is represented mathematically by a probability operator measure (POM), which consists of a Hermitian operator applied to the state. This process comports a degree of randomness in the interpretation, thus effectively it behaves as a quantum noise [52, 53, 54]. The complete theory to describe quantum transmission can be found in the work of *Hausladen* et al. [55], *Holevo* [56] and *Schumacher* et al. [57]. A fundamental result is known as Holevo's theorem, the extension of channel capacity to the case of quantum symbols [54]. CC is again obtained maximizing the signal entropy as done for the derivation of (9). However, maximization is performed not only over the probabilities of sent symbols, but also over the set of detecting operators (POMs). Consequently, a fundamental quantum limit of channel capacity is found, even in absence of classical noise contribution. In practice, quantization can be disregarded in the case of electrical interconnect, because quantum effects are less important than thermal noise. Hence, classical Shannon's theorem as expressed by (9) is directly applicable to electrical interconnect [15,48]. In optical transmission, the effect of quantization is recognizable, due to the larger photon energy $E = h\nu$. For example, *Sasaki* et al. Ihave developed a detecting scheme approaching the fundamental limit using a He-Ne laser as light source ($\lambda = 632.8\,\text{nm}$), and a light power of order 100 fW, corresponding to about 3×10^5 photon/s, and to a beam containing about 10^{-3} photons in one meter [53]. This experimentally proves that quantization becomes relevant at low signal power P_S. Power is low to this respect when the number of quanta/(s-Hz) (given by $P_\text{S}/hf\Delta f$) is close to 1 [54]. Assuming Δf=100 GHz, $f = 3 \times 10^{14}$ Hz, one obtains that quantization starts affecting channel capacity below some tens of nW. When $P_\text{S}/hf\Delta f < 1$, the transmission is in a full particle-like regime and the channel capacity has a quite different expression from (9) [54].

4.2 Material Limits

One of the most interesting feature of signal transmission using electromagnetic waves is that there is actually no need of a medium. No medium means no energetic loss. However, if oscillations are generated in space regions which are small compared to the wavelength, radiation is necessarily emitted with large angular divergence. Equation (7) shows that the beam divergence is inversely proportional to the source size. The same equation also indicates that the problem is more severe in the case of electrical interconnect, since the electrical wavelength (at least of the order of cm) is larger than the optical one. An ideal conductor – i.e., a material in which charges can move under no electrical field – transforms the spherical wave in a plane, transversal-field wave. The Poynting vector steers towards the receiver, as shown in Fig. 7. The picture also emphasizes that the wave propagates inside the dielectric between the ideal metals, rather than in the metals themselves. Dielectric properties control the propagation characteristics, including the speed. Equation (8) must be modified with substitution of ϵ with $k\epsilon_0$, μ_0 with $\mu\mu_0$, c_0 with $c_0/\sqrt{k\mu} \simeq c_0/\sqrt{k}$, where k (μ) is the relative dielectric constant (magnetic permeability).

This kind of ideal conductors should be imagined as containing free charge carriers with zero mass. Charge carriers would act as "slaves" of electromagnetic fields in the dielectric, adjusting themselves, at the metal surface, as fast as required to make the fields transversal. In this sense, metallic "waveguiding" is achieved, by letting the charge carriers inside the metal (the clad-

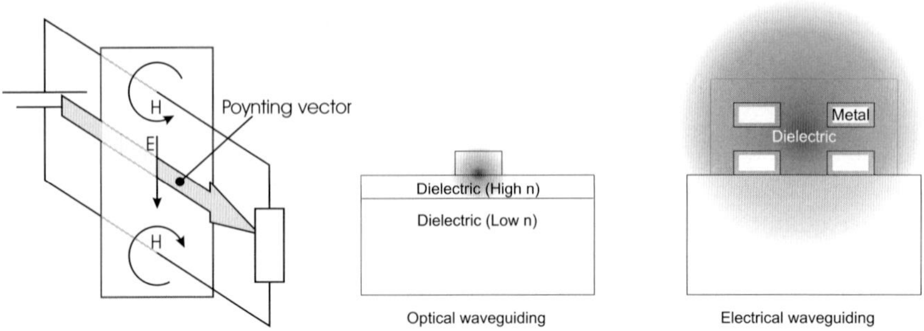

Fig. 7. Direction of Poynting vector in an electrical communication (*top*) and waveguiding (*bottom*). E and H symbols indicate, respectively, the electric and magnetic field

ding) to screen the field, confining it into the dielectric (the core). Figure 7 also suggests the physical origin of electromagnetic cross-talk: in the "open waveguide" layout, fields couple to all metallic conductors. Only the coaxial waveguide is a zero cross-talk waveguide.

Metals, however, are no ideal conductors. The first reason is that charge carriers have a nonzero mass. Free carriers are resonantly excited only at zero frequency, and their ability to follow electromagnetic fields drops with frequency. Beyond plasma frequency

$$\omega_{\mathrm{P}} = \sqrt{\frac{4\pi n q^2}{\epsilon_0 m_e}} \tag{10}$$

electrons loose any capability of following the radiation, which freely enters the metal without being absorbed. The second nonideality is damping, a mechanism which transfer mechanical energy from carriers to the solid lattice. Damping allows penetration of waves at frequencies below plasma frequency, where absorbtion is still significant. This is mechanism is the origin of the "skin effect", hinting at a thin sheet at the metallic inner surface, of thickness δ, in which carriers move for a given frequency.

$$\delta = \sqrt{\frac{1}{2\pi f \sigma \mu}} . \tag{11}$$

In ideal metals, this thickness would be 0 at all frequencies, as one understands by the inverse-root dependence on σ ($\sigma \to \infty$ in ideal metals). In real metals, however, the skin effect is normally associated with the idea of a reduction of the metal effective cross-section. This latter picture implicitly takes into account that at zero frequency the resistive behavior (which is due to damping) totally dominates over the wave-propagating picture.

Non-zero, finite resistivity leads to power dissipation (Joule effect) and to speed limiting effects, the most known being the RC delay. It is therefore advantageous to employ low resistivity materials as conductors (resistivity of aluminum and copper are, respectively, $0.0277\,\Omega\cdot$cm and $0.0172\,\Omega\cdot$cm). The effect of scaling on the resistivity has been discussed in [48]. The metal *effective* resistivity $\rho = \rho_{\mathrm{in}} + \rho_{\mathrm{ex}}$ includes both a material (intrinsic) ρ_{in} term, and an extrinsic ρ_{ex} term, which is due to scattering both from distribute impurities and from surfaces [48]. The effective resistivity due to surface scattering, in particular, is larger in smaller wires [48].

4.3 Devices Limits

With the labelling convention of Fig. 3, the time constant $\tau = RC$ associated to the resistive charging of capacitive elements is quadratically dependent on the wire length L [58, 4, 48]

$$\tau = \frac{\rho k}{(D_\rho \times D_k)} L^2 . \tag{12}$$

where ρ is the resistivity of the conductor, k is the relative permittivity of the insulator, $(D_\rho \times D_k)$ is either $(W \times L_S)$ or $(H \times X_{\mathrm{OX}})$, respectively, for IMD or ILD capacitance (see Fig. 3 at page 127 for the meaning of W, L_S, H and X_{OX}). The bandwidth $f_{\max}$ is inversely proportional to τ, thus $f_{\max} \propto A/L^2 \simeq 10^{16} A/L^2$, where A is an effective wire area [4]. Bandwidth is limited also when inductive effects dominate. Typically, inductance leads to overshoots and oscillations [48]. Although the result is less intuitive than (12), it is also $f_{\max} \propto A/L^2$ for inductive limited bandwidth. Numerically, it is $f_{\max} \simeq 10^{15} A/L^2$, since R is larger due to skin effect (11) [4, 29, 48]. Equalization can improve the figures but not the functional dependence.

Therefore, the urge of large bandwidth in long wires forces a large wire cross-section. *Svensson* [29] has proposed a graph, reported in Fig. 8, which allows evaluation of the required cross section. The graph of Fig. 8 has two regions. The top-left region is limited by line RC, whereas the bottom-right region is skin-effect limited: the transmission speed basically approaches the speed of light in the medium (full light speed is not achievable because of skin effect) [29].

In optical interconnects, on the other hand, bandwidth limitations are much less important. In principle, in free-space optical transmission, the full electromagnetic band up to optical frequency is transmittable since free-space transfer function is only a delay element.

Optical transmission in waveguides is limited by material dispersion and nonlinear effects. Therefore, it is interesting to look at such effects in materials used or compatible with Si technology. *Tsang* et al. [59] have considered the impact such effects in a 17 mm long Si rib waveguide, embedded in SiO_2 cladding. Optical pulses $\simeq$ 300-fs to 400-fs long were subject to time broadening and compression around $\simeq 50$ fs. The authors suggest (conservatively) that pulse rates of 40 Gb/s can be thus supported by the waveguide, although

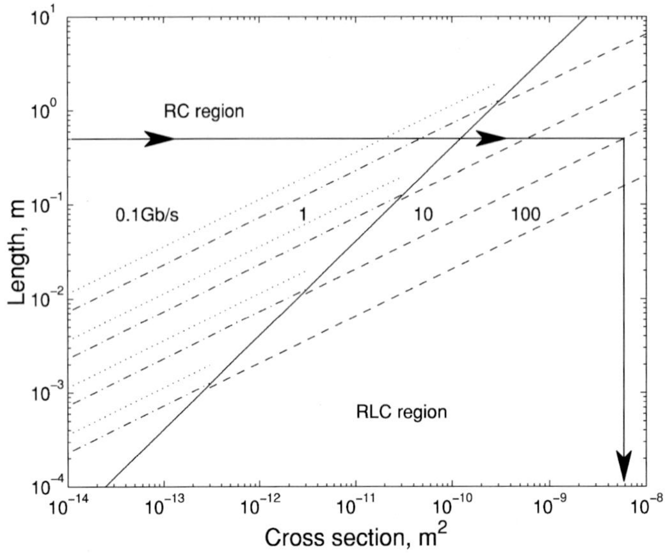

Fig. 8. Wire length versus cross-section for different data rates (copper conductor with square cross-section assumed). *Dotted*: RC region with 0% eye opening; *dashed dotted*: RC region with 64% eye opening; *dashed*: RLC region with 0% eye opening. The *solid line with no arrows* demonstrates the border between RC and RLC regions. The *line with arrows* illustrate a sample use of the graph: if one wants to interconnect electrically two points, say, 0.5 m apart, allowing 10 Gb/s transmission, a cross-section of at least about $77 \times 77\,\mu\text{m}^2$ is required. Being well inside the RLC region, the delay of such wire would be comparable to the time-of-flight. Adapted from [29]

much higher rates (THz range) are in principle compatible with their measurement[14]. Other materials which can be used in waveguiding structures are not expected to behave with significant difference as far as the bandwidth is concerned. The optical delay time can be therefore safely considered equal to time-of-flight in the medium.

Dispersion characteristics of photonic band-gap (PGB) waveguides have much more interesting features. PGB materials – also called photonic crystals (PC) – are materials with periodic modulation of the dielectric constant. Properly designed PCs exhibit photonic band-gaps: light whose photon energy lays within such gaps cannot propagate inside the material. If line defects are introduced in PC with photonic gaps, propagation might be allowed within the defect line, as in the case of Fig. 9. PC are discussed in detail in Chapter by *Geppert* et al. in this book. The reader is also directed to [61] for a recent review on PC waveguides. In PC waveguides, dispersion can be en-

[14] Measured dispersion and losses in the waveguide were, respectively, 9.1 fs/(nm to cm) and 0.1 dB/cm.

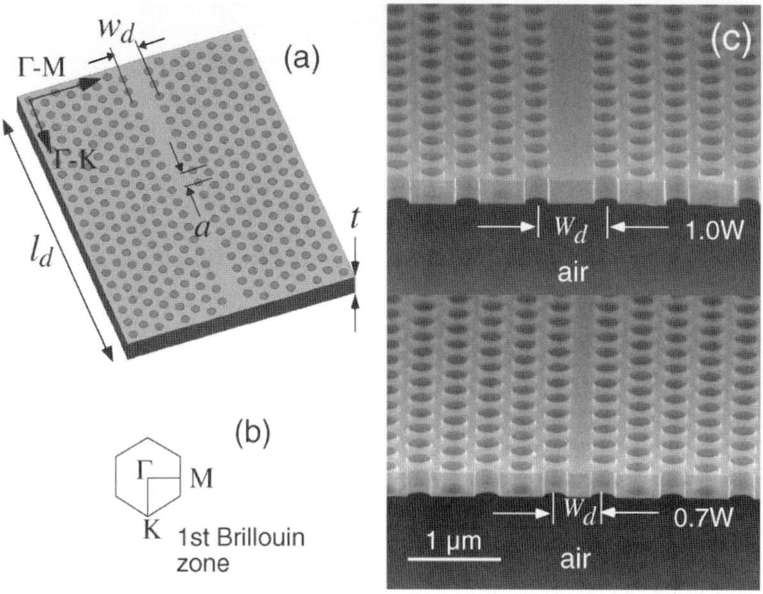

Fig. 9. Single missing-hole line defects of Si hexagonal airhole PC slabs. The structure forms a PC waveguide in the plane of periodicity. Confinement in the normal direction is ensured by the air cladding. The slab thickness is 0.2 μm. The lattice constant a is 0.39 μm ($W = a\sqrt{3}$). (**a**) Schematic of samples, (**b**) reciprocal-space representation, (**c**) scanning electron micrographs of fabricated samples. After [60]

gineered precisely. One can obtain highly dispersive waveguides, where light effectively slows down significantly. In [60], a drop of almost two orders of magnitudes has been demonstrated, as shown in Fig. 10 (right side of plot, graph a), where $n_g \simeq 100$. Slowing down the light can be interesting, in practice, when enhancements of light-matter interactions are desired (for instance, in waveguide amplifiers). However, pulse travelling with no dispersion – i.e., at speed $\simeq c_0/n$, where n is the effective refractive index of the volumetric mixture of air and Si – can also be achieved (non dispersive section, Fig. 10). It should be remembered that the y-axis is normalized: the scale in actual energy can be adjusted with the absolute value of parameter a. Therefore, any wavelength range can freely be inserted in a dispersive or in a nondispersive region. A still open problem with 2D PC waveguides is loss. Albeit the air cladding, waveguides such as the one of Fig. 9 are subject to out-of-plane scattering loss – which is their dominant loss mechanism [61] – as well as to roughness scattering. The 2D crystal is in fact not perfect (it should be infinite in the direction normal to the 2D plane). It is expected that out-of-plane losses might be reduced to values around 10 dB/cm [61].

In conclusion, in electrical interconnect, finite cross-section wires always lead to significant bandwidth limitation. For distances above 1 cm, bit rates

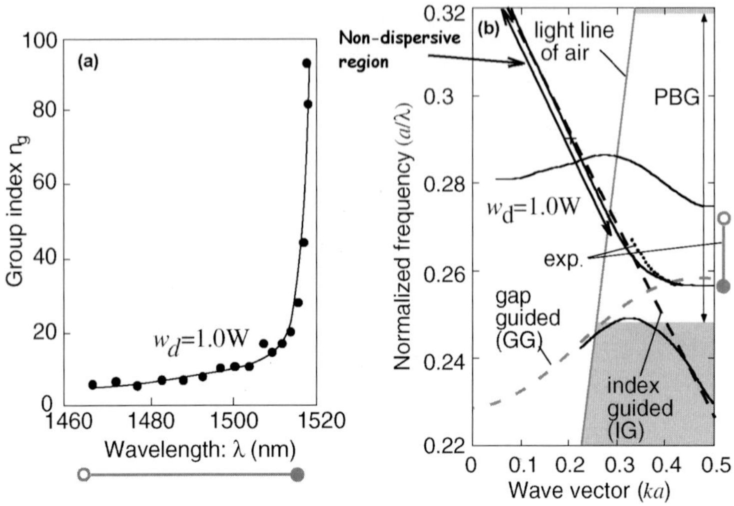

Fig. 10. (**a**) Measured group index $n_{\mathrm{g}} = n - \lambda(\mathrm{d}n/\mathrm{d}\lambda)$ versus wavelength λ for $w_{\mathrm{d}} = 1.0\,\mathrm{W}$ (see Fig. 9). (**b**) Theoretical dispersion curves for $w_{\mathrm{d}} = 1.0\,\mathrm{W}$. *Solid lines* are 3D FDTD calculations. *Broken lines* illustrate hypothetical contributions to the overall dispersion, as given from a refractive-index guiding (IG) mode (*linear curve*) and from a PBG guiding (GG) mode (*S-shaped curve*). *Solid red dots* are experimentally determined dispersion from the result in (**a**). Adapted from [60]

above 10 GHz impose wire cross sections larger than $10 \times 10\,\mu\mathrm{m}^2$. There is no such limitation for optical interconnect.

4.4 Circuit Limits

The general elementary architectures are represented schematically Fig. 11, where electrical wires and units are indicated as dashed lines, whereas optoelectronic devices and optical pathways are shown as solid lines.

The systems represented in Fig. 11 can have different realizations. For example, the laser could be replaced by light emitting diode (LED) and/or could be directly modulated – thus eliminating the need of a stand-alone modulator. The optical paths could be optical fibers, or even free-space instead of integrated waveguides. There are, however, two relevant differences between the two approaches, which are insensitive to the details of the realizations. First, optical transmission needs devices capable of handling both electrical and optical signals. Second, although optics requires more elaborated transmitting and receiving units, the signal has better chances to be smoothly transmitted through the medium. With careful design of the transmitting medium, the time width of signal pulses and the signal-to-noise ratio do not degrade appreciably for any interconnect scale considered in Fig. 12 below the km range. Simply stated, optical interconnect bottleneck is located

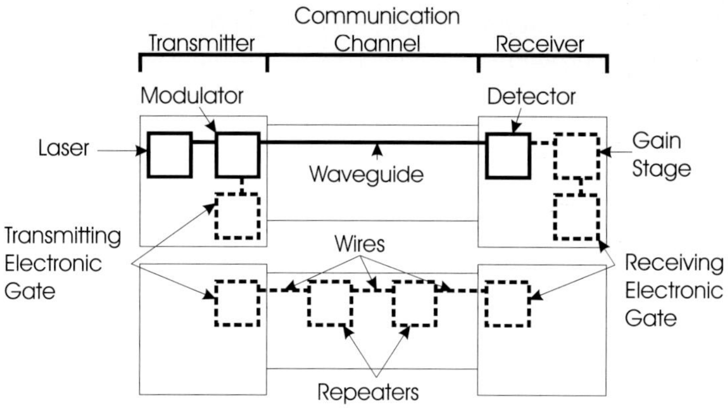

Fig. 11. Schematic comparison of typical components of an optical elementary interconnect versus an electrical one. The figure assumes that optical transmission must be preceded by an electro-optical conversion and followed by optical-electrical re-conversion, on the grounds that signal processing is performed electrically

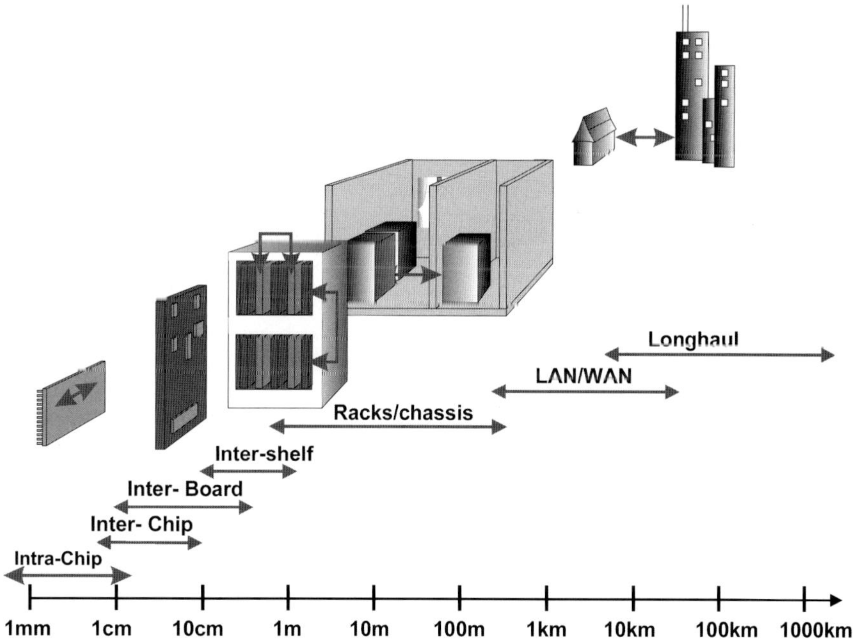

Fig. 12. Connectivity levels according to distance

at the front and end side, (i.e. at drivers and detectors, in terms of costs, chip real estate, speed, power consumption); in electrical interconnect, the bottleneck is located at the middle (wire). The optical bottleneck is exacerbated at intra-chip level because optical and electrical devices require different material properties, and bulk Si has poor optical efficiency. Integration of suitable optical devices, especially light sources, in Si is *the* critical issue of optical interconnect. Since in electrical interconnect the performance improves at shorter length, a *break-even* length can be defined as the length where optical and electrical performance are equal. The break-even length depends on which performance is considered. It can be speed, power consumption, cost and real estate. It should be emphasized, however, that even at short distances it is difficult to obtain time-of-flight limited speed performance in electrical wires because of driver output impedance.

Figure 13 gives a qualitative picture of break-even lengths for speed performance. In Fig. 13, optical interconnect has a fixed delay given by light source and receiver delays even at 0 distance, but is otherwise time-of-flight limited. Electrical interconnect has also a fixed delay given by the driver self-loading term ($R_{\mathrm{dr}}C_{\mathrm{dr}}$). Internal driver impedance R_{dr} also determines the slope of the driver limited delay. Only global interconnect with large drivers (small R_{dr}) exhibit a limited interval where speed is time-of-flight limited. At larger length, both local and global interconnect are limited by RC wire constant (12). In local wires, cross sections are kept small to allow a large number of connections. The larger resulting values of R are not critical for local wires, whose length is short and speed is usually driver limited speed. When optical fixed delay is larger than electrical fixed delay[15], as in the case shown in Fig. 13, break-even length crosses the parabolic RC limit, thus optical does not improve performance for local connections. Global wires are usually designed with large (fat) cross section to have low R. However, the parabolic dependence on length can lead to break-even length of the order of chip size. The exact evaluation of break-even length requires accurate knowledge of source and receiver latencies.

Insertion of repeaters in electrical lines transforms the parabolic dependence on interconnect length into linear dependence. In fact, given a long line with length L and repeaters with intrinsic delay τ_{rep}, if a repeater is associated to each of n sections (L/n long), the former line delay $\tau_{\mathrm{old}} = K \times L^2$ is transformed into $\tau_{\mathrm{new}} = \left[K \times (L/n)^2 + \tau_{\mathrm{rep}}\right] * n$, where K is a constant. If the number of repeaters per unit length ($N_r = n/L$) is constant, $\tau_{\mathrm{new}} = \left[K \times N_r^{-1} + \tau_{\mathrm{rep}} \times N_r\right] \times L$. Nevertheless, the overall speed will be slower

[15] This assumption might be questioned. For example, *Yayla* et al. [22] have assumed that superbuffers (tapered buffer chains) are required as electrical driving stages for long wires, and simultaneously have considered optoelectronic latency times comparable to single electrical inverter's delay (around 100 ps). In this case, break-even length for speed might even be undefinable, since optical interconnect would always result faster.

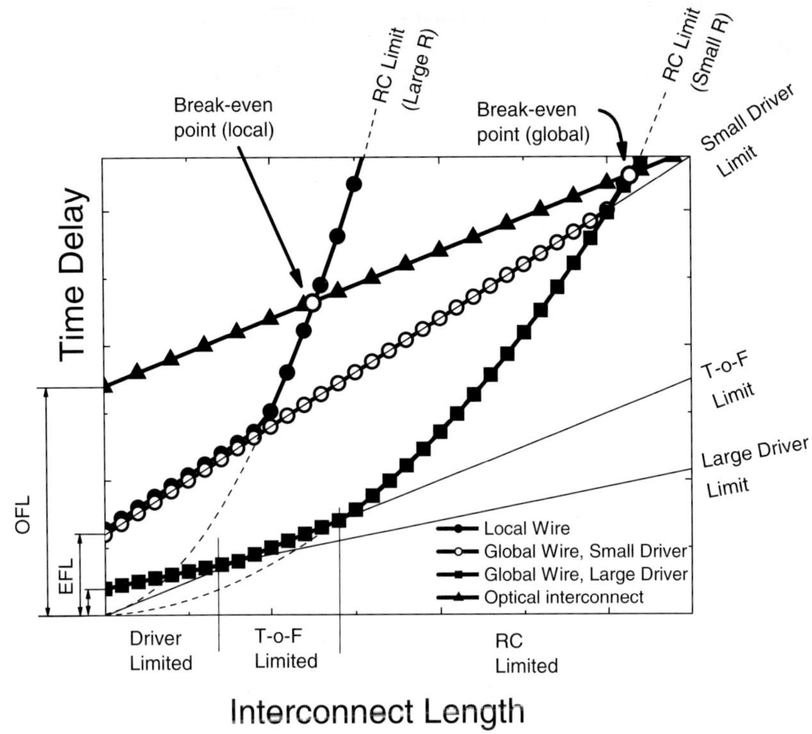

Fig. 13. Qualitative Time Delay versus Interconnect length. Both optical and electrical interconnect are represented. OFL stands for Optical Fixed Latency, and is due to the delays of light sources and receivers. EFL stands for Electrical Fixed Latency, and is due the $R_{dr}C_{dr}$ delay of the electrical driver. A large *small* driver has low *high* R_{dr}. Actual numerical values and crossing points depend on drivers' delay, on drivers' output impedance, on line's capacitance, resistance and inductance per unit length

than light. The break-even length for speed performance gets shifted towards longer values, but the effect might not be radical (for example, in Fig. 14, insertion of repeaters does not change the break-even length much). A disadvantage of repeaters insertion is a larger power consumption.

4.4.1 Break-even Length for Speed Performance

Kapur and *Saraswat* have simulated the optical transmission through a Si waveguide, whose thickness was ranging from 0.3 μm to 1 μm, at various height/width aspect ratios. 50 nm technology was assumed for the electronics CMOS circuitry. A 75 ps delay and a 250 pF capacitance were assumed, respectively, as transmitter delay and as output detector capacitance. Several simulations were performed at various values of input optical power

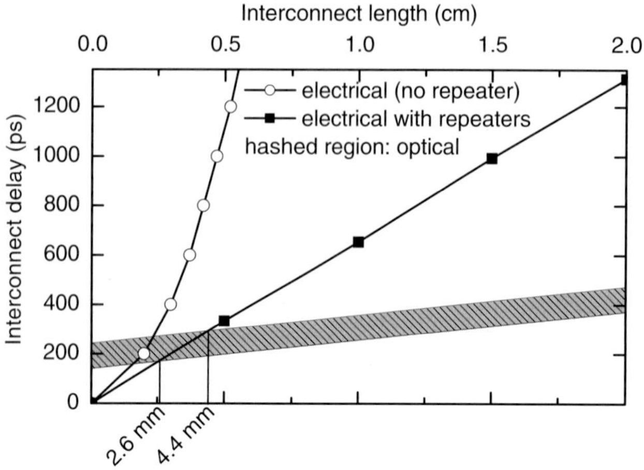

Fig. 14. Break-even length according to [62]. For the case discussed in [62], optical interconnect is predicted to be faster than electrical interconnect with repeaters above 4.4 mm

(IOP) and receiver power dissipation (RPD). In Fig. 14, the hashed region is bounded by simulations using IOP = 75 μW and RPD = 1.8 μW (lower curve), and IOP = 240 μW and RPD = 5.3 μW (upper curve). The delay decreases if larger RPD is tolerated and larger IOP is used for the transmission. The simulation reveals that in the conditions of the study, the optical transmission is advantageous for speed performance, compared to electrical transmission even with repeaters inserted, for distances over about 5 mm [62].

The first attempt to measure the latency of an optical interconnect has been reported by *Keeler* et al. [63]. The test circuit was an hybrid chip. The optical chip (GaAs-AlGaAs technology) was containing multi quantum well (MQW) p–i–n devices serving as both as input detectors and output modulators. Modulation was based on quantum confined Stark effect [64, 65, 66]. The electronic Si CMOS chip had a receiver circuit and a transmitter driver. The chips were flip-chip bonded to each other. 0.25 μm technology was used.

In the measurement scheme, an optical beam was sent the receiver, whose output was fed into the light modulator driver. The data latency of this setup is the sum of the optical-to-electrical conversion of the first receiver stage and the electrical-to-optical conversion of the transmitter. In a transmission line such as the one of Fig. 11, the stages would be ordered as 1. modulator driver, 2. optical modulator, 3. optical beam transmission, 4. detector, and 5. receiver circuit. However, the scheme proposed by [63] allows the measurement of the contribution to total latency due to the two signal conversions only. The total latency of a complete transmission should include also the time-of-flight of the optical signal travelling between transmission and reception.

The measured latency of optical-electrical-optical (O-E-O) conversion was ranging from 400 ps to 650 ps. The authors, however, suggest that a critical

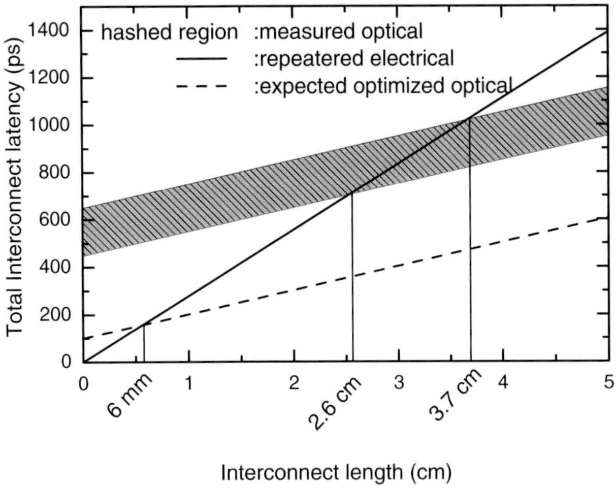

Fig. 15. Estimated break-even length for speed performance according to measurements reported in [63]. The actual measured performance of the optical connection led to numerical values between 2.6 cm and 3.7 cm. However, according to estimated optimization of the parameters, the length drops to 6 mm. Above the break-even length, the optical interconnect is expected to have better speed performance

performance bottleneck was originated by the unnecessarily high output capacitance of the detector (260 fF measured). According to authors, the use of an integrated detector would reduce the capacitance to 40 fF, and the total O-E-O latency to about 100 ps. An optimized (including repeaters) electrical *RC* connection would have a speed of $0.12 \times c_0$ [4], where c_0 is the velocity of light in free space, whereas technologically available waveguides could let the speed of the optical connection reach a value of $0.33 \times c_0$ [63], implying, respectively, latency times per unit length of 278 ps/cm and 101 ps/cm. In the plot of Fig. 15, we report various estimations of latency time as a function of the interconnect distance, both for the electrical (solid line) and the optical interconnect. The hashed band is the locus of obtainable optical latency times, allowing variations of parameters such as optical power, or DC power supply voltage of electronic stages, and considering the measured detector capacitance of (260 fF). The dashed line is the estimated performance obtainable with 40 fF detector capacitance.

The above two independent studies point out that the break-even length for speed performance under reasonable assumptions, but without including integrated light sources, is in the range 2.6 mm to 6 mm.

4.4.2 Break-even Length for Speed by Imposing Maximum Delay

Mulé et al. have considered then problem of substituting only the wires whose total latency break a ceiling acceptable maximum [67]. They proposed the following criterion to substitute electrical interconnect. An electrical wire should be replaced by an optical interconnect only when its RC delay is longer than the time-of-flight τ_{limit} associated to the *longest possible on-chip optical waveguide*. The rationale is that there is no point on having a faster interconnect than τ_{limit}, which is going to be the the chip bottleneck in any case. The longest waveguide in a Manhattan layout is twice as long as the chip side. According to this strategy, one makes the longest interconnect as fast as possible using an optical link, and then substitutes all (and only) those electrical connections which are slower than this optical one. Break-even lengths found with this criterion are longer than lengths found previously. However, this criterion defines a chip area A in which the interconnect time between any couple of devices is less than or equal to τ_{limit}. The clock rate that can be applied to the chip area A after this strategy is the maximum achievable, because it is limited by the time-of-flight of the longest interconnect. This criterion minimizes the optical interconnect under the constrain of achieving the maximum clock rate. In using this criterion, the authors did not compute the delays associated with optoelectronic circuitry.

Several break-even lengths were found, depending on the waveguide materials. The minimum value was 32 mm – obtained using SiO_2 as core material and air and methylsilsesquioxane (MSQ)[16] as cladding. The assumed CMOS technology was 45 nm generation, and the die size was 572 mm^2. The substitution led to a 34% improvement in the chip speed, using 32% of chip area for optical waveguides. The envisioned layout, shown in Fig. 16, closely resemble an electrical back-end layout. However, an hetero-epitaxial GaAs layer is inserted as hypothetical optically active layer, and an optical interconnect layer – with Manhattan layout – is shown as top-most interconnect [67].

4.4.3 Speed and Power Performance of Speed-Optimized Interconnect

Yayla et al. have calculated and compared the speed performance and the power consumption of electrical and optical interconnect optimized for speed performance [22]. The very comprehensive analysis took into consideration a variety of configurations, ranging from intra-chip one-to-one (repeatered) interconnect to off-chip 1-to-N (repeatered) connections and to H-tree[17] (repeatered) clock networks. All the cases were discussed both at intra-chip

[16] MSQ is a Si-based low-k dielectric. The reader is directed to [17] for an excellent review on low-k dielectrics.

[17] The characteristic of the H-tree network is the invariance of path length between the root and the leaf nodes. Such invariance eliminates the clock skew (footnote 13, p. 132) due to path length differences of other configurations. Clock distributions – including H-tree examples – are further discussed in Sect. 5.2.2 at p. 155.

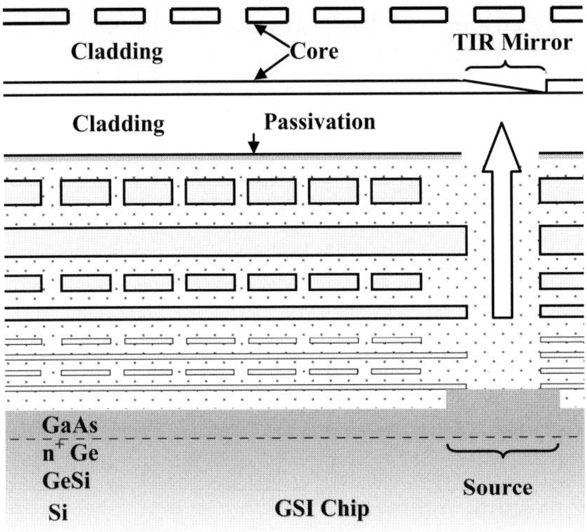

Fig. 16. Suggested interconnect back-end structure including an optical interconnect as last wiring level. The hypothetical substrate is an heterostructured Si wafer, with a SiGe buffer layer to allow GaAs heteroepitaxy. After [67]

and inter-chip level. optical interconnect included both solutions based on light generation with VCSELs and solutions based on light modulation of externally provided light. Two possible modulation schemes were considered. The first was by quantum confined Stark effect, assuming a multi quantum well (MQW) modulator, whereas the second was by electro-optical effect assuming a lead lanthanum zirconium titanate (PLZT) modulator. All the effects due to hybrid technology (e.g., the capacitance of solder bump connections of flip-chip optoelectronic components) were also considered. Typical VLSI (0.25 μm to 0.35 μm generation) as well as optoelectronic parameter were used for the devices. Under the assumptions taken, the conclusion of [22] is that optical interconnect for global wires might be faster than electrical interconnect even at short distances. The reason is that global wires require cascade buffers and large repeaters. However, *total* power consumption is generally larger in optical interconnect, although in the case of light modulators the *on-chip* consumption is expected to be actually lower. In fact, with modulators, the optical power for signals is supplied by external sources.

4.5 System Limits

A key design strategy at the system level is partitioning. System performance critically depends on how a sea-of-gates [30] is distributed in physical blocks[18]. Such partitioning is driven by the empirical Rent's rule connecting the number P of connections of a block to the outside world with the

[18] The *functional* partitioning is done at previous design levels, but that does not necessarily translate into the best *physical* partitioning.

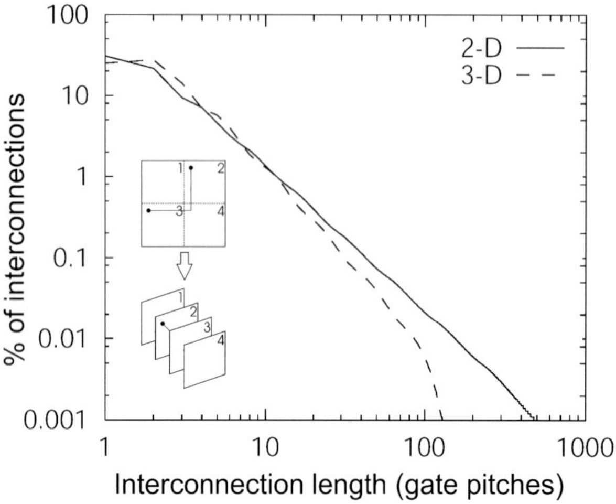

Fig. 17. *Main plot*: the interconnect length distribution in a 2D and 3D layout. The advantage of 3D layout in terms of wire shortening is pictorially shown in the *inset*. Adapted from [30]

number B of its basic circuital elements [68]

$$P(B) = FB^r \, , \tag{13}$$

where F is the average connections per basic element, and r is the Rent exponent. Regular structures such as RAMs have low complexity ($r \simeq 0.47$). Fast, full-custom VLSI design can have vales of r up to 0.75. Average-complexity circuits have $F = 2.5$ and $r = 0.6$ [30].

When the total chip area is limited by the computing elements (gates) rather than the wires, Donath's partitioning [69] gives the wire distribution $n(l)$ for the wire length l

$$n_{2\mathrm{D}}(l) = Cl^{2r-3}; \qquad n_{3\mathrm{D}}(l) = Cl^{3r-4} \tag{14}$$

Figure 17 shows a typical result obtained form (14). The 3D distribution is markedly different for larger (global) interconnect – the critical connections – which are fewer and shorter than in 2D design. Intuitively, the 3D architecture is attractive for optical interconnect because it is the natural setup for networks between sources and detectors arranged in planar circuits, using free-space interconnect (as depicted, for example, in Fig. 29 at p. 164). *Ozaktas* [70] has shown that the 3D Rent's rule applies to free-space optical interconnect similarly to electrical wires, a result which is not so obvious, since optical beams – as opposed to electrical wires – can freely cross each other. The result originates from the general fact that the minimum total communication volume required for an optical system whose total interconnection length is l_{total} is given by $l_{\mathrm{total}}\lambda^2$ [70]. Thus, even with the greatest possible amount of overlap and space sharing, the global result is that each channel requires a cross-sectional area (λ^2), as it was a solid wire [70].

Moreover, provided the interconnections are allowed to be routed through three-dimensional space, there is no disadvantage in restricting the active devices to a plane [71].

Ozaktas [71] has quantitatively compared 3D architectures for electrical (both simple and repeatered), optical and superconducting interconnect. Optical and superconducting interconnect were found superior, in that they allow arbitrary increase of bandwidth and bisection-bandwidh[19] with increasing system size. In electrical interconnect the bisection-bandwidth product $H \times f_{\max}$ is constant: if larger bandwidth is desired, fatter wires are needed, thus the number of wires per arbitrary system cross section diminishes [71].

5 Optical Interconnect Technology

5.1 Architectures

In systems' performance, the interconnect limitations can be more or less critical bottlenecks, depending on the architecture. For example, in today's monoprocessor computers (common personal computers, PCs), intrinsic memory latency still dominates over transfer latency between caching levels. Insertion of optical interconnect would not improve significantly the memory access latency, unless the architecture is redesigned. In present design, optics might be advantageous between the last caching level and the main memory, although it appears unlikely before 2010 [72].

In the shorter term, optics appears readily attractive for parallel computers. Figure 18 shows two such architectures, the symmetric processors architecture and the distributed memory architecture. The common feature is the presence of a fast bus. A symmetric processor architecture is a physically shared-memory machine with a uniform memory-access time. In distributed memory architecture, each processor has its own memory and the fast communication is conducted between processors. Substitution of shared buses with optical buses would be interesting because of potential clock performance, high degree of parallelism, and skew-free connections [72]. The reader is directed to [73] for a review of optical buses for parallel computing, based on waveguides. Optical buses based on slab waveguides have been discussed in [74].

Optics could have a major role in field programmable gate arrays (FPGAs). Compared to Application Specific integrated circuits (ASICs), FPGA have

[19] The *bisection* H is the number of connections crossing an imaginary plane dividing the system into two roughly equal parts. The bisection-bandwidth product $H \times f_{\max}$ and the bisection-inverse delay product $H \times S$ are often more meaningful measures of performance or throughput than aggregate bandwidth $N \times f_{\max}$ or $N \times S$ (where N is number of devices or computing elements) [71].

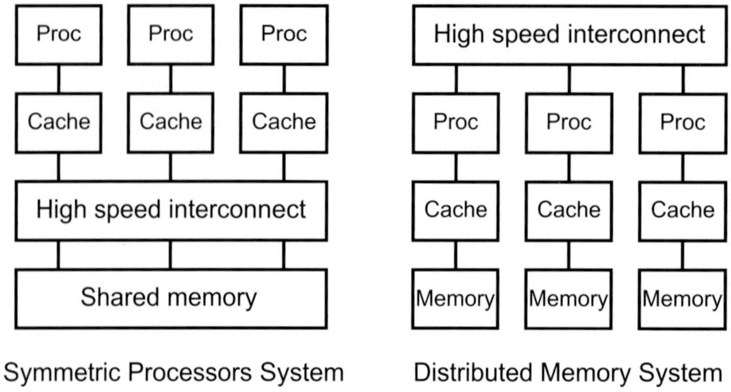

Fig. 18. Architectural layout of a Symmetric Processors System (*left side*) and of a Distributed Memory System (*right side*)

connections which can be programmed. However, the cost of such flexibility is that the volume dedicated to interconnect routing is considerable. Thus, chip area devoted to the logic is limited, and sometimes several interconnected FPGAs might be necessary for particular applications. The high density of slow and reconfigurable interconnections is a known bottleneck of FPGA [72].

Dedicated Optoelectronic Processors, which are not von Neumann architectures, and do not execute stored programs, constitute a more disruptive field where optical interconnect is also very attractive [72]. Generally, von Neumann architectures are not well suited for exploiting optical interconnect at full potentiality. On the contrary, architectures which can take advantages of locality and regularity of imaging systems – such as neural and reconfigurable computing systems – can benefit of the high degree of parallelism available in optical free-space interconnects. Implementation of this kind of optical architectures by planar free-space interconnect has been pioneered by *Jahns* [75], whose group has recently reported a sample demonstrator of an optically interconnected memory [76] (Fig. 19).

A similar architecture is the Smart Pixel Array. Figure 20 illustrates the smart pixel concept, which has been used to realize a Cross-Bar[20] switch nominally capable of handling an aggregate input of over 1 Tb/s [77, 78].

[20] A cross-bar network is a non-blocking network: a connection of one transmitter to a given receiver does not block a connection of another transmitter to a different receiver, as opposed to bus-based networks, which consist of a shared medium common to all nodes. A von Neumann architecture is a typical example of bus-based network.

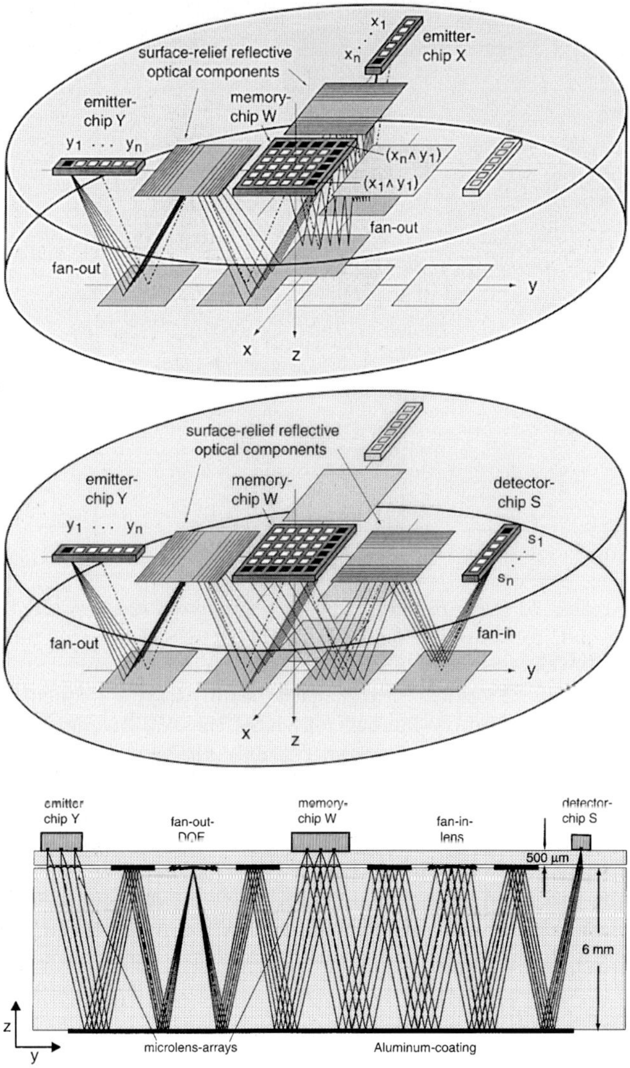

Fig. 19. Planar free space optically interconnected memory. The architecture is a neural associative memory: writing a vector x into memory w (*top*) and reading it from w (*middle*) is achieved in one shot by suitably activating the key vector y (the pixel y_1 only, in this case). The required fan-out is provided by diffractive optical elements (DOEs). The purpose of this structure is only demonstrative. One could picture massive simultaneous reading and writing of several vectors according to the idea shown in Fig. 31. It should be noted that in this demonstrator the memory pixel was a *learning* pixel (i.e., the behavior of pixels is locally organized) based on self-electrooptical effect device (SEED) modulator technology (GaAs diodes working both as photodiodes and as light modulators, flip-chip bonded on Si-CMOS circuit) [79]. The bottom graph shows the side view with the diffractive elements. The medium of the demonstrator was a quartz substrate. After [76]

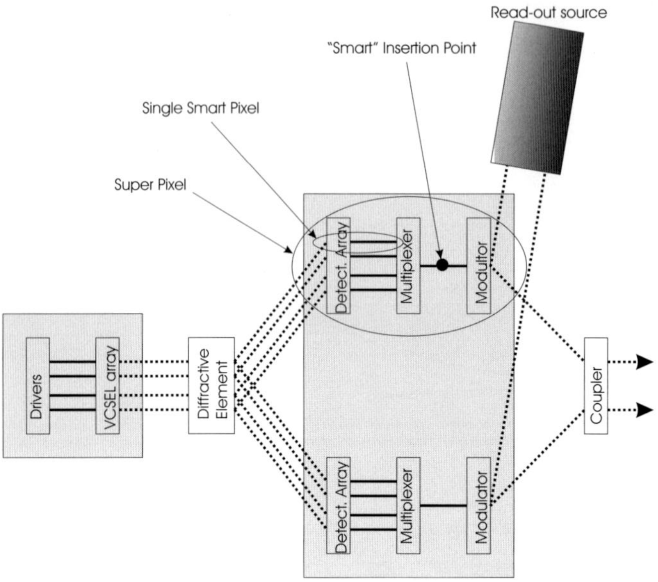

Fig. 20. Schematic operation of a smart pixel. The *gray shadowed box* represents the optoelectronic front-end, for example realized by III–V VCSEL arrays, flip-chip bonded onto a Si CMOS substrate. The array is schematically represented with $N_{\text{in}} = 4$ output modulated optical beams, each beam carrying sequences of signal packets. An optical diffractive fan-out replicates the N_{in} beams N_{diffr} times ($N_{\text{diffr}} = 2$ in the picture; it is different from N_{in} only for generality). Each of the N_{diffr} replicas impinges onto a superpixel, composed by N_{in} smart pixels. The smart pixel array is the *large shadowed box*, and includes detectors, logic, and output modulators. Inside, the multiplexer is driven directly by the beams (a routing header is supplied at the beginning of each beam packet). The selected packet can be electrically processed in the smart point (*solid circle*), and then it is converted to a modulation signal and read by a laser. If $N_{\text{in}} = N_{\text{diffr}}$, the smart pixel array processes all the input beams and re-emit them at the end (Cross-Bar switch). All the routing is directly controlled optically. In [77], a demonstrator with $N_{\text{in}} = N_{\text{diffr}} = 64$ has been fabricated

5.2 Functions

An interconnect has different design constrain depending on its function. In digital signal processing one can have interconnect for

- intra-chip single channel digital data transmission;
- inter-chip single channel digital data transmission;
- power supply;
- clock distribution networks;
- integrated optical circuits for WDM networks.

Single channel digital data transmissions constitute the mainstream framework of the whole discussion of this chapter, thus are not explicitly dealt here. In the rest of this section, some peculiarities of the last three cases are addressed.

5.2.1 Power Supply

In traditional ICs schemes, the power is carried and supplied to devices by an electrical wire network. However, that is not a necessary choice. Power can be delivered optically – at least partially. This can be done, for example, in smart pixel circuits, for the readout power: in the final stage of smart pixels suggested by *Walker* et al. [77] an external optical power supply was required, since the final stages of pixels were light modulators (not sources). An interesting implication of this design is that a large fraction of the required signal to noise ratio is not charged as a direct *electrical* power consumption. A complete power supply design for smart pixel applications, its realization and characterization has also been reported [80].

The "Si Integration Technology Working Group" – a Technical Working Group (TWG) of the Communication Technology Roadmap (CTR)[21] – has recognized that optical power supply in a general sense is to be considered as an important option for Si optoelectronics, due to present lack of efficient light sources [82].

5.2.2 Clock Distribution Networks

Today's electrical clock lines are by far the longest interconnect lines in IC's, and power dissipation due to clock distribution constitute a substantial fraction of the total IC's power, as already shown in Fig. 4 [83].

In clock distribution networks, the main characteristics are that 1. there is one centralized source and 2. the clock must be consistently distributed among synchronously working units. In fact, clock skew and jitter (footnote 13, p. 132) in synchronous architectures – i.e. the dominant strategy for commercial microprocessors – are key challenges for high-speed circuit designers because they can degrade performance and cause chip failures [84]. The problem of clock skew and jitter have been worsening as clock speed increased.

Designers have introduced symmetric networks to eliminate skew originated by different path lengths, such as the H-tree, X-tree, grid and serpentine networks. The reader is directed to the review by *Friedman* for electrical clock design strategies [85]. Wire symmetrization, however, is not a satisfactory solution, because skew can also originate from line loading asymmetries, by local differences in wire floor-plans which result in asymmetrical cross-talk

[21] The CTR (Web Page Address: `http://mph-roadmap.mit.edu`) was created in 2000 by a consortium at the Massachusetts Institute of Technology, with the goal of "providing technology development targets for the long-term evolution of photonic component integration in the optical communication industry" [81].

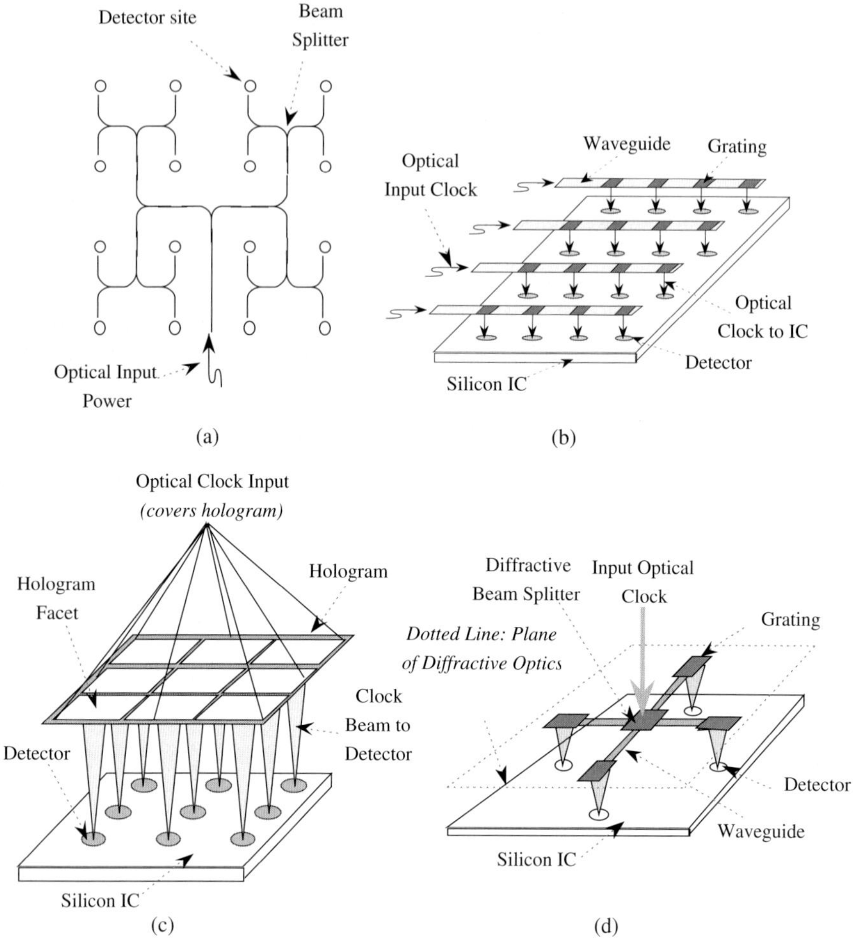

Fig. 21. General approaches for optical clock distribution over the area of an IC. (**a**) Waveguide emulating H-tree. (**b**) Tapped waveguides. (**c**) Hologram distribution of clock. (**d**) Planar diffractive optics providing vertically optical clocks. From [83]

characteristics, and by temperature differences, which affect copper resistivity and repeaters delay [86, 84, 85, 83]. optical clock distribution is an interesting perspective, since it is expected to be less sensitive to temperature and cross-talk. Another appealing feature is its insensitivity of required power to the propagation length [4]. Optical clock distribution has been recently reviewed by *Tewksbury* and *Hornak* [83]. Sample architectures for optical clock distribution at intra- and inter-chip level are shown in Fig. 21.

Bhatnagar et al. [87] have shown that optical clock distribution is very effective for jitter reduction. The suggested ideas was a receiverless circuit, in which the processing of optically delivered clock was minimized, in order

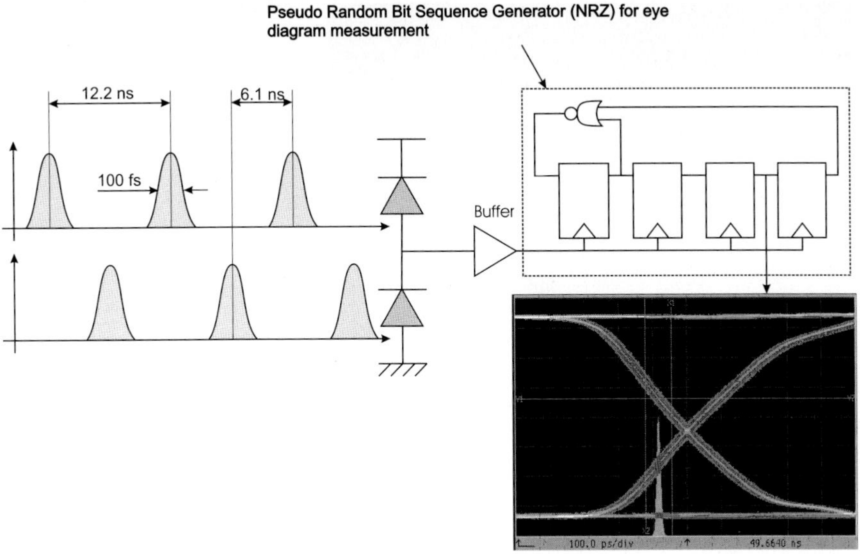

Fig. 22. Totem-pole connected photodiode detectors, converting two phase-shifted pulsed laser beams into a 50% duty cycle electrical square wave for circuit clocking. The clocked circuit (inside the box) was a pseudo-random generator, whose output was a OOK/NRZ signal, used to plot the eye-diagram, as discussed in Sect. 3.1. A 6 ps jitter was measured, to be compared with 30.5 ps from an electrical clock measurement. The resulting clock period was 82 MHz. Eye-diagram (the total shown time window is 1 ns) is from [88]

to take the maximum advantage of short light pulses. The circuit is schematically reported in Fig. 22. The clock signal was delivered optically in the format of two pulsed beams. Pulses time separation was 12.2 ns, and the two beams were time shifted by 6.1 ns. Pulses were generated by a mode-locked laser and their time width (100 fs) was very narrow compared to any electrical time constant of the circuit. The beams were sent to a couple of photodiodes connected in totem-pole configuration. The photodiodes were $12 \times 12\,\mu m^2$ $GaAs/Al_{0.3}Ga_{0.7}As$ multiple quantum well p–i–n diodes flip-chip bonded to the CMOS chip containing the pseudorandom generator. The middle point of the totem-pole was therefore oscillating as a square wave at 50% duty cycle. Since the time location of rising and falling front of the square wave was controlled by the pulse position – i.e., by the timing of the mode locked laser, which is far more accurate by any electrical alternative – the jitter of such clock was very low. *Miller* as reported an updated (March 2003) jitter value of less than 6 ps [88]. The jitter obtained by clocking the same circuit electrically was 30.5 ps.

A possible implementation of an optical clock distribution network at intra-chip level, with H-tree structure, has been proposed by *Mulé* et al. [89]. The structure was based on a single-mode rib guide, for target wavelength

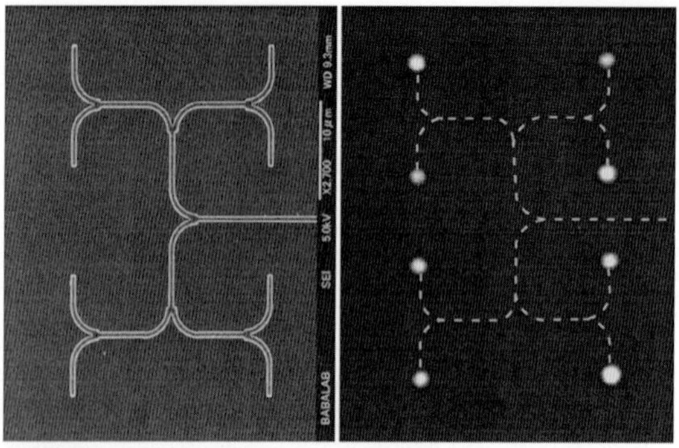

Fig. 23. Top view of H-tree circuit (*left*) and corresponding near field pattern of light output for TM-polarized light at $\lambda = 1.55\,\mu m$. *Dashed curves* have been added for better understanding of waveguide position. After [91]

$\lambda = 0.85\,\mu m$, and tested with computer simulations. Simulated rib height and width were, respectively, 0.5 and $2\,\mu m$. Assumed refractive index of core, cladding and substrate were $n_{core} = 1.54$, $n_{clad} = 1$, and $n_{sub} = 1.51$. The simulations resulted in negligible cross-talk when adjacent waveguides were spaced $4.5\,\mu m$ apart. The radii of curvature of last level (with the highest bending) was about $450\,\mu m$. Light was assumed to enter the structure by a grating coupler, whose design allowed some tens of μm misalignment with few percent drop in coupling efficiency [89].

Optical H-tree structures have been experimentally demonstrated. *Chen* has patented a polymeric technology which is both printed-board and Si-CMOS compatible. He has shown a H-tree prototype waveguide made of Ultradel 9120 polyimide, with fan out 1-to-48, directly fabricated onto a Si wafer [90]. The optical distribution was tested using $\lambda = 632.8\,nm$ He-Ne laser wavelength. According to the authors, the process is fully CMOS compatible, and although their demonstrator was fabricated at larger scale, in principle it is claimed to be applicable to chip-level interconnect.

Fukazawa et al. [91] have demonstrated the first ultra small (covering an area of $35 \times 25\,\mu m^2$) H-tree network with fan out 1-to-8 using Silicon on Insulator (SOI) technology, which is shown in Fig. 23. The network was designed for wavelength $\lambda = 1.5\,\mu m$. The waveguide had a cross-section $0.5 \times 0.32\,\mu m^2$. Combined measurements on the H-tree network and on simple waveguide structures allowed detailed estimation of losses. Any path from the root to the leaf level had three branches. In a H-tree network, there is clearly a power drop of 3 dB per branching, due to the signal split. Thus, a 8-leaf H-tree has a root-to-leaf splitting drop of $3 \times \log_2 8 = 9\,dB$. After taking this drop out, additional losses were found: 6 dB at input butt-coupling, 1.2 dB in the root-to-leaf waveguided path, a total additional 2 dB at the branches, and 3 dB as reflection at the output.

5.2.3 Integrated Optical Circuits for WDM Networks

Point-to-point WDM fixed optical transmission lines are conceptually simple. Typically, they might require fiber amplifiers to regenerate attenuated signals. When one moves to shorter distance applications, however, networks might involve many nodes, mutually connecting each other. Efficient *routing* and *wavelength assignement* schemes are then required. The traditional approach in such case is the use of opaque networks[22] [92]. There is strong motivation in finding solutions to increase the network transparency, in order to limit or eliminate the delays from O-E-O conversions. Fully optical networks are also more easily reconfigurable[23] [93]. Unfortunately, all-optical switching technology is not yet a reality. One of the difficulties is the lack of suitable all-optical wavelength conversion devices [92]. However, optical wavelength conversion is an active field of research [94, 93, 95, 96].

Integrated novel optical routing systems, which can also include O-E-O conversions, are currently under investigation. An example of such routing technology is the All Optical Label Swapping (AOLS). In AOLS, the routing is controlled by putting optical labels in front of data packets. When a packet enters the router, the routing function computes a new label and wavelength from an internal routing table given the current label, current wavelength, and fiber port [97].

Single Si-based devices, for all-optical WDM applications, have also been proposed. For example, *Gerken* and *Miller* have demonstrated a fully optical, non periodic dielectric thin-film demultiplexer, based on SiO_2 and Ta_2O_5 materials [98] (Fig. 24). *Nelson* et al.have also demonstrated chip-to-chip transmission and wavelength demultiplexing using GaAs optoelectronic devices, flip-chip bonded onto Si ICs [99]. *Ghulinyan* et al. have fabricated porous silicon Fabry–Perot microcavities with transmission resonances in the wavelength region of 1.55 µm, with quality factors up to 3380 [43]. *Prather* et al. have proposed an electro optical switch for photonic band-gap waveguides [100].

5.3 Physical Partitioning

Market-driven packaging choices strongly affect all level of Si technology, including interconnect. Many designers are shifting from System-on-Chip (SoC) solutions towards System-in-Package (SiP) alternative because the costs of the former product are daunting, and do not motivate the advantage of including all the functionality at the chip level [58, 101]. SiP is based on an intermediate level of interconnection between the thin films of ICs and the thick films of Printed Circuit Boards (PCBs). This approach has emerged in the last decade, though the idea is not new. Its original application was

[22] An optical network is said to be *opaque* (*transparent*) if it involves (does not involve) optical-to-electrical-to-optical (O-E-O) processes at routing nodes.

[23] A network is *reconfigurable* if its physical links between nodes can be changed.

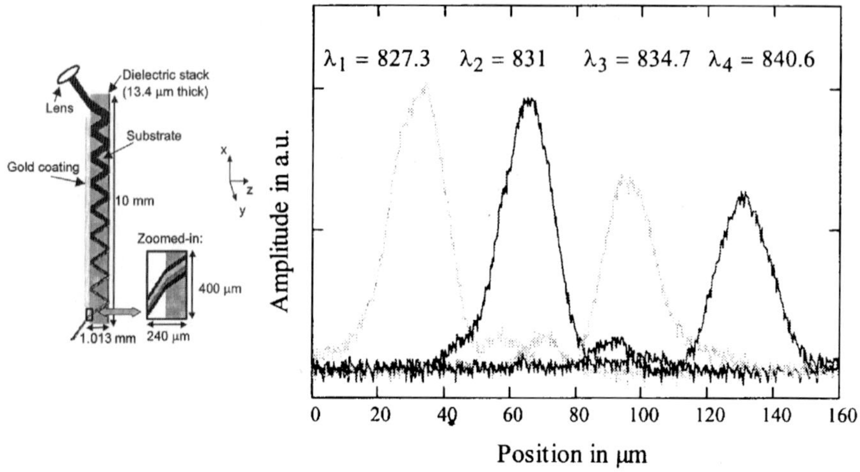

Fig. 24. The fully optical demultiplexer device proposed by *Gerken* and *Miller* is schematically shown at the *left side*. The channel separation ability has been demonstrated: the plot at the *right side* shows the channel separation in the x-direction (wavelength λ in nm [98])

for high performance supercomputers [102]. The technology is referred to as Multi-Chip-Module (MCM), and it consists on more than one bare die on a common substrate. The basic advantage is that packaging is not required at the level of one single chip. Thus, overall system size is smaller, costs are reduced, and yet more flexibility, diverse functionality and higher performance is achieved, compared to solutions involving single chip packaging.

According to the Institute for Interconnecting and Packaging Electronic Circuits (IPC), there are three main fabrication technologies for MCM substrates (Table 5):

- Laminate MCM (MCM-L), which derive from printed wiring board technology (PWB), and uses a laminate substrate such as FR-4, and copper as conductor;
- Ceramic MCM (MCM-C),which uses a cofired ceramic-based, multilayer substrate, either HTCC or LTCC, but not thick film;
- Deposited MCM (MCM-D), which uses thin-film deposition technologies both for conductors and dielectric, resulting in multilayer, deposited substrates. Starting bases for the film deposition can be silicon, ceramic or aluminum.

Among the three major categories of MCMs , the MCM-D are more expensive, but particularly suited for extending ICs interconnects and for insertion of optical functionality. In fact, deposited and etched films are achieved using techniques adapted from IC microfabrication, combined with new technologies (e.g., polymer dielectrics or copper interconnects). MCM-D technology

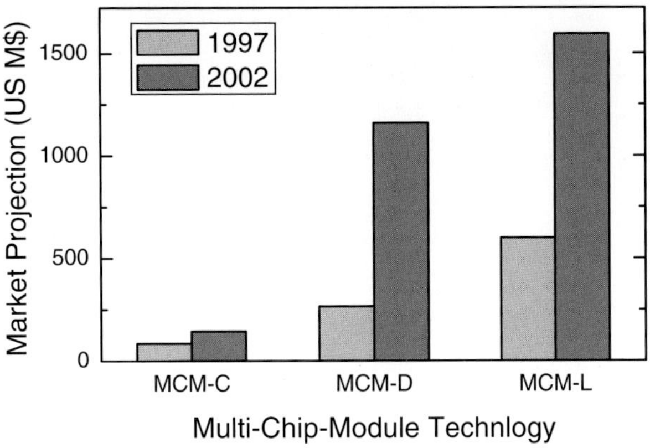

Fig. 25. MCM Market Projections, not including components. (Source: Integrated Circuit Engineering Corp., Scottsdale, AZ; `http://www.ice-corp.com`

is expected to rapidly gain market in the IC industry (Fig. 25). IC thin film technologies usually use films with thickness of about 1 μm, whereas the MCM-D film thicknesses are about 10–20 times thicker. MCM-D technology might provide effective answers for global electrical interconnect (where fat wires are required) as well as for optical interconnect, where the additional degree of freedom in the choice of materials provided by MCM-D technology is interesting, for example, for insertion of optical sources or for waveguides. MCM-D technology can be also be used for 3-D modules [102].

Table 5. Multi-Chip-Module technologies (Source: Integrated Circuit Engineering Corp., Scottsdale, AZ; `http://www.ice-corp.com`)

Characteristics	MCM-C	MCM-D	MCM-L
Pitch per layer	10 mils (254 μm)	1 mil (25 μm)	8 mils (200 μm)
No. of Layers	> 60	5	5–25
Materials	Alumina, Aluminum Nitride, Beryllium Oxide	Silicon, Alumina, Glass	FR-4, Polyimide
Power Dissipation	High	Medium	Low
Cost	Medium	High	Low
Speed Performance	Medium	High	Low

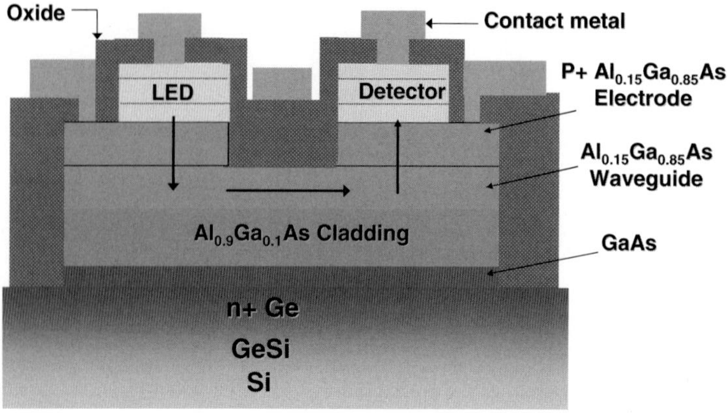

Fig. 26. Monolithic optical interconnect concept, based on silicon substrate and epitaxial growth of III–V semiconductors for the light source and the detector. The optical path is provided by a III–V based waveguide. From [108]

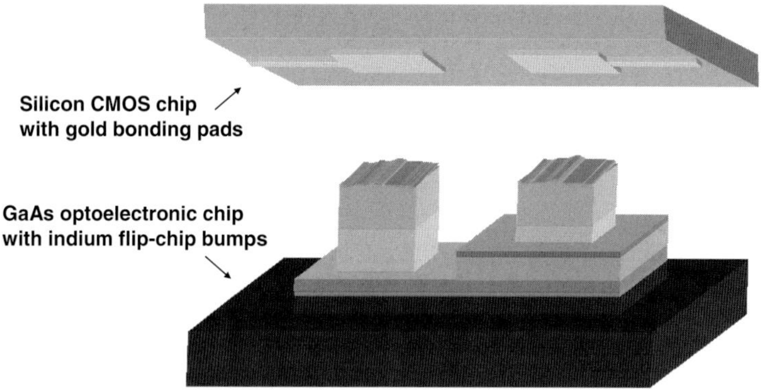

Fig. 27. Schematic illustration of the concept of hybrid integration for optical functionality. From [108]

The difficulties in achieving efficient optoelectronic Si devices – especially light sources[24] – have motivated the development of mixed technologies in which optical functions are delegated to non-Si materials. Insertion of optically active materials in Si technology – typically, III–V semiconductors such as GaAs – has been first pursued by either *monolithic* (Fig. 26) or *hybrid* (Fig. 27) integration, and only more recently MCM technology has been suggested as an additional approach, which has been termed *polylithic* [18] integration (Fig. 28). Polylithic technology allows high complexity in circuits, with ultra high density wafer level I/O metallic connections ("Sea-of-leads"

[24] It should be remember that the research for fully monolithic silicon optical sources is very active and remarkable results have been found in the last three years [103, 104, 105, 106, 107].

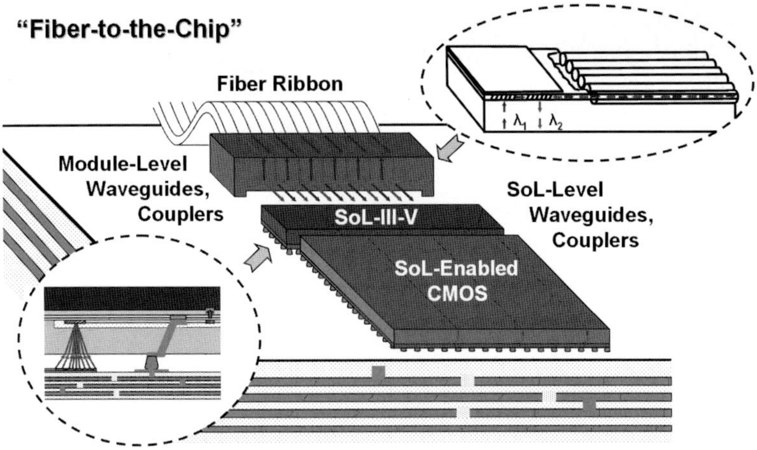

Fig. 28. Polylithic integration concept, to conjugate electrical and optical functionality by supporting them in the multi-chip module (MCM, see also Table 5 at p. 161). Chips are designed with Sea-of-Leads I/O technology, and are subsequently bonded over substrates carrying both electrical and optical busses. From [108]

technology, 10^4 leads/cm^2), and connectivity to virtually any kind of substrates, even with significantly different thermal coefficients. In fact the leads can be designed to be flexible in all three space dimensions [109].

Hybrid and polylithic approaches have gained attention not only because they fit the packaging trends, but also because the question has shifted from "how" optics is to be integrated with Si to "why" optical integration is needed. In fact, optical interconnect is mostly attractive for global and clock wires, whereas it is not advantageous for sub mm distances, as discussed in Sect. 4.4.1. Hybrid and polylithic technologies naturally suggest structures with vertical emission and detection, which could be implemented using vertical cavity surface emitting lasers (VCSELs) and taking advantage of free-space interconnect (Fig. 29). Moreover, off-chip optical interconnects are an interesting alternative to electrical solutions, as the off-chip insertion of repeatered lines is more complicate than in fully Si monolithic ICs [28].

Figure 30 shows some other examples of opportunities MCM technology can offer towards the insertion of optical functionality.

5.4 Free-Space Optical Interconnect

In off-chip interconnects, the circuit sizes are larger than the wavelength. At such feature size, beam divergence (7) is negligible, making free-space interconnect an intriguing option for off-chip interconnect. optical interconnect provides effectively a wireless technology.

A first consideration is that free-space interconnect lets take advantage of the third space dimension, leading to opportunity of massive parallelism [110,

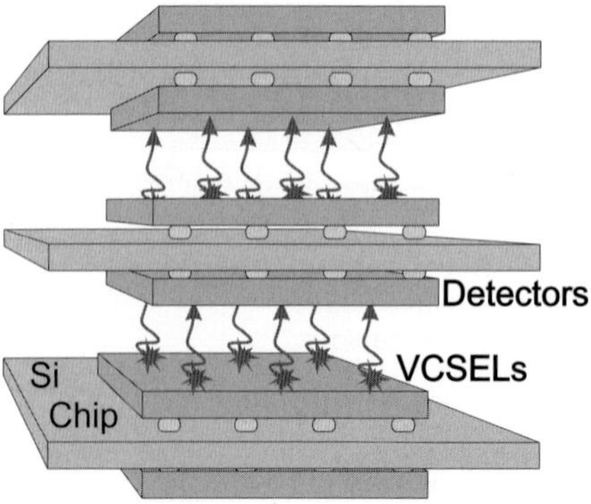

Fig. 29. Sample structure for free-space interconnect, compatible with both hybrid and polylithic technology

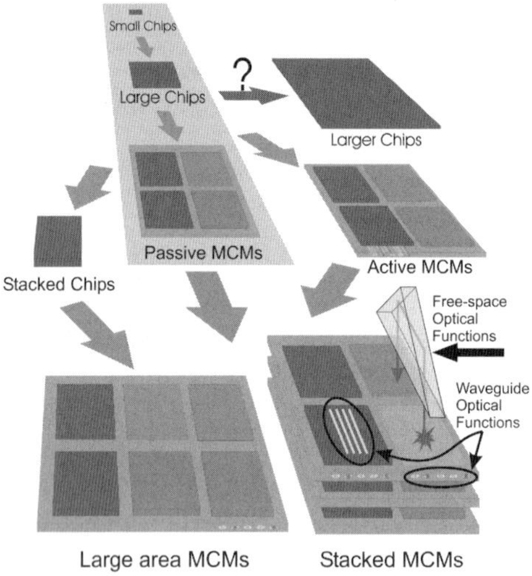

Fig. 30. Evolution of current Si chip technology towards larger area chips and passive Multi Chip Modules (MCMs). Potential future structures could be larger, more complex, passive and active substrates MCMs, also including 3D opportunities, both at chip and MCM levels. Solutions already present on the market lay on the grey background. The picture also show an alternate design to the structure suggested in Fig. 29 for free-space interconnect, which allows a flat rather than a 3D layout

111,28]. Examples of free-space optical interconnects have been already introduced in Sect. 5.1 (Fig. 19 and Fig. 20). An interesting aspect of free-space interconnect is the sharing of memories. For example, with reference to Fig. 31, a memory pixel, constituted by a LCD shutter (on-state: shutter transparent, off-state: shutter closed) in front of an inverting prism, can be read by any source-detector unit, which will receive back the emitted light only if the memory is in the on-state. Such memory pixel "knows" which reading unit is sending the read request and replies to that unit, the reading being available to any potential reader at any time with no interference between readers. Concurrent-write concurrent-read techniques, potentially interesting for parallel computing, are thought to be hardly realizable with electrical connections, but not with optical [112].

Typically, free-space communication routing is achieved by diffractive optical elements (DOE). Essentially, DOEs are generalization of diffraction gratings. They act on beams' wavefronts, redirecting fragmented beams' portions

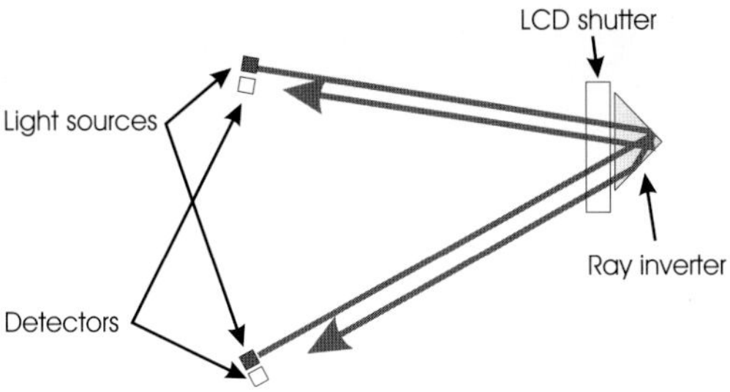

Fig. 31. Concept of memory sharing with optical free-space interconnect. A memory pixel (*right side*) is constituted by a LCD screen and an inverting mirror. Reading elements (*left side*) are composed by ba light source and a detector. The read operation involves sending a beam to the pixel and reading the back-reflected light. The pixel is available in parallel to the readers with no restriction

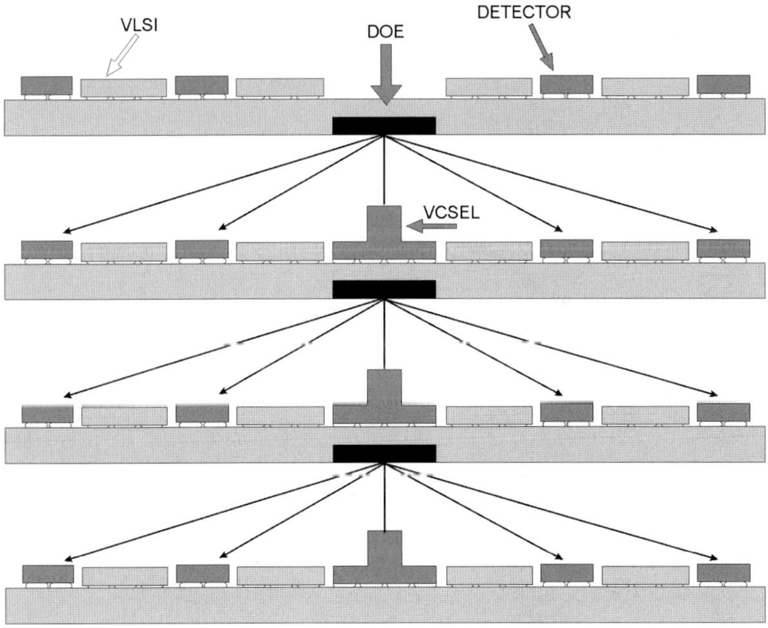

Fig. 32. Concept of insertion of Diffractive Optical Elements in 3D architectures, as discussed by Prather [115]. DOE provide fan-out, and can be integrated into Si wafers

in such a way that designed interference patterns are achieved [113]. The interest in DOEs is in their ability to provide fan out, achieving beam splitting (for instance as in [114]) and 3D advanced interconnections [115] (Fig. 32). The structure proposed in Fig. 19 also has DOEs. A popular technique to fabricate DOEs is by holography. An holographic optical element is a DOE *fabricated* by using the interference of two interfering beams [113].

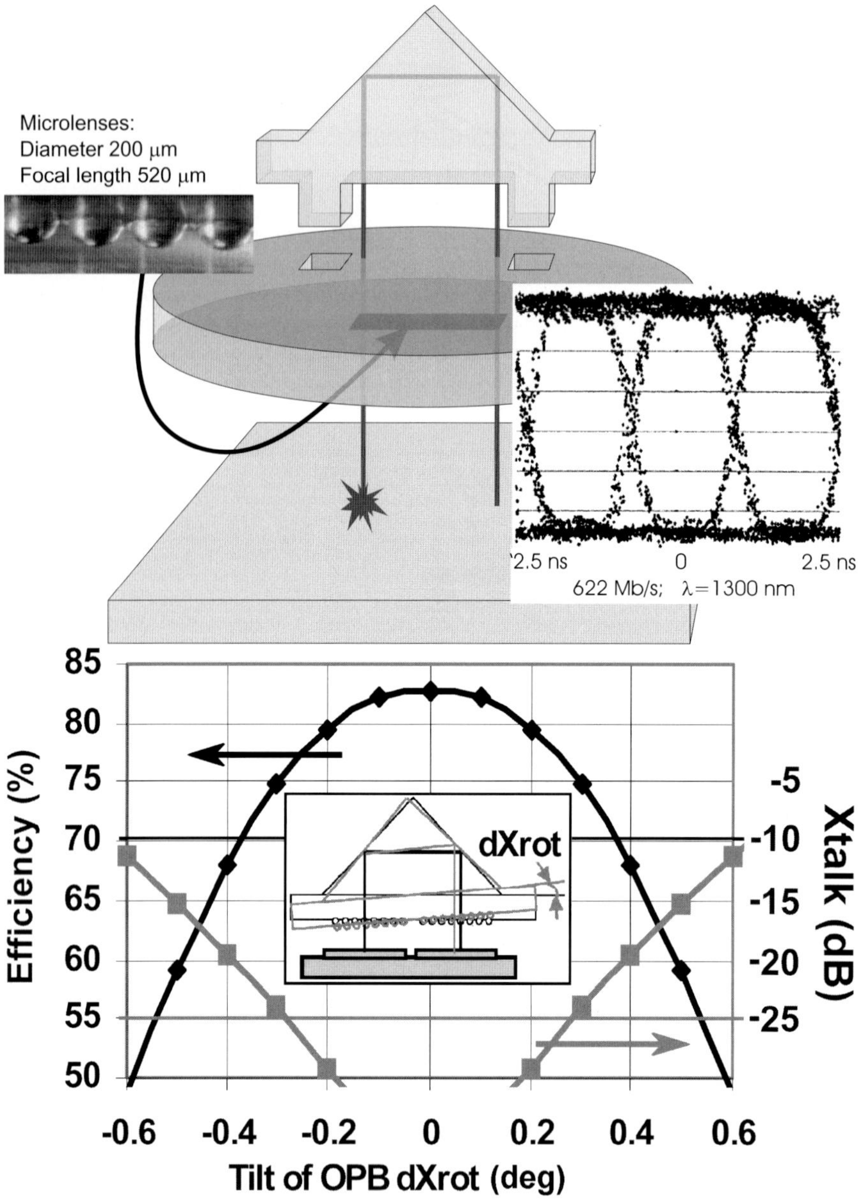

Fig. 33. Schematic realization of a prism-bridge with alinement features. Both the prism and the baseplate have been fabricated with PolyMethylMetAcrylate. The device operation is of the same type of the one shown in Fig. 30 (p. 164). The photograph shows a detail of the microlenses. The eye diagram was obtained by mimicking a VCSEL emission with a single mode fiber (NA = 0.11), and collecting the return beam with a multimode fiber (core diameter 50 μm, NA = 0.2). Neighboring channel crosstalk was between −22 dB and −27 dB. The *bottom plot* shows the calculated sensitivity of optical throughput and of crosstalk to angular misalignments. Adapted from [116, 117]

5.5 Guided Optical Interconnect

Integrated optical waveguides are reviewed in the Chapter by *Bozeat* et al. in this book. The large refractive index contrast ($\Delta n \simeq 2$) between Si ($n_{\mathrm{Si}} \simeq 3.45$) and SiO_2 ($n_{\mathrm{SiO_2}} \simeq 1.46$) has suggested the possibility of fabrication of integrated waveguides on Si chips for several years [118]. Large contrast is a necessary ingredient to allow small bending radii. However, there is a trade-off in using large index contrasts: as Δn increases, scattering losses from edge roughness become a major concern [119]. *Lee* et al. have shown a possible technique to reduce such kind of scattering losses [120]. In the case of Si/SiO_2 waveguides, the achievable smallest radius is of the order of 1 μm. Since both Si and SiO_2 are fairly transparent to the wavelength range 1.3 μm to 1.5 μm, the system Si/SiO_2 is appealing for integrated waveguide for communication wavelengths [121].

Silicon-On-Insulator (SOI) technology is interesting for waveguide fabrication because the Si layer constitutes a waveguide (cladding layers are provided by the SiO_2 layer underneath and air above). The reader is directed to [122] for an introduction on SOI and to [123, 121] for reviews on SOI optoelectronics. Other promising Si-based materials for integrated waveguides are Si nitride (Si_3N_4) and Si oxynitride (SiON) [124].

The requirement of Si-based materials for the waveguides for integrated optical circuits is relaxed in SoP and MCM technologies. In particular, polymer-based waveguides have attracted much attention because of their easy processing, compatibility with inorganic substrates and moderate losses. For a comprehensive starting point, the reader is directed to the recent review by *Ma* et al. [125]. The use of polyimide for H-tree clock distribution has already been mentioned in Sect. 5.2.2. Other used materials are polyurethane, epoxy, photoresist, polymethyl methacrylate (PMMA), polycarbonate, cyclobutene and polyacrylate. Processing usually involves spin or dip coating of a substrate, followed by lithography to define the core layer, and by the eventual deposition of other overcladding materials.

5.6 Alignment in Integrated Optical Circuits

One of the main challenges in hybrid and free-space optical interconnect is represented by the sensitivity of architectures to misalignment, which must typically be in length within few μm and in angles within fraction of degrees, and must be maintained in spite of temperature variations and vibrations [126].

For the case of free-space interconnect, *Thienpont* et al. have studied possible structures which can be alignment tolerant [116] (Fig. 33).

An interesting strategy which can be robust to misalignment is light emission and collection *through* the substrate. On example of such idea is shown in Fig. 34 [127]. The key idea is the integration of coupling optical elements (microlenses) on the back side of the substrate, which keeps the designed

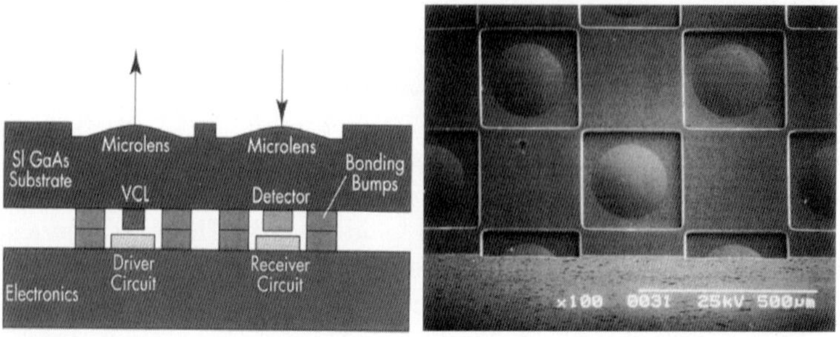

Fig. 34. Concept of through-the-substrate optical interconnect (*left*) and SEM micrograph of microlenses (*right*). The emitting vertical laser and the detector were epitaxially grown onto a semi-insulating GaAs substrate. The laser emission was red-shifted with respect to the GaAs band-gap ($E_{\mathrm{GaAs}} \simeq 1.42\,\mathrm{eV}$) by growing InGaAs quantum wells as active material, and tuning the cavity to $E_{\mathrm{laser}} \simeq 1.25\,\mathrm{eV}$. Substrate was therefore transparent to laser emission. After [127]

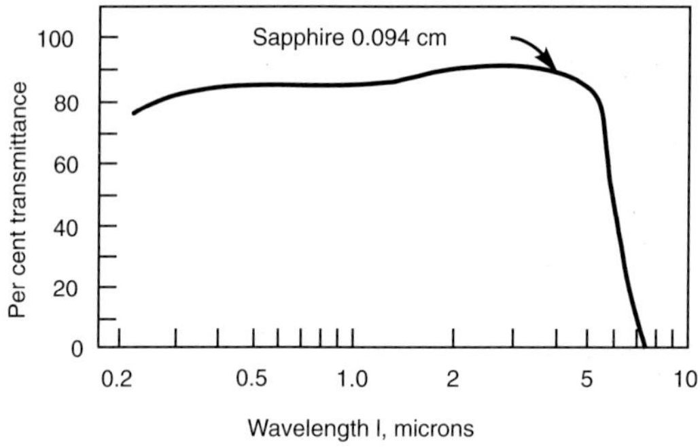

Fig. 35. Transmission spectrum of sapphire over wavelength range from UV to IR (*bottom*). From [128]

distance simultaneously avoiding most of misalignment problems of separate coupling designs. In the case of Fig. 34, microlenses were fabricated achieved using mask-shape transfer in chlorine reactive ion etching.

The recent introduction of Si-on-Sapphire (SOS) technology is particularly intriguing for through-the-substrate interconnects, because of the wide transmission window of sapphire (Fig. 35).

6 Conclusions

The appealing feature of optical interconnect is the very large bandwidth potentiality. In order to take the optical leap, however, the ability of efficient handling of optical signal at low cost is required.

At present, such an objective seems quite ambitious for Si technology. Optical functionality, however, is sneaking in, though not at the transistor scale. According to common understanding, as discussed in this chapter, such extreme scale is still penalized by electro-optical conversions. On a positive perspective, however, the situation is reminiscent of early modern Si electronic industry, when the devices were first born as bulky stand-alone elements. Integration started up then, and within few years Moore's law was stated. Perhaps this historical analogy can let us figure a restatement of such law for photonics in the near future.

Acknowledgements

I acknowledge the help of Prof. Jeffrey Davis for his advices, and the professional and cheering guidance of my research director, Prof. Lorenzo Pavesi.

References

1. G. E. Moore: Cramming more components onto integrated circuits, Electronics **38**, 114–117 (1965)
2. G. E. Moore: No exponential is forever – But we can delay 'forever'. Keynote presentation, in *International Solid-State Circuits Conference (ISSCC)* (San Francisco 2003)
3. G. Marcyk: Breaking Barriers to Moore's Law, in *Intel Developer Forum* (Intel Corp., Web address: http://www.intel.com/labs/microarch/publications.htm 2002)
4. D. A. B. Miller: Rationale and Challenges for Optical Interconnects to Electronic Chips, Proc. IEEE **88**, 728–749 (2000)
5. International Technology Roadmap for Semiconductors, 2001 EDITION, INTERCONNECT, http://public.itrs.net/
6. D. Sylvester, K. Keutzer: Impact of Small Process Geometries on Microarchitectures in Systems on a Chip, Proc. IEEE **89**, 467–489 (2001)
7. R. H. Havemann, J. A. Hutchby: High-Performance Interconnects: An Integration Overview, Proc. IEEE **89**, 586–601 (2001)
8. A. F. J. Levi: Optical Interconnects in Systems, Proc. IEEE **88**, 750–757 (2000)
9. B. Warneke, M. Last, B. Liebowitz, K. S. J. Pister: Smart Dust: Communicating with a Cubic-Millimeter Computer, Computer **34**, 41–51 (2001)
10. T. A. Schmedake, F. Cunin, J. R. Link, M. J. Sailor: Standoff Detection of Chemicals using Porous Silicon "Smart Dust" Particles, Advanced Materials **14**, 1270–1272 (2002)
11. J. R. Link, M. J. Sailor: Smart dust: Self-assembling, self-orienting photonic crystals of porous Si, Proceedings of the National Academy of Sciences **100**, 10607–10610 (2003)
12. S. John, M. Florescu: Photonic bandgap materials: towards an all-optical micro-transistor, J. Opt. A: Pure Appl. Opt. **3**, S103–S120 (2001)
13. D. Sylvester, C. Hu: Analytical Modeling and Characterization of Deep-Submicrometer Interconnect, Proc. IEEE **89**, 634–664 (2001)
14. K. Buchanan: The evolution of interconnect technology for silicon integrated circuitry, in *GaAsMANTECH Conference* (2002)

15. J. A. Davis: *A Hierarchy of Interconnect Limits and Opportunities for Gigascale Integration (GSI)*, Ph.D. thesis, Georgia Institute of Technology (1999)
16. J. D. Meindl, J. A. Davis, P. Zarkesh-Ha, C. S. Patel, K. P. Martin, P. A. Kohl: Interconnect opportunities for gigascale integration, IBM Journal of Research & Development **46**, 245–263 (2002)
17. K. Maex, M. R. Baklanov, D. Shamiryan, F. Iacopi, S. H. Brongersma, Z. S. Yanovitskaya: Low dielectric constant materials for microelectronics, Journal of Applied Physics **93**, 8793–8841 (2003)
18. J. D. Meindl: Interconnect Opportunities for Gigascale Integration, IEEE Micro **23**, 28–35 (2003)
19. R. H. Dennard, F. H. Gaensslen, H.-N. Yu, V. L. Rideout, E. Bassous, A. R. LeBlanc: Design of Ion-Implanted MOSFETs with Very Small Dimensions, IEEE J. Solid-State Circuits **SC-9**, 256–268 (1974)
20. P. Pezzati: DMS Interconnect Effects on System-on-Chip Design Flows, Electronics Systems and Sofware **1**, 14–19 (2003)
21. T. Sakurai, A. R. Newton: Alpha-power law MOSFET model and its application to CMOS inverter delay and other formulas, IEEE Journal of Solid State Circuits **25**, 584–594 (1990)
22. G. I. Yayla, P. J. Marchesand, S. C. Esener: *Optical Interconnections and Parallel Processing: Trends at the Interface* (Kluwer 1998), chap. 3, pp. 49–116
23. H. J. M. Veendrick: Short-circuit dissipation of static CMOS circuitry and its impact on the design of buffer circuits, IEEE Journal of Solid State Circuits **SC-19**, 468–474 (1984)
24. C. Hu: Device and technology impact on low power electronics, in J. Rabaey (Ed.): *Low Power Design Methodologies* (Kluwer, Boston, MA 1996), p. 2135
25. J. R. Black: Electromigration: A brief survey and some recent results, IEEE Trans. Electron Devices **ED-16**, 338–347 (1969)
26. K. Banerjee, M. Pedram, A. H. Ajami: Analysis and Optimization of Thermal Issues in High-Performance VLSI, in *ISPD01, Sonoma, California, USA* (Sonoma, CA 2001), pp. 230–237
27. G. Chandra, P. Kapur, K. Saraswat: Scaling Trends for the on chip Power Dissipation, in *IEEE* (2002), pp. 170–172
28. D. A. B. Miller: Optical interconnects to silicon CMOS, in *Device Research Conference, 25–27 June 2001* (2001)
29. C. Svensson: Electrical Interconnects Revitalized, IEEE Trans. on VLSI systems **10**, 777–788 (2002)
30. J. M. V. Campenhout: Solving the interconnect bottleneck Optoelectronic FPGAs (invited), in *Proc. of the 1998 IEEE/LEOS Summer Topical Meeting on Smart Pixels* (IEEE, Monterey 1998), pp. 21–22
31. D. Gajski, R. Kuhn: New VLSI tools, IEEE Computer **6**, 11–14 (1983)
32. A. Naeemi, J. D. Meindl: An optimal partition between on-chip and on-board interconnects, in *Proc. IEEE International Interconnect Technology Conference* (San Francisco, CA 2001), pp. 131–133
33. G. Tröster: Potentials of Chip-Package Co-Design for High-Speed Digital Applications, in *Proc. DATE99* (Association for Computing Machinery, Inc (ACM) 1999)
34. W. J. Dally, B. Towles: Route Packets, Not Wires: On-Chip Interconnection Networks, in *DAC 2001* (Las Vegas 2001), pp. 684–689

35. B. A. Floyd, C.-M. Hung, K. K. O: Intra-Chip Wireless Interconnect for Clock Distribution Implemented With Integrated Antennas, Receivers, and Transmitters, IEEE Journal of Solid State Circuits **37**, 543–552 (2002)
36. M. El-Sayed, J. Jaffe: A View of Telecommunications Network Evolution, IEEE Communications Magazine **40**, 74–81 (2002)
37. R. Jordan, C. T. Abdallah: Wireless communications and networking: An overview, IEEE Antenna's and Propagation Magazine **44**, 185–193 (2002)
38. A. C. Davies: An overview of Bluetooth Wireless Technology^{TM} and some competing LAN standards (Invited Plenary Lecture) (2002), pp. 206–211
39. A. Papoulis: *The Fourier integral and its applications*, Sciences (McGraw-Hill, New York, NY 1962)
40. T. Sakurai: Closed-form expression for interconnection delay, coupling, and crosstalk in VLSI's, IEEE Transactions on Electron Devices **40**, 118–124 (1993)
41. T. Dhaene, D. D. Zutter: Selection of lumped element models for coupled lossy transmission lines, IEEE Trans. Computer-Aided Design **11**, 805–815 (1992)
42. M. Ismail, N. Tan: Modeling Techniques for Energy-Efficient System-on-a-Chip Signaling, IEEE Circuits and Devices Magazine **19**, 8–17 (2003)
43. M. Ghulinyan, C. J. Oton, G. Bonetti, Z. Gaburro, L. Pavesi: Free-standing porous silicon single and multiple optical cavities, J. Appl. Phys. **93**, 9724–9729 (2003)
44. D. A. B. Miller: Optics for Digital Information Processing, in D. A. Miller, M. Ebrahimzadeh, D. M. Finlayson (Eds.): *Semiconductor Quantum Optoelectronics* (Institute of Physics, Bristol 1999), Proc. of the Fiftieth Scottish Universities Summer School in Physics, St. Andrews, 1998, pp. 433–461
45. D. A. Miller: Physical reasons for optical interconnection, Int. J. Optoelectron (1997)
46. O. Svelto: *Principles of Lasers*, 2 edn. (Plenum Press, New York 1982)
47. J. Meindl. Interconnect limits on ultra-large-scale integration, VLSI 1985 pp. 13–19 (1985)
48. J. A. Davis, R. Venkatesan, A. Kaloueros, M. Beylansky, S. Shukri J, K. Banerjee, K. C. Saraswat, A. Rahman, R. Reif, J. D. Meindl: Interconnect Limits on Gigascale Integration (GSI) in the 21st Century, Proc. IEEE **89**, 305–324 (2001)
49. V. W. S. Chan: Optical Space Communications, IEEE Journal on Selected Topics in Quantum Electronics **6**, 959–975 (2000)
50. C. E. Shannon: A mathematical theory of communication, The Bell System Technical Journal **27**, 379–423 and 623–656 (1948)
51. A. Papoulis: *Probability, Random Variables, and Stochastic Processes* (McGraw-Hill 1984)
52. V. A. Vilnrotter, C.-W. Lau: Quantum Detection Theory for the Free-Space Channel, The Interplanetary Network Progress Report **42**, 146 (2001)
53. M. Sasaki, J. Mmizuno, M. Fujiwara: Quantum detection circuit for quantum channel coding, Journal of the Communications Research Laboratory **49**, 105–118 (2002)
54. C. M. Caves, P. D. Drummond: Quantum limits on bosonic communication rates, Reviews of Modern Physics **66**, 481–537 (1994)
55. P. Hausladen, R. Jozsa, B. Schumacher, M. Westmoreland, W. K. Wootters: Classical information capacity of a quantum channel, Phys. Rev. A **54**, 1869–1876 (1996)

56. A. S. Holevo: The capacity of the quantum channel with general signal states (LANL e-print quant-ph/9611023, 1996.), IEEE Trans. Inf. Theory **44**, 269–273 (1998)
57. B. Schumacher, M. D. Westmoreland: Sending classical information via noisy quantum channels, Phys. Rev. A **56**, 131–138 (1997)
58. R. C. Frye, K. L. Tai, M. Y. Lau, T. J. Gabbara: Trends in Silicon-on-Silicon Multichip Modules, IEEE Design and Test of Computers **10**, 8–17 (1993)
59. H. K. Tsang, C. S. Wong, T. K. Liang, I. E. Day, S. W. Roberts, A. Harpin, J. Drake, M. Asghari: Optical dispersion, two-photon absorption and self-phase modulation in silicon waveguides at 1.5 μm wavelength, Applied Physics Letters **80**, 416–418 (2002)
60. M. Notomi, K. Yamada, A. Shinya, J. Takahashi, C. Takahashi, I. Yokohama: Extremely Large Group-Velocity Dispersion of Line-Defect Waveguides in Photonic Crystal Slabs, Physical Review Letters **87**, 253902–1 253902–4 (2001)
61. T. F. Krauss: Planar photonic crystal waveguide devices for integrated optics, Phys. Stat. Sol. (a) **197**, 688–702 (2003)
62. P. Kapur, K. Saraswat: Comparisons Between Electrical and Optical Interconnects For On-Chip Signaling, in *Proc. IEEE Interconnect Technology Conference* (2002), pp. 89–91
63. G. A. Keeler, D. Agarwal, C. Debaes, B. E. Nelson, N. C. Helman, H. Thienpont, D. A. B. Miller: Optical Pump-Probe Measurements of the Latency of Silicon CMOS Optical Interconnects, IEEE Photonics Technology Letters **14**, 1214–1216 (2002)
64. D. A. B. Miller, D. S. Chemla, S. Schmitt-Rink: Electric Field Dependence of Optical Properties of Semiconductor Quantum Wells: Physics and Applications, in H. Haug (Ed.): *Optical Nonlinearities and Instabilities in Semiconductors* (Academic Press, Inc., Boston 1988), pp. 325–359
65. S. Schmitt-Rink, D. S. Chemla, D. A. B. Miller: Linear and nonlinear optical properties of semiconductor quantum wells, Adv. Phys. **38**, 89–188 (1989)
66. D. A. B. Miller: Integrated Quantum Well Switching Devices, in H. Haug, L. Banyai (Eds.): *Optical Switching in Low-Dimensional Systems* (Plenum Press, New York 1989), pp. 1–8
67. A. V. Mulé, A. Naeemi, E. N. Glytsis, T. K. Gaylord, J. D. Meindl: Towards a Comparison Between Chip-level Optical Interconnection and Board-level Exterconnection, in *Proc. IEEE Interconnect Technology Conference* (2002), pp. 92–94
68. B. S. Landman, R. L. Russo: On a pin versus block relationship for partitioning logic graphs, IEEE Transactions on Computers **20**, 1469–1479 (1971)
69. W. E. Donath: Wire length distribution for placements of computer logic, IBM Journal of Research and Development **25**, 152–155 (1981)
70. H. M. Ozaktas: Connectivity Models for Optoelectronic Computing Systems, in *Workshop on Optics and Computer Science (WOCS), in International Parallel and Distributed Processing Symposium (IPDPS)* (Cancun, Mexico 2000)
71. H. M. Ozaktas: Comparison of Fully Three-Dimensional Optical, Normally Conducting and Superconducting Interconnections, in P. Berthomé, P. Marchand (Eds.): *Second Workshop on Optics and Computer Science (WOCS)* (Geneve (Switzerland) 1997), vol. 11th International Parallel Processing Symposium (IPPS)

72. J. H. Collet, D. Litaize, J. M. V. Campenhout, C. Jesshope, M. Desmulliez, H. Thienpont, J. Goodman, A. Louri: Architectural approach to the role of optics in monoprocessor and multiprocessor machines, Appl. Opt. **39**, 671–682 (2000)
73. S. Sahni: Models and algorithms for optical and optoelectronic parallel computers, International Journal Foundations in Computer Science **12**, 249–264 (2001)
74. M. Feldman, R. Vaidyanathan, A. El-Amawy: High Speed, High Capacity Bused Interconnects Using Optical Slab Waveguides, in *Proc. 1999 Workshop on Optics in Computer Science* (Springer-Verlag 1999), Lect. Notes Comput. Sc., vol. 1586, pp. 924–937
75. J. Jahns, A. Huang: Planar integration of free-space optical components, Appl. Opt. **28**, 1602–1605 (1989)
76. D. Fey, W. Erhard, M. Gruber, J. Jahns, H. Bartelt, G. Grimm, L. Hoppe, S. Sinzinger: Optical Interconnects for Neural and Reconfigurable VLSI Architectures, Proc. IEEE **88**, 838–848 (2000)
77. A. C. Walker, M. P. Y. Desmulliez, M. G. Forbes, S. J. Fancey, G. S. Buller, M. R. Taghezadeh, J. A. B. Dines, C. R. Stanley, G. Pennelli, A. R. Boyd, P. Horan, D. Byrne, J. Hegarty, S. Eitel, H.-P. Gauggel, K.-H. Gulden, A. Gauthier, P. Benabes, J.-L. Gutzwiller, M. Goetz: Design and construction of an optoelectronic crossbar switch containing a terabit/s free-space optical interconnect, IEEE J. Sel. Top. Quantum Electron. **5**, 236–249 (1999)
78. M. Forbes, J. Gourlay, M. Desmulliez: Optically interconnected electronic chips: a tutorial and review of the technology, Electronics and Communication Engineering Journal **13**, 221–232 (2001)
79. A. V. Krishnamoorthy, A. L. Lentine, K. W. Goossen, J. A. Walker, T. K. Woodward, J. E. Ford, G. F. Aplin, L. A. D'Asaro, S. P. Hui, B. Tseng: 3-D integration of MQW modulators over active sub-micron CMOS circuits: 375 Mb/s transimpedance receiver-transmitter circuit, EEE Photon. Technol. Lett. **7**, 1288–1290 (1995)
80. R. Iyer, Y. S. Liu, G. C. Boisset, D. J. Goodwill, M. H. Ayliffe, B. Robertson, W. M. Robertson, D. Kabal, F. K. Lacroix, D. V. Plant: Design, implementation, and characterization of an optical power supply spot-array generator for a four-stage free-space optical backplane, Applied Optics **36**, 9230–9242 (1997)
81. E. Bruce: The Communications Technology Roadmap, in *MIT Microphotonics Center* (2000)
82. L. Kimerling, E. Bruce, M. Beals: Optoelectronic Integration in Silicon, Technical Working Group (TWG) Meeting, Summary Document (2002)
83. S. K. Tewksbury, L. A. Hornak: Optical Clock Distribution in Electronic Systems, The Journal of VLSI Signal Processing–Systems for Signal, Image, and Technology **16**, 225–246 (1997)
84. D. Harris, S. Naffziger: Statistical Clock Skew Modeling With Data Delay Variations, IEEE Transaction on VLSI **9**, 888–898 (2001)
85. E. G. Friedman: Clock Distribution Networks in Synchronous Digital Integrated Circuits, Proc. IEEE **89**, 665–692 (2001)
86. P. J. Restle, T. G. McNamara, D. A. Webber, P. J. Camporese, K. F. Eng, K. A. Jenkins, D. H. Allen, M. J. Rohn, M. P. Quaranta, D. W. Boerstler, C. J. Alpert, C. A. Carter, R. N. Bailey, J. G. Petrovick, B. L. Krauter, B. D. McCredie: A Clock Distribution Network for Microprocessors, IEEE Journal of Solid State Circuits **36**, 792–799 (2001)

87. A. Bhatnagar, C. Debaes, R. Chen, N. C. Helman, G. A. Keeler, D. Agarwal, H. Thienpont, D. A. B. Miller: Receiver-less Clocking of a CMOS Digital Circuit Using Short Optical Pulses, in *Lasers and Electro-Optics Society. LEOS 2002. The 15th Annual Meeting of the 10–14 november* (2002), vol. 1, pp. 127–128
88. D. A. B. Miller: Optics joined with silicon?, in *Photonic Interconnect Seminars* (Stanford Photonics Research Center 2003), Public Lecture, Topics in International Advanced Technology Research
89. A. V. Mulé, S. M. Schultz, E. N. Glytsis, T. K. Gaylord, J. D. Meindl: Input Coupling and Guided-wave Distribution Schemes for Board-level Intra-chip Guided-Wave Optical Clock Distribution Network Using Volume Grating Coupler Technology, in *IITC* (2001)
90. L. Wu, B. Bihari, J. Gan, R. T. Chen: Fabrication and characterization of a 1-to-48 Fanout H-tree Structure for Clock Signal Distribution System, OSA Topical meeting on Integrated Photonics Research **4**, 282–284 (1998)
91. T. F. A. Sakai, T. Baba: H-Tree-Type Optical Clock Signal Distribution Circuit Using a Si Photonic Wire Waveguide, Jpn. J. Appl. Phys. (2) **41**, L1461–L1463 (2002)
92. J. F. Labourdette: Opaque and Transparent Networking, Optical Networks Magazine **4**, 3– (2003)
93. G. N. Rouskas, L. Xu: Chapter 1: Optical Packet Switching, in K. Sivalingam, S. Subramaniam (Eds.): *Optical WDM Networks: Past Lessons and Path Ahead* (Kluwer, Norwell, Massachusetts 2003)
94. T. Durhuus, B. Mikkelsen, C. Joergensen, S. L. Danielsen, K. E. Stubkjaer: All-optical wavelength conversion by semiconductor optical amplifiers, IEEE Journal of Lightwave Technology **14**, 942–954 (1996)
95. D. J. Blumenthal, J. E. Bowers, L. Rau, H.-F. Chou, S. Rangarajan, W. Wang, H. N. Poulsen: Optical signal processing for optical packet switching networks, IEEE Communications Magazine **41**, S23–S29 (2003)
96. F. Ahmed, N. Kishi: All-fiber wavelength conversion of ultra-fast signal with enhanced extinction ratio using stimulated Raman scattering, Optical Review **10**, 43–46 (2003)
97. D. J. Blumenthal, B.-E. Olsson, G. Rossi, T. E. Dimmick, L. Rau, M. M. sanović, O. Lavrova, R. Doshi, O. Jerphagnon, J. E. Bowers, V. Kaman, L. A. Coldren, J. Barton: All-optical label swapping networks and technologies, IEEE Journal of Lightwave Technology **18**, 2058–2075 (2000)
98. M. Gerken, D. A. B. Miller: Wavelength Demultiplexer Using the Spatial Dispersion of Multilayer Thin-Film Structures, IEEE Photonics Technology Letters **15**, 1097–1099 (2003)
99. B. E. Nelson, G. A. Keeler, D. Agarwal, N. C. Helman, D. A. B. Miller: Demonstration of a wavelength division multiplexed chip-to-chip optical interconnect (2002), paper CTuF6
100. D. W. Prather, A. S. Sharkawy, S. Shi, R. A. Soref: Electro-optical 2×2 switching in a photonic bandgap waveguided coupler, Proc. SPIE **4653**, 45–48 (2002)
101. J. McGrath: Sorting Stacked-die. The Challenges of Building SiPs, Advanced Packaging (http://ap.pennnet.com/home.cfm) (2003)
102. S. K. Tewksbury, L. A. Hornak, H. E. Nariman, S. M. Langsjoen, S. P. McGinnis: Cointegration of Optoelectronics and Submicron CMOS, Proc., Fifth Annual IEEE International Conference on Wafer Scale Integration **93CH3227-6**, 358–367 (1993)

103. L. Pavesi, L. D. Negro, C. Mazzoleni, G. Franzò, F. Priolo: Optical gain in silicon nanocrystals, Nature **408**, 440–444 (2000)
104. W. L. Ng, M. A. Lourenço, R. M. Gwilliam, S. Ledain, G. Shao, K. P. Homewood: An efficient room-temperature silicon-based light-emitting diode, Nature **2001**, 192–194 (2001)
105. M. A. Green, J. Zhao, A. Wang, P. J. Reece, M. Gal: Efficient silicon light-emitting diodes, Nature **412**, 805–808 (2001)
106. F. Iacona, D. Pacifici, A. Irrera, M. Miritello, G. Franzò, F. Priolo, D. Sanfilippo, G. D. Stefano, P. G. Fallica: Electroluminescence at 1.54 μm in Er-doped Si nanocluster-based devices, Appl. Phys. Lett. **81**, 3242–3244 (2002)
107. M. E. Castagna, S. Coffa, L. Caristia, A. Messina, C. Bongiorno: Quantum Dot Materials and Devices for Light Emission in Silicon (2002), p. 439
108. A. Mulé, M. Bakir, J. Jayachandran, R. Villalaz, H. Reed, K. Martin, P. Kohl, E. Glytsis, T. Gaylord, J. Meindl, N. Agrawal, S. Ponoth, J. Plawsky, P. Persans: Optical Waveguides Integrated Within a Sea-of-Leads (SoL) Wafer-Level Package, in *IEEE International Interconnect Technology Conference* (2002)
109. M. S. Bakir, H. A. Reed, A. V. Mulé, J. P. Jayachandran, P. A. Kohl, K. P. Martin, T. K. Gaylord, J. D. Meindl: Chip-to-Module Interconnections Using "Sea of Leads" Technology, MRS Bullettin **28**, 61–67 (2003)
110. D. A. B. Miller: Optical interconnect technologies for Si ULSI, in *Electron Devices Meeting, 1997. Technical Digest.* (Washington, DC 1997), pp. 343–347
111. M. Gruber, J. Jahns, S. Sinzinger: Planar-integrated optical vector-matrix-multiplier, Appl. Opt. **39**, 5367–5373 (2000)
112. L. Rudolph: Do parallel computers really need optical interconnection networks?, in *Proc. Fifth International Conference on Massively Parallel Processing* (1998), pp. 176–183
113. D. O'Shea, T. Suleski, A. Kathman, D. Prather: Diffracting Light, SPIE OE Magazine pp. 35–37 (2001)
114. G. A. Keeler, D. Agarwal, B. E. Nelson, N. C. Helman, D. A. B. Miller: Performance Enhancement of an Optical Interconnect Using Short Pulses from a Modelocked Diode Laser, Lasers and Electro-Optics (CLEO 2002) p. 163 (2002)
115. D. W. Prather: Three Dimensional VLSI-Scale Interconnects, in *Workshop on Optics and Computer Science (WOCS), in International Parallel and Distributed Processing Symposium (IPDPS)* (Cancun, Mexico 2000)
116. H. Thienpont, V. Baukens, C. Debaes, H. Ottevaere, P. Vynck, P. Tuteleers, G. Verschaffelt, B. Volckaerts, A. Hermanne, M. Hanney, M. Brunfaut, J. M. V. Campenhout, I. Veretennicoff: Scalable Micro-Optical Modules for Short Distance VLSI-Photonic Interconnections, in E. Towe (Ed.): *Heterogeneous integration* (2000), vol. SPIE Critical review CR76, pp. 251–256
117. H. Thienpont: Optical Interconnects to Silicon Chips, in *Photonic Interconnect Seminars* (Stanford Photonics Research Center 2003), Public Lecture, Topics in International Advanced Technology Research
118. R. A. Soref, J. Schmidtchen, K. Petermann: Large single-mode rib waveguides in GeSi and Si-on-SiO2, IEEE J. Quantum Electron. **27**, 1971–1974 (1991)
119. L. C. Kimerling: Silicon microphotonics: the next killer technology, in L. Pavesi, S. Gaponenko, L. D. Negro (Eds.): *Towards the first silicon laser* (Kluwer, Dordrecht 2003), Mathematics, Physics and Chemistry, vol. 93, pp. 465–476

120. K. K. Lee, D. R. Lim, L. C. Kimerling, J. Shin, F. Cerrina: Fabrication of Ultra-low Loss Si/SiO_2 Waveguides by Roughness Reduction, Optics Letters **26**, 1888 (2001)
121. L. C. Kimerling: Silicon Microphotonics, Appl. Surf. Sci. **159–160**, 8–13 (2000)
122. S. Cristoloveanu: SOI: a metamorphosis of silicon, IEEE Circuits and Devices Magazine **15**, 26–32 (1999)
123. B. Jalali, S. Yegnanarayanan, T. Yoon, T. Yoshimoto, I. Rendina, F. Coppinger: Advances in Silicon-on-Insulator Optoelectronics, IEEE Journal of Selected Topics in Quantum Electronics **4**, 938–947 (1998)
124. G.-L. Bona, R. Germann, B. J. Offrein: SiON high refractive- index waveguide and planar lightwave circuits, IBM Journal of Res. & Dev. **47**, 239–249 (2003)
125. H. Ma, A. K.-Y. Jen, L. R. Dalton: Polymer-Based Optical Waveguides: Materials, Processing, and Devices, Advanced Materials **14**, 1339–1365 (2002)
126. F. A. P. Tooley: Challenges in Optically Interconnecting Electronics, IEEE Journal on Selected Topics in Quantum Electronics **2**, 3–13 (1996)
127. D. A. Louderback, O. Sjölund, E. R. Hegblom, S. Nakagawa, J. Ko, L. A. Coldren: Modulation and Free-Space Link Characteristics of Monolithically Integrated Vertical-Cavity Lasers and Photodetectors with Microlenses, IEEE Journal of Selected Topics in Quantum Electronics **5**, 157–165 (1999)
128. A. G. Andreou, Z. K. Kalayjian, A. B. Apsel, P. O. Pouliquen, R. A. Athale, G. Simonis, R. Reedy: Silicon on sapphire CMOS for optoelectronic microsystems, IEEE Circuits and Systems Magazine **1**, 22–30 (2001)

Monolithic Silicon Light Sources

Philippe M. Fauchet

Department of Electrical and Computer Engineering, University of Rochester
Rochester NY 14627-0231, USA
fauchet@ece.rochester.edu

Abstract. Monolithic silicon light sources (LEDs and lasers) could have a significant impact when integrated on silicon chips. After a general introduction to light emission in silicon, this chapter discusses the mechanisms of light emission in bulk silicon, silicon quantum dots, silicon alloyed with germanium, and silicon doped with specific impurities such as erbium. Then, the fabrication and performance of LEDs using these materials are discussed. The chapter ends with a discussion of the prospects for a silicon laser.

1 Introduction

Light emission occurs when an electron recombines with a hole to produce a photon. The photon energy is equal to the band-gap energy if the electron is in the conduction band and the hole in the valence band. If both electron and hole are located at the same point in the Brillouin zone, the radiative recombination rate is large and the radiative lifetime τ_{rad} is short (typically a few ns). This is the case in direct gap semiconductors, such as GaAs. In indirect gap semiconductors such as Si, the bottom of the conduction band is not at the same location in reciprocal space as the top of the valence band. Radiative recombination across the band-gap thus requires the participation of a phonon, making it a second-order process. As a result, the radiative lifetime is much longer, typically in the msec regime.

The internal quantum efficiency is given by

$$\eta_{\text{in}} = \frac{\tau_{\text{non-rad}}}{\tau_{\text{non-rad}} + \tau_{\text{rad}}} \tag{1}$$

where $\tau_{\text{non-rad}}$ is the lifetime for the recombination pathways that do not involve the production of photons. Non-radiative pathways include recombination at bulk or surface defects and Auger recombination. In direct gap semiconductors, η_{in} can be very large, typically of the order of 10% and may even approach 100%, because the radiative lifetime can be shorter than the non-radiative lifetime [1]. However, in indirect gap semiconductors such as Si, the non-radiative lifetime is usually much shorter than the radiative lifetime and η drops by several orders of magnitude and typically is as low as 10^{-4}% to 10^{-5}%.

Several approaches have been proposed to improve the quantum efficiency of silicon. These approaches can be grouped into two categories. In the first

L. Pavesi, D. J. Lockwood (Eds.): Silicon Photonics, Topics Appl. Phys. **94**, 177–198 (2004)

category, non-radiative recombination pathways are suppressed. To achieve this goal, bulk and surface defects must be eliminated. In bulk silicon, this can be done by carefully passivating the surface [2] and by using high purity material [3]. In addition, the injected carrier concentration must remain low to eliminate Auger recombination, as the Auger rate increases with the third power of the injected carrier density [4]. In the second category, the radiative recombination rate is enhanced. This can be achieved by localizing the electron and hole wavefunctions in the same region of space, for example using quantum dots [5] or isoelectronic impurities [6].

Other possibilities include combining silicon with another elemental semiconductor and using specific light emitting centers. Alloying is typically done with germanium [7], although alloys of Si with C and Ge have also shown promise [8]. Silicon-germanium quantum wells superlattices have been successfully employed to achieve luminescence below the band-gap energy of silicon [9]. Transfer of an electron and a hole from silicon to a specific center can lead to efficient sub-band-gap luminescence. Erbium is by far the most widely used light emitting center in bulk silicon [10] as well as in nanocrystalline silicon [11].

Multiple studies of the photoluminescence (PL), especially its origin, in all these forms of silicon have been published. Very high PL efficiencies have been reported. However, light emitting devices must be driven by current and achieving efficient electroluminescence (EL) has proven more challenging. We need to distinguish three efficiencies: the internal quantum efficiency η_{in}, the external quantum efficiency η_{ext}, and the power efficiency η_{p}. The internal quantum efficiency expresses the potential for light emission. It is defined as the fraction of electron–hole pairs that recombine radiatively (see (1)). The external quantum efficiency takes into account the fact that not all the photons generated inside the structure can be detected outside it. For example, total internal reflection prevents photons emitted at an angle to the sample surface from escaping. This is especially a problem when the index of refraction is large, as the fraction of the light that can escape is approximately given by $1/2n^2$ [12]. Solutions to this problem include texturing the surface [3], adding an antireflection coating [13], or decreasing the average index of refraction by using silicon oxide instead of silicon. In addition, in a device structure, the internally emitted photons may be partially absorbed in heavily doped region or at metal contacts. Note that both the internal quantum efficiency and the external quantum efficiency can be measured in PL and EL experiments. Finally, the power efficiency is defined as the ratio of the output light power over the input electrical power. This is the most important quantity for evaluating how useful an LED might be. In most cases, it can simply be obtained by multiplying η_{ext} by the ratio of the emitted photon energy over the applied voltage.

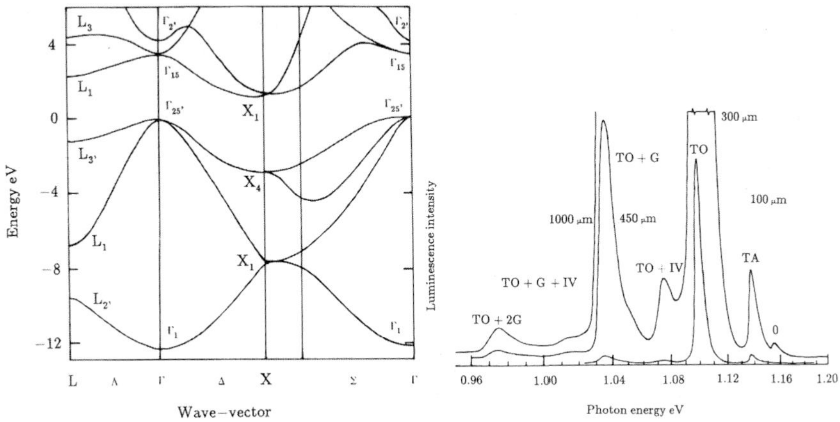

Fig. 1. Silicon band-structure and low-temperature photoluminescence spectrum. Radiative recombination without the participation of a phonon (peak labeled 0) is much weaker than radiative recombination involving at least one phonon. After *Davies* [15]

2 Mechanisms of Silicon Light Emission

Silicon is an indirect band-gap semiconductor [14]. The radiative recombination of an electron near the X-point and a hole at the Γ-point requires the participation of a phonon [15]. This has two major consequences. Because the process is a second order process, the recombination rate is orders of magnitude less than in a typical direct gap semiconductor such as GaAs. In addition, the luminescence lineshape reflects the participation of phonons as illustrated in Fig. 1.

At room temperature, light emission across the band-gap of bulk silicon results from the recombination of a free electron with a free hole. At lower temperatures however, electrons and holes are bound by Coulombic attraction and they form excitons. The volume of space through which an exciton can move before it recombines can be approximated as a sphere whose radius can exceed 100 μm in high quality samples. If this volume contains a non-radiative defect, the exciton recombines without emitting light.

By confining the exciton (or the electron and the hole) in a small volume that is free of defects, it is possible to enhance the quantum efficiency. The most common way to achieve such a confinement is to use small silicon nanocrystals. When the size of the nanocrystal drops below the Bohr exciton diameter ($\sim$ 10 nm), quantum confinement increases the value of the band-gap. Such small objects are called quantum dots. The simplest method to calculate the opening of the band-gap with decreasing size is the effective mass approximation, in which the bands are approximated by parabolas. Because of the complexity of the silicon band-structure however other methods such as tight-binding, pseudopotential, or ab initio local density approxima-

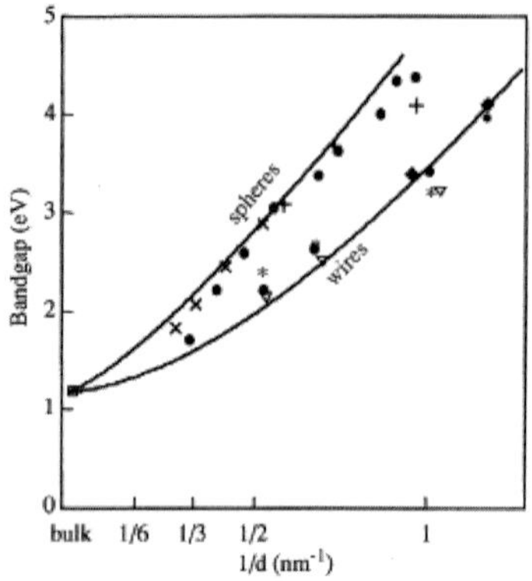

Fig. 2. Silicon quantum dot and quantum wire band-gap versus size calculated using an empirical tight binding approximation (*lines*) and other techniques. The silicon surface is passivated with hydrogen. After *Delerue* [17]

tions must be used to obtain more accurate values [16]. Figure 2 shows the calculated opening of the band-gap, defined as the energy separating the lowest conduction band state from the highest valence band state [17]. The band-gap increases as $d^{-1.3}$, where d is the quantum dot diameter, whereas the effective mass approximation would have predicted a d^{-2} dependence. The localization of the electron and hole wavefunctions in a dot of decreasing size also increases their spread in reciprocal space [5], hence increasing their overlap and the radiative recombination rate. Figure 3 shows how the recombination rate increases with decreasing size/increasing band-gap. Eventually, for very small dots, radiative recombination without participation of phonons should become more likely than radiative recombination involving a phonon. This cross-over is thought to occur for sizes ≤ 2 nm.

Quantum dots have a large surface to volume ratio, with up to 50% of atoms at or near the surface. It is thus not surprising that the quantum dot surface must be well passivated: even one non-radiative recombination center, such as a dangling bond, prevents luminescence (one defect per quantum dot corresponds to an equivalent bulk defect concentration of $\sim 10^{20}\,\text{cm}^{-3}$!). What would happen if the surface contained one center that is capable of producing light? If the states involved in the intra-center recombination are lined up inside the band-gap of the quantum dots, then the electron-hole pair is quickly captured by the center and recombination takes place at the specific energy of the intra-center transition. This is the case with some silicon-oxygen bonds and with erbium. Consider a H-passivated silicon quantum dot. If one Si = O bond forms on its surface, the states associated with this bond appear inside the band-gap provided that the dot size is small enough. Figure 4 shows that when this takes place, the measured luminescence energy increases very slowly and eventually saturates [18]. Further studies have shown that adding more Si = O bonds on the surface does not alter this picture [19].

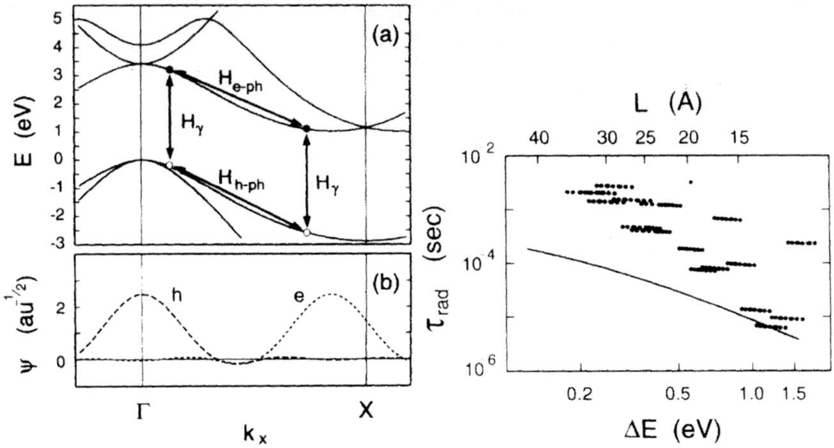

Fig. 3. Spread of the electron and hole wavefunction is a quantum dot and calculated radiative lifetime versus size for silicon quantum dots. The *line* corresponds to phonon-assisted recombination and the points to phonon-less recombination. After *Hybertsen* [5]

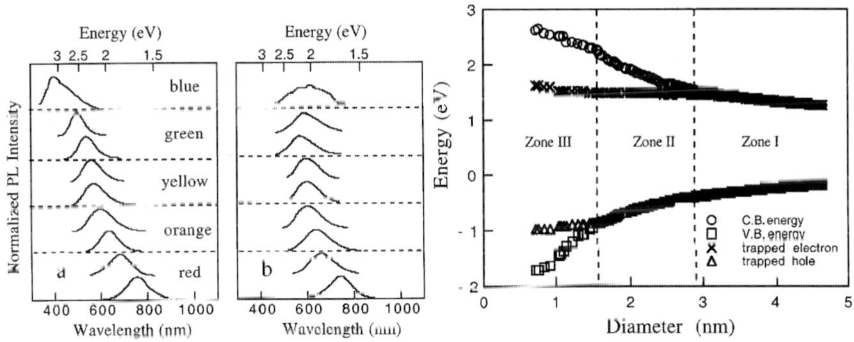

Fig. 4. Measured photoluminescence spectra in various porous silicon samples before exposure to any oxygen (*top left*) and after exposure to oxygen (*top right*). The samples are labeled according to their peak PL wavelength. The average quantum dot size decreases from $\geq 4\,\text{nm}$ (*red*) to $< 2\,\text{nm}$ (*blue*). Calculated conduction band and valence band energy levels for H-passivated Si quantum dots and calculated energy levels associated with a trapped electron and a trapped hole at a Si = O bond at the surface. As the size decreases and the band-gap increases by quantum confinement, the Si = O levels appear inside the band-gap. The expected PL energy difference between H-passivated Si quantum dots and quantum dots containing one Si = O surface bond increases as the size decreases below 2.5 nm to 3 nm, in agreement with the PL spectra. After *Wolkin* [18]

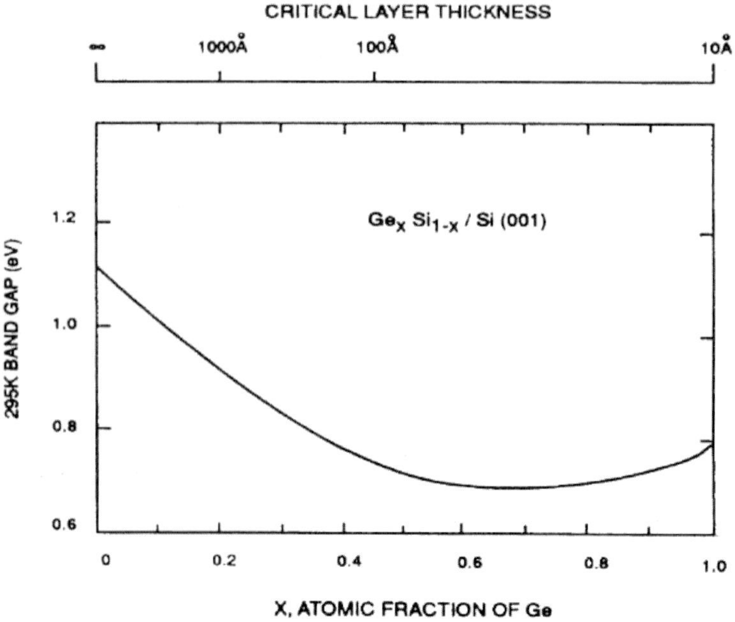

Fig. 5. Calculated band-gap energy for strained SiGe layers grown on silicon. Because the lattice mismatch between Si and the alloy increases with increasing Ge content, the critical thickness above which dislocations occur and strain is relaxed decreases dramatically. After *Pearsall* [22]

When erbium is on the surface or in close proximity to the quantum dot, energy transfer takes place between the quantum dot and erbium, resulting in efficient luminescence near 1.5 μm [20, 21].

Alloying silicon with germanium has also yielded interesting results. One approach consists of using thin (quantum wells) or thick silicon-germanium layers sandwiched between silicon barriers [22]. Luminescence is produced at the band-gap of the alloy, which can in principle be tuned from the band-gap of Si to that of Ge (Fig. 5). Because of the 4% lattice mismatch between Si and Ge, the thickness of the layers must remain small. Another approach, called Brillouin zone folding, is an attempt at modifying the band structure by using alternating layers of Si and Ge with a thickness of the order of a few atomic layers [9]. Both of these band-gap engineering approaches have so far failed to produce quantum efficiencies comparable to what is done routinely with silicon quantum dots, especially at room temperature.

3 Bulk Silicon LEDs

Silicon LEDs with high power efficiencies have been demonstrated [3]. To reach this value, it is necessary to take several steps. First, one uses silicon

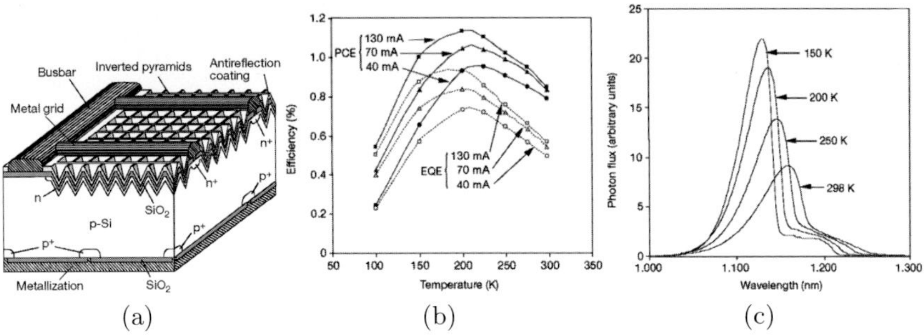

Fig. 6. (**a**) Schematic of a high-efficiency Si LED, showing the all the technological steps taken to maximize the efficiency, including the use of surface texturization. (**b**) Measured external quantum efficiency (EQE) and power efficiency (PCE) as a function of temperature for 3 current diodes. (**c**) Temperature dependence of the EL spectrum. After *Green* [3]

material with a long carrier lifetime indicative of a low defect density, such as wafers prepared by the float-zone method. Second, surface recombination is eliminated by using a high-quality surface oxide coating. Both of these steps increase η_{in}. Third, recombination and contact resistance are minimized by using small heavily-doped silicon/metal contact regions. This step increases η_{ext}. Fourth, the surface is textured in a way that is similar to the way the surface of highly efficient silicon solar cell is textured. Surface texturization improves the light extraction efficiency, thus minimizing the difference between η_{in} and η_{ext}. When these conditions are met, a power efficiency of 1% near room temperature has been demonstrated. Figure 6 shows how the surface of the LED is textured and the output spectrum and the efficiency versus temperature.

The advantage of this approach (compared to LEDs that use silicon quantum dots) is that electrical transport takes place in bulk silicon. Potential limitations of this approach can be summarized as follows. First, the maximum brightness of the device remains limited as the carrier lifetime is long and the injected carrier density must remain low enough to avoid Auger recombination. Second, direct modulation of the light output will be limited to speeds in the kHz regime as the carrier lifetime is long. Note that these two limitations are usually shared with all other monolithic silicon LEDs. Third, the requirements of high purity silicon, surface texturization, and relatively large device size make it difficult to integrate these LEDs into silicon microelectronic chips.

Another approach eliminates the need for high purity silicon. It consists of confining electrons and holes in small regions of space within the bulk silicon LED by introducing a strain field that locally modifies the band-gap [23]. This is achieved by introducing dislocation loops via ion implantation (boron

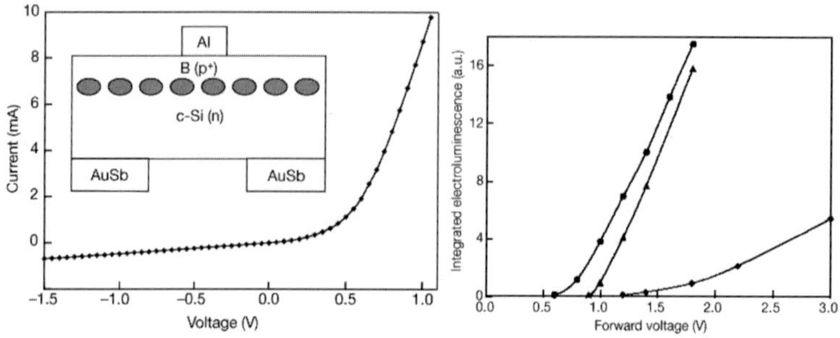

Fig. 7. Current-voltage characteristic and schematic of a Si LED containing dislocation loops in the active region. The integrated EL is plotted versus forward bias at three temperatures: from *left* to *right*, 300 K, 180 K, and 80 K. After *Ng* [23]

in n-type silicon) followed by thermal processing at 1000 °C in nitrogen. Each dislocation loop is tens of nm in size and calculations of the stress just outside the dislocation loop suggest that local band-gap around the loop can be increased by hundreds of meV [24]. This would be more than enough to confine the carriers inside the loop where they presumably will see no or very few defects and hence recombine efficiently at the silicon band-gap energy. Figure 7 shows the device cross-section and the dependence of the electroluminescence intensity on the forward voltage. Note that the EL spectra were similar to those of Fig. 6.

The advantage of this approach is that these LEDs can be fabricated using microelectronic-grade silicon. The potential limitations of this approach can be summarized as follows. First, the measured power efficiency is less 0.1% or less, although improvements would be expected with surface texturizing. Second, the brightness and modulation speeds are limited as in all monolithic Si LEDs. Third, an exact description of the microscopic phenomena is still lacking, and device-to-device reproducibility and device stability are not known.

4 Silicon Quantum Dot LEDs

Porous silicon containing silicon quantum dots has been widely used to make LEDs. Porous silicon is formed by the electrochemical dissolution of a silicon wafer (the anode) in a solution containing HF [25, 26]. When F^- and a hole injected from the wafer are both present at the interface, silicon dissolution takes place. If the current density is high or the HF concentration low, electropolishing takes place. Porous silicon is obtained by decreasing the current density or increasing the HF concentration such that the holes are the limiting species at the interface. The size of the pores and the silicon remnants can be tuned from the nanometer regime to beyond 1 μm [27]. Light emission is especially strong in "spongy" porous silicon that contains typical feature sizes in the nanometer range [28, 29]. Figure 8 shows the strong

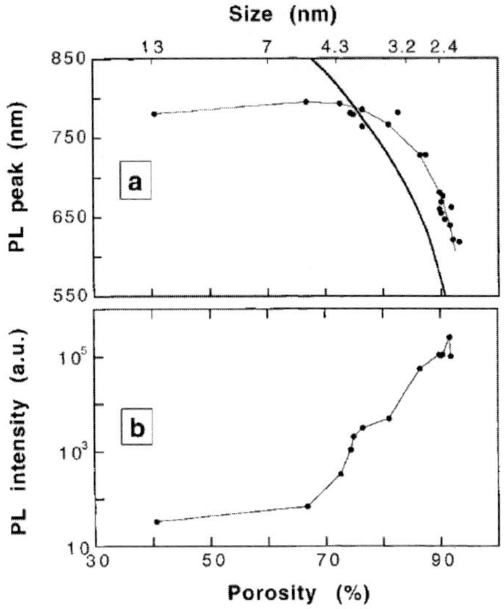

Fig. 8. Measured PL peak wavelength and PL intensity on various oxidized porous silicon films (the *full line* in graph a is a theoretical prediction). The porosity was measured gravimetrically and the average quantum dot size determined by Raman spectroscopy. The PL peak is in broad agreement with the results shown in Fig. 4 except for sizes above ~ 4.5 nm when Pl is due to defects inside the oxide. The PL peak intensity increases by several orders of magnitude as the average size decreases to ~ 2.5 nm. After *Fauchet* [30]

increase in PL efficiency as the porosity increases/average quantum dot size decreases [30]. Porous silicon is naturally passivated by Si H bonds. While these bonds provide excellent surface passivation, they are very fragile and susceptible to dissociation by many external stimuli such as heat or UV light exposure. Surface treatments such as oxidation [31] or formation of silicon-carbon bonds [32] improve the stability of the optical properties.

After the first demonstration of porous silicon LEDs, the power efficiency was quickly improved to ~ 0.01% [33]. This progress was made possible by better control of the materials science, especially the stability of silicon surface, and by improved device engineering to decrease the problems associated with charge transport in porous silicon and with the interface between porous silicon and a metal contact. After additional device optimization, power efficiencies in the 0.1% range were reported and stability in excess of 100 h was achieved [31,34]. It remained to be shown that porous silicon LEDs could be fabricated and integrated with silicon microelectronic circuitry. Integration of a porous silicon LED with a transistor was first reported in 1996 [35]. Figure 9 shows the device structure and its stability.

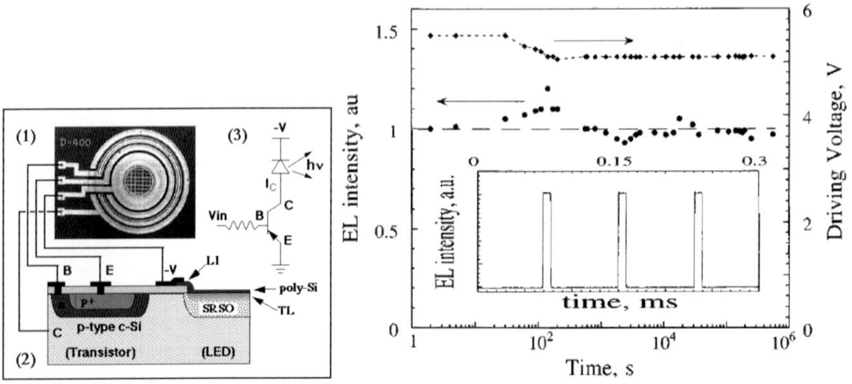

Fig. 9. Top-view picture, cross-sectional schematic, and equivalent circuit representation of a porous silicon LED integrated with a pnp transistor. The electroluminescence of a similar LED showed no degradation over days of operation under pulsed operation. After *Hirschman* [35] and *Tsybeskov* [31]

The major problems specific to porous silicon LEDs are the long-term stability, the achievable power efficiency, and the broad emission spectrum. Let us consider recent progress in each of these areas.

Steady work has continued to increase the efficiency. The external quantum efficiency of these devices was first brought up close to 1% but because the LEDs required high applied voltages ($> 10\,\mathrm{V}$), the power efficiency was still $\sim 0.1\%$ [36] With better device engineering power efficiencies very close to 1% have been achieved [37]. An important requirement for high efficiency is to keep the thickness of the porous layer to 1 μm or less. How the external quantum efficiency can be further increased remains a research topic.

In the early porous silicon LED work, little attention was paid to the passivation of the internal surface. Hence, the stability of the devices was poor. Significant progress was achieved via oxidation but the price to be paid is that transport is made more difficult as oxide barriers may be introduced. Alternative surface stabilization techniques based on silicon-carbon and silicon-carbon-oxygen bonds have been developed [38] and used to produce stable LEDs [39]. It will be interesting to see if this approach can also lead to higher efficiencies and if these LEDs can be integrated successfully with microelectronic fabrication technology.

The PL and EL spectrum of all samples containing silicon quantum dots, including porous silicon, is broad. The spectral width can be due to the quantum dot size distribution but it may also arise from the intrinsically broad luminescence linewidth of small silicon quantum dots [40, 41]. For many applications, a narrow linewidth is necessary. This can be achieved by using a microcavity. The porosity can be controlled over a wide range by changing several parameters during etching, especially the current density. Since etching requires holes at the Si/electrolyte interface and transport of holes

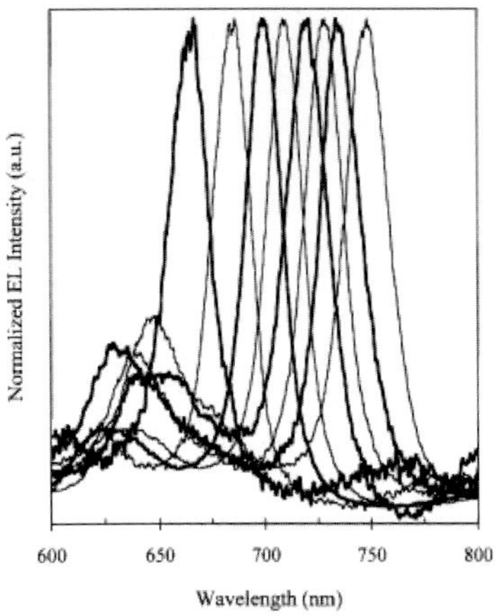

Fig. 10. Electroluminescence spectra measured with 8 porous silicon microcavity LEDs consisting of a thin light-emitting layer sandwiched between two Bragg reflectors. The porosity of the active layer and hence its refractive index was changed to achieve tuning over a 100 nm range. After *Chan* [44]

from the silicon wafer into the porous layer is preferentially to the pore tips, the properties of an existing porous layer are unaffected by further etching. Layers of different porosities can be produced by changing the current density during etching. This approach has led to anti-reflection coatings [13], Braggreflectors [42], and microcavities [43]. Microcavity LEDs made of a high porosity, light emitting layer sandwiched between two porous Braggreflectors have been demonstrated. Their EL spectra are shown in Fig. 10 [44]. Note that the PL spectrum of porous silicon microcavities can be narrowed to below 1 nm [45,46]. The LEDs can be made to cover a large region of wavelengths by simply changing the porosity and/or the thickness of the layers. They are tunable in the same sense as $Al_xGa_{1-x}As$ LEDs are tunable (changing the alloy content is equivalent to changing the porosity or the layer's thickness). The main limitation of this approach is that the power efficiency of these microcavities is very low because charge transport over several microns of porous silicon is difficult.

There are many methods to produce silicon quantum dots and all of them can lead to efficient PL [47]. Two of the most popular approaches besides porous silicon are annealing of silicon-rich SiO_2 and annealing of amorphous Si/SiO_2 superlattices. Silicon-rich SiO_2 films can be produced by implantation of SiO_2 with high doses of silicon ions or by directly depositing a substoichiometric oxide for example by plasma enhanced chemical vapor deposition. In both cases, annealing at high temperature (typically 1000 °C) is required to form silicon clusters, crystallize them, and passivate them. There are many similarities between annealed silicon-rich SiO_2 and porous silicon. The average size of the silicon quantum dots can be varied from $\sim$ 2 nm

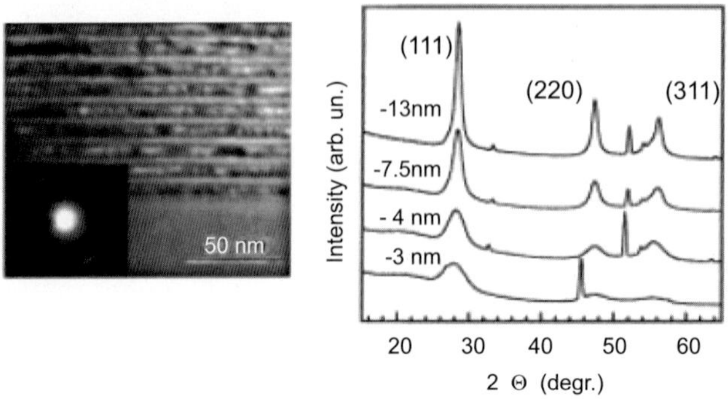

(a) (b)

Fig. 11. Cross-sectional TEM picture of an annealed and crystallized a-Si/SiO_2 superlattice. The SiO_2layers appear as the continuous *bright* layers, while the crystallized silicon layers contain *bright* and *dark* spots, corresponding to quantum dots of various orientations. The X-ray data shown on the *right*, taken on similar structures with different thicknesses of the Si layers, confirms the formation of quantum dots whose sizes are given by the thickness of the a-Si layers. After *Tsybeskov* [48]

to sizes where quantum confinement stops being important. The quantum dots are produced with a range of sizes characterized by a standard deviation that is 20% to 50% of the average. In contrast, annealing of amorphous Si/SiO_2 superlattices grown by techniques such as sputtering produces multiple layers made of a densely-packed array of silicon quantum dots whose diameters are uniform, equal to the thickness of the a-Si layers [48, 49]. Figure 11 shows a typical structure. By substituting non-stoichiometric oxide for the a-Si layer, it is possible to increase the dot-to-dot separation in each layer of these nanocrystalline Si superlattices [50].

LEDs have been made using either annealed silicon-rich SiO_2 [51, 52] or nanocrystalline Si superlattices (with CaF_2) [53]. Figure 12 shows the room temperature EL spectrum and the dependence of the EL intensity on applied voltage for an MOS-like device with a very thin (25 nm) silicon-rich SiO_2 layer [54]. To achieve threshold for electroluminescence at a voltage as low as ~ 4 V, the thickness of the silicon-rich SiO_2 must remain very small (much less than 100 nm) and the distance between adjacent silicon quantum dots small enough to allow reasonable charge transport under bias. Limitations with this approach include the low brightness achievable (due to the thin light-emitting layer) and the relatively high leakage current (due to the high density of quantum dots). In addition, direct modulation is limited to low frequencies as in all monolithic Si LEDs.

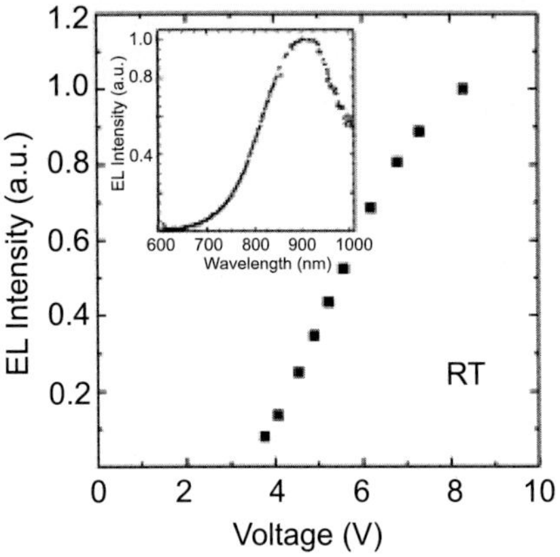

Fig. 12. Room-temperature electroluminescence intensity versus forward bias for an LED where the active layer is a 25 nm thick silicon-rich oxide layer annealed to produce ∼ 2 nm silicon quantum dots embedded in SiO_2. The *inset* shows the EL spectra measured at 4 V with a current density of 0.4 mA/cm^2. After *Irrera* [52]

5 SiGe LEDs

Bandgap engineering of III–V semiconductor structures started 30 years ago and today many commercial devices exist that use this approach. These include LEDs and lasers. It is thus reasonable to expect that band-gap engineering involving elements of the 4th column of the periodic table would produce useful devices, including light sources. The SiGe alloy system is by far the most studied system [9]. Because of the large (∼ 4%) lattice mismatch between Si and Ge, it is not possible to grow thick layers of Ge on Si without introducing dislocations. The strain present in SiGe structures is thus a limitation that does not exist in many III–V semiconductor systems. It can also be thought of as an additional degree of freedom as various degrees of strain can change the band-gap.

SiGe layers can be grown by molecular beam epitaxy [54] or low-pressure chemical vapor deposition [55]. strain in the active layers can be managed through the use of a thick alloy layer grown directly on the silicon substrate. This thick layer is relaxed and thus subsequent layers grown on it respond to the natural lattice constant of the relaxed layer, not that of the Si substrate. However, by using small growth areas, it is possible to grow thicker SiGe layers that remain strained and do not contain dislocations [56].

SiGe LEDs have been demonstrated using Si_nGe_m multilayers, SiGe quantum wells, and thick SiGe layers. Figure 13 shows the EL spectrum of devices in which the light is emitted by a thick SiGe layer [57]. All SiGe LEDs suffer from the same strong decrease in efficiency with increasing temperature. In fact, the best SiGe LEDs have a power efficiency that is at least one order of magnitude below that of good Si LEDs. Two main reasons explain this

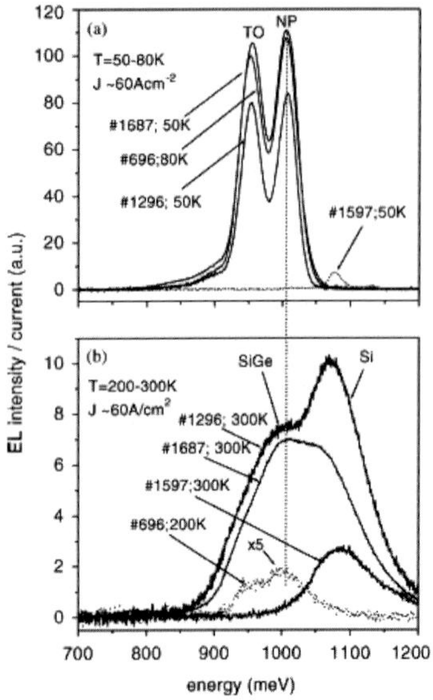

Fig. 13. Measured low-temperature and high-temperature electroluminescence spectra for LEDs in which the active layer is a strained SiGe layer. The thickness of this layer is 60 nm, 260 nm, and 445 nm in samples 696, 1296, and 1687 respectively. At low temperature, two peaks are measured, corresponding to recombination without phonon participation (NP) and with phonon participation (TO). At high temperature, the *lines* broaden and emission is also measured in silicon, as confimed by the measurement performed on a Si LED (sample 1597). After *Stoica* [57]

low efficiency. First, SiGe is still an indirect gap semiconductor, making it imperative to eliminate non-radiative centers. Second, the small band-gap discontinuity between SiGe and Si (compared to that between Si and SiO_2 for example) makes it difficult to prevent the carriers from escaping the active layer.

Recently, the activity in this material system has moved to emission from Ge islands or quantum dots [58,59] and to quantum cascade lasers that involve intraband transitions [60, 61]. The first approach is related to the on-going work on III–V semiconductor quantum dot laser and electroluminescence is only starting to emerge. The second approach does not involve transitions across the fundamental band-gap, leads to light emission in the mid to far infrared, and is thus beyond the scope of this chapter.

6 Erbium-doped Silicon LEDs

Specific light-emitting centers can be introduced into silicon. In this case, silicon itself is not directly involved in the light emission but rather serves as a host in which charge carriers are produced. These charges then are transferred to the light-emitting centers that produce sub-band-gap luminescence. This section focuses on erbium as the luminescent center, as it appears to be the most promising choice.

Erbium is widely used as a light source at the important optical wavelength of 1.54 µm. Emission at this wavelength is due to an intra-center transition between level $^4I_{13/2}$ and level $^4I_{15/2}$. The radiative lifetime is long ($\sim$ ms) and the oscillator strength of the transition is weak, not unlike in silicon itself. In silica optical fibers, erbium-doped amplifiers are used to periodically boost the optical signal. These amplifiers are long because of the weak transition rate and because the maximum density of erbium atoms is limited in order to avoid clustering of the erbium atoms that would quench the light emission.

Luminescence at 1.5 µm was first achieved in erbium-doped bulk silicon [62]. LEDs were then demonstrated and efficiencies up to near 0.1% achieved at room temperature [63]. In these devices, it is desirable that the Er ions be excited by impact excitation. It is unlikely that the performance of erbium-doped bulk Si LEDs can be improved for the following fundamental reasons: the limited maximum energy of hot electrons in silicon, efficient energy losses such as Auger recombination with free carriers in silicon or energy back transfer [64], and the limited maximum density of Er atoms to avoid clustering. At least a partial solution to these problems has been made possible by using silicon quantum dots embedded in SiO_2 instead of bulk silicon. The larger band-gap of silicon quantum dots minimizes energy back transfer to Si from Er. The use of the SiO_2 matrix guarantees that there are no free carriers that can participate in Auger recombination. More than one active Er can be excited per nanocrystal leading to a large concentration of active Er centers. Since transport takes place in an oxide matrix containing silicon quantum dots, efficient LEDs require a thin oxide layer and a large concentration of silicon quantum dots.

Figure 14 shows the cross-section of an Er-doped LED. It is an MOS structure in which the Er atoms (and the silicon quantum dots) are in the thin oxide layer. Figure 14 also shows the room temperature EL spectrum of a device that does not contain silicon quantum dots (i.e., contains a stoichiometric oxide) together with a plot of the emitted optical power as a function of the drive current [65]. Even in the absence of silicon quantum dots, the optical output of the LED is comparable to that of a III–V semiconductor LED. However, because the current-voltage curves indicate that transport takes place in the Fowler–Nordheim regime, wearout of the oxide is significant and the voltage needed to achieve a given current is very high, which limits the power efficiency. Figure 15 shows the EL spectra for a related device containing silicon quantum dots only and silicon quantum dots with erbium [66]. At low injection current density, the efficiency of the devices with and without erbium is comparable. The reported external quantum efficiency

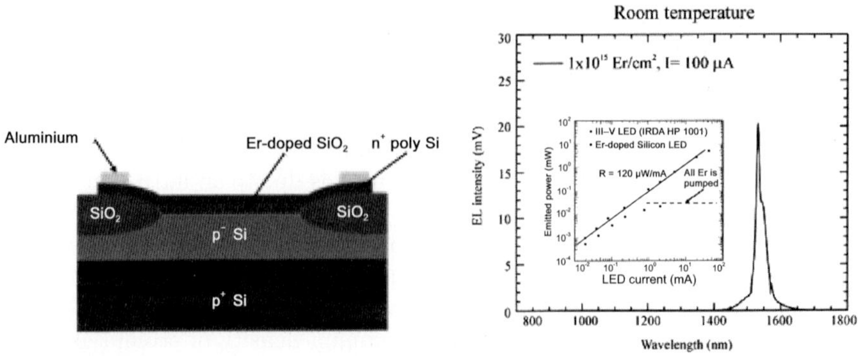

Fig. 14. Schematic cross-section of an MOS-like LED in which light emission takes place in a 70-nm-thick erbium-doped SiO_2 or silicon-rich SiO_2 layer. Measured room-temperature electroluminescence intensity spectrum for a stoichiometric oxide. The *inset* shows how the EL power increases with current. Until all the erbium atoms are pumped, this Si-based LED is comparable to a commercial III–V semiconductor LED. After *Castagna* [65]

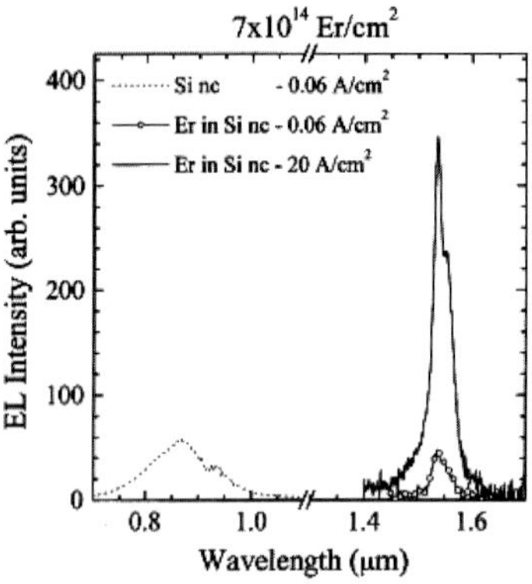

Fig. 15. Room-temperature electroluminescence spectra for an MOS-like Si LED that contain a 200-nm-thick silicon-rich SiO_2 layer. The average quantum dot size is 3.4 nm. Samples without Er showed weak EL in the near infrared whereas samples doped with erbium (6.5×10^{20} cm^{-3}) showed strong EL near 1.5 µm. After *Pacifici* [66]

of some of the Er-doped LEDs is 10% or greater [65] and it is anticipated that the power efficiency of these devices will be able to make them commercially viable. Note that using the same approach EL from other rare earths such as ytterbium and terbium has been achieved [65].

The advantages of erbium-doped LEDs include the emission wavelength (1.5 µm light is not absorbed by silicon) and the record efficiencies that make them competitive with III–V semiconductor LEDs. Their potential limita-

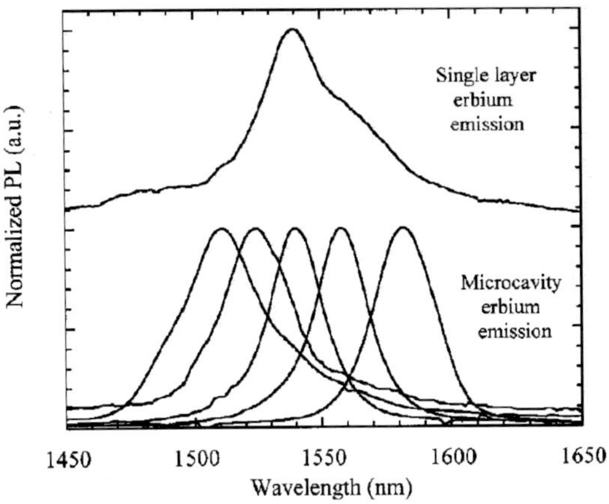

Fig. 16. Measured photoluminescence spectrum in an erbium-doped oxidized porous silicon film and measured spectra when the layer is embedded in different microcavities. After *Lopez* [67]

tions are the long-term stability of the devices and the limited speed of operation.

The luminescence spectrum of erbium-doped silicon is much narrower than that produced by silicon quantum dots. However, the emission is not really monochromatic. As mentioned earlier, the luminescence linewidth from a material can be narrowed by inserting the light emitting layer inside a microcavity. This has also been accomplished with erbium-doped porous silicon [67]. Figure 16 shows tuning of the luminescence near 1.5 μm for porous silicon microcavities doped with erbium. Because the thickness of the entire microcavity makes charge transport very inefficient [68], the achievable efficiencies of microcavity LEDs would make them unacceptable, unless an alternate electrical pumping method can be demonstrated.

7 Prospects for a Silicon-based Laser

Could a silicon laser ever be made? And if it can be made, can it be pumped electrically and can it become efficient enough to be put in use? These questions were not even raised until 2000 and yet the situation has changed drastically ever since. Entire conferences have even been devoted to the topic [69]. There are several fundamental differences between a laser and an LED. The output of a laser is coherent, of tentimes narrow band and directional. Unlike an LED, a laser is capable of producing short pulses whose duration is not related to the radiative lifetime. Thus a silicon laser could circumvent most of the limitations associated with silicon LEDs.

To achieve lasing, one first needs to achieve population inversion. In practice, this means that the upper lasing level is rapidly filled from some higher-

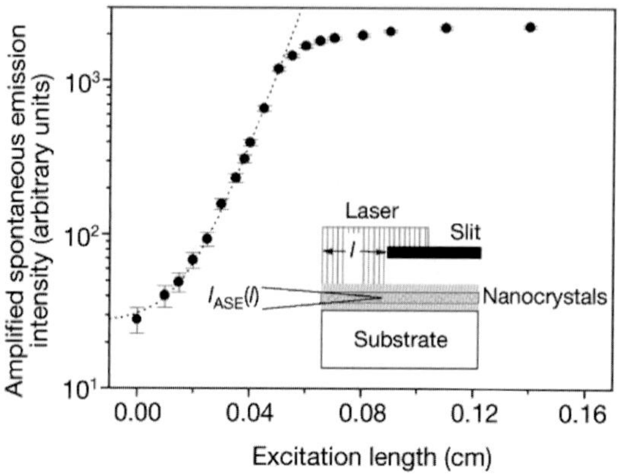

Fig. 17. Exponential growth of the amplified spontaneous emission measured by the variable stripe length method (see *inset*) on a sample containing silicon quantum dots embedded in a SiO_2 layer. The exponential growth is indicative of gain as the signal should increase as $\exp gl$, where l is the excitation length and g is the gain coefficient. After *Pavesi* [70]

lying level, and the lower lasing level is rapidly depleted of its population to some lower-lying level. This is very hard, if not impossible, to achieve in bulk silicon, where excited state absorption (i.e., free carrier absorption) and Auger recombination compete with stimulated emission. In silicon quantum dots however, the situation may become different. The radiative recombination rate increases and while pumping may involve silicon quantum dot states, lasing may involve surface states, which makes it easier to achieve population inversion. Lasing also requires that the achievable optical gain exceeds the losses associated with the cavity (i.e., mirror reflectivity, absorption losses, scattering losses).

Observations of optical gain have made using a single-pass amplifier geometry in different samples containing silicon quantum dots [70,71,72]. Figure 17 shows results from the first publication on this topic [70]. An exponential growth of the signal is observed under constant optical pumping intensity as the pumping length increases, in agreement with expectations for optical gain. Figure 18 shows that as the pumping intensity is increased, switching from losses to gain is observed [72]. In these experiments, gain was observed only under pulsed pumping and not cw excitation, presumably because pulsed excitation produces a larger density of photoinjected carriers. The origin of the gain and even its existence are still under debate. It appears unlikely that gain involves recombination of an electron and a hole within silicon quantum dots. Silicon-oxygen bonds may be involved but presently there is no confirmation of this. Other experiments have also failed to exhibit gain [73] casting doubt on the existence of gain itself.

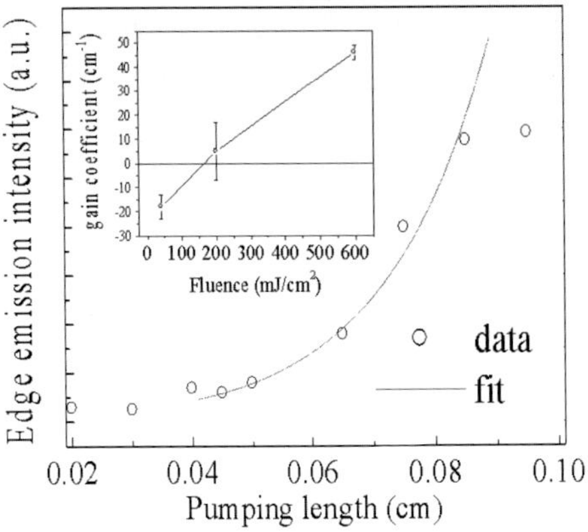

Fig. 18. Exponential growth of the amplified spontaneous signal measured using the variable stripe length method on a crystallized Si/SiO$_2$ superlattice similar to that in Fig. 11. The *inset* shows how the behavior can switch from loss to gain when the pump laser fluence increases. After *Ruan* [72]

These uncertainties should be settled after more focused work. The real test is whether a laser cavity can be build and lasing achieved. Once a laser is demonstrated, it will be possible to measure and optimize the performance of a silicon-based laser and to evaluate whether such a laser can have attractive properties and become viable. The major potential obstacle to the realization of a practical silicon-based laser is the need to pump it electrically, not optically. As we have already seen in other sections, electrical pumping is very often much less efficient than optical pumping, raising the prospect that lasing threshold may not be achievable. Note that erbium-doped samples also appear to be attractive candidates for achieving lasing.

Acknowledgements

Work in the author's laboratory related to silicon light sources has been supported in part by the National Science Foundation, the Army Research Office, and the Semiconductor Research Corporation. Contributions from the following present and past group members are gratefully acknowledged: Selena Chan, Hui Chen, Karl Hirschman, Herman Lopez, Cheng Peng, Jinhao Ruan, Leonid Tsybeskov, and Michal Wolkin.

References

1. I. Schnitzer, E. Yablonovitch, C. Caneau, T. J. Gittmer: Appl. Phys. Lett. **62**, 131–133 (1993)
2. E. Yablonovitch, D. L. Allara, C. C. Chang, T. J. Gmitter, T. B. Bright: Phys. Rev. Lett. **57**, 249 (1986)

3. M. A. Green, J. Zhao, A. Wang, P. J. Reece, M. Gal: Nature **412**, 805–808 (2001)
4. E. Yablonovitch, T. Gmitter: Appl. Phys. Lett. **49**, 587 (1986)
5. M. S. Hybertsen: Phys. Rev. Lett. **72**, 1514 (1994)
6. T. G. Brown, D. G. Hall: in *Light Emission in Silicon: From Physics to Devices*, ed. by D. J. Lockwood, Semicond. Semimet. **49** (Academic Press, San Diego 1997), chap. 3
7. G. Abstreiter: Phys. World **5**, 36 (1992)
8. B. A. Orner, J. Olowolafe, K. Roe, J. Kolodzey, T. Laursen, J. W. Mayer, J. Spear: Appl. Phys. Lett. **69**, 2557 (1996)
9. G. Abstreiter: In: *Light Emission in Silicon: From Physics to Devices*, ed. by D. J. Lockwood, Semicond. Semimet. **49** (Academic Press, San Diego 1997), chap. 2
10. J. Michel, L. V. C. Assali, M. T. Morse, L. C. Kimerling: In: *Light Emission in Silicon: From Physics to Devices*, ed. by D. J. Lockwood, Semicond. Semimet. **49** (Academic Press, San Diego 1997), chap. 4
11. S. Lombardo, S. U. Campisano, G. N. van den Hoven, A. Cacciato, A. Polman: Appl. Phys. Lett. **63**, 1942 (1993)
12. J. Gowar: *Optical Communication Systems* (Prentice-Hall, London 1984), p. 245
13. C. C. Striemer, P. M. Fauchet: Appl. Phys. Lett. **81**, 2980 (2002)
14. M. L. Cohen, J. Chelikowsky: *Electronic Structure and Optical Properties of Semiconductors*, 2nd edn. (Springer, Berlin 1989)
15. G. Davies: Phys. Rep. **176**, 83 (1989)
16. C. Delerue, G. Allan, M. Lannoo: In: *Light Emission in Silicon: From Physics to Devices*, ed. by D. J. Lockwood, Semicond. Semimet. **49** (Academic Press, San Diego 1997), chap. 7
17. C. Delerue, G. Allan, M. Lannoo: J. Lumin. **80**, 65 (1999)
18. M. Wolkin, J. Jorne, P. M. Fauchet, G. Allan, C. Delerue: Phys. Rev. Lett. **82**, 197 (1999)
19. M. Luppi, S. Ossicini: J. Appl. Phys. **94**, 2130 (2003)
20. W. Jantsch, S. Lanzerstorfer, L. Palmetshofer, M. Stepikhova, H. Preier: J. Lumin. **80**, 9 (1999)
21. G. Franzo, F. Priolo, S. Coffa: J. Lumin. **80**, 19 (1999)
22. T. P. Pearsall: Prog. Quant. Opt. **18**, 97 (1994)
23. W. L. Ng, M. A. Lourenco, R. M. Gwilliam, S. Ledain, G. Shao, K. P. Homewood: Nature **410**, 192–194 (2001)
24. M. A. Lourenco, M. S. A. Siddiqui, G. Shao, R. M. Gwilliam, K. P. Homewood: In: *Towards the First Silicon Laser*, ed. by L. Pavesi, S. Gaponenko, L. Dal Negro, NATO Science Series (Kluwer Academic, Dordrecht 2003), pp.11–20
25. R. L. Smith, S. D. Collins: J. Appl. Phys. **71**, R1 (1992)
26. A. G. Cullis, L. T. Canham, P. D. J. Calcott: J. Appl. Phys. **82**, 909 (1997)
27. L. T. Canham (Ed.): *Properties of Porous Silicon*, Electronic Materials Information Service Datareviews Series **18** (INSPEC, The Institution of Electrical Engineers, London 1997)
28. R. T. Collins, M. A. Tischler, P. M. Fauchet: Physics Today **50**, 24 (1997)
29. P. M. Fauchet: In: *Light Emission in Silicon: From Physics to Devices*, ed. by D. J. Lockwood, Semicond. Semimet. **49** (Academic Press, San Diego 1997), chap. 6

30. P. M. Fauchet, J. von Behren: Phys. Stat. Sol. (b) **204**, R7 (1997)
31. L. Tsybeskov, S. P. Duttagupta, K. D. Hirschman, P. M. Fauchet: Appl. Phys. Lett. **68**, 2058 (1996)
32. J. M. Buriak, M. J. Allen: J. Lumin. **80**, 29 (1999)
33. W. Lang, P. Steiner, F. Kozlowski: J. Lumin. **57**, 341 (1993)
34. A. Loni, A. J. Simmons, T. I. Cox, P. D. J. Calcott, L. T. Canham: Electron. Lett. **31**, 1288 (1995)
35. K. D. Hirschman, L. Tsybeskov, S. P. Duttagupta, P. M. Fauchet: Nature **384**, 338 (1996)
36. K. Nishimura, Y. Nagao, N. Ikeda: Jpn. J. Appl. Phys. **37**, L303 (1998)
37. B. Gelloz, N. Koshida: J. Appl. Phys. **88**, 4319–4324 (2000)
38. J. M. Buriak, M. J. Allen: J. Am. Chem. Soc. **120**, 1339 (1998)
39. B. Gelloz, H. Sano, R. Boukherroub, D. D. M. Wayner, D. J. Lockwood, N. Koshida: Appl. Phys. Lett. **83**, 2342–2344 (2003)
40. J. Valenta, R. Juhasz, J. Linnros: Appl. Phys. Lett. **80**, 1070 (2002)
41. C. Delerue, G. Allan, M. Lannoo. Phys. Rev. B **64**, 193402 (2001)
42. M. G. Berger, C. Dieker, M. Thonissen, L. Vescan, H. Luth, H. Munder, W. Theiss, M. Wernke, P. Grosse: J. Phys. D: Appl. Phys. **27**, 1333 (1994)
43. L. Pavesi: La Rivista del Nuovo Cimento **20**, 1 (1997)
44. S. Chan, P. M. Fauchet: Appl. Phys. Lett. **75**, 274 (1999)
45. P. J. Reece, G. Lerondel, W. H. Zheng, M. Gal: Appl. Phys. Lett. **81**, 4895 (2002)
46. M. Ghulinyan, C. J. Oton, G. Bonetti, Z. Gaburro, L. Pavesi: J. Appl. Phys. **93**, 9724 (2003)
47. P. M. Fauchet: J. Lumin. **70**, 294 (1996)
48. L. Tsybeskov, K. D. Hirschman, S. P. Duttagupta, M. Zacharias, P. M. Fauchet, J. McCaffrey, D. J. Lockwood: Appl. Phys. Lett. **72**, 43 (1998)
49. G. F. Grom, D. J. Lockwood, J. P. McCaffrey, H. J. Labbe, P. M. Fauchet, B. White, Jr., J. Diener, D. Kovalev, F. Koch, L. Tsybeskov: Nature **407**, 358 (2000)
50. M. Zacharias, J. Heitmann, R. Scholz, U. Kahler, M. Schmidt, J. Blasing: Appl. Phys. Lett. **80**, 661 (2002)
51. N. Lalic, J. Linnros: J. Lumin. **80**, 263 (1999)
52. A. Irrera, D. Pacifici, M. Miritello, G. Franzo, F. Priolo, F. Iacona, D. Sanfilippo, G. Di Stefano, P. G. Fallica: Physica E **16**, 395 (2003)
53. V. Ioannou-Sougleridis, A. G. Nassiopoulou, T. Ouisse, F. Bassani, F. A. d'Avitaya: Appl. Phys. Lett. **79**, 2076 (2001)
54. G. Abstreiter, K. Eberl, E. Friess, W. Wegscheider, R. Zachai: J. Cryst. Growth **95**, 431 (1989)
55. T. Stoica, L. Vescan, M. Goryll: J. Appl. Phys. **83**, 3367 (1998)
56. L. Vescan, T. Stoica: J. Lumin. **80**, 485 (1999)
57. T. Stoica, L. Vescan, A. Muck, B. Hollander, G. Schope: Physica E **16**, 359 (2003)
58. A. Beyer, E. Muller, H. Sigg, S. Stutz, D. Grutzmacher: Appl. Phys. Lett. **77**, 3218 (2000)
59. N. N. Ledentsov: In: *Towards the First Silicon Laser*, ed. by L. Pavesi, S. Gaponenko, L. Dal Negro, NATO Science Series (Kluwer Academic, Dordrecht 2003) pp. 281–292

60. D. J.Paul, S. A. Lynch, R. Bates, Z. Ikonic, R. W. Kensall, P. Harrison, D. J. Norris, S. L. Liew, A. G. Cullis, P.Murzyn, C.Pidgeon, D. D. Arnone, D. J. Robbins: Physica E **16**, 309–314 (2003)
61. L. Diehl et al.: Physica E **16**, 315 (2003)
62. H. Ennen, J. Schneider, G. Pomrenke, A. Axmann: Appl. Phys. Lett. **43**, 943 (1983)
63. G. Franzo, F. Priolo, S. Coffa, A. Polman, A. Carnera: Appl. Phys. Lett. **64**, 2235 (1994)
64. G. Franzo, F. Priolo, S. Coffa, A. Carnera: Phys. Rev. B **57**, 4443 (1998)
65. M. E. Castagna, S. Coffa, M. Monaco, L. Caristia, A. Messina, R. Mangano, C. Bongiorno: Physica E **16**, 547 (2003)
66. D. Pacifici, A. Irrera, G. Franzo, M. Miritello, F. Iacona, F. Priolo: Physica E **16**, 331 (2003)
67. H. A. Lopez, P. M. Fauchet: Appl. Phys. Lett. **77**, 3704 (2000)
68. H. A. Lopez, P. M. Fauchet: Mat. Sci. Eng. B **81**, 91 (2001)
69. L. Pavesi, S. Gaponenko, L. Dal Negro (Eds.): *Towards the First Silicon Laser* (Kluwer, Dordrecht 2003)
70. L. Pavesi, L. Dal Negro, C. Mazzoleni, G. Franzo, F. Priolo: Nature **408**, 440 (2000)
71. L. Khriachtchev, M. Rasanen, S. Novikov, J. Sinkkonen: Appl. Phys. Lett. **79**, 1249 (2001)
72. J. Ruan, P. M. Fauchet, L. Dal Negro, M. Cazzanelli, L. Pavesi: Appl. Phys. Lett. **83**, 5479 (2003).
73. R. G. Elliman, M. J. Lederer, N. Smith, B. Luther-Davies: Nucl. Instr. Meth. B **206**, 427 (2003)

Light Generation, Amplification, and Wavelength Conversion via Stimulated Raman Scattering in Silicon Microstructures

Bahram Jalali, Ricardo Claps,
Dimitri Dimitropoulos, and Varun Raghunathan

Department of Electrical Engineering, University of California
Los Angeles, CA 90095-1594, USA
jalali@ucla.edu

Abstract. This chapter is organized in two parts. In part one, we present the theory of Spontaneous and Stimulated Raman Scattering (SRS), as well as that of Coherent Anti Stokes Raman Scattering (CARS) in silicon. The treatment of these phenomena in silicon is more complex than that in silica fiber, due to crystal symmetry considerations. We show that, because of the intrinsically large Raman coefficient in silicon, and the tight optical confinement in silicon-on-SiO_2 waveguides, both amplification and wavelength conversion can be achieved in waveguides with millimeter scale lengths. In part two, we review our recent demonstration of spontaneous Raman emission, optical gain, and wavelength conversion in silicon waveguides. The chapter concludes by discussing practical issues and methods for realizing commercially viable devices.

1 Introduction

Silicon is the ideal platform for integrated optics and optoelectronics. Driven by the electronic IC industry, the quality of commercial silicon wafers continues to improve while the cost continues to decrease. This, along with compatibility with silicon IC manufacturing and silicon MEMS technology, is fuelling interest in silicon photonics [1]. Compared to other integrated optics platforms, a distinguishing property of silicon is the tight optical confinement made possible by the large index mismatch between the silicon and SiO_2. This renders silicon ideal for realization of photonic band structures as it minimizes radiation into the substrate even when extreme index modulation is performed. While a myriad of high performance passive devices has been realized [1], creation of active devices has been an elusive goal. In particular, the near-absence of nonlinear optical properties caused by the symmetric crystal structure, and the lack of efficient optical transitions due to the indirect band structure, have contributed to the premature perception of silicon being useful only as an optically passive material.

The need for active devices has stimulated significant amount of research, mostly in light generation in silicon. While the progress was slow in the early

L. Pavesi, D. J. Lockwood (Eds.): Silicon Photonics, Topics Appl. Phys. **94**, 199–238 (2004)

years, several breakthroughs have been made in the last three years. Various approaches that have been investigated for light generation consist of: 1. silicon nanocrystals, 2. Erbium doped silicon-rich oxide, 3. surface-textured bulk silicon, and 4. Si/SiGe quantum cascade structures. Silicon nanocrystals have received the most attention, and indications of stimulated emission have been observed in these structures [2, 3, 4, 5, 6, 7, 8, 9]. The nanocrystal approach, in which silicon nano-particles are formed within an SiO_2 film, has been motivated by earlier observation of light emission in porous silicon [10]. Porous silicon suffers from stability due to the high reactivity of exposed silicon surfaces, rendering it non-ideal for device applications. This problem is mitigated when the nanocrystals reside within the oxide layer. The structures are formed by high temperature anneal of a nonstochiometric silicon-rich oxide. The latter can be formed by 1. implantation of silicon into SiO_2, 2. Plasma Enhanced Deposition (PECVD), or 3. sputter deposition. The latter two approaches may be preferred over implantation since they produce a more uniform distribution of silicon within the oxide and hence, a narrower size distribution of nanocrystals.

The main limitation of silicon nanocrystals is that the emission occurs at wavelengths below 1000 nm, away from the important bands for telecommunication applications (1300 nm and 1550 nm). This problem has been mitigated by incorporating rare-earth dopants into the system [10, 11]. The emission wavelength can be engineered by choosing the particular dopants (Er, Yb, Pr, etc.) [12]. Additionally, structures can be electrically pumped by impact ionization, caused by hot electrons injection, realized in a Metal Oxide Semiconductor (MOS) diode. Using these two approaches, silicon LEDs with $\sim 10\%$ efficiencies and output powers approaching 100 μW have recently been demonstrated [12, 13]. These devices would be competitive with III–V LEDs once reliability issues are resolved.

By using high purity silicon, minimizing recombination at surfaces and electrical contacts, and texturing the surface to enhance the escape of photons, LEDs with efficiencies of 1%–10% have been demonstrated [14]. The measured spontaneous emission is due to the band-to-band transitions in bulk silicon and is centered at approximately 1100 nm wavelength. The main limitation of this approach is its slow modulation speed, limited by the long (ms) lifetime in high purity silicon. Also, stimulated emission is yet to be demonstrated.

Another approach to light generation in silicon is through intra-band transitions in Si/SiGe quantum cascade systems [15, 16, 17]. The approach is similar to celebrated quantum cascade lasers in III–V semiconductors capable of producing coherent radiation in the mid-IR spectrum [18]. In silicon, the concept is implemented with holes, since most of the band offset occurs in the valence band in strained layer Si/SiGe epitaxial layers. The strain caused by the lattice mismatch between Si and SiGe limits the number of layers that can be grown defect free. This, along with a high effective mass of holes makes realization of silicon quantum cascade devices more challenging than their III–V counterparts. Nonetheless, excellent progress is being made and electroluminescence from such devices has been demonstrated [17, 19].

At UCLA, we have taken an entirely different approach to achieving optical functionality in silicon, namely through Raman scattering [20, 21, 22, 23, 24, 25, 26, 27, 28, 29, 30, 31]. The Raman effect is successfully exploited in fiber amplifiers and lasers; however, several kilometers of fiber are required to create a useful device. Fortunately, the gain coefficient for Stimulated Raman Scattering (SRS) is approximately 10^4 times higher in silicon than in silica. Additionally, SOI waveguides can confine the optical field to an area that is approximately 100 times smaller than the modal area in a standard single mode optical fiber. The combination of these properties will make SRS observable over the millimeter-scale interaction lengths encountered in an integrated optic device. A desirable feature of Raman scattering is that light generation and amplification can be achieved at any wavelength (as long as the corresponding pump laser wavelength is available). Further, it does not require special impurities; therefore, the resulting devices are truly compatible with silicon manufacturing.

In addition to light generation and amplification, we have shown that the Raman effect can also perform wavelength conversion [28,29,30,31]. The conversion is achieved by the Stimulated Raman Scattering (SRS) contribution to the third order nonlinear susceptibility. Stokes to Anti-Stokes coupling occurs through coherent interaction of the two waves with a high intensity pump. Wavelength conversion is of paramount importance in optical networks because it enables optical packet switching, a milestone in the realization of all-optical networks. A key issue is to realize high conversion efficiencies. Since the efficiency in typical waveguides is limited by phase mismatch, the crucial problem is how to engineer the Group Velocity Dispersion (GVD) inside the waveguide in order to attain phase matching. The last section in this chapter describes how this can be achieved using birefringence to compensate for material dispersion in silicon waveguides.

2 Theory of Raman Scattering in Silicon

2.1 Spontaneous Raman Emission

Raman scattering in silicon was the subject of extensive research in the early 1970s as a means to study optical phonons in the material [32]. Measurements were performed on bulk silicon and at visible wavelengths. As expected, first-order Raman scattering in silicon is due to the scattering of light by the threefold degenerate zone-center optical phonons in the crystal. It corresponds to the largest feature observed in the Raman spectrum of silicon, shown in Fig. 1, where second-order Raman scattering modes are also shown. These second order modes are nearly 2 orders of magnitude weaker than the fundamental mode and will not be considered any further in the

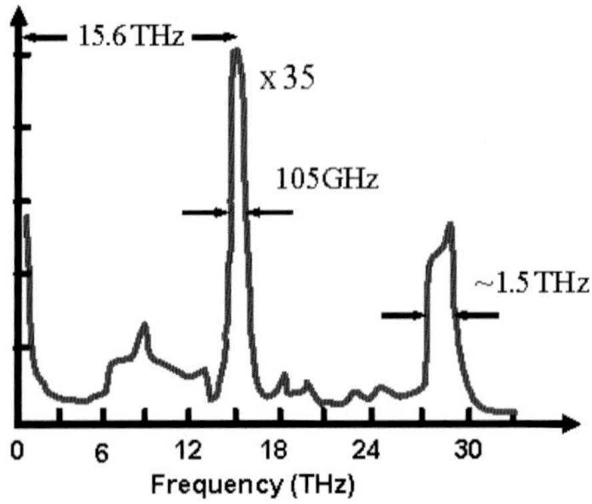

Fig. 1. Raman Spectrum of Silicon (from [32])

present manuscript. Throughout this chapter, the first-order Raman mode will be referred to as the "Stokes" scattered field, whereas the incident field that generates the process will be referred to as the "pump" field. The Stokes frequency, ω_s, is $\sim 15.6\,\mathrm{THz}$ down-shifted from the pump frequency, ω_p, this shift corresponds to the frequency of the zone center optical phonon in silicon, from now on denoted as Ω. The FWHM of the spectra is $\sim 105\,\mathrm{GHz}$ and gives rise to a response time of $\sim 10\,\mathrm{ps}$ for the Raman process in silicon. The amount of Raman radiation obtained from a given material is determined by the spontaneous scattering efficiency, S ($\mathrm{cm}^{-1} \cdot \mathrm{sr}^{-1}$), which relates the pump intensity, I_p ($\mathrm{W/cm}^2$) to the total Stokes–Raman power, P_R (W), measured over a solid angle, Φ (sr), scattered from a volume, V, of the material,

$$P_\mathrm{R} = (\Phi \times V) \times S \times I_\mathrm{P}\,. \tag{1}$$

A more detailed analysis of the Raman scattering process needs to take into account the crystal symmetry of silicon. Furthermore, it is necessary to pave the way for introducing nonlinear effects in the picture (which will be detailed in the following section). The most convenient way to do this, is through the induced-polarization vector, $\boldsymbol{P}(\omega_\mathrm{s})$, and the pump electric field, $\boldsymbol{E}(\omega_\mathrm{p})$, responsible for the Stokes radiation associated with the phonon displacement, $\boldsymbol{Q}$. The induced polarization is related to the incident electric field as [33, 34]:

$$P_i(\omega_\mathrm{s}) = R_{ijk} Q_k^* \times E_j(\omega_\mathrm{p})\,. \tag{2}$$

Equation (2) defines the Raman tensor, $\boldsymbol{R}$ Its components, R_{ijk}, are proportional to $\partial a_{ij}/\partial Q_k$, where a is the polarizability of the material. The term, Q_k, in (2) is the kth component of the phonon mode, $\boldsymbol{Q}$, and ω_p and ω_s are the pump and Stokes frequencies, respectively. In the linear domain, where the pump power levels are below a threshold value, the tensor $\boldsymbol{R}$ is independent of the pump electric field, and can be defined in terms of the scattering efficiency, S, as follows [35]

$$S = S_0 \sum_{n=1,2,3} \left|\hat{e}_s R_n \hat{e}_p\right|^2 . \tag{3}$$

The vectors $\hat{e}_p$ and $\hat{e}_s$ denote the polarization of the pump and Stokes electromagnetic fields. S_0 is an intrinsic microscopic property of silicon and has the same dimensions as S. It contains specific information about the derivatives of the polarizability in silicon, and the absolute amplitude of the displacement of the zone-center optical phonons in the crystal. Another parameter that is usually seen in the literature is the linear Raman susceptibility, χ_R, which is related to S_0 through the following expression [33],

$$S_0 = \frac{k_S^4}{32\pi^2 n} V \chi_R^2 , \tag{4}$$

where k_s is the wavevector of the Stokes wave, $k_s = 2\pi n_s/\lambda_S, n_s$ is the index of refraction, and V is the same as in (1). Given the scattering efficiency for a particular wavelength, S_0 can be calculated for other wavelengths by taking into account the λs^{-4} dependence given by (4). The value of S_0 at $\sim 1550\,\mathrm{nm}$ is estimated to be $8.4 \times 10^{-7}\,\mathrm{cm}^{-1} \cdot \mathrm{sr}^{-1}$ [20].

In a coordinate system oriented along the crystallographic axes of silicon, the Raman tensor components are given as [23]:

$$\overleftrightarrow{R_1} = \begin{bmatrix} 0 & 1 & 0 \\ 1 & 0 & 0 \\ 0 & 0 & 0 \end{bmatrix}, \quad \overleftrightarrow{R_2} = \begin{bmatrix} 0 & 0 & 1 \\ 0 & 0 & 0 \\ 1 & 0 & 0 \end{bmatrix}, \quad \overleftrightarrow{R_3} = \begin{bmatrix} 0 & 0 & 0 \\ 0 & 0 & 1 \\ 0 & 1 & 0 \end{bmatrix} . \tag{5}$$

Table 1 shows the relative spontaneous Raman intensities obtained for different scattering configurations in silicon. The table is obtained by using the tensor components in (5) into (3) and solving for the absolute scattering intensity. In the table, the vectors $\hat{k}_p$ and $\hat{k}_s$ denote the pump and Stokes field propagation, respectively, and for the case of forward scattering are identical.

Table 1. Spontaneous Scattering efficiency for various wavevector and polarization directions of Pump and Stokes fields

$\hat{k}_p$ & $\hat{k}_s$	$\hat{e}_p$	$\hat{e}_s$	Relative efficiency
$[1\bar{1}0]$	[110]	[110]	S_0
$[1\bar{1}0]$	[110]	[001]	S_0
$[11\bar{2}]$	[111]	[111]	$4S_0/3$
[111]	$[1\bar{1}0]$	$[11\bar{2}]$	$2S_0/3$
[111]	$[11\bar{2}]$	$[11\bar{2}]$	S_0
$[0\bar{1}1]$	[011]	[100]	S_0
$[0\bar{1}1]$	[011]	[011]	S_0

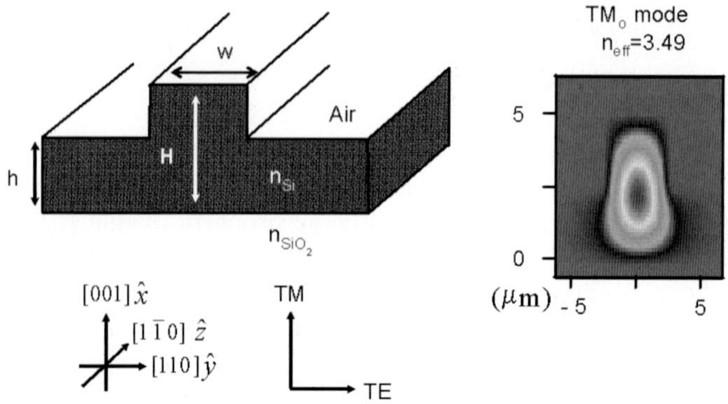

Fig. 2. SOI waveguides used. The *coordinate axes* are labeled in the usual crystallographic notation. The (x, y, z) labels appear as they will be used throughout the text. The resulting TM_0 mode ($\lambda = 1.54\,\mu m$) from a BPM calculation is depicted to the *right*, for the case $H = w = 5.0\,\mu m, h = 2.2\,\mu m$. The effective index of refraction is calculated assuming $n = 3.5$ for bulk silicon

From Table 1, it is seen that the configuration oriented along the $[1\bar{1}0]$ direction has the property that the scattering intensity is polarization independent, with the pump field oriented along the $[110]$ direction. This, together with the favorable cleaving properties of silicon in this orientation, makes the afore-mentioned configuration highly convenient if one is thinking in terms of a polarization-independent device. For this reason, the theoretical and experimental work presented here is based on rib waveguides fabricated parallel to the $[1\bar{1}0]$ direction on a $[001]$ silicon surface. The cross-section of the rib structure and a typical mode profile associated with it, are shown in Fig. 2, together with a convenient selection of a coordinate axes system (x, y, z) that will be used for the remainder of this chapter.

2.2 Stimulated Raman Scattering

When the intensity of the pump field inside the material surpasses certain threshold value, the linear regime in which (3) applies is no longer valid, and the process of Stimulated Raman Scattering (SRS) takes place. The nonlinear effect concerns the electric field dependence of the induced polarization. This dependence can be obtained by solving for the phonon displacement, $\boldsymbol{Q}$. The equation of motion for $\boldsymbol{Q}$ is governed by the differential equation for a forced-harmonic oscillator with damping coefficient, $\varGamma$, and resonance frequency, $\varOmega$ [34]:

$$\frac{\mathrm{d}^2 Q_k}{\mathrm{d}t^2} - 2\varGamma \frac{\mathrm{d}Q_k}{\mathrm{d}t} + \varOmega^2 Q_k = R_{kmn} E_m(-\omega_\mathrm{p}) E_n(\omega_\mathrm{s}) \,, \tag{6}$$

where use has been made of the fact that the force on the Q_k phonon component goes as $F_{Q_k} = -\partial\varepsilon/\partial Q_k$, with $\varepsilon = -\boldsymbol{P} \times \boldsymbol{E}$, and $\boldsymbol{P}$ as given by (2).

Equation (6) is solved for Q in terms of the pump and Stokes electric fields. It is customary to define the nonlinear susceptibility, χ^R_{ijmn}, by the following relation:

$$P_i^{NL}(\omega_s) = \varepsilon_0 \chi^R_{ijmn} E_j(\omega_p) E_m(-\omega_p) E_n(\omega_s) \,. \tag{7}$$

By introducing the solution for (6) into (2), and comparing with the definition given in (7), we arrive at the following expression for χ^R_{ijmn}

$$\chi^R_{ijmn} = \frac{1}{\varepsilon_0} \frac{R_{ijk} R_{kmn}}{(\Omega^2 - (\omega_p - \omega_s)^2 - 2i\Gamma(\omega_p - \omega_s))} \,, \tag{8}$$

The pump and signal frequencies, ω_p and ω_s, respectively, need not satisfy the 'Raman-resonance' condition: $\omega_p - \omega_s = \Omega$. It will be demonstrated below that the nonlinear optical constants for bulk silicon can be used to accurately describe the rib-SOI waveguides, shown in Fig. 2 with negligible correction factors for Raman-induced cross-polarization effects. Equation (8) will be useful in the next section; there, a full vectorial treatment of the electric fields in the waveguide will be necessary to carry out Coupled-Mode calculations of the SRS effect in SOI waveguides. For the moment, we will proceed with an approach that considers bulk propagation. In this regime, the gain coefficient, g_R, defined through the relation, $\mathrm{d}P_R/\mathrm{d}z = g_R P_R I_p$ (cf 1), is obtained from the spontaneous scattering efficiency, S, as [35]:

$$g_R = \frac{8\pi c^2 \omega_p}{\hbar \omega_s^4 n^2(\omega_s)(N+1)\Delta\omega} S \,, \tag{9}$$

The factor N is the Bose occupation number (0.1 at room temperature), and $\Delta\omega$ is the FWHM of the spontaneous Raman spectrum. The relation between g_R and χ^R_{ijmn} will be made explicit below. Substituting the appropriate values, the gain coefficient is obtained as 0.076 cm/MW. When compared to silica (0.93×10^{-5} cm/MW) [36,37], the Raman gain in silicon is approximately 10^4 times larger! Such a large difference has its origin in the linewidth of the Raman spectrum being much narrower in crystalline silicon compared to the amorphous fiber. The relation between g_R and χ^R_{ijmn} can be obtained from Maxwell's equations and (7) as [27]

$$\chi^R_{2222}(\omega_p - \omega_s = \Omega) = \frac{i}{(\mu_0/\varepsilon_0)^{1/2}} \frac{2ncg_R}{\omega_s} = 11.2 \times 10^{-14} i \frac{\mathrm{cm}^2}{\mathrm{V}^2} \,.$$

Following the same derivation for the total Raman amplification in optical fibers [21], the amplified Stokes signal power after a length, L, in the rib waveguide, with an initial input stokes power of $P_R(0)$ is given as [21]:

$$P_R(L) = P_R(0) \exp\left[-\gamma L + \frac{g_R I_P(0)}{\gamma}(1 - \exp(-\gamma L))\right] \,, \tag{10}$$

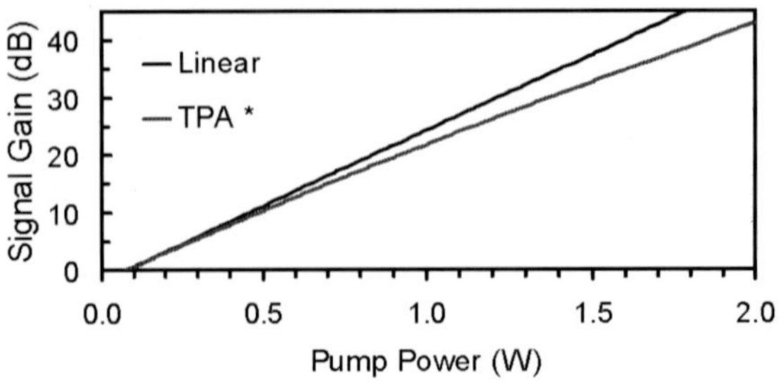

Fig. 3. Stimulated Raman gain as a function of the pump power with no other nonlinear contribution and with two-photon-absorption (TPA). $L = 1.9\,\text{cm}$, $\gamma = 1\,\text{dB/cm}$, $g_\text{R} = 0.076\,\text{cm/MW}$, $a = 2\,\mu\text{m}^2$, $P_\text{R} = 1\,\mu\text{W}$

where γ is the propagation loss (cm^{-1}), $I_\text{p}(0) = P_\text{p}(0)/A$ is the input pump intensity, and A is the modal overlap between the propagating TE_0 and TM_0 waveguide modes. The plot of the net Stokes gain as a function of the pump power for a waveguide of length, $L = 1.9\,\text{cm}$, is shown in Fig. 3. The plot shows that more than 10 dB gain can be obtained with a pump power of less than 50 mW. Moreover, the threshold power, which is defined as the pump power required to achieving transparency, is relatively small in comparison to fibers, because of the tight confinement of the optical fields in the rib structure. The use of low-loss waveguides with efficient coupling schemes, and antireflection coated facets, can reduce the threshold power even further. The pump is assumed to be a monochromatic source. The finite linewidth of the pump laser will result in a lower gain than what is predicted in Fig. 3. The effect of the pump linewidth is discussed in Sect. 3.2.1. At high enough pump powers, two photon absorption (TPA) may contribute significantly towards pump depletion. Including TPA in (10) involves reduction of the pump power with distance, as given below [20]:

$$P_\text{p}(z) = \frac{P_\text{p}(0)e^{-\gamma z}}{(1 + B\,z\,I_\text{p}(0))}\,, \tag{11}$$

where B (cm/W) is the TPA coefficient for silicon. The effect of TPA is a negligible reduction in gain, as shown in Fig. 3.

It is clear that for amplification levels of up to 10 dB, the gain reduction produced by TPA is below 10%. The value of B used to plot Fig. 3 was obtained from experiments described in Sect. 3.2.2. Another nonlinear effect in silicon that may affect the Raman gain coefficient is Coherent Anti-Stokes Raman Scattering (CARS). As will be described in Sect. 3.3, this effect can actually be exploited in performing wavelength conversion when a phase matching condition is satisfied. The results presented up to this point start from the basis that there is no coupling between the TE_0 and TM_0 modes of propagation along the waveguide. This simplifies the analysis and eventually leads to (10). In single-mode silica fibers, this assumption is justified because the Raman signal has the same polarization as the pump signal. As we have seen (cf Table 1) this is not the case for crystalline silicon, where the pump and Stokes fields have crossed polarization, and therefore the fields couple to different modes in the waveguide. In the next section, we will use a detailed Coupled-Mode-Calculation theory (CMC) to investigate the issue of Raman-induced cross-coupling between TE_0 and TM_0 field modes, and demonstrate the validity of (10) and the results derived from it (Fig. 3).

2.3 Coupled-Mode Theory Analysis of Raman Scattering

In the rib SOI waveguide, with the pump field being launched in the TE_0 mode and the Stokes field assumed to be in both TE_0 and TM_0 modes, these fields can be represented mathematically as follows:

$$\begin{aligned} \boldsymbol{E}(\omega_\mathrm{s}) &= \sum_{\mu=1,2} A_{s,\mu} \in_\mu (\omega_\mathrm{s}, x, y) e^{i\beta_\mu z} \,, \\ \boldsymbol{E}(\omega_\mathrm{p}) &= A_{p,1} \in_1 (\omega_\mathrm{p}, x, y) e^{i\beta_1 z} \,, \end{aligned} \tag{12}$$

where μ denotes the polarization of the field (1 for TE_0, and 2 for TM_0), $A_{s,1} = \sqrt{aP_\mathrm{s}}$, $A_{s,2} = \sqrt{(1-a)P_\mathrm{s}}$, a is the fractional power in the TE_0 polarization, $A_{p,1} = \sqrt{P_\mathrm{p}}$, P_p denotes the coupled pump power (W), $\in$ denotes the normalized modal profile, and β_μ (cm^{-1}) is given by $\beta_\mu = 2\pi\, n_\mu/\lambda_\mu$, as usual ($n_\mu$ is the effective index of refraction of the mode in question). The coupled mode equations relating the Stokes field in TE_0 and TM_0 polarizations, with no pump depletion, are given as follows [27,34]:

$$\frac{\mathrm{d}A_{s,1}}{\mathrm{d}z} = i\kappa_{11} P_\mathrm{p} A_{s,1} + i\kappa_{12} P_\mathrm{p} A_{s,2} \exp(i(\beta_2 - \beta_1)z) \,, \tag{13a}$$

$$\frac{\mathrm{d}A_{s,2}}{\mathrm{d}z} = i\kappa_{22} P_\mathrm{p} A_{s,2} + i\kappa_{21} P_\mathrm{p} A_{s,1} \exp(-i(\beta_2 - \beta_1)z) \,. \tag{13b}$$

The coupling coefficients $\kappa_{\mu\mu'}$ are given by:

$$\kappa_{\mu\mu'} = -i \sum_{i,j,m,n=1,2,3} \frac{2g_\mathrm{R}}{Z^2} \xi_{ijmn} \times \iint \in_\mu^{*i}(\omega_\mathrm{s}) \in_1^j (\omega_\mathrm{p}) \in_1^{*m} (\omega_\mathrm{p}) \in_{\mu'}^n (\omega_\mathrm{s}) \mathrm{d}x\mathrm{d}y \,, \tag{14}$$

where $Z = Z_0/n_{Si}$ is the impedance in the waveguide ($Z_0 = (\mu_0/\varepsilon_0)^{1/2} = 377\,\Omega$ [43]), $\xi_{ijmn} = \chi^R_{ijmn}/\chi^R_{1111}$ and $\in^i_\mu$ is the mode profile of the ith component of the μ polarization-mode vector. In the case of weak coupling the gain, G (dB/cm), will be given by,

$$G_\mathrm{TE(TM)} = 4.34 \times 2 \left|\kappa_{11(22)}\right| P_\mathrm{p} \,. \tag{15}$$

For the spontaneous Raman Effect, the coupled-mode equations as mentioned above are modified into the following form:

$$\frac{\mathrm{d}A_{s,\mu,\nu}}{\mathrm{d}z} = i\bar{\kappa}_{\mu\nu} A_{p,1} \,. \tag{16}$$

The index ν refers to the optical phonon component. The coupling coefficients in this case are

$$\bar{\kappa}_{\mu,\nu} = \omega_{\mathrm{s}}\varepsilon_0\chi_{\mathrm{R}} \iint \in^*_{\mu}(\omega_{\mathrm{s}})\boldsymbol{R}_{\nu}\in_1(\omega_{\mathrm{p}})\mathrm{d}x\mathrm{d}y\,. \tag{17}$$

The scattering efficiency for a Stokes mode μ, S_{μ}, is defined as

$$S_{\mu} = \frac{|A_{s,\mu}|^2}{|A_{p,1}|^2 l} = l\left(|\bar{\kappa}_{\mu,1}|^2 + |\bar{\kappa}_{\mu,2}|^2 + |\bar{\kappa}_{\mu,3}|^2\right)\,. \tag{18}$$

Comparing (3), (17) and (18), S_{μ} can be rewritten as:

$$S_{\mu} = S_0\frac{32\pi^2}{k_0^4}\omega^2\varepsilon_0^2\frac{\sum\limits_{\lambda=1,2,3}\left|\iint(\varepsilon_{\mu}(\omega_{\mathrm{s}}))^* \times \overleftrightarrow{R}_{\lambda} \times \varepsilon_1(\omega_{\mathrm{p}})\mathrm{d}x\mathrm{d}y\right|^2}{A_{\mathrm{eff}}}\,, \tag{19}$$

with the scattering volume, $V = A_{\mathrm{eff}}l$, and A_{eff} taken as the effective area of the TE_0 pump mode:

$$A_{\mathrm{eff}} = \frac{\left(\iint|\varepsilon_1(\omega_{\mathrm{p}})|^2\,\mathrm{d}x\mathrm{d}y\right)^2}{\iint|\varepsilon_1(\omega_{\mathrm{p}})|^4\,\mathrm{d}x\mathrm{d}y}\,. \tag{20}$$

Table 2 lists the coupling coefficients and the modal scattering efficiency, S_{μ}, for varying rib widths, with the rib height and slab height varied proportional to the widths [27]. The mode profiles that were used in computing the coefficients in (14), (19) and (20) were evaluated using a finite-difference time domain method [38]. Now it is possible to define an effective area, $A_{\mathrm{R}}^{\mathrm{eff}}$, by relating the effective gain obtained from (15), to the bulk Raman gain, g_{R}, as expressed in (9) above. This relation can be written as

$$2\kappa_{11(22)}P_{\mathrm{P}} = g_{\mathrm{R}}I_{\mathrm{p}} = g_{\mathrm{R}}P_{\mathrm{P}}/A_{\mathrm{R}}^{\mathrm{eff}}\,. \tag{21}$$

Using (21), it is seen that the value of $A_{\mathrm{R}}^{\mathrm{eff}}$ decreases linearly as the waveguide dimensions are reduced, whereas the spontaneous efficiency (19) and the gain coefficient (15) are found to increase, as expected. The modal spontaneous efficiencies are larger for TE_0 than for TM_0 modes, but the gain coefficients follow the opposite trend. Though this contradicts the Einstein relation between spontaneous efficiency and gain, this trend is in fact real. The validity of Einstein's relation is limited to electric fields with spatial homogeneity and isotropic energy density [39], and as the waveguide dimensions are reduced, this assumption is no longer valid and the effect mentioned above is observed. It is also clear that the inter-modal coupling coefficients are less than two orders of magnitude smaller than the self coupling coefficients. This implies that the two Stokes modes propagate independent of each other, confirming the validity of the approach leading to (10). Furthermore, the CMC analysis proves that simultaneous amplification of TE_0/TM_0 Stokes modes is possible through polarization multiplexing of the signals.

Table 2. Effective area, coupling coefficients and spontaneous efficiency for a rib waveguide with $w/h = 2/1.4$ and $w/H = 2/2.15$

w (µm)	A_R^{eff} TE (µm²)	A_R^{eff} TM (µm²)	κ_{11} TE (cmW)$^{-1}$	κ_{22} TM (cmW)$^{-1}$	κ_{12} (10^{-3}) (cmW)$^{-1}$	S_{TE} (10^{-9} cm^{-1})	S_{TM} (10^{-9} cm^{-1})
4	16.2	15.3	−j0.215	−j0.23	−0.32 − j0.58	19.6	19.2
3	8.9	8.25	−j0.39	−j0.42	−2.5 + j1.36	35.8	35.2
2	4.3	3.9	−j0.82	−j0.89	−11.9 + j13.3	69.2	67.0
1.2	1.6	1.4	−j2.16	−j2.45	−284 + j89.7	192	183
0.8	1.1	0.9	−j3.1	−j3.88	212 − j69.5	271.2	222.0
0.4	0.42	0.36	−j8.28	−j9.7	−58.9 − j18	787.0	540.0

2.4 Coherent Anti-Stokes Raman Scattering (CARS) in Silicon

In the previous analysis of the Raman process in silicon, no mention was made of the anti-Stokes field (also denoted as a-Stokes, or simply aS). This field is generated simultaneously to the Stokes radiation being scattered by the interaction between the pump field and the optical phonons in silicon. In the case of the anti-Stokes wave, the phonon vibration in the material transfers energy to the electromagnetic field. The resulting wave then has a higher frequency than the pump wave, shifted by the same amount as the Stokes wave (15.6 THz for optical phonons in bulk silicon). Under no external influence on the vibronic state of the crystal, and standard equilibrium temperature conditions (up to 100 °C), the amount of a-Stokes radiation relative to the Stokes radiation is governed by a Boltzmann distribution, and is negligible for all practical purposes. However, the extent to which this assertion is true and the nature of the Raman process that occurs in the waveguide depend upon the phase mismatch between the pump (β_P), Stokes (β_S) and anti-Stokes (β_{aS}) waves, which is given as follows [40]:

$$\Delta\beta = 2\beta_p^{\mu_p} - \beta_s^{\mu_s} - \beta_{aS}^{\mu_{aS}} \,, \tag{22}$$

where β_λ^μ is the wave-vector for the given wavelength and the corresponding mode of polarization μ (1 for TE_0 and 2 for TM_0). As $\Delta\beta$ approaches zero, pump, Stokes and anti-Stokes waves experience a coherent interaction. This phenomenon is well known in the literature as Coherent anti-Stokes Raman Scattering (CARS) and can be explained in simple terms as the creation of anti-Stokes (Stokes) photons as a result of the coupling of two pump photons and one Stokes (anti-Stokes) photon, with the in-phase creation and annihilation of one zone-center optical phonon, as illustrated in Fig. 4. This interaction of the three fields, E_p, E_S and E_aS, is a Four-Wave-Mixing (FWM) process, and is governed by the third order nonlinear susceptibility in silicon, $\chi^{(3)}$. In general, the $\chi^{(3)}$ tensor has two contributions. The first is the "electronic", nonresonant component, $\chi_{NR}^{(3)}$, which is nearly instantaneous in response and hence broadband, and which accounts for such effects as self-phase modulation (SPM) and FWM. The second is the Raman component, $\chi_\mathrm{R}^{(3)}(\Omega = \omega_\mathrm{p} - \omega_\mathrm{S})$ which accounts for SRS and CARS. The $\chi_\mathrm{R}^{(3)}(\Omega)$ component has a strong spectral dependence, with a resonant peak at $\Omega = \Omega_0$, the frequency of oscillation of zone-center optical phonons in silicon. The element $\chi_\mathrm{R}^{(3)}(\Omega)$ has therefore the spectral signature of the Raman-active phonon modes of vibration of the crystal lattice. This implies that there is a response time associated with this component, which corresponds to the phonon de-phasing lifetime. In the case of first order Raman scattering from silicon at room temperature, where only zone-center optical phonons are involved, the bandwidth is $\sim 105\,\mathrm{GHz}$. This corresponds to a response time of about 10 ps. The details of CARS in silicon and how it applies to SOI waveguides will be worked out theoretically in the next two sections. Experimental demonstration of this phenomenon appears in Sect. 3.2.3.

2.5 Coupled-Mode Analysis of CARS in Silicon Waveguides

The evolution of the Stokes and the anti-Stokes waves coupled by the third-order susceptibility can be described according to Maxwell's equations by the following differential equations [30, 41]:

$$\frac{\mathrm{d}E_\mathrm{S}}{\mathrm{d}z} = \eta_{11} E_\mathrm{S} + \eta_{12} E_\mathrm{aS}^* e^{i\Delta\beta z}\,, \tag{23a}$$

$$\frac{\mathrm{d}E_\mathrm{aS}^*}{\mathrm{d}z} = -\eta_{22} E_\mathrm{aS}^* - \eta_{21} E_\mathrm{S} e^{-i\Delta\beta z}\,, \tag{23b}$$

where E_S and E_aS refer to the electric fields of the Stokes and anti-Stokes waves, respectively. The coupling coefficients can be written in terms of the nonlinear susceptibility and the pump electric field (E_p) as:

$$\eta_{11} = i\frac{k_\mathrm{S}^2}{2\beta_\mathrm{S}}\left(\chi_\mathrm{NR}^{(3)} + \chi_\mathrm{R}^{(3)}\right)|E_\mathrm{p}|^2\,, \qquad \eta_{12} = i\frac{k_\mathrm{aS}^2}{2\beta_\mathrm{S}}\left(2\chi_\mathrm{NR}^{(3)} + \chi_\mathrm{R}^{(3)}\right)E_\mathrm{p}^2\,, \tag{24a}$$

$$\eta_{22} = i\frac{k_\mathrm{aS}^2}{2\beta_\mathrm{aS}}\left(\chi_\mathrm{NR}^{(3)} + \chi_\mathrm{R}^{(3)}\right)|E_\mathrm{p}|^2\,, \qquad \eta_{21} = i\frac{k_\mathrm{S}^2}{2\beta_\mathrm{aS}}\left(2\chi_\mathrm{NR}^{(3)} + \chi_\mathrm{R}^{(3)}\right)E_\mathrm{p}^2\,, \tag{24b}$$

where $k_{s,\mathrm{aS}} = \beta_{s,\mathrm{aS}}/n_{s,\mathrm{aS}}$. The steady state solution for the fields, in terms of their initial value at $z = 0$, can be written in the following form:

$$\begin{pmatrix} E_{\mathrm{S}}(z) \\ E_{\mathrm{aS}}(z) \end{pmatrix} = \begin{pmatrix} T_{11} & T_{12} \\ T_{32} & T_{33} \end{pmatrix} \begin{pmatrix} E_{\mathrm{S}}(0) \\ E_{\mathrm{aS}}(0) \end{pmatrix}, \tag{25}$$

where the transmission coefficients are:

$$T_{11} = e^{\eta_{11} z} \frac{s^- e^{s^+ z} - s^+ e^{s^- z}}{s^- - s^+}, \tag{26a}$$

$$T_{12} = \eta_{12} e^{\eta_{12} z} \frac{e^{s^- z} - e^{s^+ z}}{s^- - s^+}, \tag{26b}$$

$$T_{22} = e^{-\eta_{22} z} \frac{\lambda^- e^{\lambda^+ z} - \lambda^+ e^{\lambda^- z}}{\lambda^- - \lambda^+}, \tag{26c}$$

$$T_{21} = -\eta_{21} e^{-\eta_{21} z} \frac{e^{\lambda^- z} - e^{\lambda^+ z}}{\lambda^- - \lambda^+}, \tag{26d}$$

with

$$\lambda^{\pm} = s^{\pm} - i\Delta\beta + \eta_{22} + \eta_{33}, \tag{27a}$$

$$s^{\pm} = \frac{i\Delta\beta - \eta_{22} - \eta_{33} \pm \sqrt{(i\Delta\beta - \eta_{22} - \eta_{33})^2 - 4\eta_{23}\eta_{32}}}{2}. \tag{27b}$$

Exactly at the Raman resonance, $\chi_{\mathrm{R}}^{(3)}$ is imaginary and $\chi_{NR}^{(3)}$ can be neglected. Then $\eta_{11} = g_{\mathrm{R}}/2$, where g_{R} is the steady-state Raman gain, and to a good approximation $\eta_{11} \approx \eta_{22} \approx \sqrt{\eta_{12}\eta_{21}}$. The dispersion regime for which $2\eta_{22} \ll |\Delta\beta|$, is the region in which the Stokes wave is amplified and the Stokes anti Stokes coupling is highly suppressed. As $\Delta\beta \to 0, s^{\pm} \to -g_{\mathrm{R}}$, and the Stokes and anti-Stokes fields can be approximated in the following form:

$$\begin{aligned} E_S(z) &= E_S(0) + (E_S(0) + E_{\mathrm{aS}}^*(0))g_{\mathrm{R}} z/2, \\ E_{\mathrm{aS}}^*(z) &= E_{\mathrm{aS}}^*(0) - (E_S(0) + E_{\mathrm{aS}}^*(0))g_{\mathrm{R}} z/2. \end{aligned} \tag{28}$$

The solution predicts no change in the Stokes and anti-Stokes fields if the initial fields are equal in amplitude and opposite in phase. The above equation predicts a linear increase in the fields with distance, but this is valid under the assumption that $s^+ \approx s^-$, which holds true only for small propagation lengths. At lengths $z \sim 1/(s^+ - s^-)$, the difference becomes significant and leads to saturation of the fields. This characteristic length corresponds to tens of cm in silicon; therefore, this regime is not expected to occur in chip-scale devices. The physics of this saturation is the cancellation of the phonon oscillation amplitude, due to dephasing of the creation and annihilation events depicted in Fig. 4.

To illustrate the physical process described in (26d)–(28), Fig. 5 depicts the effective amplification coefficient of the Stokes (or a-Stokes) wave,

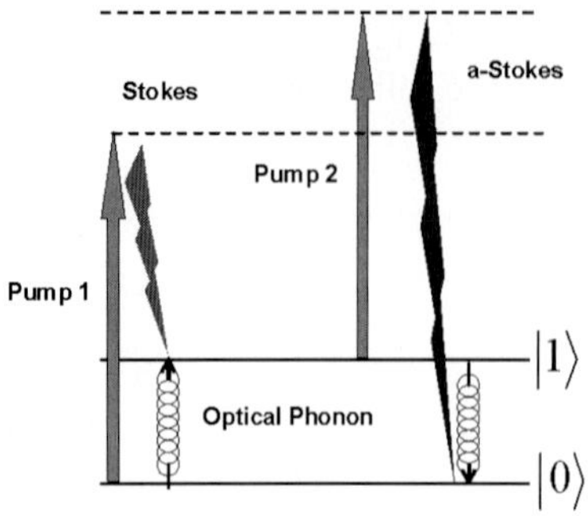

Fig. 4. Schematic energy-level diagram of the CARS process. Notice that the four photons involved are synchronized with the phonon oscillation that is created and reabsorbed in the process

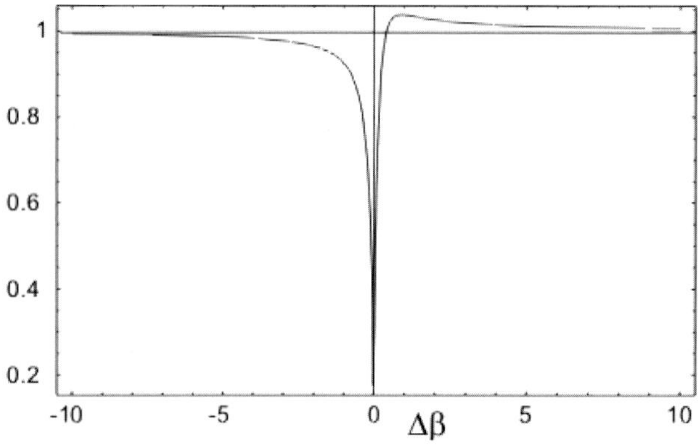

Fig. 5. The variation of normalized Raman gain as a function of the phase mismatch ($\Delta\beta\,\mathrm{cm}^{-1}$)

$A = \mathrm{Im}(s^{\pm})$, normalized to the bulk SRS gain coefficient, g_{R}, as a function of the phase mismatch, $|\Delta\beta|$ [28]. At large values of $|\Delta\beta|$, SRS is the predominant effect and leads to the amplification of the Stokes signal with an effective gain that is very close to g_{R}. At small, negative values of $\Delta\beta$, the Stimulated Raman effect is considerably suppressed and the parametric coupling effects dominate. In this region, Stokes, anti-Stokes and pump fields are strongly coupled and parametric scattering, enabled through the Raman susceptibility, dominates. For small, positive values of $\Delta\beta$, the normalized gain slightly exceeds unity due to modulation instability. This effect has also been predicted and observed in optical fibers [40, 42].

2.6 Parametric Wavelength Converter

The power conversion efficiency between the Stokes and anti-Stokes waves, $|T_{12}|^2$ in (26d) as a function of the phase mismatch, $\Delta\beta$, is shown in Fig. 6,

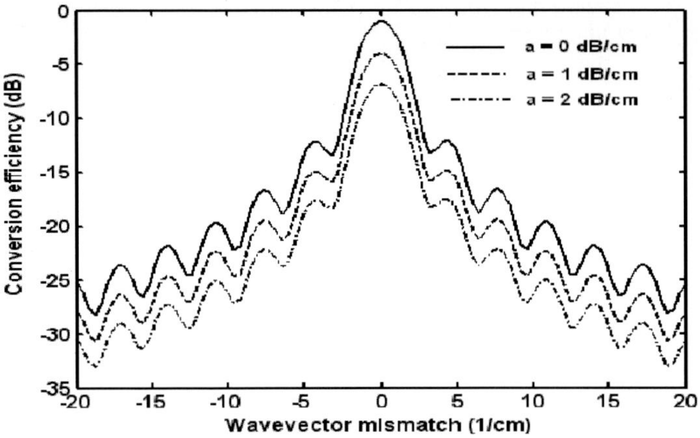

Fig. 6. Power conversion efficiencies (in dB) vs. wavevector mismatch for a 2 cm long waveguide of cross-section $4\,\mu m^2$ and pump power of 400 mW

with 400 mW of pump power effectively coupled into a waveguide of modal cross-section, $A = 4\,\mu m^2$. The efficiency is plotted for different values of the propagation loss. In the following paragraph, we will consider the possibility of using the CARS effect to transfer information between the Stokes and anti-Stokes waves. The conversion efficiency of the wavelength converter is strongly dependent on the phase mismatch, as evident in Fig. 6.

From Fig. 6, the conversion efficiency is within 3 dB of the maximum value for a range of values of $\Delta\beta_{3\,\mathrm{dB}} = \pm 1.3\,\mathrm{cm}^{-1}$, around the phase matched condition.

An important parameter is the utility bandwidth, defined as the 3-dB bandwidth over which efficient conversion occurs. Clearly, an upper limit for the utility bandwidth is given by the Raman bandwidth itself, $2\Gamma = 105$ GHz. The utility bandwidth will depend on the bandwidth dictated by phase matching. To determine this, we need to introduce the detuning parameter, $\Delta\varpi$, defined in Fig. 7. The value of $\Delta\beta$ between the pump $(\beta_\mathrm{p}, \omega_\mathrm{p})$, Stokes $(\beta_\mathrm{S}, \omega_\mathrm{S} = \omega_\mathrm{p} - \Omega - \Delta\varpi)$ and anti-Stokes $(\beta_\mathrm{aS}, \omega_\mathrm{aS} = \omega_\mathrm{p} + \Omega + \Delta\varpi)$ signals, with differing polarizations and detuning, $\Delta\varpi$, around the Raman resonant frequency, will be calculated next. The dependence of $\Delta\beta$ on $\Delta\varpi$ will yield the utility bandwidth. Assuming that $\Delta\varpi \ll \Omega$ (which is quite reasonable for $\Delta\varpi < 105$ GHz, the total $\Delta\beta$ can be written as [30]:

$$\Delta\beta = 2\left(\beta_\mathrm{p}^{\mu} - \beta_\mathrm{p}^{*\mu}\right) - \left[\left(\frac{\mathrm{d}^2\beta_\mathrm{p}^{*\mu}}{\mathrm{d}\omega^2}\right)_\mathrm{mat} + \left(\frac{\mathrm{d}^2\beta_\mathrm{p}^{*\mu}}{\mathrm{d}\omega^2}\right)_\mathrm{wg}\right]\Omega(2\Delta\varpi - \Omega) + \Delta\beta_\mathrm{Raman}\,, \quad (29)$$

where β_p^{μ} and $\beta_\mathrm{p}^{*\mu}$ take into account the most general configuration between the polarization of the Stokes, pump, and a-Stokes fields. The effect of bire-

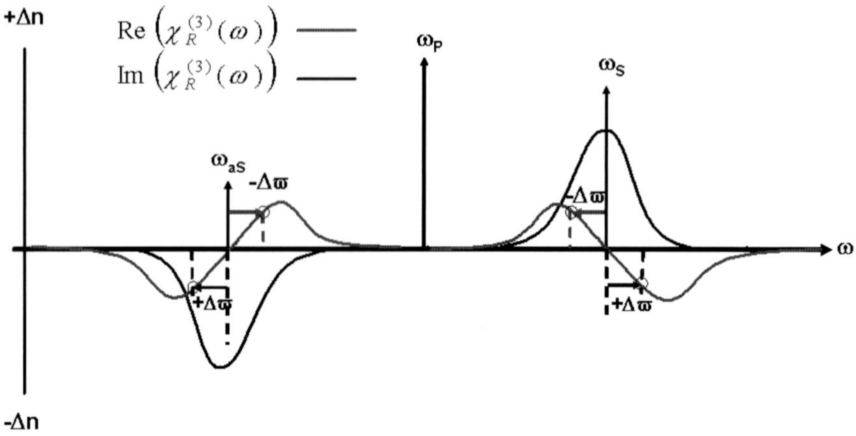

Fig. 7. Real and imaginary part of $\chi_{\mathrm{R}}^{(3)}(\omega)$. The fact that $\mathrm{Im}(\chi_{\mathrm{R}}^{(3)}(\omega_{\mathrm{S}})) = -\mathrm{Im}(\chi_{\mathrm{R}}^{(3)}(\omega_{\mathrm{aS}}))$, expressed in (28), gives rise to a nonzero Δn for a finite value of $\pm\Delta\varpi$. See text for details

fringence is taken into account in the first term of (23a). The second term groups together the effect of material and waveguide dispersion around the Raman-resonant frequency. The last term denotes the excess phase-mismatch arising from the real part of the Raman susceptibility. The real part of the Raman susceptibility leads to changes in the refractive index, Δn, when the Stokes and a-Stokes frequencies are slightly detuned off the Raman resonance. Both changes (Stokes and a-Stokes index) have the same sign [43], and therefore, their contributions to $\Delta\beta$ add up, to effectively introduce a total nonzero phase-mismatch around the resonant wavelength; this is clearly depicted in Fig. 7. The term $\Delta\beta_{\mathrm{Raman}}$ can be expressed as:

$$\Delta\beta_{\mathrm{Raman}} = \frac{2\omega_{\mathrm{p}}\Delta n}{c}, \tag{30}$$

$$\Delta n = \frac{\chi_{\mathrm{real}}^{R}(\Omega - \Delta\varpi)\,|E_{\mathrm{p}}|^2}{2n} = -\frac{\xi_{\mathrm{R}} I_{\mathrm{p}}}{2cn^2\varepsilon_0}\left(\frac{\Delta\varpi/\Gamma}{1+(\Delta\varpi/\Gamma)^2}\right), \tag{31}$$

where $|E_{\mathrm{p}}|^2 = \frac{I_{\mathrm{p}}}{2cn\varepsilon_0}$ [43], and $\xi_{\mathrm{R}} = -11.2\times10^{-14}\,\mathrm{cm}^2/\mathrm{V}^2$ (cf Sect. 2.2). The bandwidth of the process can be defined in terms of $\Delta\beta_{3\,\mathrm{dB}}$ by the condition $|\Delta\beta| < \Delta\beta_{3\,\mathrm{dB}}$. The contribution of detuning, $\Delta\varpi$, to $\Delta\beta$, via material and waveguide dispersion, contained in the second term of (29), has the following values:

$$(\mathrm{d}^2\beta_{\mathrm{TM}}/\mathrm{d}f^2)_{\mathrm{wg}} = -3.5\times10^{-2}\,\mathrm{THz}^{-2}\cdot\mathrm{cm}^{-1}, \tag{32}$$

$$(\mathrm{d}^2\beta_{\mathrm{TM}}/\mathrm{d}f^2)_{\mathrm{mat}} = 0.48\quad\mathrm{THz}^{-2}\cdot\mathrm{cm}^{-1}. \tag{33}$$

These values were obtained from simulations performed with commercial mode-solving software for the specific SOI waveguide described in the previous paragraphs (Fimmwave^{TM} Version 4.1.6). It is clear from (32) and (33)

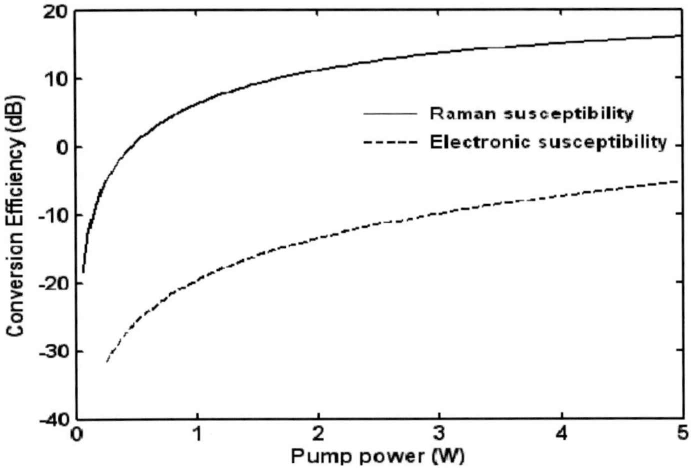

Fig. 8. Power conversion efficiency (in dB) for the Stokes to Anti-Stokes conversion via the Raman susceptibility and Electronic susceptibility for a waveguide 2 cm long and with 4 μm^2 cross-sectional area

that, for $\Delta\varpi < \Gamma$, the contribution of the material and waveguide dispersion to $\Delta\beta$ is much smaller than $\Delta\beta_{3\,\mathrm{dB}}$. Then, the only contribution to $\Delta\beta$ that needs to be considered is that coming from $\Delta\beta_{\mathrm{Raman}}$, (30) and (31). The result is a utility bandwidth that is limited by the input pump intensity (and therefore, conversion efficiency). For values of pump intensity, $I_{\mathrm{p}} < 35\,\mathrm{MW/cm^2}$ (1.4 W effectively coupled into a waveguide of the dimensions considered here), $\Delta\beta_{\mathrm{Raman}}$ is negligible and the utility bandwidth is still $2 \times \Gamma$ (= 105 GHz). At higher pump intensities, $\Delta\beta_{\mathrm{Raman}}$ becomes the limiting factor, e.g. at $100\,\mathrm{MW/cm^2}$ the total utility bandwidth has been reduced to 18 GHz.

At this point, it is instructive to compare the relative contributions to the conversion process due to each of the terms, $\chi_{NR}^{(3)}$ and $\chi_{R}^{(3)}$, in silicon. Figure 8 shows the Stokes-to-a-Stokes conversion efficiency for each component, and it is clear that, in silicon waveguides, the Raman susceptibility dominates the process by approximately 20 dB, regardless of the pump power value.

Wavelength conversion can also be performed using the nonresonant third-order susceptibility in an optical fiber, as shown in Fig. 9a. However, the nonresonant susceptibility, since it is a broadband contribution, also causes cross talk between adjacent WDM channels. In contrast, the conversion in silicon, shown in Fig. 9b, is performed by the Raman susceptibility which, in addition to being much more efficient than its nonresonant counter part, does not result in cross talk. This represents a fundamental difference between conversion in silicon and that in fiber.

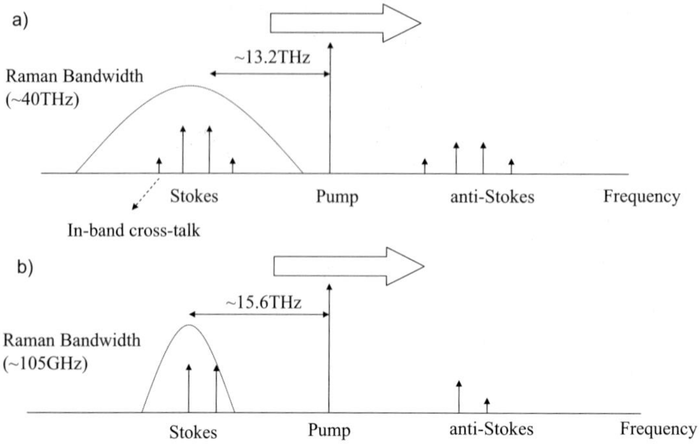

Fig. 9. Pictorial description of the parametric conversion process and the effect of cross-talk. (**a**) In silica fibers, FWM is dominant over CARS. FWM leads to cross-talk between adjacent channels. (**b**) In silicon waveguides, CARS is the dominant effect. The CARS process leads to parametric conversion to the anti-Stokes spectrum, with negligible cross-talk due to CARS and/or FWM. The laser sources are assumed to be monochromatic and the frequency axis is not drawn to scale

2.7 Phase Matching by Dispersion Engineering in Silicon Waveguides

The variation of the power conversion efficiency with phase mismatch, as depicted in Fig. 6, clearly shows that the conversion efficiency of the wavelength converter is strongly dependent on the phase mismatch. The various contributions to the phase mismatch have been considered in (29). In this section, we will focus on the design of the rib waveguide dimensions and choose the pump, Stokes and anti-Stokes wavelengths in such a way that perfect phase matching is achieved and the conversion efficiency is maximized. Material dispersion can be modelled based on the refractive index variation with wavelength for silicon, as given by the following expression [44]:

$$n^2 = \varepsilon_1 + \frac{A}{\lambda^2} + \frac{B\lambda_1^2}{\lambda^2 - \lambda_1^2}, \tag{34}$$

where $A = 0.939816, B = 8.10461 \times 10^{-3}$ and $\lambda_1 = 1.1071\,\mu\text{m}$. The third term is the contribution of the indirect band-gap and the first two represent contributions from all other resonances. The phase mismatch contribution due to material dispersion is graphically represented in Fig. 10. To a good approximation, the phase mismatch is dependent only on $\mathrm{d}^2\beta/\mathrm{d}\omega^2$. Since we are concerned with wavelengths below the band-gap, the dispersion is normal and phase mismatch is negative.

The effect of waveguide dispersion was characterized by numerical simulations, and the phase mismatch is calculated to be within $+5$ and $-15\,\text{cm}^{-1}$

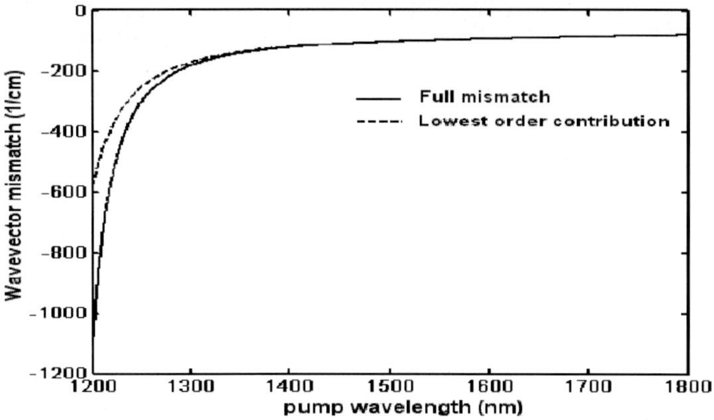

Fig. 10. Phase mismatch contribution (cm^{-1}) due to material dispersion, as a function of wavelength

for the range of waveguide dimensions described below (see Fig. 12). These values are insufficient to cancel the effect of material dispersion. As mentioned earlier, the phase mismatch due to SPM and XPM is also very small and insignificant compared to material dispersion. When the pump is launched in orthogonal polarization relative to the signal (both Stokes and anti-Stokes), the contribution of the birefringence term can be in the range $-600\,cm^{-1}$ to $600\,cm^{-1}$, depending on the waveguide dimensions and the polarization configuration. However, the birefringence of the waveguide, $D = \beta^{TM} - \beta^{TE}$, can be conveniently engineered by adjusting the waveguide dimensions. Figure 11 shows a schematic representation of the dispersion curves for the TE_0 and TM_0 modes of a typical silicon waveguide. The three wavelengths, λ_p, λ_{aS} and λ_s, are indicated. In general, D is wavelength dependent, but it can be assumed constant for the sake of simplicity. The value of $\Delta\beta$ for the configuration in which the pump field is coupled in the TE_0 mode, with the Stokes and a-Stokes fields coupled into the TM_0 mode, is labelled $\Delta\beta_1$, and its different components are given by the orange marks in the figure. The opposite configuration, with the pump field coupled into the TM_0 mode and the Stokes, a-Stokes fields coupled into the TE_0 mode, has a total phase mismatch $\Delta\beta_2$, and its components are marked in blue.

It can be easily seen that, in general, $\Delta\beta_2 < \Delta\beta_1$, and the amount by which these two values are different is $\sim 4D$. Since the value of D depends on the waveguide design, it is possible to design a waveguide such that $\Delta\beta_1 = 2D$, and $\Delta\beta_2 = -2D$. Under these conditions, the device will be a polarization-independent wavelength converter, as can be deduced by the symmetry of both SRS and CARS effects, relative to $\pm\Delta\beta$ (see Figs. 5 and 6). The conversion efficiency of the device can be adjusted by choosing a small value of D. Achieving this phase matched condition requires careful design of the waveguide dimensions. Figure 12 shows the plot of total

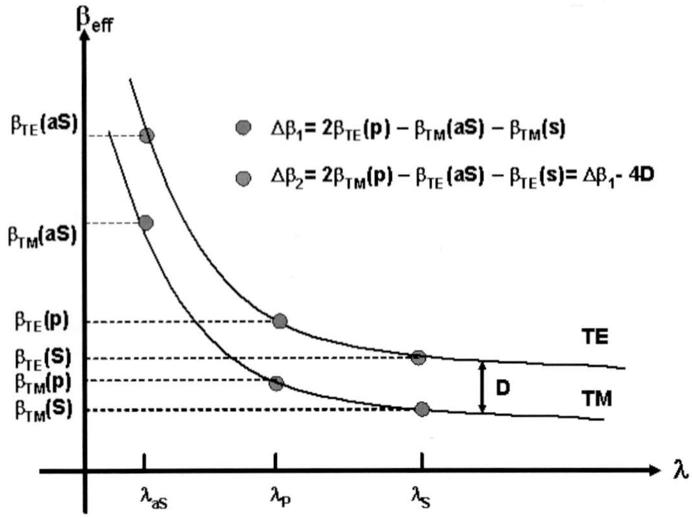

Fig. 11. Schematic representation of birefringence in the CARS effect. The indexes are: P-pump field, S-Stokes field, and aS-anti-Stokes field. See text for details

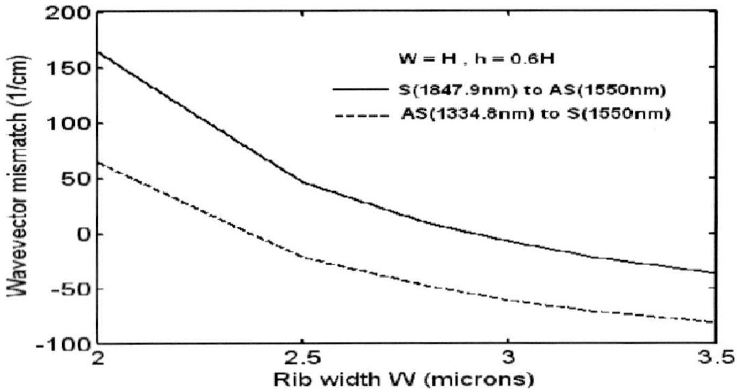

Fig. 12. Total momentum mismatch vs. dimension increase for both Stokes (1847.9 nm) to Anti-Stokes (1550 nm) (*solid line*) and Anti-Stokes (1334.8 nm) to Stokes conversion (1550 nm) (*dashed line*)

phase mismatch as a function of one of the rib dimensions, obtained by simulation [29, 30]. These results were obtained for waveguides with $w = H$ and $h = 0.6H$, with the rib width varied from 2 μm to 3.5 μm and for two different cases of wavelength combinations. These dimensions satisfy the single mode propagation condition. The solid line shows the phase mismatch for Stokes to anti-Stokes conversion from 1847.9 nm to 1550 nm respectively, and the dashed line shows the phase mismatch for anti-Stokes to Stokes conversion from 1550 nm to 1334.8 nm. Phase matching is achieved at an optimum rib width of 2.9 μm and 2.35 μm in the above cases, respectively.

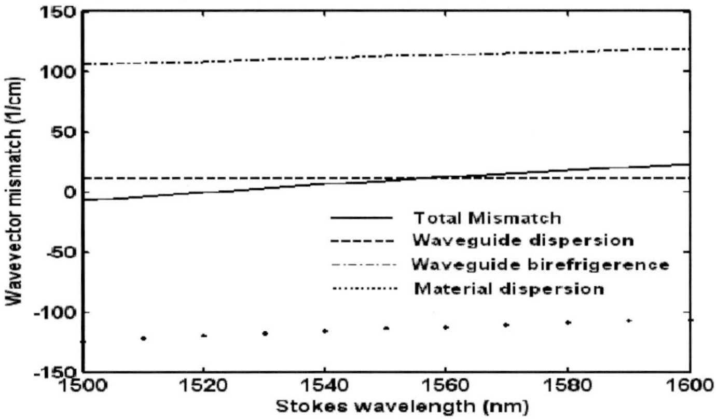

Fig. 13. Momentum mismatch for waveguide $w = H = 2.3\,\mu\text{m}, h = 0.6H$ versus wavelength in CARS: total mismatch (*solid line*), material dispersion contribution (*dotted line*), waveguide dispersion contribution (*dashed line*) and birefringence contribution

The variation of the total phase-mismatch as a function of the Stokes wavelength is shown in Fig. 13, for a waveguide with $w = H = 2.3\,\mu\text{m}$ and $h = 0.6H$. The calculations were performed by adjusting the pump and a-Stokes wavelength such that, for every value of the Stokes frequency, the following is true: $\omega_\text{p} - \omega_\text{s} = \omega_\text{aS} - \omega_\text{p} = \Omega$ (= 15.6 THz). The Stokes wavelength is varied between 1500 nm–1600 nm, and phase matching is achieved around 1520 nm, with the corresponding anti-Stokes wavelength around 1312 nm. The contribution of each of the components to the total phase mismatch is shown in the figure. The calculated waveguide dispersion is found to be almost constant over the entire wavelength range. It is important to analyze the effect of deviations of the waveguide dimensions due to fabrication errors on the phase matching condition. Deviations in waveguide width can arise due to error in the mask features and the improper transfer of the pattern during lithography. Uncertainties in etching and undercutting could lead to error in the rib/slab heights. The effect of these uncertainties on the phase mismatch is graphically depicted in Fig. 14. For the mismatch to remain within $\pm 1.3\,\text{cm}^{-1} (= \Delta\beta_{3\,\text{dB}})$, it is required that the variations in waveguide width and slab height be within 70 nm and 10 nm, respectively. As a benchmark, the transistor gate length in CMOS circuits going into production next year is 70 nm, with ±7 nm of accuracy. Therefore, the dimensional accuracy required for achieving phase matching in silicon waveguides is within the tolerance of the silicon manufacturing process.

In conclusion, waveguide engineering provides a powerful tool to maximize the efficiency of nonlinear optical phenomena in silicon microstructures. In the following section, the experimental realization of the ideas discussed up to now, will be described.

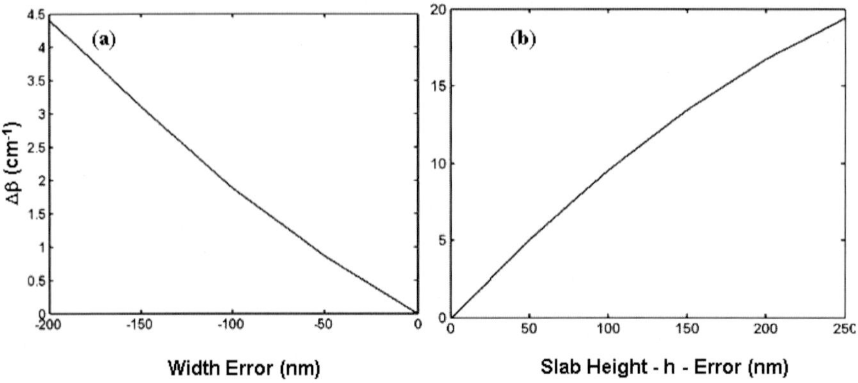

Fig. 14. Variation of phase mismatch with deviation in waveguide dimensions (**a**) Rib width (nm) (**b**) Slab height (nm)

3 Experimental Results

3.1 Spontaneous Raman Scattering Observed in Silicon Waveguides

The experimental setup used to observe Spontaneous Raman Scattering in silicon waveguides is shown in Fig. 15. The pump laser is a high power fiber laser that delivers CW, randomly-polarized light, at 1427 nm, with a 2 nm linewidth at maximum power.

The selected scattering configuration combines a large relative Raman efficiency (see Table 1), with orthogonal pump and signal polarizations. The orthogonal configuration satisfies the Raman selection rules for scattering from a silicon crystal oriented with its $[1\bar{1}0]$ axis along the beam propagation direction, and is helpful for signal-pump isolation at the output. In order to

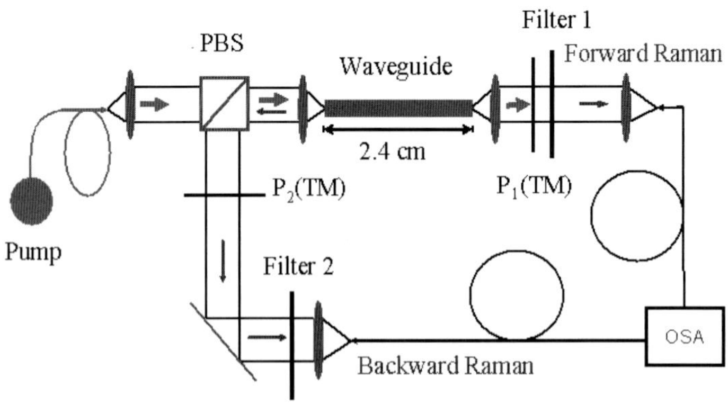

Fig. 15. Experimental setup for measurement of spontaneous Raman scattering from silicon waveguides

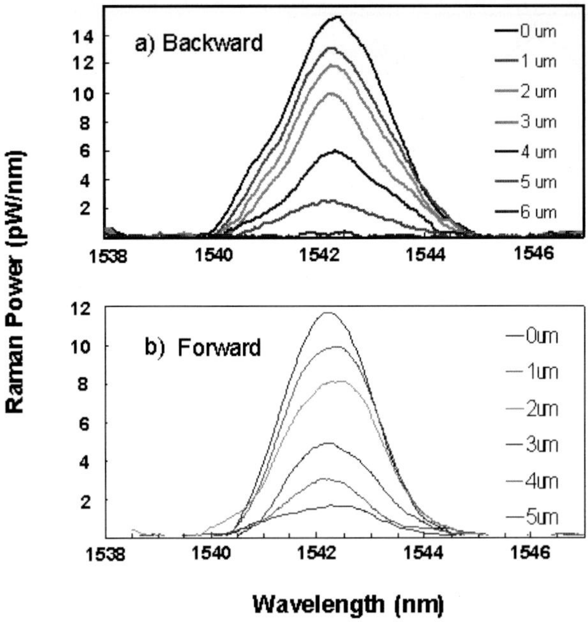

Fig. 16. Raman spectra from silicon obtained for different horizontal offsets of the waveguide relative to the optical axis. (**a**) shows the back-scattered spectra, and (**b**) shows the forward-scattered spectra

realize this configuration, a Polarization Beam Splitter (PBS) was used to split the pump beam into S and P polarized beams. The P-polarized beam is sent into the waveguide and acts as the TE_0-mode Raman pump. The Raman signal backscattered from the waveguide (TM_0 mode) is collected from Port 2 of the PBS. For further signal filtering and background rejection, a polarizer (P_1), and a 1530 nm–1550 nm bandpass filter (Filter 1) are placed before the collection fiber port in Port 2. A polarizer (P_2) and a bandpass filter (Filter 2) remove the pump signal and pass the backward scattered radiation into another collection fiber port. The silicon rib waveguide used is shown in Fig. 2. The waveguide for this particular experiment has a 5 μm width (w) and 2.5 μm rib height (h), with a 5 μm total thickness (H).

Figure 16 shows the experimental results obtained in both forward and backward scattering configurations. All the spectra are shifted by 15.6 THz from the pump laser line (at 1427 nm), and therefore, belong to the spontaneous Raman signal from zone-center optical phonons in bulk silicon. The different spectra on each of the Fig. 16a and b correspond to displacements of the waveguide in the plane perpendicular to the optical axis of the experimental setup shown in Fig. 15. The rapid decrease in the signal as the waveguide is displaced by 2 μm or 3 μm, either laterally or vertically, proofs that the signal collected comes directly from the bulk of the SOI waveguide,

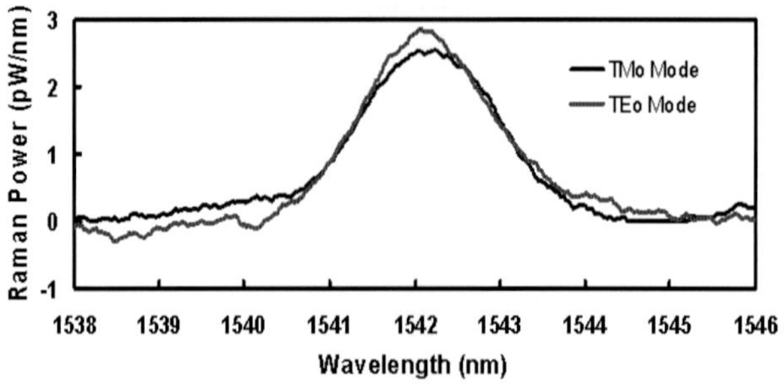

Fig. 17. Raman spectra from SOI waveguides obtained by rotating the polarizer P1 (Fig. 15) by 90°. The scattering intensity is essentially the same, as expected from the selection rules

and not from the front-end facet of the silicon chip. The selection rules, discussed in Sect. 2.1, can also be easily verified using the experimental setup shown in Fig. 15. The Raman scattering intensity, collected from the TM_0 and the TE_0 modes of the waveguide, can be obtained by simply rotating the polarizer P_1 over 90°. The result is shown in Fig. 17, where it can be seen that the scattering intensity is the same in both polarizations.

The fact that the absolute intensity of Forward and Backward Raman scattering are comparable, as shown in Fig. 16, is expected from bulk Raman scattering. Other authors have found a surprising asymmetry between the two processes in GaP waveguides, and attributed it to phonon confinement effects introduced by the waveguide structure [45]. These effects should have no role in the case of silicon waveguides, because in silicon the three optical modes participate equally in the first-order Raman process. However, if the Raman intensities are plotted as a function of the pump power intensity, a clear asymmetry is found between Forward and Backward scattering, as shown in Fig. 18. The origin of this asymmetry has to be related to the waveguide propagation losses, rather than coupling losses, since the latter affect both directions of propagation in a symmetric fashion. In the case of spontaneous Raman scattering, the equations that govern the propagation of pump ($P_\mathrm{P}(z)$) and signal ($P_\mathrm{R}(z)$) field power along the waveguide are obtained in a manner similar to [46], but with the effect of the stimulated emission term omitted. This is validated since the low value of the Raman signal measured clearly shows that the experiment was performed well within the linear regime in silicon. The CMC theoretical approach discussed in Sect. 2.3, demonstrates that there is no inter-modal coupling within the waveguide; therefore, in the following derivation only the intensity of the fields will be considered, in the usual manner. The expressions for $P_\mathrm{P}(z)$ and $P_\mathrm{R}(z)$ are then,

$$P_\mathrm{P}(z) = P_\mathrm{P}(0)e^{-\gamma z}\,, \tag{35}$$

$$\frac{\mathrm{d}P_\mathrm{R}(z)}{\mathrm{d}z} = \mp\gamma P_\mathrm{R}(z) \pm \alpha P_\mathrm{P}(z)\,, \tag{36}$$

where the upper (lower) sign in (36) applies in forward (backward) scattering. Here, γ is the propagation loss of the waveguide in units of cm^{-1}, and α is the spontaneous Raman coefficient for the silicon waveguide, defined as:

$$\alpha = S \Delta \Omega \,. \tag{37}$$

with S the Raman scattering efficiency ($\mathrm{cm}^{-1} \cdot \mathrm{sr}^{-1}$) in silicon at 1542 nm, as defined in Sect. 2.1. The factor $\Delta \Omega$ is the effective solid angle of collection for the TM_0 waveguide mode, and is measured to be $\Delta \Omega = 0.013\,\mathrm{sr}$. The model used to obtain (35) and (36) is actually a simplification of the real optical field propagation along the waveguide, and assumes that the field has already been coupled into a single-mode. In reality, the situation is quite different, since the parameter γ turns out to be a function of the propagation length, $\gamma(z)$. Modeling of this effect is out of the scope of the present work, but measurements can be accurately done using well known, infrared scattering measurement techniques [47]. For the purpose of the present discussion, we will keep the model from (35), in the understanding that the parameter γ is rather an "effective" or averaged value of the propagation loss along the waveguide. The total Raman signal power, P_R, measured either in forward configuration (at the output end of the waveguide) or in backward configuration (from Port 2 in the PBS) is related to the pump power at the input facet of the waveguide (P_P) by integrating (35) and (36), with the following result:

$$\text{Backward: } P_\mathrm{R} = \alpha \left(\frac{1 - e^{-2\gamma L}}{2\gamma} \right) P_\mathrm{P} \,, \tag{38}$$

$$\text{Forward: } \quad P_\mathrm{R} = \alpha L e^{-\gamma L} P_\mathrm{P} \,, \tag{39}$$

where L is the length of the waveguide. From (38) and (39), it is seen that the ratio between the slopes of the lines in Fig. 18, m, can be related toγ (in the low propagation loss limit) by

$$m = \frac{\sinh(\gamma L)}{\gamma L} \,, \tag{40}$$

with the value of m obtained from Fig. 17 and using (40), we get an estimate for the propagation loss, $\gamma = 0.64\,\mathrm{cm}^{-1} = 2.8\,\mathrm{dB/cm}$. Note that the value thus obtained is independent of the Raman scattering coefficient and the optical coupling efficiency. It is expected that the value for propagation losses obtained using the Raman method would give an over-estimate. The reason for this is that the mode-coupling loss mentioned above also contributes to a larger backward scattering efficiency. This is because the scattered Raman light is more likely to be collected in the backward direction, since it is not effectively coupled to the waveguide in the forward direction. Other traditional methods to measure propagation loss of the single-mode along the waveguide, like cut-back techniques, result in lower values. Including this value of γ into (39) and using the measured slope in Fig. 18, with further

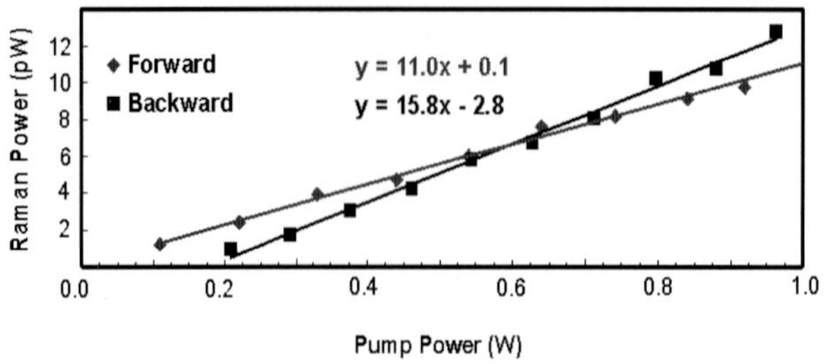

Fig. 18. Spontaneous Raman intensity as a function of pump power. Forward and backward scattering cases are shown. The pump power was measured between the PBS and the input coupling lens

correction for the back-scattering collection losses and waveguide coupling losses (20 dB total), a value is obtained of $\alpha = (5.3 \pm 3.2) \times 10^{-9}\,\mathrm{cm}^{-1}$. This is remarkably close to the values obtained above using the CMC formulation (cf the last two columns in Table 2). Notice that the values listed there correspond to waveguides with a total rib height of $H = 4.3\,\mu\mathrm{m}$, as compared to the $5\,\mu\mathrm{m}$ rib waveguide used in this experiment. With the above value for $\Delta\Omega(0.013\,\mathrm{sr})$, the scattering efficiency in silicon is obtained to be $S = (4.1 \pm 2.5) \times 10^{-7}\,\mathrm{cm}^{-1} \cdot \mathrm{sr}^{-1}$. The difference (a factor of 2), with the value obtained by *Ralston* and *Chang* for bulk silicon [35], should be related to the modal coupling losses in the waveguide, as discussed above. Indeed, as the high power pump laser impinges on the front facet, it propagates with high losses through the first few millimeter inside the silicon waveguide, before coupling into the single TE_0 mode. This means that, for most of the length of the waveguide, the effective pump laser power is less than assumed in the calculations, therefore reducing the value of the measured scattering coefficient.

3.2 Third-order Optical Nonlinearities in Silicon Waveguides

In bulk silicon, due to the crystal symmetry, second-order nonlinearities are highly suppressed, but third order nonlinearities may be observable if the field intensities used to generate them are high enough. In what follows, the three third-order optical nonlinearities in silicon, which have been treated theoretically above, will be demonstrated experimentally in silicon waveguides. These processes are: Stimulated Raman Scattering (SRS), Two-Photon Absorption (TPA), and Coherent Anti-Stokes Raman Scattering (CARS). The results confirm the values obtained from third-order nonlinear properties of bulk silicon.

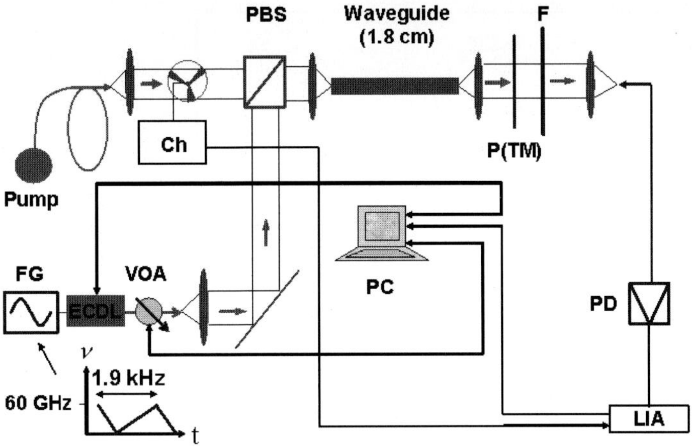

Fig. 19. Experimental setup: Pump-CRC fiber laser; Ch-Chopper; PBS-Polarization Beam Splitter; LIA-Lock-in amplifier; ECDL-External cavity diode laser (tunable); FG-Function generator (60 GHz freq. range); VOA-Variable optical attenuator; PD-optically-broadband photodetector. The *thick lines* represent electrical connections and wiring, the *thin lines* represent free-space optical beams, and the *colored lines* represent optical fiber. The SOI rib waveguides used for this experiment have a length, $L = 1.8\,\text{cm}$, and a measured TM_0 modal area of $a = 5.4\,\mu\text{m}^2$

3.2.1 Observation of Stimulated Raman Scattering in Silicon Waveguides

Figure 19 shows the experimental setup used to measure SRS in SOI waveguides. The pump laser is the same as described in the previous section. In the present case, the S-polarized input of the PBS is used to couple the signal beam into the TM_0 mode of the waveguide. The signal laser is an External Cavity Diode Laser (ECDL) with a linewidth of $< 300\,\text{kHz}$. The laser wavelength was scanned (0.2 nm/step) through a wavelength range from 1537.0 nm to 1547.0 nm. A variable optical attenuator (VOA) was used to regulate the amplitude of the signal beam. A polarizer (P) and a high pass filter are used to collect the field from the TM_0 mode coupled out of the waveguide.

To isolate the pump and signal interaction, a low frequency amplitude modulation is introduced in the pump beam, using a chopper (Ch) operating at 100 Hz. In the presence of a nonlinear interaction between pump and signal beams in the waveguide, this modulation will be transferred onto the CW signal beam. The interaction should possess the spectral characteristics of the Raman Effect. Given the 1427 nm pump used in our experiment, and the 15.6 THz optical phonon frequency, the gain bandwidth is expected to be centered at 1542.3 nm. An optically-broadband photodetector (PD) is used to measure the signal modulation. The signal from the PD is processed by a lock-in amplifier (LIA), which is locked to the chopper frequency.

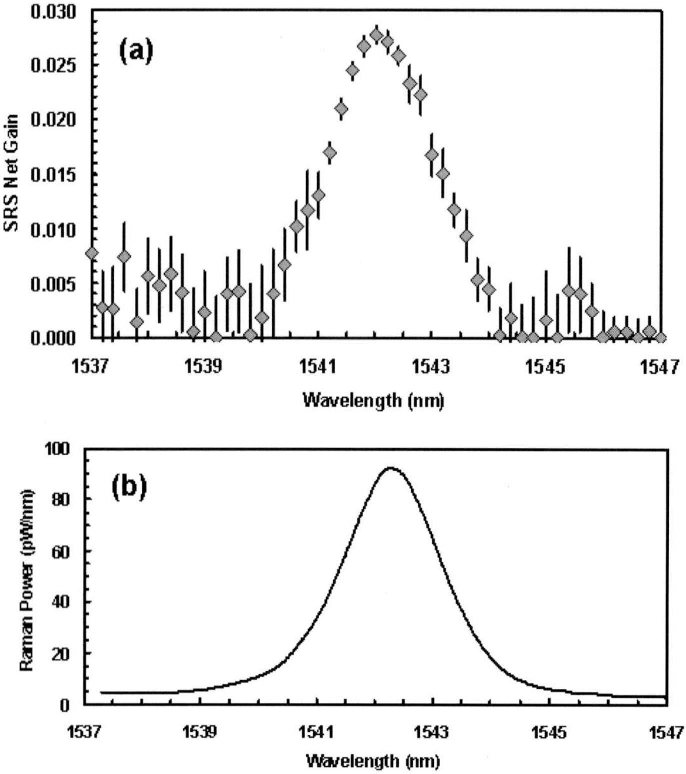

Fig. 20. (**a**) Measured spectral characteristic of the Stimulated Raman Scattering (SRS) in the silicon waveguide. The *error bars* are the standard deviation from 10 averages per point. The pump power was 0.64 W at the front facet of the waveguide. SRS Net gain is the ratio of the amplitude of the LIA output to the average signal power throughput. (**b**) Spontaneous Raman Spectra of the same waveguide with the same pump power as in (**a**), collected with an OSA

A function generator (FG) was used to modulate the ECDL frequency. The purpose of this is to average out the Fabry–Perot (FP) fringes produced by the waveguide facets, which are not AR coated. Shown in Fig. 20a is the measured signal power gain as a function of signal laser wavelength. The pump power, measured before the waveguide, was 0.64 W. The characteristic wavelength dependence of the Raman process is clearly shown, and a signal amplification of 3% is obtained. For comparison, shown in Fig. 20b is the measured spontaneous emission. The gain peak in the stimulated emission occurs at 1542.3 nm, as expected. Additionally, the FWHM linewidth of 310 GHz is consistent with the measured value for spontaneous emission.

In Fig. 21, the maxima of the signal wavelength scans are plotted versus the effective pump power. The maximum signal gain obtained is 0.25 dB, corresponding to $\sim$ 6% signal amplification. The slope of the curve is approx-

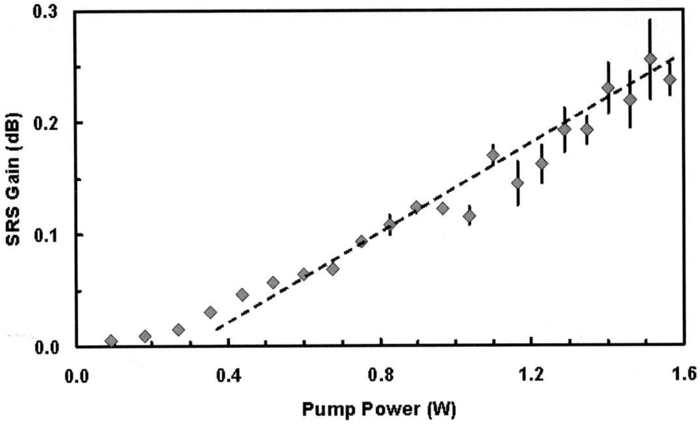

Fig. 21. The maxima from each spectral scan are plotted against effective pump power coupled into the front facet of the waveguide. A maximum of 0.25 dB (6%) amplification is obtained

imately linear, as expected for the gain of a Raman amplifier as a function of pump power.

The amplified signal power expected from a waveguide of length, L, given by (10), should be adjusted to accommodate for the pump broadening of the Raman signal, as follows,

$$P_S(L) = P_S(0)\ \exp\left(-\gamma\, L + \frac{g_\mathrm{R}\, P_\mathrm{p}(0)}{A \times (1 + \Delta\,\nu_\mathrm{p}(P_\mathrm{p})/\Delta\,\nu_\mathrm{R})} L_\mathrm{eff}\right) . \tag{41}$$

The factor $(1 + \Delta\,\nu_\mathrm{p}/\Delta\,\nu_\mathrm{R})$ is an approximation describing the increase in Raman threshold due to the finite linewidth of the pump laser. The dependence of threshold on pump linewidth is well known in Stimulated Brillouin Scattering (SBS) in optical fibers [40]. Because of the extremely large Raman bandwidth in fibers (5 THz–10 THz), this effect is neglected in modelling SRS in fibers. However, due to the narrow Raman linewidth in silicon ($\Delta\,\nu_R = 105$ GHz), the dependence of threshold on pump linewidth must be considered here. Additionally, the linewidth, $\Delta\,\nu_\mathrm{p}$, of the pump laser increases with output power. Based on the specifications provided by the manufacturer [48], the dependence of linewidth with pump power can be assumed to be linear, with a rate of 70 GHz/W. By using the total Stokes amplification obtained experimentally, in (41), a value of $g_\mathrm{R} = 2 \times 10^{-8}$ cm/W is found for the SRS gain coefficient. As described in Sect. 2.1, the value obtained by using the measured waveguide spontaneous scattering efficiency and (9), is $g_\mathrm{R} = 3.7 \times 10^{-8}$ cm/W, in good agreement with the present measurement. Other nonlinear optical effects may be detrimental to SRS. A serious problem for fiber-based Raman amplification is presented by SBS. However, the Brillouin scattering coefficient for silicon is about two orders of magnitude smaller than the Raman coefficient [49]. Furthermore, as has been discussed above, the pump broadening reduction of the effective SBS gain is more pronounced than in the SRS case, due to the smaller bandwidth of the Brillouin signal. In conclusion, the possibility of SBS pump depletion in SOI-based Raman amplification schemes should be discarded. The hypothesis of two-photon-absorption quenching the SRS amplification measured above, as outlined in Sect. 2.2, should be ruled out by direct measurement of TPA in SOI waveguides. This will be the topic of the following section.

3.2.2 Two-Photon Absorption in Silicon Waveguides

Two-Photon Absorption has been recognized as a detrimental factor in the operation of SOI optical devices in the telecommunications C-band (1.52 μm–1.58 μm), particularly causing pump-power depletion in waveguide-Raman amplifiers (see Sect. 2.2), and also through induced cross-talk in DWDM devices [50]. However, TPA in silicon has been effectively used at these wavelengths for diverse purposes, like implementing optical-beam-induced-current imaging techniques, used in IC failure analysis [51], and ultrafast sampling systems [52, 53]. The TPA coefficient has been measured in silicon at 1.06 μm [54, 55], but there have been very few reports in the 1.5 μm wavelength region [50, 56]. Furthermore, theoretical efforts to calculate TPA at different wavelengths have been concentrated on III–V and II–VI semiconductors [57], which present a direct band-gap transition. The indirect transition involved in TPA for silicon, at the 1.5 μm region, requires the treatment of a phonon-assisted electronic process, which complicates the analysis. In this context, the availability of experimental data for TPA in silicon at 1.56 μm becomes of practical importance. It supports recently measured data in this wavelength region, providing relevant information with which theory can be compared, and settling the ground for the practical use of nonlinear effects in silicon.

The experimental setup used for this measurement is depicted in Fig. 22. A passively mode-locked laser operating at a 25 MHz pulse-repetition rate, and with a 0.9 ps pulsewidth, produces an input power level an order of magnitude higher than those used in previous reports of TPA in silicon [50, 56]. The laser operates at a center wavelength of $\lambda = 1560\,\text{nm}$, which is well below the indirect band-gap of silicon; therefore, linear absorption can be considered negligible. The regime that this experiment then explores is that of indirect, simultaneous two-photon absorption [55]. Such an effect is of interest for ultrafast photodetection in silicon.

Shown in Fig. 23 is a plot of the throughput peak power, P_{out} as a function of the input peak power, P_{in}. The effect of the nonlinear absorption is clearly visible for input power levels above 50 W. This is more than an order of magnitude higher than the power levels used to observe SRS.

Under TPA-dominating conditions, the input power at the front facet of the waveguide, P_{in}, and the signal measured on PD_2, namely, P_{out}, are related

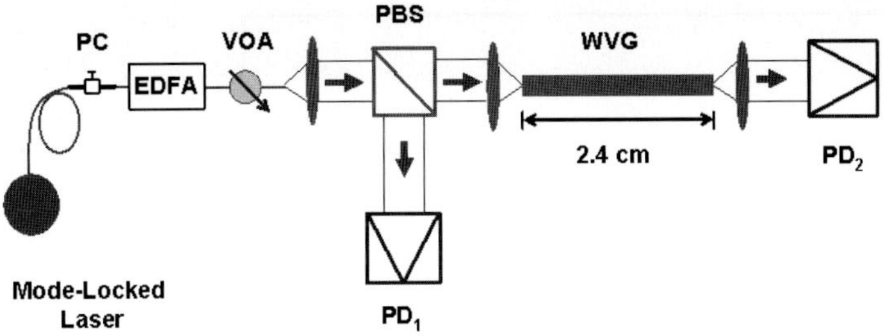

Fig. 22. Experimental setup for measuring TPA. The waveguide in this case is different from the one used for SRS. PC-Polarization controller; VOA-Variable optical attenuator; PBS-Polarization beam splitter; PD1 and PD2 photodetectors

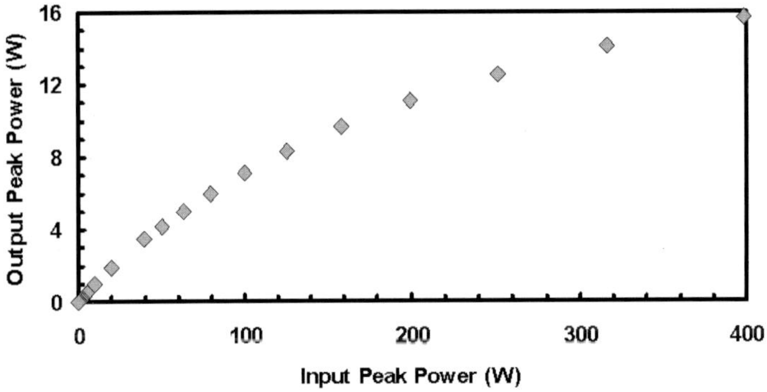

Fig. 23. Output power vs. input power results, using a mode-locked laser, and depicting a nonlinear relationship

by the following expression [55],

$$\frac{P_{\text{in}}}{P_{\text{out}}} = e^{\gamma L}\left(1 + \frac{BL_{\text{eff}}}{A}P_{\text{in}}\right), \tag{42}$$

where $\gamma = 0.46 \pm 0.1\,\text{cm}^{-1}$) is the propagation loss in the waveguide, A is the modal cross section ($= 8.1 \pm 1.0 \times 10^{-8}\,\text{cm}^2$), and B is the TPA coefficient (in cm/W). The result is plotted in Fig. 24, where the linear behavior of $P_{\text{in}}/P_{\text{out}}$ versus P_{in} is clearly demonstrated up to 400 Watt of input peak power. The fact that spontaneous TPA in silicon is the dominant absorption process up to such high power regimes, in the 1.5 μm region, is of fundamental importance in the design of an ultrafast silicon photodetector and/or imaging device. The resulting value for B is then,

$$B = -\frac{c_1 A}{L_{\text{eff}}}e^{-\gamma L} = (4.4 \pm 1.0) \times 10^{-10}\,\text{cm/W}, \tag{43}$$

where c_1 is the slope of the curve shown in Fig. 24 and the main elements in the error are the uncertainties in γ, and in A.

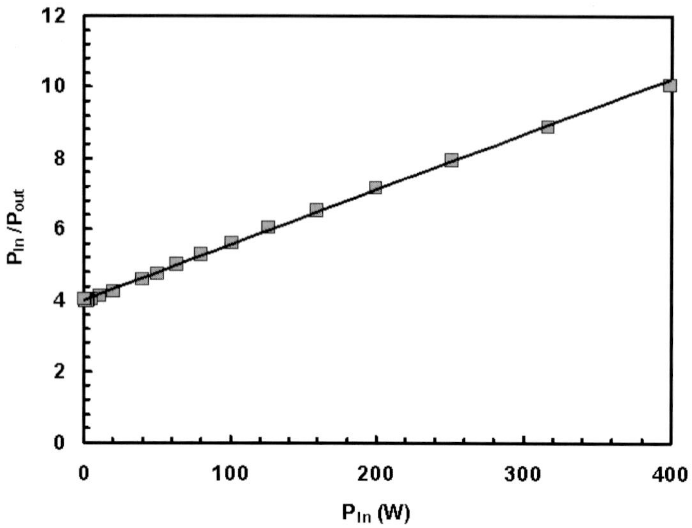

Fig. 24. TPA measurement result. The input power, P_{in}, is not corrected for coupling losses. The linear behavior is maintained up to 400 W

This value of TPA agrees with what has been reported in the literature and with a lorentzian extrapolation of values measured at 1.06 µm [55]. In summary, the value measured for B implies that no practical effect of TPA should be considered in the results of Sect. 3.2.1 for SRS measurements.

3.2.3 Demonstration of Coherent Anti-Stokes Raman Scattering (CARS) in Silicon Waveguides

Coherent Anti-Stokes Raman Scattering (CARS), Parametric Wavelength Conversion (PWC), and Optical Parametric Amplification (OPA) are nonlinear optical processes occurring in fibers and nonlinear crystals [58, 59]. PWC and OPA have been investigated in optical fibers in the context of broadband signal amplification, or as wavelength-conversion mechanisms [60, 61]. The theory that governs these phenomena has been fully developed in Sects. 2.4 and 2.5. The effect involves interaction of three fields: the pump, the Stokes, and the anti-Stokes (a-Stokes) fields, at wavelengths (frequencies) $\lambda_{\text{P}}(\omega_{\text{P}})$, $\lambda_{\text{S}}(\omega_{\text{S}})$, $\lambda_{\text{aS}}(\omega_{\text{aS}})$, respectively. The incoming fields will be the pump, $\boldsymbol{E}_{\text{P}}$ (TE_0 mode), and the Stokes, $\boldsymbol{E}_{\text{S}}$ (TM_0 mode). In this section, a demonstration of the effect of Stokes to anti-Stokes parametric conversion in silicon waveguides and its counterpart, that is, a-Stokes to Stokes conversion, will be presented. The potential for building a CARS-based, silicon wavelength-converter will be discussed in the following section. The experimental setup used for this series of experiments is similar to the one presented above to measure spontaneous Raman emission and SRS in silicon waveguides (see Figs. 15 and 19). As usual, the pump laser is coupled into the TE_0 mode,

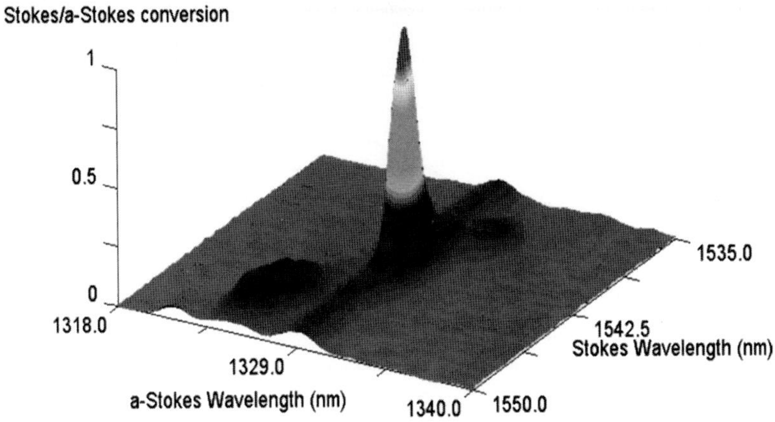

Fig. 25. Plot of the a-Stokes spectra collected for a given value of the signal (Stokes wavelength). The z-axis represents the Stokes/a-Stokes conversion efficiency, normalized to unity. Note the clear appearance of two "satellite" resonances, as mentioned in the text

and the signal laser is coupled into the TM_0 mode. The signal laser in this case consists of two different types. The Stokes signal laser can be scanned in a wavelength range from 1530 nm to 1560 nm, and the anti-Stokes signal laser is fixed at the Raman anti-Stokes wavelength, i.e. 1328.8 nm. Figure 25 shows the a-Stokes spectra measured as a function of the Stokes signal wavelength. The spectra were collected by scanning the spectrum analyzer in the range 1318 nm to 1340 nm, while the Stokes laser was set at a wavelength λ_S, and the pump laser effective power was 0.7 W. The wavelength of the Stokes laser, λ_S, was tuned from 1535 nm to 1550 nm. There is a clear peak at 1328.8 nm of anti-Stokes emission when the Stokes laser is tuned to 1542.3 nm. The latter corresponds to the peak of the spontaneous Raman emission in silicon. The nature of the weaker features in Fig. 25 is due to the wavelength-dependent character of the phase-matching parameter, $\Delta\beta$, in the SOI waveguide, and the characteristic Sinc^2 dependence of the conversion efficiency with $\Delta\beta$ (Sect. 2.6).

To obtain the down-conversion efficiency, the a-Stokes integrated power was measured as a function of the Stokes signal power, for different values of pump power. The result is shown in Fig. 26.

From the slopes in Fig. 26, the maximum Stokes/anti-Stokes power conversion efficiency measured for this particular waveguide is found to be $\sim 10^{-6}$ for an effective pump power of ~ 0.7 W. By using the signal laser at 1328.8 nm through the TM_0 port of the PBS, the opposite effect is found, namely, power conversion into Stokes wavelength. Figure 27 shows the effect for different values of the a-Stokes laser power, with the pump laser effective power set at 0.7 W. In this case it was not possible to obtain a plot like the one shown

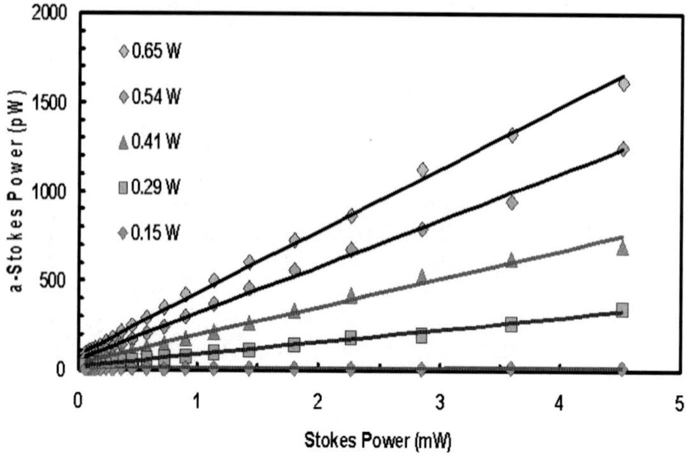

Fig. 26. Stokes to a-Stokes conversion efficiency. The Stokes power was varied using a VOA. The efficiency, defined as the slope of the plots shown, increases with pump power, as expected

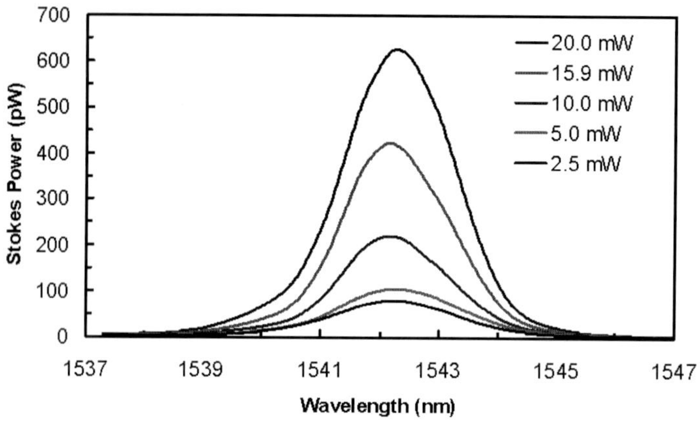

Fig. 27. Anti-Stokes to Stokes conversion. The a-Stokes power was varied using a VOA. Stokes power shown is not corrected for collection efficiency (~ 10 dB from output facet of waveguide)

on Fig. 25 because the a-Stokes laser used is not tunable across such a wide range.

3.3 Demonstration of Parametric Raman Wavelength Conversion in Silicon

The results shown up to this point have demonstrated the potential for light amplification and wavelength conversion in silicon optical circuits. The analysis carried out in Sects. 2.4 and 2.5 led us to the conclusion that the para-

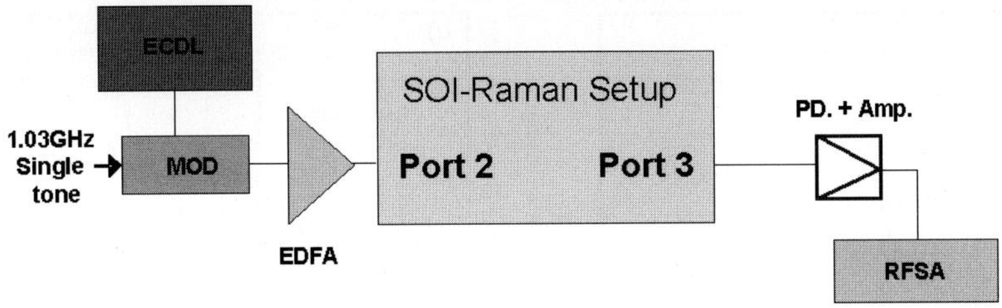

Fig. 28. CARS-based, SOI signal transmission switch

metric coupling between pump, Stokes and anti-Stokes waves, through the Raman susceptibility, can be used to transfer information between the wavelength channels. It has been demonstrated that the conversion efficiency of this process is dependent upon the phase difference between the three waves, the intensity of the Pump wave, and the Raman susceptibility of the medium, as detailed in Sect. 2.5. The experimental setup used to realize this wavelength-conversion scheme is shown in Fig. 28. The SOI-Raman setup consists of the same pump laser and waveguide as described in the previous experiments. An external cavity diode laser (ECDL) is used as the Stokes source at 1542 nm. The ECDL is externally modulated and amplified by an Erbium-doped fiber amplifier (EDFA), in order to maximize the amount of down-converted, a-Stokes signal. At the output, a BPF filter is used to extract the converted a-Stokes signal, at 1328.5 nm, before photodetection and 40 dB electrical amplification.

Figure 29a shows the converted a-Stokes signal spectrum. Note that the FWHM for wavelength conversion, which is approximately 250 GHz, is determined solely by the pump laser linewidth. Figure 29b shows the conversion of 1.03 GHz RF modulation from 1542 nm to 1328.5 nm. The input RF signal power applied to the Stokes wavelength is shown in the inset. The measured electrical Signal to Noise Ratio (SNR) is 34 dB. Figure 30a shows the wavelength-converted eye diagram when a 1 Mb/s PRBS $(2^7 - 1)$ modulation is applied to the Stokes laser. Figure 30b depicts the measured electrical SNR versus the input Stokes optical power, along with the theoretical fit. Also shown in Fig. 30b is the SNR when the conversion efficiency is increased by 10 dB.

As has been discussed in Sect. 2.7, by proper rib waveguide design, the waveguide birefringence term can be used to obtain phase matching conditions. This hypothesis has been verified using numerical simulations [30]. Under these conditions, with 0.7 W of pump power coupled into a waveguide with 1 μm^2 single-mode cross-section, the conversion efficiency would be approximately 65%. As also mentioned in Sect. 2.7, if the proper design is used for the SOI waveguide, a polarization-insensitive device can be envisioned. This would double the effective data-processing capacity of the device, by using polarization multiplexing.

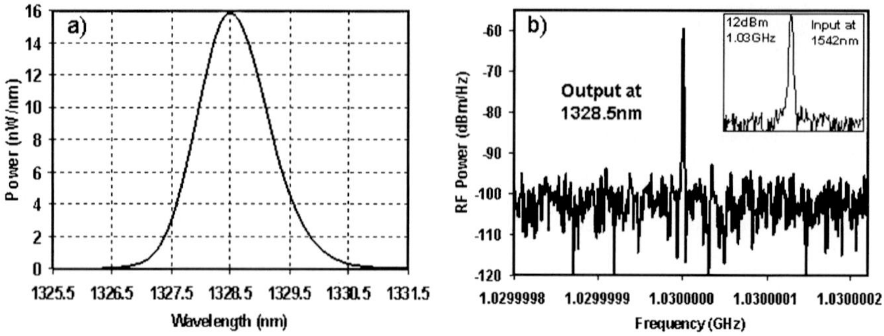

Fig. 29. (**a**) Maximum a-Stokes signal obtained. (**b**) Single-tone data conversion at 1.03 GHz. The *inset* shows the incoming signal at 1542.3 nm

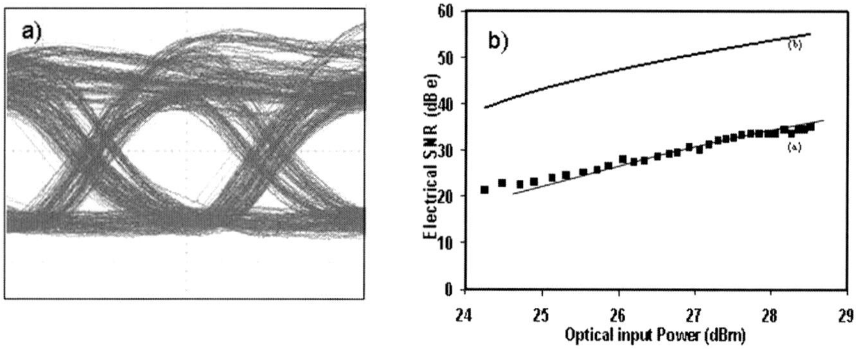

Fig. 30. Wavelength Conversion: (*left*) Eye diagram (1 Mb/s), (*right*) Electrical SNR: Measured (*squares*) and calculated (*solid lines*)

4 Summary

This chapter has outlined a method for light generation, amplification and wavelength conversion in silicon using stimulated Raman processes. These effects are routinely observed in optical fiber; however, several kilometers of length are required to do so. What enable us to achieve these processes on millimeter length scales on a chip are two fundamental differences between an optical fiber and a silicon microstructure. The first is the difference in atomic structures. An amorphous material such as the silica fiber supports many vibrational modes, hence the Raman gain has a very large bandwidth and a low peak value, requiring long interaction lengths for the effect to be observed. In contrast, silicon is single crystal and supports only three optical vibration modes. The result is a much narrower gain bandwidth but a much higher gain peak (nearly 10^4 times higher). Second, is the difference in modal areas. The large index-contrast in the silicon/SiO_2 waveguides results in a mode area that is approximately 100 times smaller (assuming $0.8\,\mu m^2$ waveguides) than in a standard single mode fiber (mode area $= 80\,\mu m^2$).

The proportionally higher power density in the silicon waveguide lowers the threshold for nonlinear optical processes.

The intrinsic Raman bandwidth in silicon is 105 GHz, and it is broadened in the experiments by the pump laser linewidth (typically $\sim$ 2 nm). The resulting bandwidth is sufficient for amplifying several WDM channels. The bandwidth can be broadened by broadening the pump linewidth even further (although this reduces the peak gain) or by using multiple pump wavelengths.

The Raman phenomenon is fully tunable; the tuning range is only limited by the available pump wavelengths. This is an advantage over the nanocrystal approach for light generation and amplification. In addition, the Raman effect can also perform wavelength conversion, a critical function needed for optical packet switching. Wavelength conversion in silicon is yet to be demonstrated with other techniques. Another important advantage of this approach is that it does not require rare-earth dopants or nanostructures. Hence, it is truly compatible with silicon IC manufacturing.

Going forward, low loss waveguides with small cross sections are required for realizing high performance devices. Surface roughness produces strong scattering and high propagation loss due to the high index contrast between the silicon waveguide core and the cladding (air or SiO_2). As such, losses of silicon waveguides tend to increase with reduction in cross section [62]. Fortunately, new waveguide fabrication processes that are in development promise low-loss waveguides with submicron cross sections [63]. Naturally, one must be able to efficiently couple light into these structures. Several novel approaches for high efficiency coupling into submicron waveguides with impressive results have been recently demonstrated [64].

References

1. B. Jalali, S. Yegnanarayanan, T. Yoon, T. Yoshimoto, I. Rendina, F. Coppinger: Advances in Silicon-on-Insulator Optoelectronics, IEEE J. of Sel. Top. in Quant. El. **4**, 938–947 (1998)
2. L. Pavesi, L. Dal Negro, C. Mazzoleni, G. Franzo, F. Priolo: Optical gain in silicon nanocrystals, Nature **408**, 440 (2000)
3. M.H. Nayfeh, N. Barry, J. Therrien, O. Akcakir, E. Gratton, G. Belomoin: Stimulated blue emission in reconstituted films of ultrasmall silicon nanoparticles, Appl. Phys. Lett. **78**, 1131 (2001)
4. P.M. Fauchet, J. Ruan: Towards the first silicon laser, NATO Series (Kluwer, New York 2003), p. 197
5. L. Khriachtcheve, M. Rasanen, S. Novikov, J. Sinkkonen: Optical gain in Si/SiO_2 lattice: Experimental evidence with nanosecond pulses, Appl. Phys. Lett. **79**, 1249 (2001)
6. K. Luterova, I. Pelant, I. Mikulskas, R. Tomasiunaset, D. Muller, J.J. Grob, J.L. Rehspringer, B. Honerlage: Stimulated emission in blue-emitting Si^+-implanted SiO_2 films, J. Appl. Phys. **91**, 2896 (2002)
7. M. Ivanda, U.V. Densica, C.W. White, W. Kiefer: Towards the first silicon laser, NATO Series (Kluwer, New York 2003), p. 191

8. L. Dal Negro, M. Gazzanelli, N. Daldosso, Z. Gaburro, L. Pavesi, F. Priolo, D.Pacifici, G. Franzo, F. Iacona: Stimulated emission in plasma-enhanced chemical vapour deposited silicon nanocrystals, Physica E **16**, 297 (2003)
9. L. Pavesi: Will silicon be the photonic material of the third millenium?, J. Phys. Condens. Matter **15**, R1169–R1196 (2003)
10. S. Coffa, G. Franzo, F. Priolo: High efficiency and fast modulation of Er-doped light emitting Si diodes, Appl. Phys. Lett. **69**, 2077 (1996)
11. S. Coffa, G. Franzo, F. Priolo, A. Pacelli, A. Lacaita: Direct evidence of impact excitation and spatial profiling of excited Er in light emitting Si diodes, Appl. Phys. Lett. **73** 93 (1998)
12. M. E.Castagna: Quantum Dot Materials and Devices for Light Emission in Silicon, Proc. 32th European Solid-State Device Res. Conf., Firenze, Italy, (University of Bologna 2002) p. 439–442
13. S. Coffa: ST sets world record for silicon light emission, ST Press Release, **3** (2002)
14. M. A. Green, J. Zhao, A. Wang, P. J. Reece, M. Gal: Efficient Silicon Light-emitting diodes, Nature **412**, 805 (2001)
15. R. A. Soref: Thin Solid Films **294**, 325 (1997)
16. G. Dehlinger, L. Diehl, U. Gennser, H. Sigg, J. Faist, K. Ensslin, D. Grutzmacher: Intersubband electroluminescence from silicon-based quantum cascade structures, Science **290**, 2277 (2000)
17. L. Diehl, D. Mentese, H. Sigg, E. Muller, D. Grutzmacher, U. Gennser, I. Sagnes, T. Fromhertz, J. Stangl, T. Roch, G. Bauer, Y. Campidelli, O. Kermarrec, D. Bensahel, J. Faist: Electroluminescence from strain-compensated $Si_{0.2}Ge_{0.8}$/Si quantum-cascade structures based on a bound-to-continuum transition, Appl. Phys. Lett. **81**, 4700 (2002)
18. S. A. Lynch, R. Bates, D. J. Paul, D. J. Norris, A. G. Cullis, Z. Ikonic, R. W. Kelsall, P. Harrison, D. D. Arnone, C. R. Pidgeon: Intersubband electroluminescence from Si/SiGe cascade emitters at terahertz frequencies, Appl. Phys. Lett. **81**, 1543–1545 (2002)
19. J. Faist, F. Capasso, D. L. Sivco, C. Sirtori, A. L. Hutchinson, A. Y. Cho: Quantum cascade laser, Science **264**, 553 (1994)
20. R. Claps, D. Dimitropoulos, B. Jalali: Stimulated raman scattering in silicon waveguides, IEE Electron. Lett. **38**, 1352–1354 (2002)
21. R. Claps, D. Dimitropoulos, Y. Han, B. Jalali: Observation of Raman emission in silicon waveguides at 1.54 μm, Opt. Express **10**, 1305–1313 (2002)
22. D. Dimitroupoulos, R. Claps, B. Jalali: Prospects for Raman amplification in silicon, Materials Research Society, MRS Annual Meeting (Boston, MA 2002)
23. D. Dimitropoulos, R. Claps, B. Jalali: Nonlinear optics in silicon waveguides: stimulated Raman scattering and two-photon absorption, Proc. SPIE **4987** (2003), pp. 140–148
24. R. Claps, V. Raghunathan, D. Dimitropoulos, B. Jalali: Raman emission in Silicon waveguides: Prospect for Silicon amplifier and Laser (CLEO, Baltimore, MD 2003)
25. R. Claps, D. Dimitropoulos, V. Raghunathan, Y. Han, B. Jalali: Observation of stimulated Raman scattering in Silicon waveguides, Opt. Express **11**, 1731–1739 (2003)
26. B. Jalali, R. Claps, D. Dimitropoulos: Integrated Photonic Research (IPR) (Washington, DC 2003)

27. D. Dimitropoulos, B. Houshmand, R. Claps, B. Jalali: Coupled-mode theory of Raman effect in silicon-on-insulator waveguides, Opt. Lett. **28** (2003)
28. R. Claps, V. Raghunathan, D. Dimitroupoulos, B. Jalali: Anti-stokes Raman conversion in silicon waveguides, Opt. Express **11**, 2862–2872 (2003)
29. V. Raghunathan, R. Claps, D. Dimitropoulos, B. Jalali: Wavelength Conversion in Silicon Waveguides using Parametric Stokes to Anti-Stokes Coupling (CLEO 2004), submitted
30. D. Dimitropoulos, V. Raghunathan, R. Claps: Silicon Optical Wavelength conversion based on Parametric Raman effect in Dispersion compensated waveguides, manuscript in preparation
31. D. Dimitropoulos, V. Raghunathan, R. Claps, B. Jalali: Phase-matching for Efficient Parametric Wavelength Conversion in Silicon Waveguides, Opt. Express (submitted)
32. P. A. Temple, C. E. Hathaway: Multiphonon Raman Spectrum of Silicon, Phys. Rev. A **7**, 3685 (1973)
33. M. Cardona, G. Guntherodt: *Light Scattering in Solids II*, Top. Appl. Phys. **50** (Springer, Berlin 1982)
34. A. Yariv: *Quantum Electronics*, 3rd edn. (Wiley, New York 1988)
35. J. M. Ralston, R. K. Chang: Spontaneous-Raman-scattering efficiency and stimulated scattering in silicon, Phys. Rev. B **2**, 1858–1862 (1970)
36. R. H. Stolen, E. P. Ippen: Raman gain in glass optical waveguides, Appl. Phys. Lett. **22**, 276–278 (1973)
37. R. H. Stolen: Nonlinear properties of optical fibers, in *Optical Fiber Telecommunications* (Academic Press, New York 1979)
38. D. H. Choi, W. J. R. Hoefer: The Finite-Difference Time-Domain method and its applications to Eigenvalue problems, IEEE Trans. Microwave Theory Tech. (1986), pp. 1464–1470
39. R. Loudon, *The Quantum Theory of Light*, 3rd edn. (Oxford University Press)
40. G. P. Agrawal: *Non Linear Fiber Optics* (Academic Press, San Diego 2001)
41. Y. R. Shen, N. Bloembergen: Theory of stimulated Brillouin and Raman scattering, Phys. Rev. **137**, A1787–A1805 (1965)
42. E. Golovchenko, P. V. Mamyshev, A. N. Pilipetskii, E. M. Dianov: Mutual influence of the parametric effects and stimulated Raman scattering in optical fibers, IEEE. J. Quant. Elec. **26**, 1815–1820 (1990)
43. R. W. Boyd: *Nonlinear Optics* (Academic Press, San Diego 1992)
44. H. H. Li: Refractive index of silicon and germanium and its wavelength and temperature derivatives, J. Phys. Chem. Ref. Data **9**, 591–658 (1980)
45. T. Saito, K. Suto, J. Nishizawa, M. Kawasaki: Spontaneous Raman scattering in [100], [110] and [11-2] directional GaP waveguides, J. Appl. Phys. **90**, 1831 (2001)
46. R. G. Smith: Optical power handling capacity of low loss optical fibers as determined by stimulated Raman and Brillouin scattering, Appl. Opt. **68**, 2489 (1972)
47. S. V. Rao, K. Moutzouris, M. Ebrahimzadeh, A. De Rossi, G. Dintz, M. Calligaro, V. Ortiz, V. Berger: Influence of scattering and two-photon absorption on the optical loss in GaAs-Al_2O_3 nonlinear waveguides measured using femtosecond pulses, IEEE J. Quant. Elec. **39**, 478–486 (2003)
48. Spectra-Physics Telecom: Model RL5 Raman Fiber Laser Specifications
49. M. Grimsditch, M. Cardona: Absolute cross-section for Raman scattering by phonons in silicon, Phys. Stat. Sol. B **102**, 155 (1980)

50. H.K. Tsang, C.S. Wong, T.K. Lang, I.E. Day, S.W. Roberts, A. Harpin, J. Drake, M. Asghari: Optical dispersion, two-photon absorption and self-phase modulation in silicon waveguides at 1.5 μm wavelength, Appl. Phys. Lett. **80**, 416 (2002)
51. C. Xu, W. Denk: Two-photon optical beam induced current imaging through the backside of integrated circuits, Appl. Phys. Lett. **71**, 2578 (1997)
52. K. Kikuchi: Optical sampling system at 1.5 μm using two photon absorption in Si avalanche photodiode, IEE Elecron. Lett. **34**, 1354 (1998)
53. T.K. Liang, H.K. Tsang, I.E. Day, J. Drake, A.P. Knights, M. Asghari et al.Silicon waveguide two-photon absorption detector at 1.54 μm wavelength for autocorrelation measurements, Appl. Phys. Lett. **81**, 1323 (2002)
54. J.M. Ralston, R.K. Chang: Optical limiting in Semiconductors, Appl. Phys. Lett. **15**, 164 (1969)
55. J.F. Reintjes, J.C. McGroddy: Indirect two-photon transitions in Si at 1.06 μm, Phys. Rev. Lett. **30**, 901 (1973)
56. M. Dinu, F. Quochi, H. Garcia: Third-order nonlinearities in silicon at telecom wavelengths, Appl. Phys. Lett. **82**, 2954 (2003)
57. M. Sheik-Bahae, E.W. Stryland: *Optical Nonlinearities in the Transparency Region of Bulk Semiconductors, Nonlinear Optics in Semiconductors I*, Semicond. Semimet. **58** (Academic Press, 1999)
58. H. Vogt: Coherent and hyper-Raman techniques, in M. Cardona, G. Guntherodt (Eds.): *Light Scattering in Solids II*, Top. Appl. Phys. **50** (Springer, Berlin 1982), chap. 4
59. M.C. Ho, K. Uesaka, M. Marhic, Y. Akasaka, L.G. Kazovsky: 200-nm-bandwidth fiber optical amplifier combining parametric and Raman gain, J. Light. Technol. **19**, 977 (2001)
60. M.E. Marhic, K.K.-Y. Wong, L.G. Kazovsky: Prospects for CW Fiber OPAs and OPOs, CLEO 2003, Baltimore, MD, USA, paper no. CTuA1 (2003)
61. J.K. Chee, J.-M. Liu: Polarization-dependent parametric and Raman processes in a birefringent optical Fiber, IEEE J. Quant. Elec. **26**, 541 (1990)
62. T. Zinke et al.: Comparison of optical waveguide losses in silicon-on-insulator, IEE Elec. Lett. **29**, 2031 (1993)
63. P. Koonath, K. Kishima, T. Indukuri, B. Jalali: SIMOX sculpting of 3-D nano-optical structures, LEOS Annual Meeting (Tucson, AZ, USA 2003)
64. IEEE LEOS Summer Topical Meetings, 14–16 July 2003, Vancouver, BC, Canada, Proceedings Session TuA2: Coupling Issues in Microphotonics, IEEE Catalog Number 03TH8701

Silicon Photo-Receivers

Horst Zimmermann

Vienna University of Technology, 1040 Vienna, Austria
hzimmermann@emst.tuwien.ac.at

Abstract. The properties of photodiodes being exploitable in standard bipolar, CMOS, and BiCMOS technologies are summarized. In addition examples of advanced photodiodes will be introduced in order to show how the properties of integrated photodiodes can be improved significantly by minor process modifications. Furthermore, examples of optoelectronic integrated circuits (OEICs) for such important applications like optical storage systems and optical fiber receivers are described. New trends for the circuit topology of digital-video-disk (DVD) and digital-video-recording (DVR) read-OEICs are covered. Progress of OEIC receivers for optical data transmission and communication as well as optical interconnects is also summarized.

1 Photodetectors in Standard Silicon Technologies

Nowadays, there are three types of silicon process technologies for the fabrication of integrated circuits: bipolar, complementary metal oxide semiconductor (CMOS), and a combination of both in BiCMOS processes. These technologies are very mature and there is a trend towards integrated sensors. The integration of optical sensors especially is very interesting. In the following, properties of photodiodes being available without process modifications are described.

1.1 Photodetectors in Bipolar Technology

Modern bipolar integrated circuits are manufactured in standard-buried-collector (SBC) technology [1]. Without any technological modifications, the buried N^+ collector can serve as the cathode (see Fig. 1), the N collector epitaxial layer can serve as the 'intrinsic' layer of a PIN photodiode, and the base implant can serve as the anode in order to integrate PIN photodiodes with a thin 'intrinsic' region in bipolar technologies [2,3]. The process of [2] was described in [4]. The only difference in Fig. 1 compared to the cross section of an NPN transistor is that the N^+ emitter (lying within the P-base) is omitted, which is a pure layout matter and does not need any technological modification.

The small thickness of the epitaxial layer of high-speed bipolar processes in the range of about 1 μm causes a low quantum efficiency in the yellow to the infrared spectral region (580 nm to 1100 nm). The rise and fall times of

L. Pavesi, D. J. Lockwood (Eds.): Silicon Photonics, Topics Appl. Phys. **94**, 239–268 (2004)

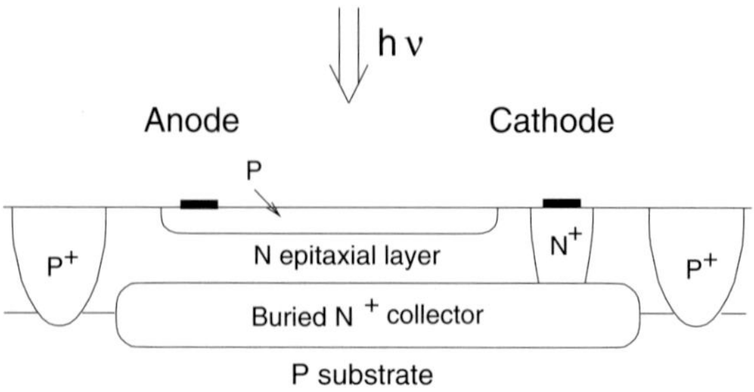

Fig. 1. Schematic cross section of a base-collector photodiode

the photocurrent for light pulses are very short due to the small epitaxial layer thickness. In [2] a bit rate of 10 Gb/s for the base-collector diode and a responsivity R of merely 48 mA/W for 840 nm were reported. This low responsivity at wavelengths from 780 nm to 850 nm, which are widely used for optical data transmission on short fiber lengths of up to several kilometers, is a major disadvantage of standard bipolar OEICs.

The base-collector diode with a sensitive area of 100 μm^2 fabricated in a 0.8 µm silicon bipolar technology worked up to 3 Gb/s for a wavelength of 850 nm [3]. A sensitivity of 0.045 A/W was reported. A phototransistor with a very small sensitive area of 10 μm^2 reached a data rate of 5 Gb/s.

1.2 Photodetectors in CMOS Technology

The simplest way to build CMOS OEICs is to use the PN junctions available in CMOS processes: source/drain-substrate, source/drain-well, and well-substrate diodes. These PN photodiodes, however, possess regions which are free of electric fields. In these regions, the slow diffusion of photogenerated carriers determines the transient behavior of such PN photodiodes. Published PN CMOS OEICs are characterized by bandwidths of less than 15 MHz [5, 6, 7]. Another example is described in [8], where the OEIC was optimized for a dynamic range of six decades in illumination.

The source/drain-substrate and source/drain-well photodiodes are more appropriate for the detection of wavelengths shorter than about 600 nm, whereas the well-substrate photodiode is more appropriate for long wavelengths like 780 nm or 850 nm. Figure 2, for instance, shows an N^+/P-substrate photodiode.

In addition to carrier diffusion, the series resistance of the photodiodes due to lateral anode contacts at the silicon surface together with the relatively large junction capacitance of the photodiode may limit the dynamical PN photodiode behavior. In an N-well process, the anode of the N^+/P-substrate photodiode has to be at V_{SS} potential, which may be a restriction for circuit design.

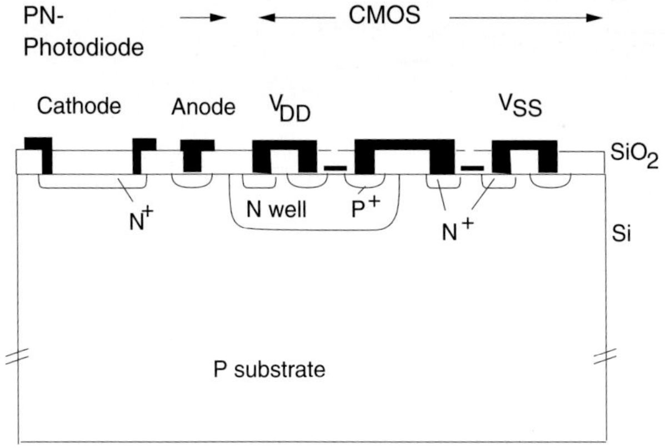

Fig. 2. Schematic cross section of a PN photodiode integrated in a one-well CMOS chip

A lateral N$^+$/P-substrate/P$^+$ photodiode was used as an optical detector, into which light was coupled by an integrated waveguide [5]. This detector was realized in a 0.8 µm N-well CMOS process. A maximum bandwidth of 10 MHz was achieved with a transimpedance amplifier for a photocurrent of 1 µA with $\lambda = 675$ nm. The speed of the detector was limited by carrier diffusion due to photogeneration outside the diode [6].

The N well/P-substrate diode in a 2 µm N-well CMOS process was used as a photodiode in [7]. A bandwidth of 1.6 MHz with a wavelength $\lambda =$ 780 nm was reported for an unoptimized system. The leakage current density of the photodiode was 15 pA/mm^2 at 5 V. The responsivity for $\lambda = 780$ nm was 0.5 A/W (quantum efficiency $\eta = 70\%$). The light was coupled into the photodiode via an integrated waveguide.

1.3 Photodetectors in BiCMOS Technology

In order to demonstrate the speed limitation caused by the rather high doping concentrations in modern standard technologies, we consider an N$^+$/P-substrate photodiode in a CMOS-based BiCMOS technology. The problem, however, is exactly the same in modern CMOS processes. Figure 3 shows the cross section of this photodiode.

Modern CMOS and BiCMOS process are optimized with respect to a minimum number of masks and, therefore, implement a self-aligned well processing scheme. Such a process scheme has the effect, that only N wells can be defined in the layout (for a process using P-type substrate) and everywhere outside of the N wells P wells are present. Therefore, there is a P well with a doping concentration exceeding 10^{16} cm^{-3} incorporated in the N$^+$/P-substrate photodiode. This leads to a thin N$^+$/P-substrate space-charge region and the effect of slow carrier diffusion is very pronounced (see Fig. 4). Rise

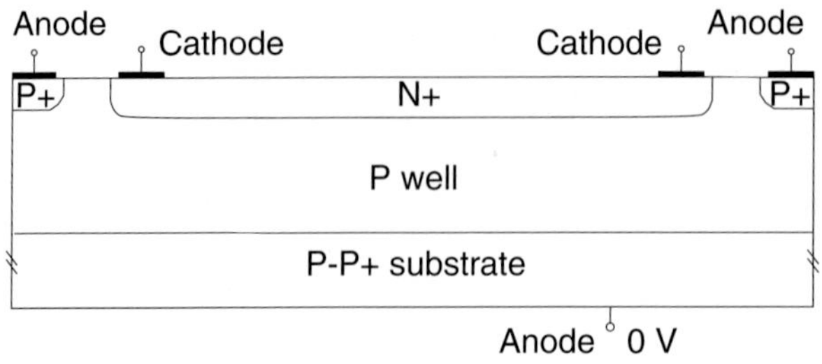

Fig. 3. Cross section of a N^+/P-substrate photodiode in self-adjusting well BiCMOS technology

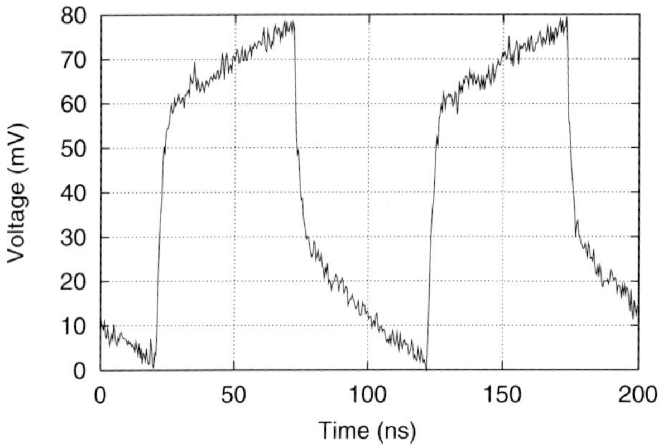

Fig. 4. Transient response of a N^+/P-substrate photodiode in a self-adjusting well BiCMOS technology

and fall times of 26 ns and 28 ns, respectively, are determined due to the pronounced diffusion tail. The measured −3 dB bandwidth is 6.7 MHz [9].

A standard BiCMOS technology with a minimum effective channel length of 0.45 μm (corresponding to a drawn or nominal channel length of about 0.6 μm) without any modifications was used [10, 11] to exploit a fast photodiode. The buried N^+-collector in Fig. 5 was used for the cathode of the PIN photodiode. The P^+-source/drain island served for the anode. The intrinsic zone of the PIN photodiode was formed by the N well and, therefore, had only a thickness of 0.7 μm. Consequently, the responsivity of the photodiode was only 0.07 A/W for a wavelength of 850 nm [10]. For a bias of 2.5 V across the $75 \times 75\,\mu m^2$ PIN photodiode with a capacitance of 1.8 pF, a −3 dB bandwidth of 700 MHz was reported. The OEIC reached a bit rate of 531 Mb/s with a bit error rate of 10^{-9} and a sensitivity of −14.8 dB · m. This bit rate was limited by the capacitance of the photodiode and the feedback resistor of 1.4 kΩ in the amplifier transimpedance input stage. The dark current of the photodiode was 10 nA for a reverse voltage of 2.5 V at room temperature.

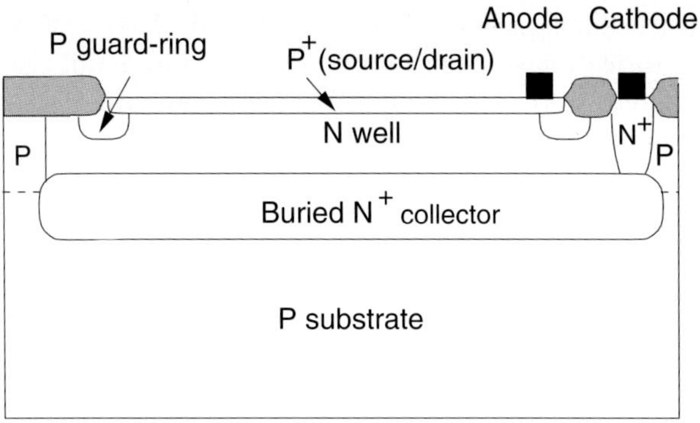

Fig. 5. PIN photodiode in a SBC-based BiCMOS technology [10]

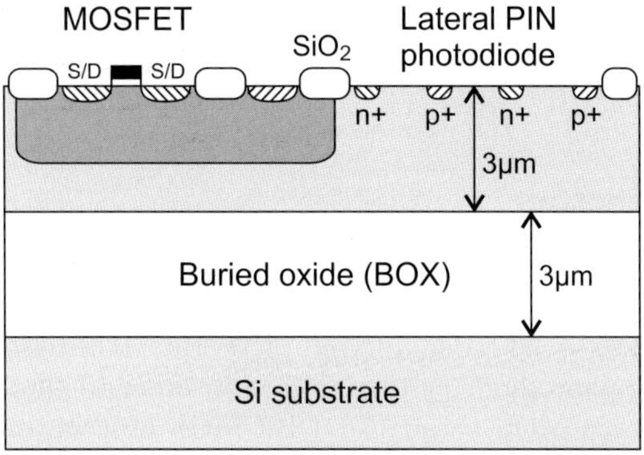

Fig. 6. Cross section of lateral SOI PIN photodiode

In [11] a laser with a wavelength of 670 nm was used for the characterization of the same OEIC as in [10]. The data rate was increased to 622 Mb/s for this wavelength. No measured result for the responsivity was given in [11]. Instead, a three times larger responsivity value of approximately 0.2 A/W for $\lambda = 670$ nm was estimated due to the lower penetration depth than for $\lambda = 850$ nm. This estimation, however, may be doubtful, because possible destructive interference in the isolation and passivation stack is neglected. The main drawback for the photodiode used in [10, 11] was its low responsivity due to the thin epitaxial layer in the SBC-bipolar based BiCMOS process.

1.4 Lateral PIN Photodiodes on SOI

Silicon-on-Insulator (SOI) was suggested to be the technology of future CMOS [12]. SOI uses a thin crystalline silicon film for the integration of devices. A cross section of the integration scheme used for SOI receivers [13] is shown in Fig. 6. The starting material was bonded and etched-back silicon-on-insulator (BESOI), and the active Si layer was N-type (100) with a resistivity of

$50\,\Omega\cdot$cm–$80\,\Omega\cdot$cm and a thickness of 3 µm. The buried oxide layer (BOX) was also 3 µm thick. The fabrication steps have been described in detail elsewhere [14]. All processing steps were fully compatible with standard CMOS processes. Indeed, this structure could easily be implemented in a standard process with no modifications other than the starting substrate.

The P- and N-fingers of the lateral, interdigitated PIN photodiode were formed during the P-well contact and source/drain implants, respectively. The finger width was 2 µm and the spacing between the fingers was 5 µm. The total photodiode area was 51 µm by 46 µm.

The dark current of the SOI device was 2.1 nA at 5 V and 2.2 nA at 20 V. The onset of breakdown occured at 40 V. The external quantum efficiency was measured as a function of wavelength using a white light source, spectrometer, chopper, and lock-in amplifier using a calibrated Si photodiode as a reference. At 850 nm, the efficiency of the SOI photodiode was 29% corresponding to a responsivity of 0.20 A/W. The −3 dB bandwidth of the SOI photodiode with a bias $|V_{PD}| = 20$ V was 2.8 GHz [13].

2 Advanced Photodetectors

2.1 Double Photodiode

In a standard CMOS or a CMOS-based triple-diffused (3-D) BiCMOS technology [9], the so-called double photodiode can be applied in order to eliminate slow carrier diffusion. In a CMOS based (3-D) BiCMOS process, an N well can be used as the collector of a bipolar NPN transistor. This N well can also be used in order to form the cathode of the double photodiode (DPD). Figure 7 shows the cross section of such a DPD. The two anodes are connected to ground. The cathode is connected to the amplifier in the OEIC. Two PN junctions are vertically arranged.

In addition to the two space-charge regions at the two vertically arranged PN junctions, an electric field is also present between the two space-charge regions due to the doping gradient of the N well. Therefore, there is no contribution of slow carrier diffusion from the region between the two space-charge regions. With an integrated polysilicon resistor of 1 kΩ, rise and fall times of

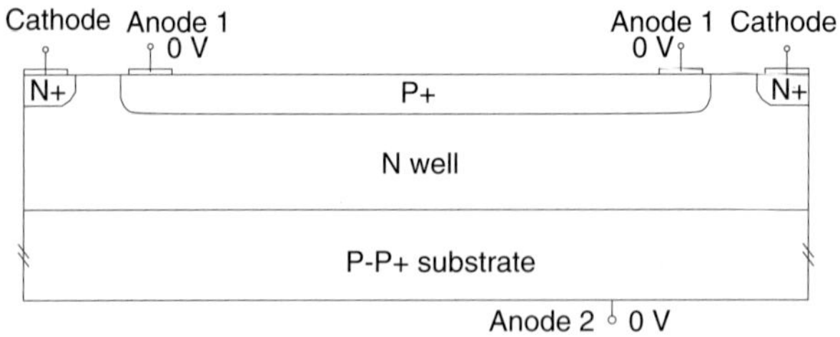

Fig. 7. Cross section of a double photodiode in BiCMOS technology [15]

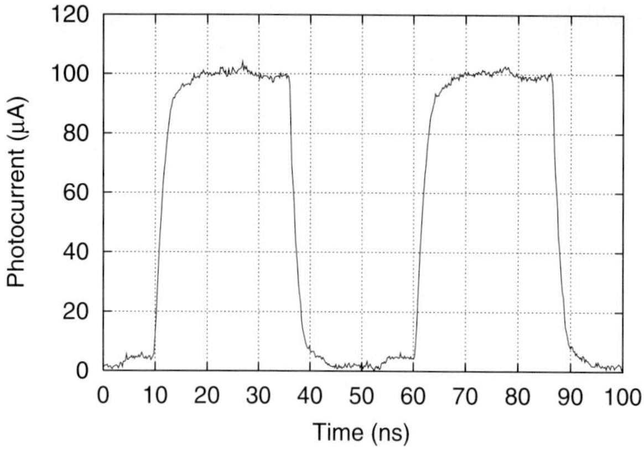

Fig. 8. Transient response of the double photodiode in BiCMOS technology shown in Fig. 7 [16]

3.2 nm and 2.8 ns, respectively, were measured for the DPD with $\lambda = 638$ nm and $U_\mathrm{r} = 2.5$ V [16]. With an integrated 500 Ω resistor, rise and fall times of 1.8 ns and 1.9 ns, respectively, were found [15]. There is no indication of a so-called diffusion tail in the photocurrent (Fig. 8).

2.2 Vertical PIN Photodiode

Figure 9 shows the structure of an N^-N^+ CMOS-OEIC in the twin-well CMOS approach [9]. Here the N^+ substrate serves as the cathode and the P^+ source/drain region forms the anode of the integrated PIN photodiode.

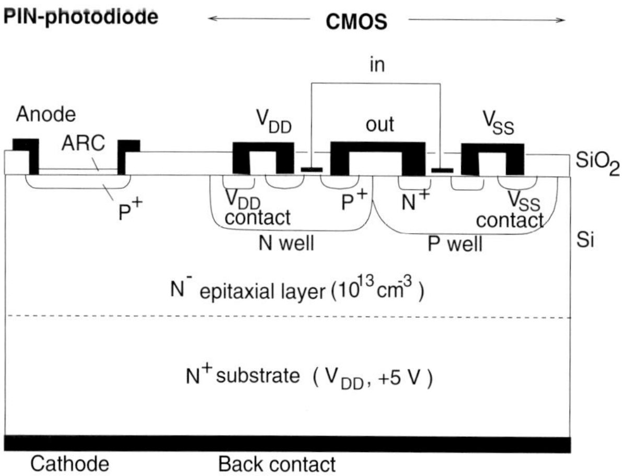

Fig. 9. Cross section of a N^-N^+ PIN-CMOS-OEIC [33]

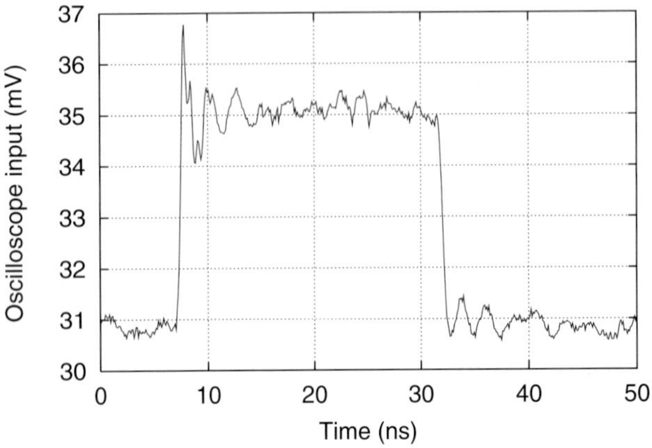

Fig. 10. Measured transient response of a CMOS-integrated PIN photodiode with an I-layer doping concentration of $2 \times 10^{13}\,\mathrm{cm}^{-3}$ for $\lambda = 638\,\mathrm{nm}$ and $|V_{\mathrm{PD}}| = 3.0\,\mathrm{V}$. The overshoot in the signal is due to the direct modulation of the laser [18]

PIN-CMOS photodiodes with an area of $2700\,\mu\mathrm{m}^2$, with a standard doping concentration of $1 \times 10^{15}\,\mathrm{cm}^{-3}$ and with reduced doping concentrations in the epitaxial layer down to $2 \times 10^{13}\,\mathrm{cm}^{-3}$, and with an integrated polysilicon resistor of $500\,\Omega$, were fabricated in an industrial $1.0\,\mu\mathrm{m}$ CMOS process [17]. For the measurements, a laser with $\lambda = 638.3\,\mathrm{nm}$ was modulated with a commercial ECL generator. The light pulses were coupled into the photodiodes on a wafer prober via a single-mode optical fiber. The rise (t_{r}) and fall (t_{f}) times of the photocurrent of the photodiodes were measured with a picoprobe (pp), which possesses a $-3\,\mathrm{dB}$ bandwidth of 3 GHz and an input capacitance of 0.1 pF, and with a 20 GHz digital sampling oscilloscope HP54750/51.

For the PIN photodiode with a doping concentration in the epitaxial layer of $2 \times 10^{13}\,\mathrm{cm}^{-3}$, the oscilloscope extracted $t_{\mathrm{r}}^{\mathrm{osc,disp}} = 0.37\,\mathrm{ns}$ and $t_{\mathrm{f}}^{\mathrm{osc,disp}} = 0.57\,\mathrm{ns}$ from the waveform shown in Fig. 10. The evaluation according $(t_{\mathrm{r}}^{\mathrm{PIN}})^2 = (t_{\mathrm{r}}^{\mathrm{osc,disp}})^2 - (t_{\mathrm{r}}^{\mathrm{laser}})^2 - (t_{\mathrm{r}}^{\mathrm{pp}})^2 - (t_{\mathrm{r}}^{\mathrm{osc}})^2$ with $t_{\mathrm{r}}^{\mathrm{laser}} = 0.30\,\mathrm{ns}$, $t_{\mathrm{f}}^{\mathrm{laser}} = 0.51\,\mathrm{ns}$, $t_{\mathrm{r/f}}^{\mathrm{pp}} = 0.1\,\mathrm{ns}$ and $t_{\mathrm{r/f}}^{\mathrm{osc}} \approx 0.02\,\mathrm{ns}$ results in $t_{\mathrm{r}}^{\mathrm{PIN}} = 0.19\,\mathrm{ns}$ and $t_{\mathrm{f}}^{\mathrm{PIN}} = 0.24\,\mathrm{ns}$ [33]. With $f_{3\,\mathrm{dB}} = 2.4/(\pi(t_{\mathrm{r}} + t_{\mathrm{f}}))$, the $-3\,\mathrm{dB}$ bandwidth can be estimated to be 1.7 GHz for $|V_{\mathrm{PD}}| = 3.0\,\mathrm{V}$. With the conservative estimate $\mathrm{BR} = 1/(1.5(t_{\mathrm{r}} + t_{\mathrm{f}}))$, a bit rate, BR, of 1.5 Gb/s results for $|V_{\mathrm{PD}}| = 3.0\,\mathrm{V}$. With an antireflection coating (ARC), the quantum efficiency η could be increased from 49% to 94%. To our knowledge this is the first time that such a high speed and such a high quantum efficiency have been achieved with an integrated silicon photodiode for a reverse voltage of only 3 V, whereby an only slightly modified standard CMOS process has been used.

In contrast to transistors in bipolar OEICs, the electrical performance of the N- and P-channel MOSFETs is not degraded when the epitaxial layer is modified. This statement was verified by measurements (Table 1). The threshold voltages of the NMOS and PMOS transistors are practically independent

Table 1. Measured threshold voltages U_{Th} and drain leakage currents $I_{\mathrm{D}}^{\mathrm{r}}$ for different doping levels in the epitaxial layer

I-Doping (cm^{-3})	$U_{\mathrm{Th}}^{\mathrm{NMOS}}$ (V)	$I_{\mathrm{D}}^{\mathrm{r,NMOS}}$ (pA)	$U_{\mathrm{Th}}^{\mathrm{PMOS}}$ (V)	$I_{\mathrm{D}}^{\mathrm{r,PMOS}}$ (pA)
Standard	0.79	2.19	−0.62	63.1
1×10^{14}	0.79	0.83	−0.60	66.1
5×10^{13}	0.79	0.81	−0.60	66.1
2×10^{13}	0.78	1.02	−0.60	67.6

of the doping concentration in the epitaxial layer, because these transistors are placed inside wells which possess a much higher doping level of several times $10^{16}\,\mathrm{cm}^{-3}$ than the standard epitaxial layer with about $1 \times 10^{15}\,\mathrm{cm}^{-3}$ and because the threshold implants produce an even higher doping level ($\approx 10^{17}\,\mathrm{cm}^{-3}$) than the wells [19].

The reverse, i.e. the leakage current of the drain to well diodes is also listed in Table 1 for the NMOS and PMOS transistors [19]. The leakage current for the NMOS transistor in an epitaxial layer with reduced doping concentrations actually seems to be smaller than for the standard concentration.

These results confirm the superiority of the PIN-CMOS integration [20] compared to the PIN-bipolar integration [21]. In contrast to the PIN-bipolar integration, the electrical transistor parameters of the standard twin well CMOS process are completely unaffected for the PIN-CMOS integration and can be used for circuit simulations within the design of OEICs.

2.3 Finger Photodiode

A new photodiode for the UV/blue spectral range down to wavelengths of 400 nm, which can be integrated monolithically with CMOS circuits was presented [22]. Such optoelectronic integrated circuits (OEICs) with a high sensitivity in the UV/blue spectral range are needed in near-future optical storage systems like DVR (Digital Video Recording) with a storage capacity of more than 30 GB compared to todays CD-ROMs with 650/700 MB, which are read with 780 nm light.

At 400 nm, where laser diodes are available [23], the so-called finger photodiode achieved a responsivity of 0.23 A/W corresponding to a quantum efficiency η of 70% (with an anti-reflection coating (ARC)) and rise and fall times of 1.0 ns and 1.1 ns, respectively. The finger photodiode can be used in the red spectral range, too, for which todays DVD optical pickups use red laser diodes with a wavelength of 630/650 nm. At 638 nm, the responsivity is 0.49 A/W ($\eta = 95\%$) and rise and fall times of less than 2.3 ns were achieved. For the integration of the finger photodiode in an industrial 1 µm twin-well CMOS process only one additional mask (beside the ARC mask) was needed in order to block out the threshold voltage implantation in the photoactive region.

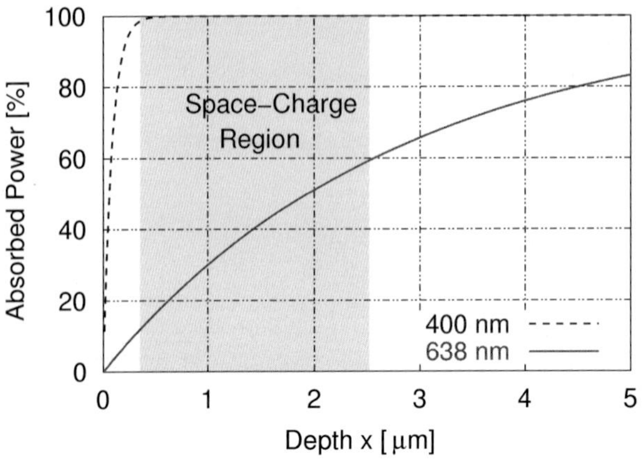

Fig. 11. Absorption of light in silicon for $\lambda = 400\,\text{nm}$ and $\lambda = 638\,\text{nm}$ (higly doped P^+-anode on n-substrate with a doping concentration of $1 \times 10^{15}\,\text{cm}^{-3}$, $V_{PD} = 3\,\text{V}$)

The absorption depth of light with a wavelength of 400 nm in silicon is approximately 0.1 µm. The absorbed optical power is given by the Lambert–Beer-Law:

$$P = (1 - R)\, P_0\, e^{-\alpha x} . \tag{1}$$

P is the absorbed optical power at the depth x in the photodiode (whereby the origin is located at the surface of the photodiode), R is the reflectivity, P_0 is the incident optical power, and α is the absorption coefficient. Considering a conventional PN photodiode with a noninterdigitated P^+-surface anode we can conclude that almost all of the incident photons are absorbed in the quasi neutral region (at the left of the space-charge region in Fig. 11) of the P^+-anode, which normally has a depth of 0.2 µm–0.4 µm.

The majority of the photogenerated carriers recombine in the P^+-anode, because of the small diffusion coefficient due to the low electron and hole mobilities (keeping the Einstein relation in mind). As a consequence of this strong recombination, the responsivity for $\lambda = 400\,\text{nm}$ with a noninterdigitated anode (or cathode for a P-substrate) is low. Furthermore, the slow diffusion processes outside the space-charge region are responsible for a limited response speed.

To avoid both, recombination and diffusion, space-charge regions at the surface have to be implemented. This can be achieved with an interdigitated structure as it is shown in Fig. 12. The N^+-substrate is used as the cathode and the P^+-fingers which are located in the N-epitaxial layer are used as the anode of the photodiode. The P^+-fingers are connected by metal outside of the photosensitive region. Photogenerated carriers can be separated in the space-charge region, electrons drift to the N^+-substrate and holes drift to the P^+-fingers. In order to find the optimum concerning response speed

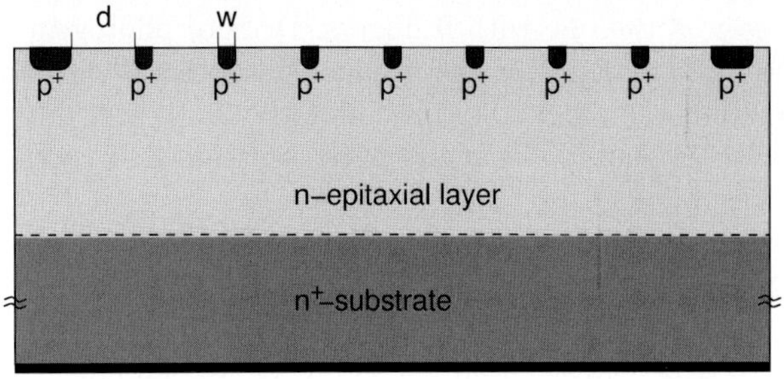

Fig. 12. Cross section of the 9-finger photodiode on n-substrate

and responsivity, several finger photodiodes on wafers with different doping concentrations in the epitaxial layer and with different numbers of fingers have been processed. An industrial 1 µm twin-well CMOS process with one additional mask to block out the threshold voltage implantation in the photo-active region was used. The antireflection coating (ARC) was realized by a stack consisting of SiO_2 and Si_3N_4 (see [19]).

For measurements, an integrated polysilicon resistor with a resistance of $R_L = 0.5\,\mathrm{k\Omega}$ for $\lambda = 630\,\mathrm{nm}$ and $R_L = 1\,\mathrm{k\Omega}$ for $\lambda = 400\,\mathrm{nm}$ (because of different optical signal power) was connected in series with the photodiode. The cathode (substrate) was connected to V_{PD} and the anode was connected to the resistor. The second pin of the resistor was connected to ground.

The laser light was coupled into the photodiode with an photo-active area of $50 \times 50\,\mathrm{\mu m^2}$) via an optical fiber. The photocurrent resulted in a voltage V_{out} across R_L, which was measured with a picoprobe (with an input capacitance of 0.1 pF and a high input resistance of 10 MΩ). For the transient measurements a sampling oscilloscope (HP 54750A) was used. The dark currents were measured with a parameter analyzer (HP 4156A). A calibrated photodiode was used to determine the incident optical power for responsivity measurements at the finger photodiodes.

In order to find the optimized interdigitated structure, the finger photodiodes were processed with different numbers of fingers. The distance d between the fingers and the finger width w (see Fig. 12) were varied. Table 2 summarizes the results for the 3-, 4-, 6-, and 9-finger photodiodes ($\lambda = 400\,\mathrm{nm}$).

Table 2. Dimensions of the finger photodiodes with different numbers of fingers [22]

Fingers	d (µm)	w (µm)	t_r (ns)	t_f (ns)
3	18.6	4.2	6.9	7.9
4	12.8	2.8	5.2	6.3
6	8.0	1.6	3.8	4.5
9	4.8	1.2	1.0	1.1

The fastest photodiode was the one with 9 fingers. Drift of photogenerated carriers leads to the fast response for the 9-finger photodiode with an epitaxial doping concentration $C_e = 1 \times 10^{15}\,\mathrm{cm}^{-3}$. For the 3-, 4- and 6-finger photodiode the distance of the anode fingers was too wide and carrier diffusion slows down the response speed.

A responsivity of 0.23 A/W at 400 nm was measured for the 9-finger photodiode compared to 0.08 A/W for a reference photodiode having a noninterdigitated anode both with the same antireflection coating. For the reference photodiode without ARC a responsivity of only 0.045 A/W was found.

For 400 nm, an SOI photodiode suggested in [24] would be a good choice. An SOI photodiode, however, shows a low responsivity for red light due to the thin SOI layer. Therefore, the finger photodiode was optimized at the same time for the UV/blue and red spectral ranges by varying the doping concentration of the epitaxial layer C_e. Four different wafers with $C_e = 2 \times 10^{13}\,\mathrm{cm}^{-3}$, $C_e = 5 \times 10^{13}\,\mathrm{cm}^{-3}$, $C_e = 1 \times 10^{14}\,\mathrm{cm}^{-3}$, and $C_e = 1 \times 10^{15}\,\mathrm{cm}^{-3}$ were used. For $\lambda = 638\,\mathrm{nm}$, the 9-finger photodiode had a responsivity of 0.49 A/W (with ARC, independent of C_e). Table 3 summarizes the rise and fall times of the 9-finger photodiode on these wafers. Figure 13 shows a transient response for $\lambda = 638\,\mathrm{nm}$ ($C_e = 2 \times 10^{13}\,\mathrm{cm}^{-3}$). Figure 14 shows the transient response for $\lambda = 400\,\mathrm{nm}$ on wafers with $C_e = 2 \times 10^{13}\,\mathrm{cm}^{-3}$.

Table 3. Rise and fall times of the 9-finger photodiode on wafers with different epitaxial doping concentrations ($R_L = 0.5\,\mathrm{k\Omega}$ for $\lambda = 638\,\mathrm{nm}$, $R_L = 1\,\mathrm{k\Omega}$ for $\lambda = 400\,\mathrm{nm}$, $V_{PD} = 3\,\mathrm{V}$) [22]

C_e (cm^{-3})	$t_r^{638\,\mathrm{nm}}$ (ns)	$t_f^{638\,\mathrm{nm}}$ (ns)	$t_r^{400\,\mathrm{nm}}$ (ns)	$t_f^{400\,\mathrm{nm}}$ (ns)
2×10^{13}	0.3	0.9	1.3	1.7
5×10^{13}	0.4	1.0	1.2	1.6
1×10^{14}	0.5	2.3	1.1	1.4
1×10^{15}	15.2	16.9	1.0	1.1

It can be seen that the photodiode on standard epitaxial wafers ($C_e = 1 \times 10^{15}\,\mathrm{cm}^{-3}$) gives the best results for 400 nm. The reason is, that the space-charge regions of adjacent fingers still meet each other for the 9-finger photodiode for $C_e = 1 \times 10^{15}\,\mathrm{cm}^{-3}$. For lower C_e, the space-charge region extends deeper resulting in a lower drift speed and a longer drift distance increasing the rise and fall times.

At 638 nm the 9-finger photodiode with $C_e = 1 \times 10^{15}\,\mathrm{cm}^{-3}$ shows a slow response due to the thin space-charge region letting room for slow diffusion. With lower doping concentration the space-charge region extends deeper into the epitaxial layer resulting in a faster response. The difference between the rise and the fall time is more pronounced in the case of red light (see Table 3). This is due to the longer drift distance of the holes generated deep in the

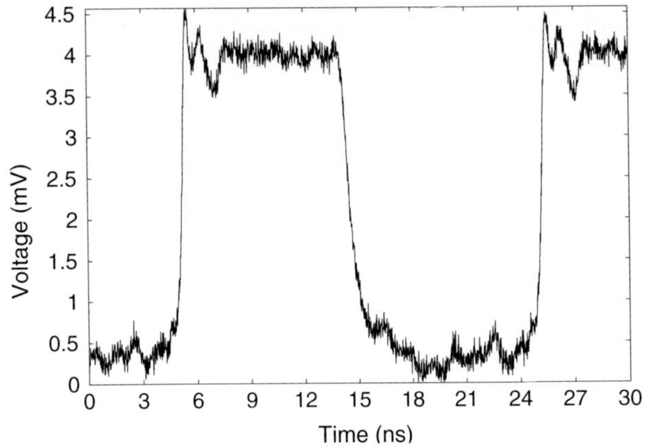

Fig. 13. Transient response of the 9-finger photodiode ($C_e = 2 \times 10^{13}\,\text{cm}^{-3}$, $\lambda = 638\,\text{nm}$, $R_L = 0.5\,\text{k}\Omega$ and $V_{PD} = 3\,\text{V}$) [22]

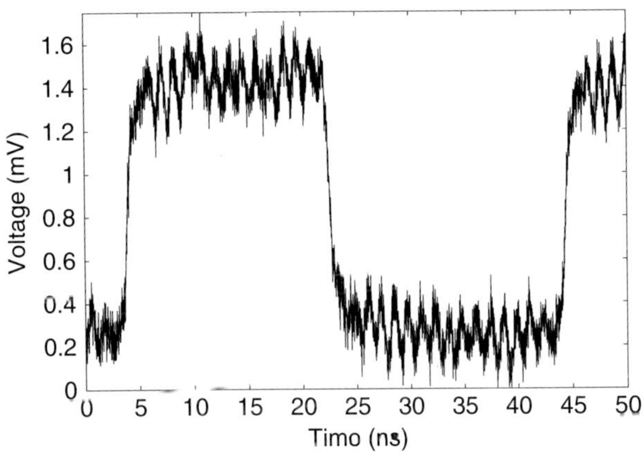

Fig. 14. Transient response of the 9-finger photodiode ($C_e = 2 \times 10^{13}\,\text{cm}^{-3}$, $\lambda = 400\,\text{nm}$, $R_L = 1\,\text{k}\Omega$ and $V_{PD} = 3\,\text{V}$) [22]

epitaxial layer to the P^+-anode compared to the lateral drift distance of the holes generated close to the Si surface for $\lambda = 400\,\text{nm}$.

The finger photdiodes processed on wafers with a doping concentration of $C_e = 1 \times 10^{14}\,\text{cm}^{-3}$ in the epitaxial layer are sensitive and fast enough for todays red and for near-future UV/blue OEICs.

In OEICs for optical storage systems a low offset voltage is required [25]. Therefore the dark current of the photodiode which is connected to the input of the amplifier has to be small. Even for reverse voltages up to 5 V the dark current remains below 1 pA at room temperature for finger photodiodes with an area of $2500\,\mu\text{m}^2$ [22]. Therefore, there is no problem with offset voltages.

Let us conclude: with the finger photodiodes the responsivity for a wavelength $\lambda = 400\,\text{nm}$ was increased by a factor of 2.8 compared to a conventional photodiode and rise and fall times of less than 1.1 ns were obtained for the standard epitaxial doping concentration. The finger photodiodes have a responsivity of 0.49 A/W (with ARC) for $\lambda = 638$ nm and rise and fall times of less than 2.3 ns for $C_e \leq 1 \times 10^{14}\,\mathrm{cm}^{-3}$. The finger photodiodes can be integrated in twin-well CMOS-processes with only one additional mask (besides the ARC mask) and are therefore well suited for UV/blue-sensitive OEICs for optical storage systems of the next generation, which are compatible with todays optical storage systems using red light.

3 Examples of Photo-Receiver Circuits

The design of microelectronic integrated circuits is a highly developed area. Circuits for advanced OEICs, however, are not yet investigated thoroughly. In the following, progress in this field will be summarized especially at the examples of OEICs for optical storage systems (OS-OEICs) like DVD and DVR as well as OEICs for receiver applications in optical interconnects, optical data transmission and optical communication.

3.1 OS-OEICS for DVD Applications

The first presented approach requires no process modification to implement a double photodiode (DPD) as shown in Fig. 7. An OEIC with the DPD with a light-sensitive area of about $50 \times 50\,\mu\mathrm{m}^2$ was fabricated in a 0.8 μm BiCMOS process [26].

The circuit diagram of a fast channel (A–D) of the OEIC for optical storage systems (OS-OEIC) is shown in Fig. 15. An integrated DPD is connected to a transimpedance amplifier using an operational amplifier to obtain a low output offset voltage compared to a 2.5 V reference voltage as is required for applications in optical storage systems.

For a universal applicability, the gain is switchable by MOS elements between high (H, R3), medium (M, R4||R3) and low (L, R5||R4||R3) with

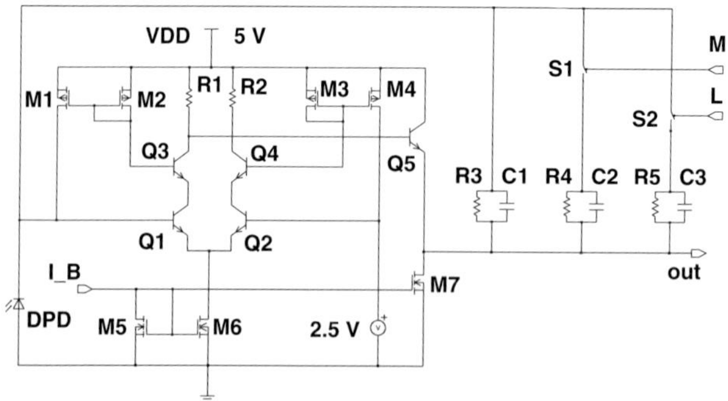

Fig. 15. Schematic of a fast channel amplifier (A–D) in a BiCMOS OS-OEIC [26]

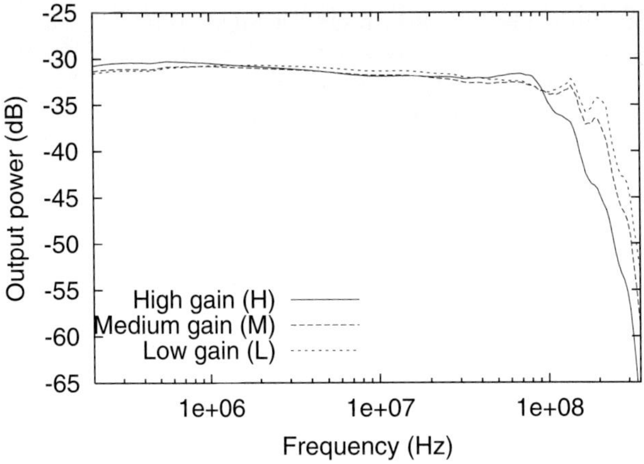

Fig. 16. Frequency response of a fast channel amplifier (A–D) in a BiCMOS OS-OEIC [26]

a ratio of approximately 1/3 each (Fig. 15). Polysilicon-polysilicon-capacitors can be produced in the 0.8 µm BiCMOS process and are used for frequency compensation with C1, C2 and C3. Only NPN transistors were applied in the signal path of operational amplifier to achieve high −3 dB bandwidths. The NPN transistors Q3 and Q4 are used to sense the base current of Q1 and Q2, respectively. These currents are mirrored by the current mirrors M1/M2 and M3/M4 back to the bases of the input transistors Q1 and Q2, respectively. This method is called bias-current cancellation. This biasing of the input transistors Q1 and Q2 reduces the systematic output offset of approximately 110 mV, which would result from the base current of Q1 across the largest resistor R3 = 20 kΩ, by more than one order of magnitude. The low frequency open-loop gain of the operational amplifier was 27 dB and its transit frequency was 870 MHz with a 1 kΩ and 10 pF load. An OEIC was packaged, mounted on a printed circuit board together with these load elements, and a probe head with an input capacitance of 1.7 pF is used to measure the frequency response shown in Fig. 16.

The measured bandwidths exceeded 92 MHz, which was much larger than the bandwidth value 7.3 MHz of the circuit for 8× speed CD-ROMs fabricated in 0.8 µm CMOS technology with off-chip photodiodes [27].

The four sensitive channels E–H have a ten times larger sensitivity for tracking control in the optical storage system. A double photodiode with approximately twice the size of the DPDs in the channels A–D was implemented here. 90 mV/µW sensitivity in combination with a low offset voltage was realized. Each amplifier of channels A–H covered an active area of approximately 0.079 mm^2. The total OEIC die area was 3.25 mm^2. The power consumption of the OS-OEIC was less than 75 mW at 5.0 V. Table 4 summarizes the properties of the OS-OEIC channels A–H.

Table 4. Measured results of the high-bandwidth OS-BiCMOS-OEIC

	H	M	L
$f_{-3\,\mathrm{dB}}$ (MHz) A–D	92.0	94.9	95.1
$f_{-3\,\mathrm{dB}}$ (MHz) E–H	5.2	8.5	14.6
Sensitivity (mV/μW) A–D	8.8	2.9	0.9
Sensitivity (mV/μW) E–H	88.1	29.3	9.1
U_{Offset} (mV) A–D	< 10.8	< 9.5	< 9.0
U_{Offset} (mV) E–H	< 7.4	< 6.4	< 6.4
Noise (dBm) at 10 MHz with 30 kHz RBW A–D	−81.5	−85.0	−85.2
Noise (dBm) E–H	−66.0	−67.5	−73.5

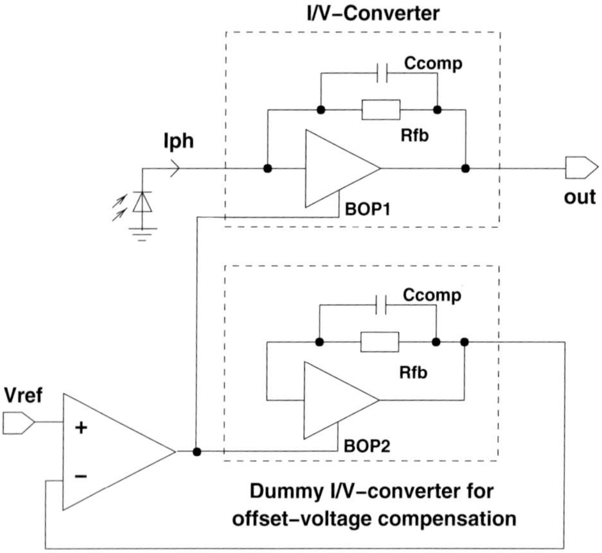

Fig. 17. Basic topology for output offset voltage reduction [28]

The next shown circuit topology implements transistor transimpedance amplifiers instead of operational amplifiers in transimpedance configuration. The innovative topology of this OEIC for optical storage systems is shown in Fig. 17. A current-to-voltage converter is connected to the integrated photodiode. The BOP1 (bias operating point) input allows to correct the operating point of this transimpedance amplifier in order to cancel the output offset voltage.

For this purpose an identical dummy current-to-voltage converter with an open input and a low-offset operational amplifier are integrated together with the transimpedance amplifier connected to the photodiode. The operational amplifier measures the output offset voltage of the dummy current-to-voltage (I/V) converter compared to $V_{\mathrm{ref}} = 2.1\,\mathrm{V}$ and changes its operation

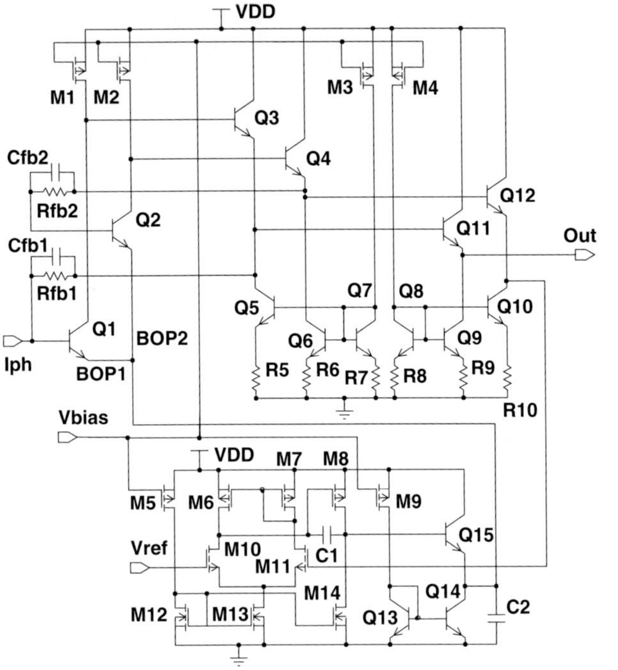

Fig. 18. Schematic diagram of one BiCMOS OS-OEIC channel

point via the BOP2 input in order to minimize the output offset voltage. The BOP1 input of the actual transimpedance amplifier is also connected to the output of the operational amplifier. In such a way, the output offset voltage of the transimpedance amplifier can be reduced provided that a good matching between the two I/V converters and a low offset voltage of the operational amplifier can be guaranteed. The first requirement can be dealt with a careful layout reducing the distance between corresponding devices in the two I/V converters. The second requirement can be fulfilled, because the operational amplifier can be designed to have a very slow response, i.e. transistors with a large emitter area or large gate lengths and widths for a good matching can be used.

An OEIC containing four fast channels (for data extraction) and four sensitive channels (for tracking control in the optical storage system) with Rfb $\approx 60\,\mathrm{k\Omega}$ values was fabricated in a $0.8\,\mu\mathrm{m}$ BiCMOS technology. The N-well/P-substrate photodiode was implemented with a light sensitive area of $2500\,\mu\mathrm{m}^2$ in the fast channels and $9100\,\mu\mathrm{m}^2$ in the sensitive channels.

The schematic diagram of the fast and sensitive channels of the BiCMOS OEIC is shown in Fig. 18. The transistor transimpedance amplifier is formed by Q1 in common-emitter configuration, the PMOS load element M1, the emitter follower Q3, and the feedback resistor Rfb1. Q3 provides a low output impedance of this input stage. The capacitor Cfb1 is used for com-

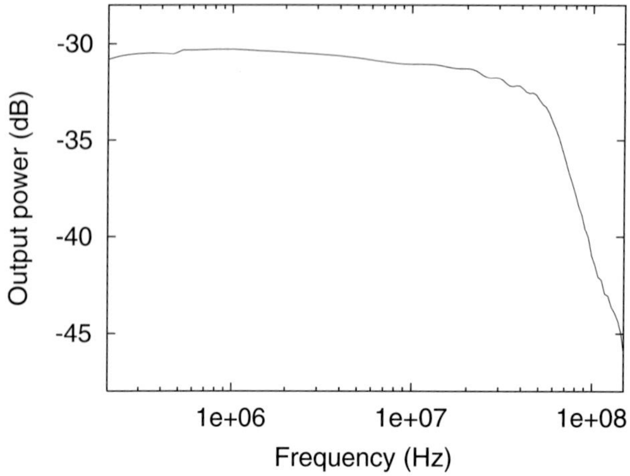

Fig. 19. Frequency response of fast channel of BiCMOS OS-OEIC [26]

pensation. The emitter follower Q11 decouples the feedback path from the output and provides a low overall output impedance. The dummy I/V converter is formed by Q2, M2, Q4, Rfb2, Cfb2 and Q12. Q5–Q10, R5–R10 and M3/M4 are used for biasing. The circuit diagram in Fig. 18 corresponds to the interlaced layout for good matching of the two I/V converters.

The BiCMOS operational amplifier is based on a CMOS Miller type topology with the NPN emitter follower Q15 at the output in order to supply a voltage source with a low output resistance for the emitter inputs BOP of the two I/V converters. The simulated open-loop gain of the operational amplifier was 77 dB and its transit frequency was 7.3 MHz.

A sensitivity for the fast channel of 25 mV/μW and of 51 mV/μW for the sensitive channels at 638 nm was found by determining the incident light power with a calibrated photodiode. A −3 dB frequency of 58.5 MHz was measured for a fast channel with a load of 10 kΩ and 10 pF (Fig. 19). The sensitive channels showed a −3 dB frequency of 26.7 MHz with the same load. All measured output offset voltages for photodiodes and OEICs in the dark were lower than 7.5 mV. This value is considerably lower than in the maximum simulated offset voltage of a reference amplifier without offset compensation of more than 72 mV for worst case transistor parameters at room temperature.

It should be mentioned that the BiCMOS operational amplifier and the emitter BOP inputs reduce the bandwidth of the I/V converters. The output resistance of the BiCMOS operational amplifier, however, is obviously rather low, since the bandwidth of 58 MHz is not considerably lower than that of a reference I/V converter of 82 MHz where the emitter of Q1 was connected to V_{ref} directly.

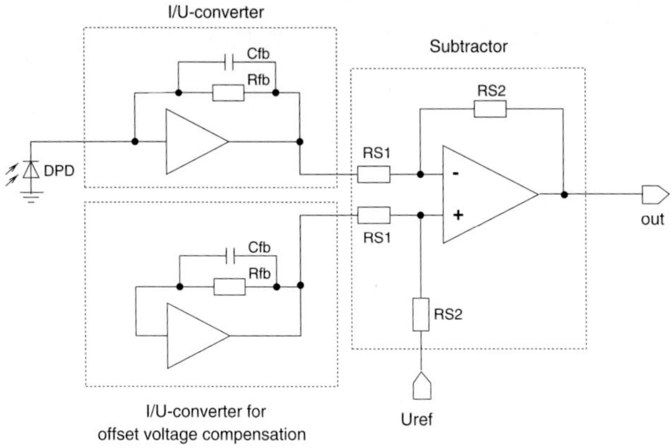

Fig. 20. Block diagram of a two-stage optical receiver for a fast channel of a high-speed OS-BiCMOS-OEIC

The bandwidth of the N-well/P-substrate photodiode may seem to be rather high compared to [7] where a bandwidth of 1.6 MHz was reported for a similar photodiode for $\lambda = 780\,\mathrm{nm}$. The high bandwidth determined here, however, can be explained by the shorter wavelength and by an electric field, i.e. by a drift zone for photogenerated carriers, due to the doping gradient of the N-well similarly to that in the N-well of a double photodiode [29].

Noise measurements at 10 MHz with a resolution bandwidth of 30 kHz revealed values of $-76\,\mathrm{dB} \cdot$ mand $-65\,\mathrm{dB} \cdot$ mfor the fast and sensitive channels, respectively. The power consumption determined by simulation was 36 mW for a fast channel at a supply voltage of 5.0 V. The active die area of each channel was approximately $0.042\,\mu\mathrm{m}^2$ and the total chip area was $3.1\,\mathrm{mm}^2$.

When a high speed and a higher sensitivity are required in addition to a lower output offset voltage for the OS-OEICs, a two-stage optical receiver may be necessary. The circuit principle of such a two-stage amplifier [30, 31] is shown in Fig. 20.

The circuit consists of a transimpedance amplifier for the photocurrent and a reference I/V converter for offset compensation plus an operational amplifier in subtractor configuration. The subtractor can also be used as a voltage amplifier with the amplification factor RS2/RS1 enabling a high overall sensitivity of the OEIC. It has be mentioned, however, that offset voltages due to mismatch of the two transimpedance preamplifiers are also amplified. The circuit diagram of the complete circuit is shown in Fig. 21.

In order to achieve a high bandwidth, only NPN transistors are used in the signal paths of the preamplifiers. Q1 is used in common-emitter configuration, Q3 is used as emitter follower, and the feedback resistor Rfb1 together with Q1 and Q3 represents a low input impedance for the photocurrent of the double photodiode (DPD).

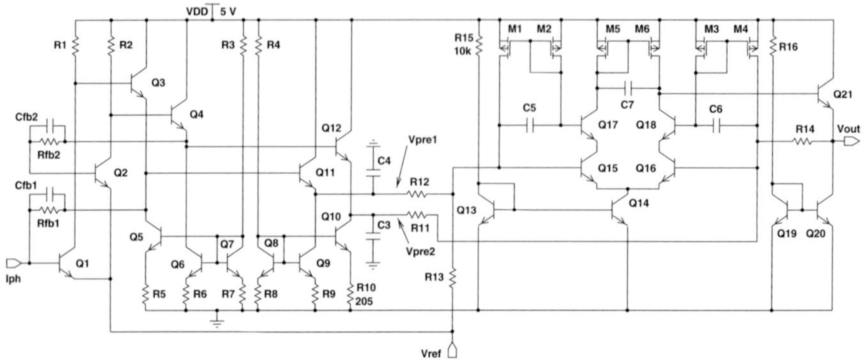

Fig. 21. Schematic of high-speed BiCMOS-OEIC for the fast channels A–D in an OS-BiCMOS-OEIC

Thereby, the effect of the DPD capacitance is minimized. The reference voltage V_{ref} is chosen as the emitter potential of Q1 in order to increase the reverse voltage of the DPD to $V_{\text{BE,Q1}} + V_{\text{ref}}$. The values for R1, Rfb1 and Cfb1 were 3 kΩ, 27 kΩ and 25 fF, respectively. A second emitter follower (Q11) is implemented for level shifting and decoupling of output and feedback path. The second preamplifier consists of transistors Q2, Q4 and Q7 and the feedback resistor Rfb2 plus the compensation capacitor Cfb2. At the outputs of the preamplifier, C3 and C4 are added as further compensation capacitors.

Perfect matching of the two preamplifiers is necessary in order to obtain $V_{\text{pre1}} = V_{\text{pre2}}$ for a dark photodiode (Fig. 21). This perfect matching of the two preamplifiers requires a careful layout to achieve a low output offset voltage.

The preamplifiers are connected to the operational amplifier in subtractor configuration via R_{11} and R_{12}. The bias current cancellation introduced in Fig. 15 is applied to reduce the input currents of the operational amplifier enabling a low output offset voltage. A PMOS current mirror load with M5 and M6 is used here in order to obtain a higher open loop gain of the operational amplifier. For $\text{R}_{11} = \text{R}_{12}$ and $\text{R}_{13} = \text{R}_{14}$, an analysis of the subtractor amplifier yields the transfer function

$$V_{\text{out}} \approx V_{\text{ref}} + \frac{R13}{R12} I\text{ph}R\text{fb1}$$

when we assume a large open-loop voltage gain of the operational amplifier. A −3 dB frequency of 189 MHz was determined by numerical prelayout simulation for the complete amplifier with a load of $\text{R}_\text{L} = 1\,\text{k}\Omega$ and $\text{C}_\text{L} = 10\,\text{pF}$. The complete two-stage amplifier was designed for a sensitivity of 10 mV/μW and an offset voltage less than 10 mV for R11 = R12 = R13 = R14.

The OEIC with the DPD and the two-stage amplifier was fabricated in a 0.8 μm BiCMOS technology. The measured frequency response of this two-stage optical receiver is shown in Fig. 22. A −3 dB frequency of 147.7 MHz

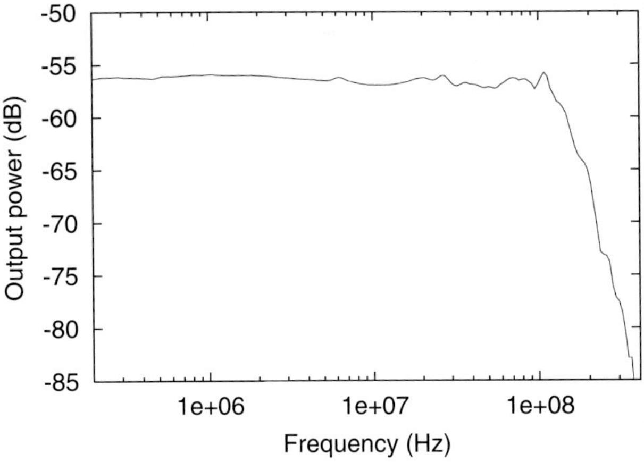

Fig. 22. Frequency response of a two-stage optical BiCMOS receiver for the fast channels A–D of a high-speed BiCMOS-OEIC for optical storage systems [28]

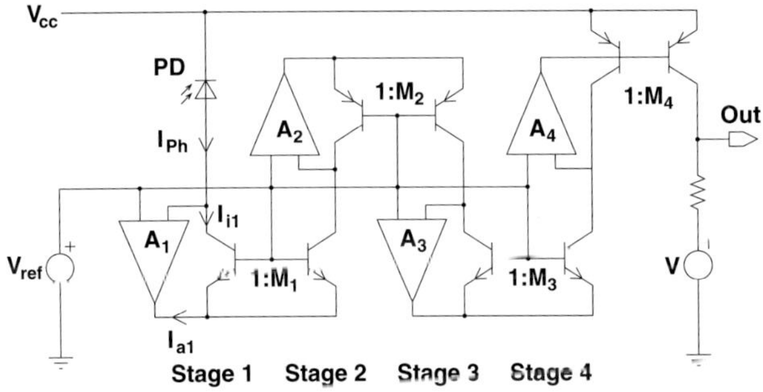

Fig. 23. Circuit topology of one segment in the fast OEIC for optical storage systems [32]

is determined from this frequency response. The power consumption of the two-stage optical receiver was 35 mW at a supply voltage of 5 V. The active die area of the two-stage optical receiver was 340 µm × 140 µm.

An even faster OEIC for optical storage systems has been presented recently. A bandwidth of 250 MHz has been achieved with a four-stage current amplifier in a standard 0.6 µm BiCMOS technology [32]. This OEIC can be used for channel data rates up to 600 Mb/s corresponding to 22× DVD speed with 12 000 rpm. The photodiode of this OEIC was optimized to a low capacitance of 0.25 pF. Nevertheless, the high bandwidth is an astonishing result when we consider that no bias currents were superimposed to the photocurrents to reduce the offset of the amplifiers. Lowest power consumption, therefore, also could be achieved. Figure 23 shows the topology of the current preamplifier used in the 8-channel OEIC.

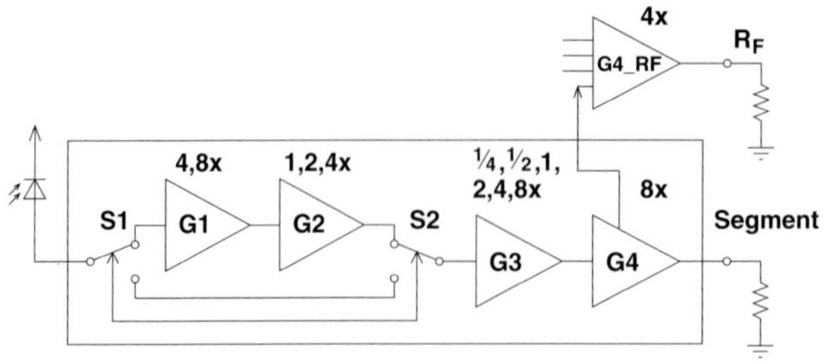

Fig. 24. Gain partitioning of one segment in the fast OEIC for optical storage systems [32]

Since the gain-bandwidth product of the BiCMOS process used was 400 GHz and a gain over 1000 was needed, a four-stage amplifier with a bandwith of at least 400 MHz for each stage was necessary. The transistors in Fig. 23 form current dividers for the output currents of the transconductance (V/I) amplifiers A1–A4. The loops are closed over one arm of the current dividers.

The gain was switchable implementing MOS switchees between and in the stages from 2 to 2048 for different laser powers during read and write as well as for different media according to the scheme illustrated in Fig. 24.

A summation amplifier (RF amplifier) for the channels A–D was implemented. The bandwidth of this RF amplifier only weakly depended on the gain. The gain of the satellite amplifiers in channels E–H was always four times higher than shown in Fig. 24.

It has to be mentioned that a trick was necessary to achieve the high bandwidth for low photocurrents, since photocurrents of 100 nA to 1 μA caused the transit frequencies of the NPN transistors T1 and T2 to drop to tens of kHz to 1 MHz [32]. The large attenuation resulting at 400 MHz hat to be eliminated. This was done by bypass capacitors, which were not shown in the circuit diagram however. The V/I amplifiers A1 to A4 used NMOS input transistors to avoid base input currents, their DC-current offset, and their base current shot noise. Gain peaking around 200 MHz was caused by crosstalk between the output drivers and the first two stages. It was stated that this problem can be solved in a redesign by separated front-end and back-end supplies.

The output stage delivered a current of 8 mA into a 150 Ω load with a one-sigma offset value of 117 μV.

3.2 Fiber Receivers

First, we demonstrate an innovative monolithic integration of vertical PIN photodiodes in a twin-well CMOS process (Fig. 9), which uses epitaxial wafers, in order to combine the high speed and the large quantum efficiency of the photodiode [33]. In contrast to the OEICs of [34,35], here, only a single

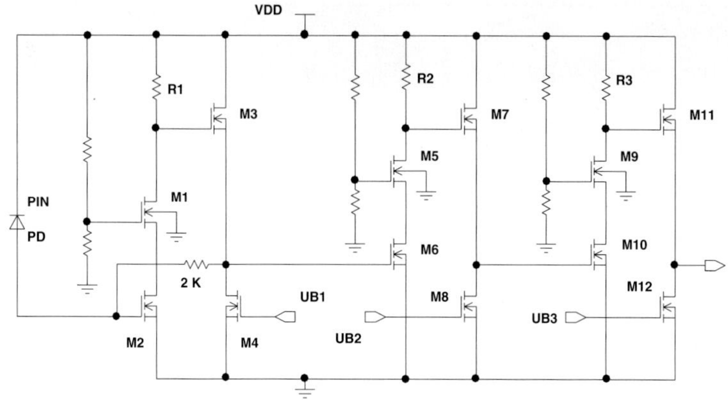

Fig. 25. Circuit of a fiber receiver OEIC [33]

power supply of 2.5 V, 3.3 V, of 5 V is needed. The integration of PIN photodiodes in a twin-well CMOS technology requires much less additional process complexity than the published approaches to standard-buried-collector (SBC) based bipolar OEICs [21, 36]. Three additional masks were necessary for the PIN-bipolar integration and for the avoidance of the Kirk effect. Only one photodiode protection mask is added for the PIN-CMOS integration in order to block out an originally unmasked threshold implantation from the photodiode area. This additional mask seems to be justified due to the expected high volumes of OEICs and OPTO-ASICs.

The PIN-CMOS photodiodes with an area of $2700\,\mu\text{m}^2$, a doping concentration C_o in the epitaxial layer of $2 \times 10^{13}\,\text{cm}^{-3}$, and an integration polysilicon resistor of $500\,\Omega$, as well as complete OEICs, were fabricated in an industrial $1.0\,\mu\text{m}$ CMOS process. Reference samples both for the photodiodes and OEICs with the standard doping concentration $C_e = 10^{15}\,\text{cm}^{-3}$ were also fabricated. The measurements on the PIN photodiodes and their results were described in the Chapter by *Paniccia* et al.

The circuit of a high-speed preamplifier with an integrated PIN photodiode is shown in Fig. 25. The cascode transistors M1, M5, and M9 reduce the Miller effect and increase the bandwidth correspondingly. The input stage with transistors M1–M4 is a transimpedance configuration, which converts the photocurrent change in the integrated PIN photodiode to a voltage change. The source followers M3, M7 and M11 as well as the current sources M4, M8 and M12 are used for level shifting. The threshold voltage of transistors M3, M7 and M11 was reduced intentionally to about 0.4 V by the photodiode protection mask in order to obtain lower V_{GS} values. Due to the feedback across the $2\,\text{k}\Omega$ resistor and to the identical dimensions of the transistors in the different stages, a good independence from process deviations within the relatively large specified process tolerances of the used digital CMOS process was obtained. Three identical biasing circuits (UB1 = UB2 = UB3)

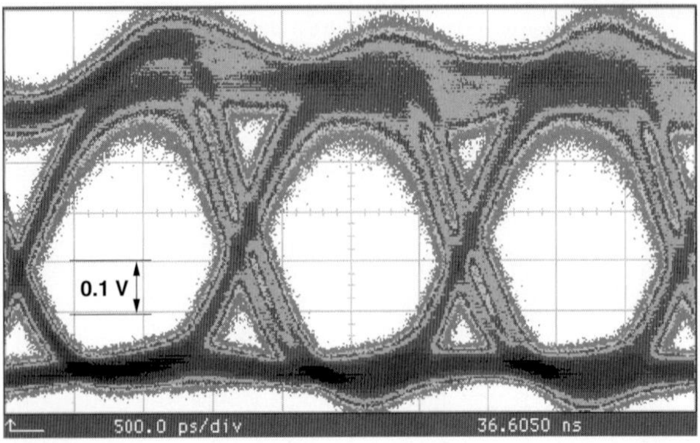

Fig. 26. Eye diagram of the CMOS preamplifier OEIC for a data rate of 622 Mb/s (500 ps/div; 0.1 V/div)

were used instead of one in order to avoid parasitic coupling between the stages. The sensitivity of the PIN amplifier OEIC was 4.7 mV/μW increasing to 9.0 mV/μW with ARC, which corresponds to an overall transimpedance of 18.4 kΩ. Its power consumption was 44 mW at 5.0 V reducing to 17 mW at 3.3 V and to 9 mW at 2.5 V. The photodiode together with its metal shield around covered an area of approximately 150 μm × 150 μm. The preamplifier occupied an active area of less than 190 μm × 200 μm.

The rise and fall times for the OEIC with the epitaxial doping concentration $C_e = 10^{15}\,\text{cm}^{-3}$ were $t_r = 15.5$ ns and $t_f = 17.6$ ns. These large values for t_r and t_f were due to the slow carrier diffusion in the standard epitaxial layer of the photodiode. The corresponding values for the doping concentration $2 \times 10^{13}\,\text{cm}^{-3}$ in the epitaxial layer, where the depletion region spread through the whole epitaxial layer and carrier diffusion in the photodiode was eliminated, were 0.53 ns and 0.69 ns, respectively. These values indicate that CMOS OEICs with a reduced doping concentration in the epitaxial layer having an appropriate output buffer can be used as receivers for optical data transmission via fibers or for optical interconnects on a board level up to a bit rate (BR) of 622 Mb/s in the non-return-to-zero (NRZ) mode, verified by a measured eye diagram with pseudo random bit sequences (PRBS) of $2^{23} - 1$ (Fig. 26).

The next fiber receiver described used PIN photodiodes in the same 1.0 μm CMOS process also with a reduced doping concentration in the epitaxial layer. The high-speed preamplifier circuit is shown in Fig. 27. Only N-channel MOSFETs with a minimum channel lengths were used in order to achieve a high bandwidth. The transistors M1–M5 operate in common-source configuration. The input stage with M1 in transimpedance configuration converts changes in the photocurrent of the photodiode to voltage changes. The feedback resistor in the input stage is formed by gate polysilicon. Two further stages with two transistors each are used as voltage amplifiers. The polysil-

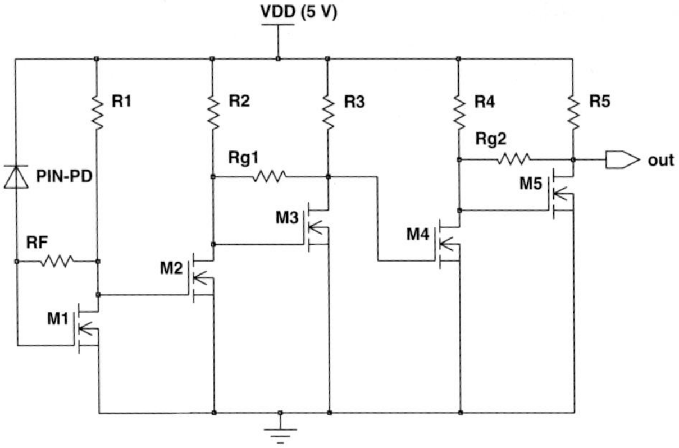

Fig. 27. Circuit of a fiber receiver OEIC [37]

icon feedback resistor in each of the two stages (Rg1, Rg2) limits the gain and therefore boosts the bandwidth. Due to the feedback resistors, a good independence from process deviations within the relatively large specified process tolerances of the used digital CMOS process is obtained. The sensitivity of the PIN preamplifier OEIC was 13.8 mV/μW without ARC, which corresponds to an overall transimpedance of 45.9 kΩ. Compared to the above-mentioned receiver, the sensitivity of the new OEIC has been increased by a factor of more than 2.5. Its power consumption is 2.5 mW at 5.0 V. The photodiode has an area of 2700 μm^2. The preamplifier occupies an active area of 245 μm × 140 μm due to the large area necessary for the resistors formed by low-resistivity (gate) polysilicon.

The rise and fall times measured at the OEIC output with the picoprobe were 0.72 ns and 0.57 ns, respectively, with a supply voltage of 5.0 V. With these values a bandwidth of $f_{-3\,\mathrm{dB}} = 2.2/(\pi(t_\mathrm{r} + t_\mathrm{f})) = 550\,\mathrm{MHz}$ can be estimated for the OEIC. The higher data rate of this OEIC is due to less transistors in the preamplifier than in the typical high-frequency circuit with cascode transistors and source followers described before. In the above-mentioned receiver, 12 NMOS transistors were used in the preamplifier, which therefore contained more parasitic capacitanes. Here, the number of parasitic capacitances is considerably lower. In addition, the gain per amplifying transistor is lower here, allowing for a higher data rate of the new preamplifier.

The so-called Q-factor is defined as $(B_1 - B_0)/(\sigma_1 + \sigma_0)$ [38], where B_1 and B_0 are the mean values and σ_1 and σ_0 are the standard deviations of the output signal for a "1" and "0". B_1, B_0, σ_1 and σ_0 can be determined from eye diagram measurements.

The bit error rate (BER) then is available from [38]: $\mathrm{BER} \approx e^{-\frac{Q^2}{2}}/Q\sqrt{2\pi}$. The BER of the OEIC has been determined with an HP 54570/51 digital sampling oscilloscope for the NRZ data rates of 622 Mb/s and 1 Gb/s in dependence on the optical input power. The results for the bit error rate

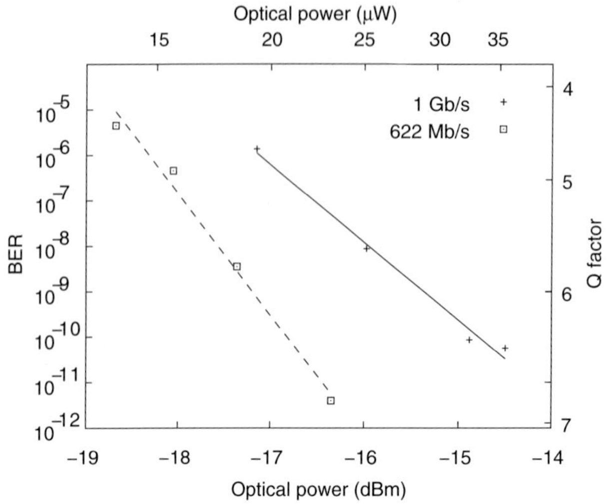

Fig. 28. BER of PIN OEIC without ARC in dependence on the optical input power at 638 nm

are shown in Fig. 28. From this diagram for $\lambda = 638\,\text{nm}$ being especially interesting for data transmission via plastic optical fibers and for optical interconnects, it can be seen that an optical input power of $-15.4\,\text{dB}\cdot\text{m}$ is necessary for a BER of 10^{-9} at a data rate of 1 Gb/s. At 622 Mb/s the sensitivity of the OEIC for the same BER is $-17.2\,\text{dB}\cdot\text{m}$. The sensitivity for a BER of 10^{-10} at a data rate of 1 Gb/s is $-14.8\,\text{dB}\cdot\text{m}$. The sensitivity for a BER of 10^{-11} at a data rate of 622 Mb/s is $-16.5\,\text{dB}\cdot\text{m}$. For $\lambda = 850\,\text{nm}$, a sensitivity of $-15.3\,\text{dB}\cdot\text{m}$ for a BER of 10^{-9} at a data rate of 622 Mb/s has been verified.

These values are very good results for an OEIC in a 1.0 μm CMOS technology. The overall transimpedance of 46 kΩ and the data rate of 1 Gb/s result in a 46 Tb·Ω/s transimpedance data rate product. With the bandwidth of 550 MHz, a 25 THz·Ω effective transimpedance product results, which exceeds the value of 18 THz·Ω reported in [39]. The sensitivity of the PIN CMOS OEIC for a data rate of 1 Gb/s and a BER of 10^{-9} with a value of $-15.4\,\text{dB}\cdot\text{m}$ for 638 nm exceeds the sensitivity of $-6.3\,\text{dB}\cdot\text{m}$ of a 0.35 μm OEIC for 850 nm [35] by a factor of 8. It shall be mentioned that the implementation of an antireflection coating will improve the sensitivity of the OEICs by a value of 2 dB–3 dB. The minimum sensitivity of $-17\,\text{dB}\cdot\text{m}$ specified in the Gigabit Ethernet networking standard, therefore, may be achievable with PIN CMOS OEICs. The data rate of the OEIC is limited by the amplifier in a 1.0 μm technology. With submicrometer PIN-CMOS-OEICs data rates in excess of 1 Gb/s are possible.

The circuit diagram for a receiver with a SOI PIN photodiode (Fig. 6) is shown in Fig. 29. Optimum results were obtained for input transistors

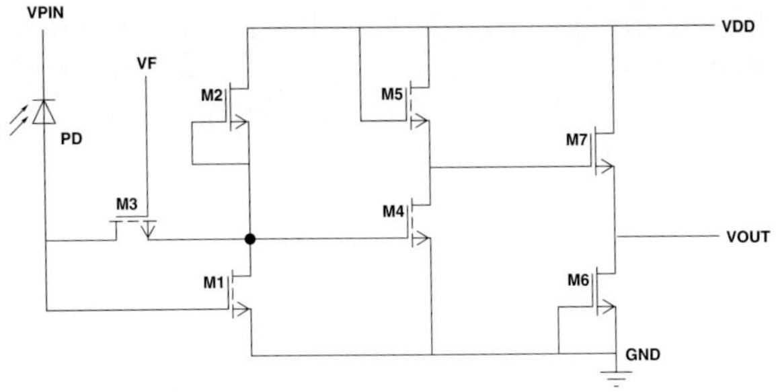

Fig. 29. Circuit diagram of the SOI receiver [13]

(M1–M2) with 50 µm widths for the SOI receiver. The functionality of the three-stage amplifier has been described in [14]. The small-signal transfer function of this design is discussed in [13].

At $V_{DD} = 5$ V and $V_{PD} = 20$ V, for 850 nm and a BER $= 10^{-9}$, the sensitivity of the SOI receiver was −26.1 dB · m at 622 Mb/s, −20.2 dB · m at 1.0 Gb/s as well as −12.2 dB · m at 2.0 Gb/s. The dependence of the sensitivity on the photodiode voltage was also investigated. The SOI receiver demonstrated a penalty that was typically less than 1 dB as the photodiode supply was reduced to 10 V. For single-supply 5 V operation, the sensitivity degraded by 2.8 dB to 3.9 dB and the maximum data rate was 1.5 Gb/s with a sensitivity of −12 dB · m.

The dynamic range of the receivers was measured with both a constant voltage supply and with a variable feedback voltage (V_F) as a simple form of automatic gain control (AGC). The photodiode bias has a large impact on the dynamic range, although the feedback voltage can be used very effectively to compensate for this degradation. At the optimum bias point and a constant V_F, the SOI receiver exhibited dynamic ranges of > 23.5 dB · m, > 17.7 dB · m, and 12.0 dB · m at 622 Mb/s, 1.0 Gb/s and 2.0 Gb/s.

This receiver was not optimized for the best sensitivity at the higher data rates. The dominant noise term was thermal noise of the feedback resistor (M_F) and not the channel noise of the input transistor. Better performance is possible with an optimized circuit design in a more advanced technology.

4 Conclusion

Double photodiode, vertical PIN photodiode, and lateral SOI photodiode avoid slow carrier diffusion effects. The performance of integrated photodiodes can be improved significantly when minor process changes are made to obtain these advanced integrated photodiodes. Finger photodiodes, another advanced approach, are very good appropriate for application together with blue light in near future DVR systems.

Advanced OEICs for optical storage systems have been realized. These low-offset solutions have bandwidths ranging from 58 MHz to 250 MHz. Furthermore, high-speed fiber receiver OEICs have been presented. PIN CMOS OEICs reach sensitivities of $-17.2\,\mathrm{dB \cdot m}$ at 622 Mb/s and $-15.4\,\mathrm{dB \cdot m}$ at 1 Gb/s (BER $= 10^{-9}$) with a single 5 V supply voltage. Using higher photodiode bias voltages, the SOI receiver reaches sensitivities of $-20.2\,\mathrm{dB \cdot m}$ at 1 Gb/s and $-12.2\,\mathrm{dB \cdot m}$ at 2 Gb/s. The speed of integrated photodiodes and of amplifiers can be increased with sub-µm technologies. Then receivers for Gigabit Ethernet and the Fiber Channel working with 850 nm light, for instance, can be realized with Si OEICs.

References

1. Landolt-Börnstein: *Numerical data and functional relationships in science and technology*, vol. 17c (Springer, Berlin 1984), p. 474
2. J. Popp, H. v. Philipsborn: 10 Gbit/s on-chip photodetection with self-aligned silicon bipolar transistors, Proc. ESSCIRC (1990), pp. 571–574
3. J. Wieland, H. Duran, A. Felder: Two-channel 5 Gb/s silicon bipolar monolithic receiver for parallel optical interconnects, Electronics Letters **30**, 358 (1994)
4. H. Kabza, K. Ehinger, T. F. Meister, H.-W. Meul, P. Weger, I. Kerner, M. Miura-Mattausch, R. Schreiter, D. Hartwig, M. Reisch, M. Ohnemus, R. Köpl, J. Weng, H. Klose, H. Schaber, L. Treitinger: A 1-µm polysilicon self-aligned bipolar process for low-power high-speed integrated circuits, IEEE Electron Device Letters **10**, 344–346 (1989)
5. E. Braß, U. Hilleringmann, K. Schumacher: System integration of optical devices and analog CMOS amplifiers, IEEE J. Solid-State Circuits **29**, 1006–1010 (1994)
6. U. Hilleringmann, K. Goser: Optoelectronic system integration on silicon: waveguides, photodetectors, and VLSI CMOS circuits on one chip, IEEE Trans. Electron Devices **42**, 841–846 (1995)
7. E. Fullin, G. Voirin, M. Chevroulet, A. Lagos, J.-M. Moret: CMOS-based technology for integrated optoelectronics: a modular approach, IEEE Int. Electron Device Meeting (1994), pp. 527–530
8. R. Kauert, W. Budde, A. Kalz: A monolithic field segment photo sensor system, IEEE J. Solid-State Circuits **30**, 807–811 (1995)
9. H. Zimmermann: *Integrated Silicon Optoelectronics* (Springer 2000)
10. P. J.-W. Lim, A. Y. C. Tzeng, H. L. Chuang, S. A. S. Onge: A 3.3 V monolithic photodetector/CMOS preamplifier for 531 Mb/s optical data link applications, Proc. ISSCC (1993), pp. 96–97
11. D. M. Kuchta, H. A. Ainspan, F. J. Canora, R. P. Schneider: Performance of fiber-optic data links using 670 nm CW VCSELs and a monolithic Si photodetector and CMOS preamplifier, IBM J. Res. Develop. **39**, 63–72 (1995)
12. J.-P. Colinge: *Silicon-on-insulator Technology: Materials to VLSI*, (Kluwer Academic Publishers, Boston 1991)
13. J. Schaub, R. Li, J. Csutak, J. Campbell: High-speed monolithic silicon photoreceivers on high resistivity and SOI substrates, J. Lightwave Technology **19**, 272–278 (2001)

14. C. Schow, J. Schaub, R. Li, J. Qi, J. Campbell: A 1-Gb/s monolithically integrated silicon NMOS optical receiver, IEEE J. Selected Topics in Quantum Electronics **4**, 1035–1039 (1998)
15. H. Zimmermann, K. Kieschnick, M. Heise, H. Pless: BiCMOS OEIC for optical storage systems, Electronics Letters **34**, 1875–1876 (1998)
16. H. Zimmermann: Full custom CMOS and BiCMOS OPTO-ASICs, Proc. 5th Int. Conf. on Solid-State and Integrated-Circuit Technology (1998), pp. 344–347
17. H. Zimmermann, U. Müller, R. Buchner, P. Seegebrecht: Optoelectronic receiver circuits in CMOS-technology, in: *Mikroelektronik'97, GMM-Fachbericht 17* (VDE-Verlag, Berlin, Offenbach 1997), pp. 195–202
18. H. Zimmermann, T. Heide, A. Ghazi, K. Kieschnick: PIN-CMOS-receivers for optical interconnects, Ext. Abstr. 2nd IEEE Workshop on Signal Propagation on Interconnects (1998), pp. 88–89
19. H. Zimmermann, A. Ghazi, T. Heide, R. Popp, R. Buchner: Advanced photo integrated circuits in CMOS technology, Proc. 49th Electronic Components and Technology Conference (ECTC) (1999), pp. 1030–1035
20. H. Zimmermann: Monolithic Bipolar-, CMOS-, and BiCMOS-receiver OEICs, Proc. Int. Semicond. Conference (CAS'96) (1996), pp. 31–40
21. M. Yamamoto, Kubo, K. Nakao: Si-OEIC with a built-in pin-photodiode, IEEE Trans. Electron Devices **42**, 1093-1099 (1995)
22. A. Ghazi, H. Zimmermann, P. Seegebrecht: CMOS photodiode with enhanced responsivity for the UV/blue spectral range, IEEE Trans. Electron Devices **49**, 1124–1128 (2002)
23. S. Nakamura, G. Fasol: *The Blue Laser Diode* (Springer 1998)
24. J.-P. Colinge: p–i–n photodiodes made in laser-recrystallized silicon-on-insulator, IEEE Trans. on Electron Devices **33**, 203–205 (1986)
25. A. Ghazi, T. Heide, H. Zimmermann, P. Seegebrecht: DVD OEIC and 1 GBit/s Fiber Receiver in CMOS Technology, Proc. 7th IEEE Int. Symposium on Electron Devices for Microwave and Optoelectronic Applications (EDMO) (2000), pp. 224–229
26. H. Zimmermann, Kieschnick, M. Heise, H. Pless: High-bandwidth BiCMOS OEIC for optical storage systems, Proc. IEEE Int. Solid-State Circuits Conference (1999), pp. 384–385
27. B. Kim, M. Jeong, D. Cho, J. Kim, J. Lee, S. Kim: 0.8 μm CMOS Analog Front-End Processor for 8X Speed CD-ROM, IEEE Trans. Consumer Electronics **42**, 826–831 (1996)
28. H. Zimmermann, K. Kieschnick: Low-offset BiCMOS OEIC for optical storage systems, Electron. Letters **36**, 1223–1224 (2000)
29. H. Zimmermann, K. Kieschnick, T. Heide, A. Ghazi: Integrated high-speed, high-responsitivity photodiodes in CMOS and BiCMOS technology, Proc. 29th European Solid-Stage Device Research Conf. (1999), pp. 332–335
30. T. Takimoto, N. Fukunaga, M. Kubo, N. Okabayashi: High speed Si-OEIC (OPIC) for optical pickup, IEEE Trans. Consumer Electronics **44** 137–142 (1998)
31. K. Kieschnick, T. Heide, A. Ghazi, H. Zimmermann, P. Seegebrecht: High speed photonic CMOS and BiCMOS receiver ICs, Proc. 25th European Solid-State Circuits Conf. (1999), pp. 398–401

32. G. de Jong, J. Bergervoet, J. Brekelmans, J. van Mil: A DC-to-250 MHz current pre-amplifier with integrated photodiodes in standard CBiMOS, for optical-storage system, Proc. IEEE Int. Solid-State Circuits Conference (2002), pp. 362–363, 474
33. H. Zimmermann, T. Heide, A. Ghazi: Monolithic high-speed CMOS photoreceiver, IEEE Photon. Technol. Lett. **9**, 254–256 (1999)
34. C. Schow, J. Qi, L. Garrett, J. Campbell: A silicon NMOS monolithically integrated optical receiver, IEEE Photon. Technol. Lett. **9**, 663–665 (1997)
35. T. Woodward, A. Krishnamoorthy: 1 Gbit/s CMOS photoreceiver with integr. detector operating at 850nm, Electron. Letters **34**, 1252–1253 (1998)
36. M. Kyomasu: Development of an integrated high speed silicon PIN photodiode sensor, IEEE Trans. Electron Devices **42**, 1093–1099 (1995)
37. H. Zimmermann, T. Heide: A monolithically integrated 1-Gb/s optical receiver in 1-μm CMOS technology, IEEE Photon. Technol. Lett. **13**, 711–713 (2001)
38. G. Agrawal: *Fiber-Optic Comm. Systems* (Wiley, New York 1997)
39. M. Ingels, G. Plas, J. Crols, M. Steyaert: A CMOS 18 THz 240Nb/s transimpedance amplifier and 155Mb/s LED-driver for low-cost optical fiber links, IEEE J. Solid-State Circuits **29**, 1552–1559 (1994)

Silicon Based Waveguides

Robert J. Bozeat, S. Day, F. Hopper, F. P. Payne,
S. W. Roberts, and M. Asghari

Bookham Technology plc, Oxfordshire, UK

Abstract. Optical waveguide based devices are key to the operation of many modern optical communication systems. The following chapter outlines the design and application of optical waveguides of the Silicon On Insulator (SOI) type. The important properties of these waveguides are described, the design choices that must be made and their implications are highlighted. The interfacing of silicon waveguides to single mode optical fibres by the use of mode transforming tapers is described and the design methodology behind the production of adiabatic tapers is discussed. Finally the design of Arrayed Waveguide Gratings (AWGs) and Variable Optical Attenuators (VOAs) in silicon waveguide technology is outlined.

1 Introduction

Devices incorporating planar optical waveguides are increasingly finding application within modern data communication and telecommunication systems. These devices may be found multiplexing and demultiplexing signals within Dense Wavelength Division Multiplexed (DWDM) systems [1], switching and routing within reconfigurable system nodes [2] or gain equalising within fibre amplifiers [3].

Much of the early research work into planar waveguide technology was focussed around the lithium niobate material system. Research interest was driven by lithium niobate's large linear electro-optic effect (Pockels effect) and hence potential for the production of electro-optic waveguide modulators [4]. Lithium niobate has subsequently developed into a commercially mature high speed modulator technology. In the 1980s research interest began to focus upon Silica-on-Silicon [5] and Silicon-on-Insulator (SOI) [6] as material systems that were perceived to offer greater potential for the fabrication of more complex waveguide based devices.

Silica-on-Silicon waveguides confine the propagating radiation within multilayer doped silica films that are deposited onto standard, microelectronics grade, silicon substrates. The composition of the waveguiding materials and the geometry of the waveguiding structure are normally chosen to enable the straightforward, low loss interfacing to standard single mode optical fibre. Low index contrast waveguides of this type do not however yield compact devices due to the large bend radii that must be employed. Higher index contrast glasses offer the potential for more compact devices [7] but interfacing these waveguides to single mode fibres is problematic.

L. Pavesi, D. J. Lockwood (Eds.): Silicon Photonics, Topics Appl. Phys. **94**, 269–294 (2004)

SOI waveguides confine the propagating radiation within an epitaxially grown, single crystal, silicon film which has an implanted lower oxide cladding. These high index contrast waveguides offer the potential for small radius waveguide bends and hence highly compact components. In addition active devices are possible because silicon can be made optically absorbent at infrared wavelengths [8] either by doping or by carrier injection. The efficient coupling of SOI waveguides to standard single mode optical fibres is possible but requires careful interface design.

The following will focus specifically upon optical waveguides fabricated in SOI materials. After a brief introduction to the material system a selective account of historical developments in the field will be given. The design of passive waveguides for the realisation of compact bends and low birefringence will be described. Finally the interfacing of SOI waveguides to single mode fibre will then be considered and some example devices will be described.

2 Background

Silicon on insulator materials have been an active area of research since the early 1960s. Research interest was initially stimulated by the potential of these materials for the production of high speed and radiation hard CMOS microelectronic circuits [9]. Early SOI substrates were largely of the Silicon-on-Sapphire (SOS) type comprising of a thin layer of silicon which was deposited onto the surface of a sapphire substrate wafer. Unfortunately SOS substrates tended to suffer from poor quality silicon layers due to the lattice mismatch between silicon and sapphire. In the late 1970s the SIMOX technique was first used for the production of SOI wafers. SIMOX wafers have a buried silicon dioxide insulating layer that is formed using a multiple oxygen implantation and annealing process. It was found to be possible to achieve high quality silicon surface layers using the SIMOX approach. More recently direct wafer bonding [10] has emerged as an attractive alternative to the SIMOX process for the production of SOI wafers with thick silicon surface layers. SOI wafers of both bonded and SIMOX types are now widely available, commercial products.

Research interest in SIMOX as a waveguiding material dates from the late 1980s. SIMOX waveguides were first fabricated from wafers with sub-micron silicon surface layers [11]. The dimensions of these waveguides were consistent with the dimensions required for single moded operation in SOI slab waveguides. In 1991 *Soref* et al. [12] described single moded SOI waveguides with much larger cross sectional dimensions where shallow etched ribs were used to provide weak optical confinement and hence single moded operation. Figure 1 is a schematic of a large cross section SOI rib waveguide as described by Soref. Low loss Waveguides of this type were soon demonstrated by several research groups [13].

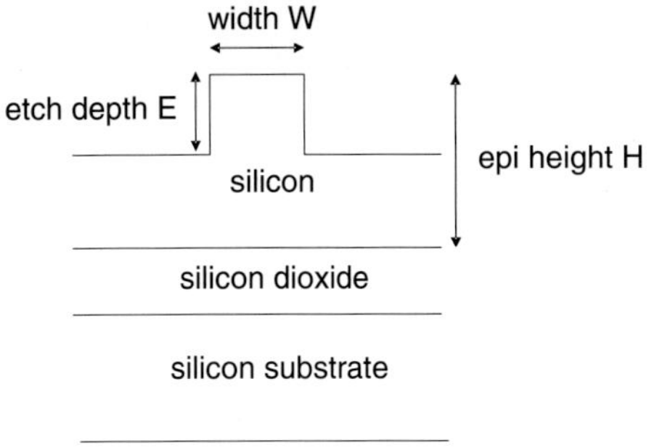

Fig. 1. The Silicon on Insulator rib waveguide geometry

Since these early demonstrations much research work has focussed upon waveguides fabricated in the SOI material system. Waveguides with dimensions varying from less than 1 µm [14] to several µm have been extensively researched. Complex active and passive devices have been demonstrated [15] and commercial products have been produced.

3 Waveguide Properties and Design

The design of optical waveguides in SOI materials requires the judicious selection of a number of structural parameters. The desired optical performance parameters, the manufacturing yield required and the cost and complexity of device packaging will all, to some extent, be influenced by these fundamental decisions. It is unlikely that there exists one best waveguide design for all applications. The intention here is not therefore to suggest practical waveguide designs but instead to consider the fundamental properties of SOI waveguides and to explain how these properties are influenced by the design decisions that are made.

The following paragraphs will first examine the modes of the silicon waveguide. The design criteria for single moded operation will be described. Design for compact waveguide bends and for the control of form birefringence will then be considered. Finally the impact of sidewall roughness upon SOI waveguides will be examined.

3.1 Modes of the SOI Waveguide

The correct operation of many waveguide device types is critically dependant upon the propagation of only a single guided optical mode. Higher order guided modes introduce cross-talk in Arrayed Waveguide Gratings (AWGs) [16], unexpected split ratios in Y-junctions and Multi-Mode-Interference devices (MMIs) and low extinction ratios in directional coupler type switches. In any complex waveguide based component light will inevitably, at

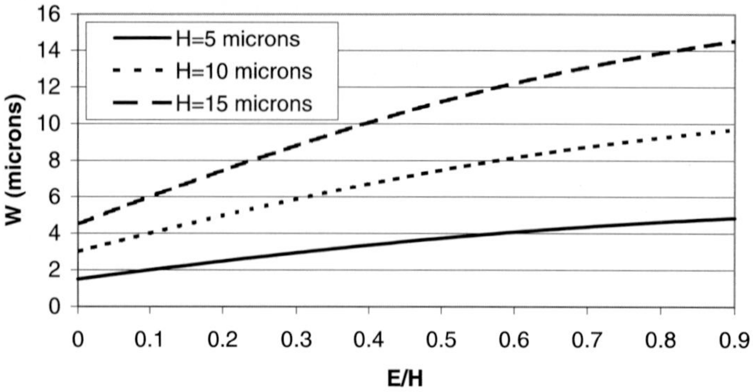

Fig. 2. The variation of single mode width with etch depth

some point, be scattered from the fundamental waveguide mode. This may occur at the optical fibre to waveguide interface, at a junction between straight and curved waveguides or in a straight waveguide itself due to scattering by lithographic roughness on the etched waveguide sidewalls. a primary design consideration is therefore that waveguides can be produced that will allow only a single guided mode to propagate with low loss.

Soref arrived at the following empirical single mode condition, valid when E/H is small, for the waveguide geometry of Fig. 1

$$\frac{W}{H} \leq 0.3 + \frac{H - E}{\sqrt{E^2 - 2HE}} . \tag{1}$$

Figure 2 plots the single mode condition for a variety of epitaxial layer thicknesses. The area below each line indicates the region of single mode operation according to the condition of (1).

From Fig. 2 it is clear that the thicker the epitaxial silicon layer that is employed the wider the range of rib widths that are available to the designer for single mode operation at the same slab to epitaxial layer ratio. Put another way as the epitaxial layer thickness increases so too does the etch depth required to achieve the same degree of optical confinement.

For most practical applications it is also important to be able to construct single mode waveguides which can achieve minimal bend radii with well controlled birefringence. Small bend radii allow for the fabrication of compact components and birefringence control is required for polarisation independent operation. The factors that influence these properties are now described.

3.2 Design of Curved Waveguides

The modal characteristics of curved SOI waveguides are, in general, quite different to those of straight waveguides of the same dimensions. The single mode condition described earlier is not appropriate for the design of curved

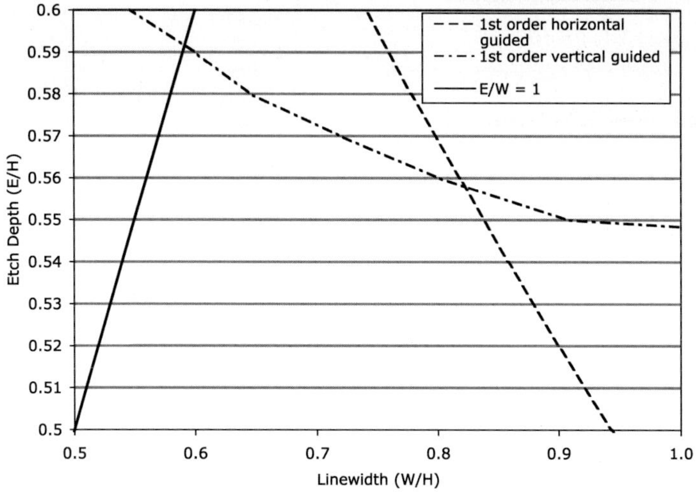

Fig. 3. The modal properties of an SOI waveguide with a 3 μm thick epitaxial layer

waveguides as the modal characteristics of these waveguides are heavily dependant upon the bend radius employed. Instead guided modes must be studied either by the use of a complex eigenmode solver or by the use of a propagation technique such as the Beam Propagation Method (BPM) [17].

Figure 3 summarises the results of TE polarised BPM simulations of a straight SOI waveguide with a 3 μm thick epitaxial layer. Figure 3 indicates the waveguide dimensions at which higher order modes begin to propagate with low loss. In addition, also shown on Fig. 3, is the condition at which the etch depth is equal to the rib width. There are manufacturing considerations that make it undesirable to fabricate guides where the etch depth exceeds the rib width. Figure 3 therefore indicates an available design region for single mode operation of a straight guide with a 3 μm thick silicon epitaxial layer.

Increasing the waveguide etch depth increases the confinement of the fundamental mode and reduces the radius of curvature that may be employed. From Fig. 3 the greatest practical etch depth for curved operation that allows for the design of single moded straight sections, in 3 μm thick epitaxial layers, can be obtained. Figure 4 below shows the radius of curvature at which the curved waveguide with this optimum etch depth begins to lose confinement of the fundamental TE mode as the waveguide width is varied. An excess bend loss of $> 0.1\,\mathrm{dB/cm}$ is used to define this condition. Also shown in Fig. 4 are similar curves for waveguides with optimum etch depth calculated for other epitaxial layer thicknesses.

The decreasing bend propagation loss with guide width displayed by the waveguide geometry studied in Fig. 4 is characteristic of curved SOI waveguides. As the rib width is increased a whispering gallery mode results [18] at which point further increases in guide width do not further reduce guide loss.

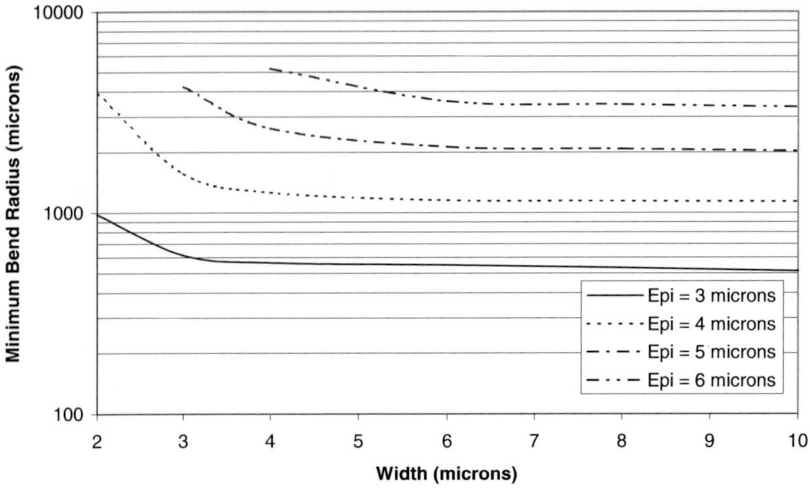

Fig. 4. Minimum practical radius of curvature waveguide bends as a function of width

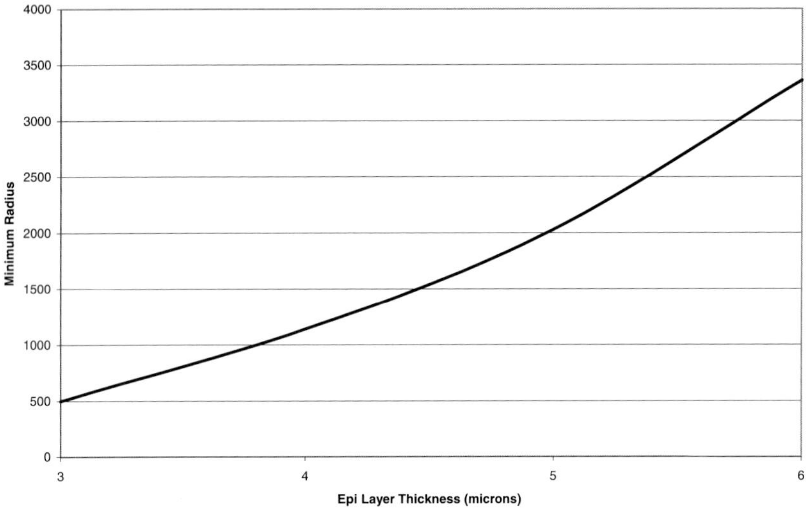

Fig. 5. The minimum achievable waveguide bend radius for a loss of less than 0.1 dB/cm as a function of epitaxial layer thickness

For a given epitaxial layer thickness and acceptable waveguide loss there is therefore a minimum limit upon the radius of curvature that may be achieved. Assuming an acceptable bend loss of 0.1 dB/cm of propagation Fig. 5 plots the radius of curvature below which this loss is exceeded as the epitaxial layer thickness is varied for the fundamental TE mode.

As the thickness of the selected silicon epitaxial layer is increased the practical minimum bend radius that can be achieved also increases.

3.3 Waveguide Birefringence Control

Most planar waveguide technologies produce waveguides with inherent birefringence. This birefringence may result from the properties of the waveguiding material itself [19], the fabrication processes used to construct the waveguide [20] or the asymmetry of the waveguide geometry [18]. As polarisation independent device operation is almost always essential it is most important to understand and control this fundamental waveguide property.

Most practically employed SOI waveguides are asymmetric leading to form birefringence. It can be shown, from electromagnetic theory [18], that form birefringence increases with the asymmetry of the waveguide geometry and also as the dimensions of the waveguide are reduced. In the following discussion the birefringence of the previously described waveguide geometry will be considered as the dimensions of the guide are varied.

Waveguide Birefringence is defined to be the difference between the modal indices of the fundamental Transverse Electric (TE) and Transverse Magnetic (TM) modes. In interferometric devices, such as arrayed waveguide gratings and Mach–Zehnder interferometers, birefringence manifests itself as a polarisation dependant shift (PDF) in the centre frequency of the device passband. Expressing birefringence as a frequency shift proves particularly useful when describing Dense Wavelength Division Multiplexed (DWDM) components. The birefringence induced frequency shift is here defined mathematically by the following equation where f_0 is the optical frequency.

$$\Delta f = 2\frac{N_{\mathrm{TM}} - N_{\mathrm{TE}}}{N_{\mathrm{TM}} + N_{\mathrm{TE}}} f_0 \, . \tag{2}$$

Consider now the birefringence of the single mode SOI waveguide described earlier. Figure 6 shows the birefringence of this waveguide, calculated using a polarised finite difference eigenmode solver, as the epitaxial layer thickness is varied.

As might be expected increasing the epitaxial layer thickness reduces the birefringence of the waveguide. In practise however it is more important to ensure that the birefringence is reproducible rather than minimal as techniques exist for on-chip birefringence compensation. In Fig. 7 the magnitude of the first derivative of the birefringence is plotted as the epitaxial layer thickness, waveguide width and etch depth are varied.

Waveguide designs with high derivatives in waveguide birefringence require tighter tolerances from fabrication processes in order to yield devices with acceptable birefringence. From the data presented in Fig. 7 it is straightforward to calculate the degree of fabrication process control required for a certain waveguide yield to a particular birefringence specification. For

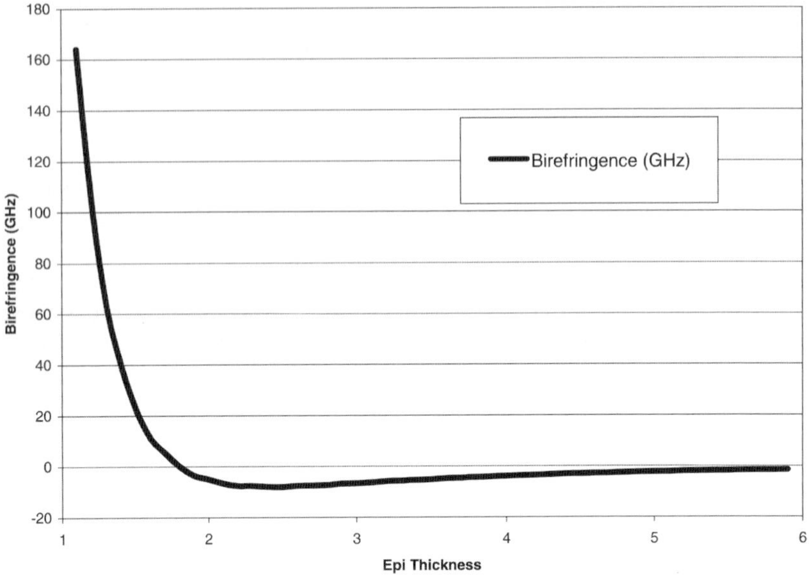

Fig. 6. Variation in waveguide birefringence with epitaxial layer thickness

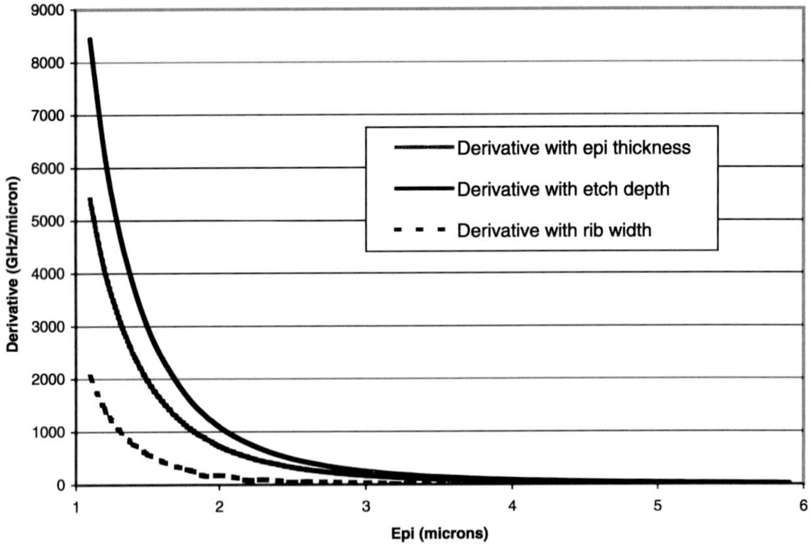

Fig. 7. Variations in the first derivative of the guide birefringence with waveguide dimensions

a possible specification consider the control of the passband centre frequency within a silicon arrayed waveguide grating. For a 100 GHz channel spacing system the useful grating passband widths are normally designed to

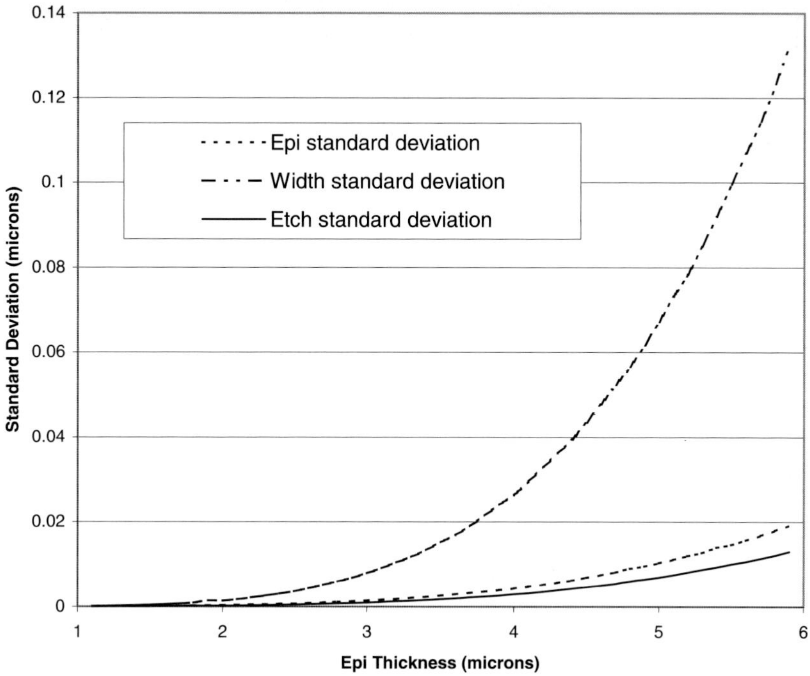

Fig. 8. Processing standard deviations required for a 0.25 GHz standard deviation in waveguide birefringence

be 25 GHz wide. The greater the variation in passband centre frequency with polarisation the smaller the usable bandwidth available within the system. System considerations suggest a reasonable standard deviation in passband centre frequency is around 1% of channel bandwidth, in this particular case 0.25 GHz. Figure 8 plots the process control required for a standard deviation in fabricated guide birefringence of 0.25 GHz.

The trend shown in Fig. 8 is generally true for single moded SOI waveguide geometries. The thinner the epitaxial silicon layer becomes the more difficult it becomes to control the birefringence of the fabricated waveguides.

3.4 Scattering Losses

Waveguide scattering losses result when propagating radiation interacts with roughly etched waveguide sidewalls. It is often assumed that scattering imposes severe limitations upon the loss performance that can be achieved in single mode high index contrast waveguides. This assumption originates from well known scattering theory which suggests that scattering losses should increase with the difference between the squared refractive indices of the waveguide core and cladding [21]. The magnitude of the scattering loss is however

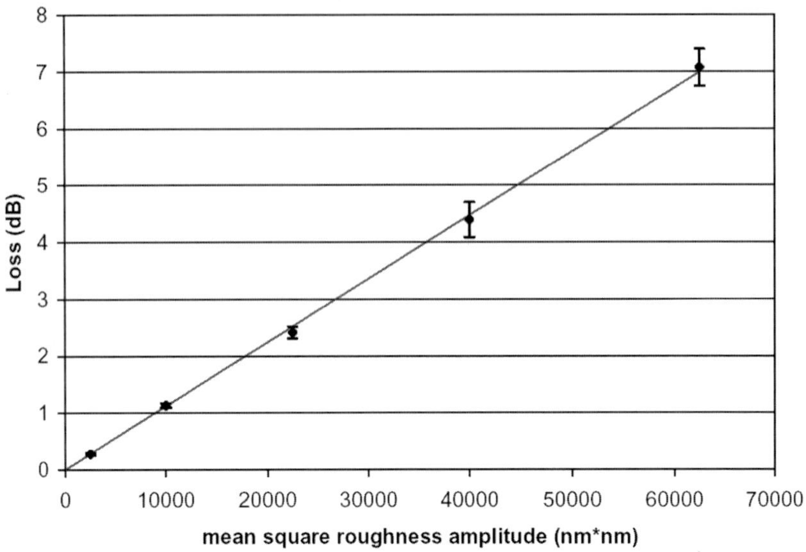

Fig. 9. Simulated scattering loss from an SOI rib waveguide as a function of the mean square roughness amplitude

also dependant upon the extent to which the guided mode interacts with the sidewall roughness.

SOI waveguides are normally designed with dimensions that are many times the wavelength of the guided radiation. The interaction of the guided mode with the etched sidewalls is therefore weak and, with typical levels of sidewall roughness, very low propagation losses are achieved. Roughness induced losses become problematic as the waveguide dimensions approach the wavelength, in silicon, of the guided radiation. Waveguides of these dimensions require special fabrication processes in order to reduce the amplitude of the roughness, and hence the loss, to acceptable levels [22].

It is possible to simulate the effects of sidewall roughness using the *Beam Propagation Method* [23]. Empirical studies have found that fabrication induced roughness displays an exponential autocorrelation function [24, 25]. Roughness of this form may be added to the sidewalls of a simulated rib waveguide and losses then calculated by averaging the losses of multiple beam propagation runs. Figure 9 shows the predicted waveguide propagation loss for an example SOI waveguide with a 4 μm epitaxial layer thickness, a 4 μm wide rib and a 2 μm deep etch as a function of the mean square sidewall roughness. a correlation length of 100 μm was assumed for the roughness used to generate Fig. 9.

The error bars on Fig. 9 indicate the range into which two thirds of the simulation results fall. As predicted by theory the propagation loss is found to be proportional to the mean square roughness amplitude. With modern silicon microfabrication technology, rms roughnesses of less than 20 nm amplitude can typically be achieved without the use of non standard processing steps. For the simulated geometry, and most geometries used in practise, scattering losses of less 0.1 dB per cm are normally achieved.

3.5 Design Conclusions

The design of SOI waveguides requires a clear understanding of both the fabrication processes that will be used to construct the device and the ultimate performance that will be required. The selection of the appropriate waveguide geometry involves a compromise between the compactness of the complete device and the degree to which it is necessary to control the form birefringence. As a general rule the smaller the guide cross section the more difficult it becomes to control the form birefringence but the easier it becomes to fabricate compact components.

4 Interfacing Silicon Waveguides to Single Mode Fibres

Due to the high dielectric constant of silicon, single mode SOI waveguides may have optical field dimensions which are small in comparison to standard single mode optical fibre. As an example consider an SOI waveguide comprising a 4 μm thick silicon epitaxial layer, a 3 μm rib width and a 2 μm etch depth. Such a waveguide would have a fundamental mode with a mode field radius of 1.6 μm, compared to the fibre mode field radius of 5.2 μm. The coupling loss between this waveguide and the optical fibre would be in excess of 4.7 dB To reduce the coupling loss of the fibre waveguide interface, it is necessary to use a mode transformer. The simplest mode transformer that can be used is to expand the waveguide horizontally. However, horizontal tapering only expands the field in one dimension and so the best coupling that can be achieved is with a 13 μm wide taper which gives a coupling loss of 3.1 dB.

To obtain a further improvement in coupling it is necessary to increase the height of the SOI waveguide. An SOI waveguide with a 12.5 μm thick silicon epitaxial layer, a 12.5 μm rib width and a 10 μm etch depth has a fundamental mode with a mode field radius of 4.8 μm and coupling loss to the fibre of 0.24 dB. Achieving such a low coupling loss enables SOI devices to be produced with losses comparable to other passive waveguide technologies such as silica on silicon, whilst still retaining all the advantages of the silicon material system. a single moded SOI waveguide which is sufficiently large to give good coupling efficiency to standard optical fibre will be unsuitable for the fabrication of compact waveguide bends and so will need to be tapered to smaller dimensions.

A schematic diagram of a taper that achieves low loss coupling is given in Fig. 10. The taper consists of two regions, there is a lower taper which is

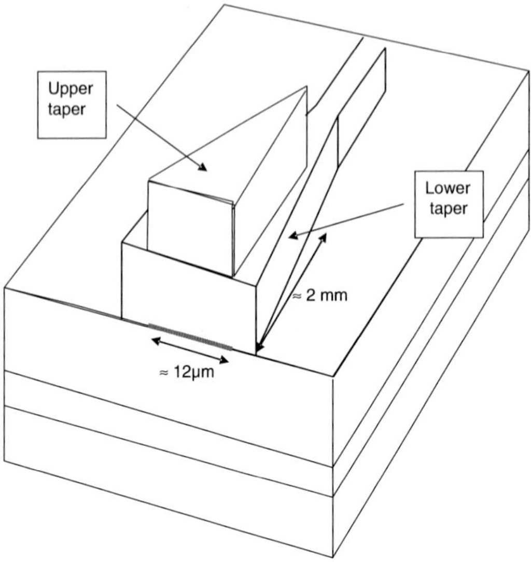

Fig. 10. Schematic of SOI taper to give low loss coupling to single mode fibre

about 20 μm wide, produced in 4 μm thick silicon etched to a depth of 2 μm. On top of this taper is an upper taper which is 12.5 μm high. Figure 11 shows how the light propagates along the taper. Initially the light is confined within the upper taper region, but at a width of around 6 μm, the light starts to transfer down and becomes a mode of the lower rib waveguide. This lower rib waveguide continues tapering beyond the end of the upper taper to produce the final single mode SOI waveguide with a 3 μm width. The total taper length is about 3 mm.

The taper is required to transform the fundamental mode of the optical fibre to that of the single mode SOI waveguide. When used in this way the taper is designed so that the fundamental waveguide mode propagates with minimal coupling to higher order modes if the waveguide is multimoded, or to radiation modes if it is single moded. A simple design approach for achieving this is to calculate the transmission of the fundamental mode through the taper as a function of the taper length L, using a beam propagation tool. The result will be a curve similar to that in Fig. 12.

From this curve the adiabatic limit is easily determined, and sets the taper length at which no significant mode coupling occurs. This approach is perfectly valid, and ensures low loss tapers. However, it will not produce a design of minimal length for a given loss. This is because a linear taper design, with constant taper angle, ignores the different rate of mode coupling along the taper, and consequently, at a given point along the taper, the taper angle may be larger than that needed to avoid unwanted mode coupling, making the taper longer than necessary. a rigorous approach to designing an optimal taper would be to solve the coupled mode equations along the taper.

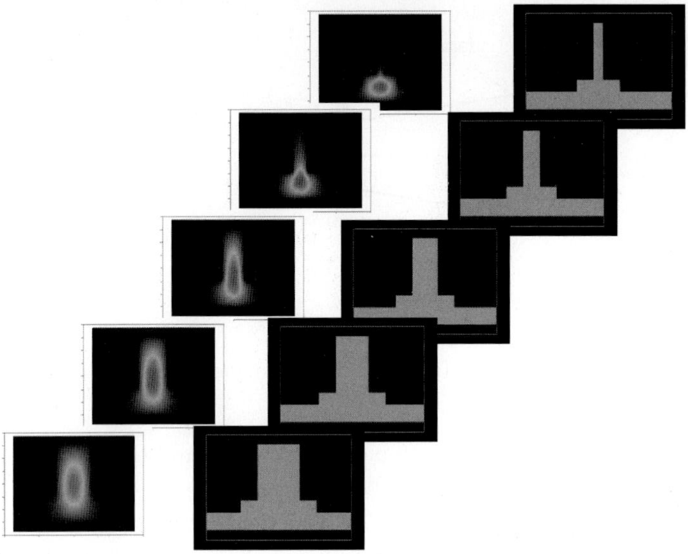

Fig. 11. Schematic of light propagation along taper

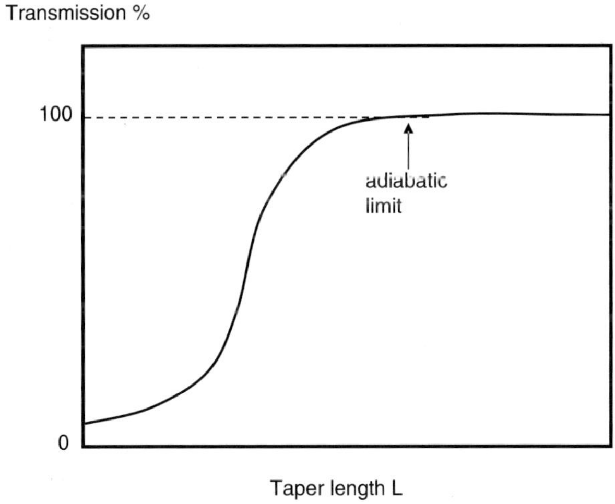

Fig. 12. Typical transmission through a taper as a function of taper length

However, this is unnecessarily complicated, and for most design purposes a simpler, more empirical approach can be used. This is based on the idea that for a taper to be adiabatic, the length scale over which its width varies should be long compared to the beat length between the fundamental mode and the nearest mode in effective index to which the fundamental can couple. Referring to Fig. 13, at a point z along the taper the waveguide half width

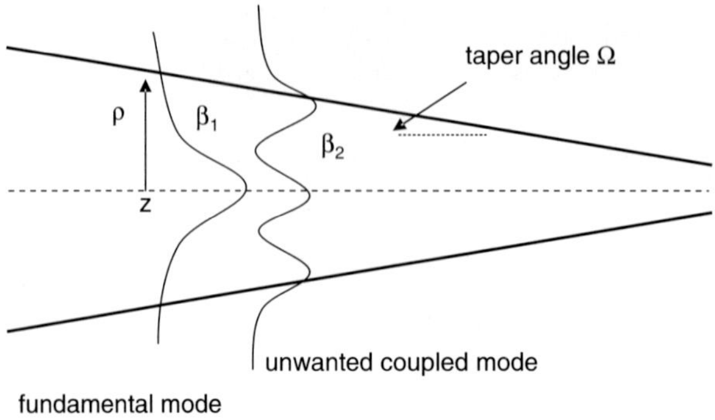

Fig. 13. Taper with coupling between fundamental mode and unwanted mode

is $\rho(z)$; the fundamental mode propagation constant is β_1, and the unwanted mode to which it can couple has propagation constant β_2.

The beat length between the modes is given by

$$L_{\mathrm{B}} = \frac{2\pi}{\beta_1 - \beta_2} \, . \tag{3}$$

The length scale over which the taper width is changing is

$$\rho \Big/ \frac{\mathrm{d}\rho}{\mathrm{d}z} \, . \tag{4}$$

The condition for an adiabatic taper is that (4) should be much greater than (3)

$$\rho \Big/ \frac{\mathrm{d}\rho}{\mathrm{d}z} \gg L_B \, . \tag{5}$$

Then from(3) we find that the local taper angle Ω must satisfy

$$\Omega = \frac{\mathrm{d}\rho}{\mathrm{d}z} \ll \frac{\rho}{L_{\mathrm{B}}} \, . \tag{6}$$

Equation (6), combined with (3), sets the adiabatic limit for a tapered waveguide. Because the beat length will vary with the waveguide half width ρ, we see from (6) that the limit on the taper angle will change with position along the taper, resulting in a taper with nonlinear profile. It should be emphasised that although the inequality in (6) is an empirical result, it can be shown that it sets an accurate limit on taper angle, and that tapers that exceed this limit will generally be lossy, and tapers which satisfy (6) will have low loss [26].

If the tapered waveguide supports only the fundamental mode, then loss will occur by coupling to radiation modes. In this case β_2 will be given by $n_2 k_0$

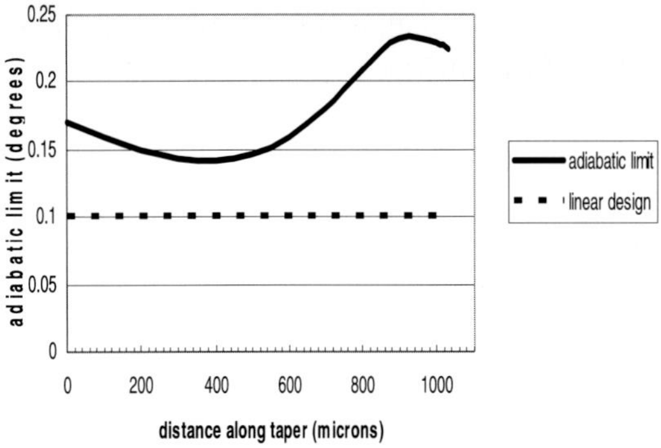

Fig. 14. Adiabatic limit

if the waveguide is surrounded by a medium of constant refractive index n_2; if it is a rib guide then the appropriate propagation constant will be that of the slab surrounding the rib.

An important point when applying (3) and (6) is understanding which modes couple. To answer this we consider the coupling coefficient C_{12} that defines mode coupling along the taper. It can be shown that C_{12} is given by [27]

$$C_{12} = \left(\frac{\varepsilon_0}{\mu_0}\right) \frac{k_0}{4} \frac{1}{\beta_1 - \beta_2} \int e_1 e_2 \frac{\partial n^2}{\partial z} \mathrm{d}A\,, \tag{7}$$

where e_1 and e_2 are the vector modal electric fields of the coupled modes, and n_2^2 is the refractive index distribution of the tapered waveguide at coordinate z along the taper. For most waveguides, n_2^2 only changes at the waveguide boundary. For example, for the tapered waveguide of Fig. 13 we see that $\partial n^2/\partial z$ is given by

$$\frac{\partial n^2}{\partial z} = \left(n_1^2 - n_2^2\right) \frac{\partial \rho}{\partial z} \delta\left(r - \rho(z)\right)\,, \tag{8}$$

where n_1 and n_2 are the core, cladding indices, and δ is the Dirac delta function. Consequently, the integral in (7) reduces to the sum of the products of the modal fields on each waveguide boundary which changes direction as a result of the taper. This means that only those modes that have the same symmetry with respect to the tapering waveguide boundaries can couple. It is important to remember that all the local modes along the taper retain their orthogonality, but can still couple because the integration in (7) reduces to the modal products on the waveguide boundary. Figure 14 shows an example of how the adiabatic limit varies along such a taper and shows that although the middle section requires an angle less 0.15°, the middle and end sections can use much larger angles, thereby resulting in a shorter taper for the equivalent loss.

Using the method described above, tapers have been fabricated which give a total pigtailed device loss less than 1 dB [28].

A second consideration at the fibre waveguide interface is back reflection from the waveguide into the fibre. There is large refractive index mismatch between the optical fibre (1.45) and the silicon (3.45), leading to both loss and back reflection. The Fresnel reflection at normal incidence is given by

$$R = \left(\frac{n_1 - n_2}{n_1 + n_2} \right)^2 . \tag{9}$$

Hence $R = -7.8\,\text{dB}$ gives the back reflection and $1 - R = 0.79\,\text{dB}$ gives the transmission loss. An anti reflection coating can be used to eliminate the transmission loss, but an angled interface is also required to reduce the back reflection to better than $-45\,\text{dB}$. Introducing an angle between the fibre and waveguide leads to an additional complication due to refraction at the interface. The facet angle for the optical fibre θ_1 is given by

$$\theta_1 = \arcsin \left(\frac{n_2 sin(\theta_2)}{n_1} \right) . \tag{10}$$

For example if the silicon waveguide is fabricated to have a facet angle of 5°, the facet angle for the optical fibre is required to be 12°.

5 Silicon Arrayed Waveguide Gratings

5.1 Introduction

The arrayed waveguide grating concept was first reported by *Smit* [29] and *Dragone* [30], and demonstrated using SOI waveguide technology by *Jalali* [31]. The general design of such components in all planar technologies still follows that of Dragone, though numerous improvements have been published. In designing for the SOI platform we concentrate on achieving low loss and PDL, high crosstalk suppression, wide, flat passbands and a small footprint. Most importantly, these device properties must be tolerant to the manufacturing variations that result from an economical fabrication process.

5.2 Function

The well-known AWG layout consists of an input and an output star coupler that are interconnected by a number (or array) of waveguides. The Star couplers are constructed in such a way that an image of the field at the end of the input waveguide is reproduced somewhere along the output focal plane. However, the interconnecting waveguides are not all of the same length, but differ in length from the first through to the last by a fixed increment. This path length difference results in a wavelength dependence (or dispersion) of

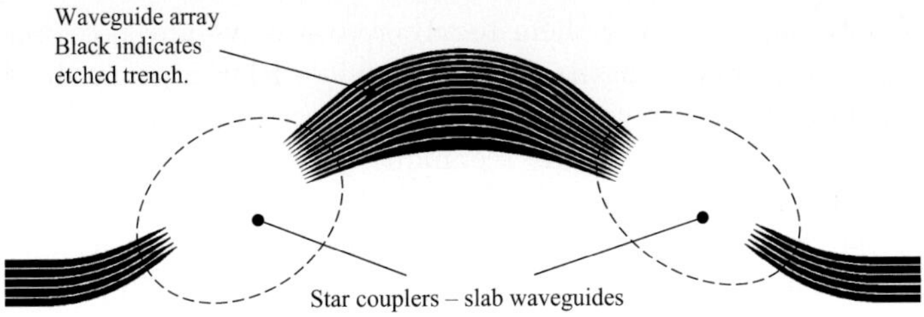

Fig. 15. AWG Schematic

the image position along the focal plane and allows the collection of different wavelength lights into different output waveguides. Hence the device functions as a wavelength demultiplexer or, if used in reverse, a multiplexer.

Rather than review the detailed design of SOI AWGs here, we will instead highlight the main ways in which they differ from those reported for the silica technology.

5.3 Loss Due to the Array-Slab Interface

The majority of the loss in the basic AWG device occurs at the two interfaces between the array waveguides and the slab waveguides. This is due to a mismatch between the array waveguide mode fields and the planar slab modes that travel parallel to them. In silica devices, the loss may be reduced by bringing the array waveguides close together where they join the slab waveguide and reducing their confinement [32]. This forces their optical modes to overlap each other, producing a resultant field that more closely matches a flat phase-front within the slab waveguide. This technique is less straightforward for SOI waveguides because the high refractive index contrast prevents the array waveguide modes overlapping within the etched regions. However, reduction of the ridge waveguide etch depth near to the interfaces allows the optical modes to overlap further and the loss to be substantially reduced [33,34,35]. By using a variation on the double-etch-depth scheme we are able to achieve losses of less than 0.5 dB and 1.0 dB for the centre and end channels respectively in a forty channel device.

5.4 PDF Control

A major advantage of the SOI waveguide technology is that birefringence is repeatable and can be controlled. In the absence of a waveguide cover material, modelling accurately predicts the measured value of the form birefringence. However, if a silicon dioxide cover layer is thermally-grown on the wafer, a strain is induced in the waveguide as the structure cools [36]. The birefringence is then modified and can even take the opposite sign to the form birefringence. Hence by patterning a film grown using a standard silicon oxidation process it is possible to create waveguides that have precise birefringences of both signs and use them together to remove any polarisation-dependence in optical path lengths. In AWG devices PDF is maintained below 2 GHz by leaving the oxide cover over a part of each array waveguide and removing it elsewhere. However, the technique is only successful if the waveguide cross-sectional dimensions are relatively large (Sect. 3.3). This is because the form birefringence is greater and becomes far more sensitive to exact dimensions in smaller waveguides. Hence PDF variation due to process fluctuation is likely to become prohibitive if there is an attempt to manufacture devices constructed of low-dimensional waveguides.

5.5 Single Mode Design

It has been stated previously that relatively large waveguide cross-sections are required if the waveguide birefringence, and hence PDF, are to be controlled within acceptable limits. However, these dimensions result in slab waveguides that support the propagation of many guided modes with low loss. The presence of higher-order modes in a star coupler slab waveguide can lead to undesirable effects if the coupling to and from the ridge waveguides is not highly efficient. For instance, if the ridge-to-slab fundamental mode power coupling efficiency is 99% at each junction, then it might be assumed that the only consequence is a small loss. However, if the other 1% of power is not lost to radiation modes but transmitted across the coupler as the guided first-order slab mode, then it will interfere with the fundamental at the output. The slab waveguide birefringence will cause the result to differ between the TE and TM polarisations and there will be a Polarisation-Dependent Loss (PDL). In this example the PDL could be up to 0.175 dB, depending on the length and thickness of the slab waveguide and the wavelength of the light. Hence ridge-to-slab waveguide junctions must be designed very carefully in SOI technology if PDL is to be minimised. Fortunately it is possible to arrange for fundamental mode power coupling efficiencies in the region 99.7% by adiabatic tapering of the waveguide width before joining the slab waveguide. This guarantees that the PDL from this source is kept below approximately 0.05 dB.

5.6 Example Results

If the design measures described above are all taken, then SOI AWG devices can be manufactured that are entirely competitive with those in other technologies such as silica-on-silicon. Figure 16 shows the transmission spectrum of such an AWG. The device is designed to have clear passbands that are $>$ 35 GHz wide and centred on 100 GHz-spaced ITU grid frequencies. This allows a large margin to guarantee that the product has 25 GHz-wide clear channels at the end-of-life under all ambient conditions. The on-chip loss at the edge of the worst band is less than 4.5 dB while the 1 dB bandwidths are

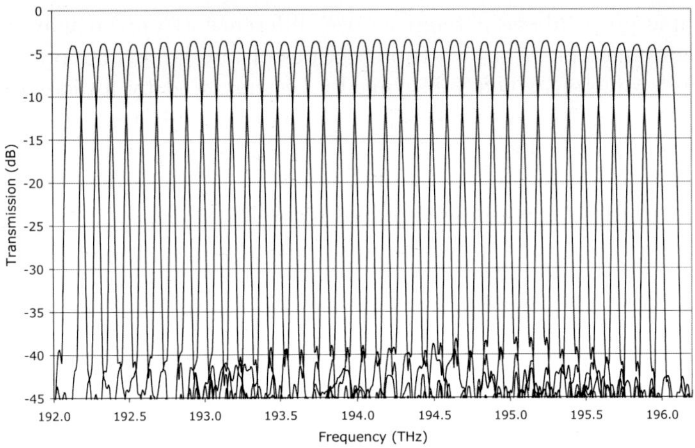

Fig. 16. Measured silicon AWG spectrum

all greater than 50GHz. The PDL across the bands is less than 0.3 dB, and the summed, worst-possible-case crosstalk < -24 dB (-29 dB typical).

6 The Silicon Variable Optical Attenuator

6.1 Mechanisms

Silicon has a cubic crystal structure, and therefore has no first order electrooptic effect. The second order electrooptic effect is small, as is the Franz Keldish effect (change of band-gap with electric field). This leaves two practical options for bringing about a change in the propagation characteristics of light in silicon waveguides. Firstly, it is possible to exploit the dependence of refractive index on temperature, or thermooptic effect. Secondly, the refractive index and absorption coefficient of silicon may be changed by altering the concentration of electrons and holes, making use of the contribution made by free carriers to the optical constants of the semiconductor.

6.2 Thermooptic Effect

The thermooptic coefficient of silicon is $1.86 \times 10^{-4}\,°\mathrm{C}^{-1}$. Thus a temperature rise of 5 °C for example will increase the refractive index by $\Delta n \sim 10^{-3}$. The resulting phase shift per unit length of waveguide is $2\pi\delta n/\lambda$, where λ is the free-space wavelength, so that a heated waveguide length of the order of 1mm will provide a phase shift of π radians. By arranging that this phase shift takes place in one arm of an interferometer, an attenuator can be produced. a suitable configuration is the Mach–Zehnder interferometer, using Y-branches, evanescent couplers or multimode interference couplers to provide the functions of path splitting and combining.

In the example given above a temperature difference of approximately 5 °C must be maintained between two 1mm sections of waveguide in opposite arms of the interferometer, which will typically be separated by a distance of the order of 100 µm. The power required to maintain this temperature differential is approximately 300 mW, and the response time is in the range of tens of microseconds. a rough figure of merit, the product of power dissipation and response time, is approximately a hundred times lower than that for the corresponding silica waveguide system. As in silica technology it is possible to reduce the power requirement by improving the thermal isolation, for example by increasing the thickness of the buried oxide layer.

6.3 Free Carrier Effects

In general the optical properties of a semiconductor in the visible and infrared regions of the spectrum are dominated by absorption due to bound electrons at photon energies below the band-gap, which for silicon corresponds to a wavelength of 1.1 µm. At wavelengths shorter than 1.1 µm the optical absorption coefficient is therefore very high, while at higher wavelengths there is no longer any contribution from bound electrons. There is still, however, a contribution from free carriers, which may be visualised as follows.

In the simplest, classical approximation, free carriers in the semiconductor can be considered to respond to the accelerating force provided by an electric field, and also to the viscous damping, or momentum relaxation, experienced as the aggregate result of the various carrier scattering mechanisms. This is the well-known Drude approximation, which leads to the following expression for conductivity:

$$\sigma = \frac{Ne^2\tau/m^*}{1+\omega^2\tau^2}(1+i\omega)\,, \tag{11}$$

where the symbols have the following meaning:

σ	electrical conductivity	N	carrier concentration
e	electronic charge	τ	momentum relaxation time
m^*	carrier effective mass	ω	angular frequency

In this convention the real part of the conductivity is derived from the component of carrier velocity which is in phase with the applied electric field, and which leads to familiar expressions for dc conductivity and carrier mobility. It is this component of velocity which is responsible for the absorption of energy from the field, giving rise to electrical resistance at low frequencies and to attenuation at optical frequencies.

At frequencies comparable to the momentum relaxation rate, $1/\tau$, the carrier velocity begins to lag behind the driving field, and a quadrature component of velocity develops, which is represented by the imaginary part of the conductivity.

If the displacement of carriers is considered instead of their velocity, then corresponding expression is obtained for the polarisation generated by the free carriers, and hence the dielectric constant:

$$\varepsilon_r = 1 - \frac{Ne^2\tau^2/\varepsilon_0 m^*}{1+\omega^2\tau^2}\left(1+\frac{i}{\omega\tau}\right). \tag{12}$$

Since the carrier velocity and displacement are in quadrature, the real component of the conductivity, which is associated with attenuation, corresponds to the imaginary component of dielectric constant, and vice-versa. As the frequency increases and the carrier velocity begins to lag behind the driving field, the component of displacement which develops is in antiphase with the field. In other words the effect of carrier motion is to reduce the real part of the dielectric constant.

The momentum relaxation time, τ lies in the picosecond range, so that in the infrared and optical regions of the spectrum the condition $\omega\tau \gg 1$ holds, from which the following approximations for refractive index and absorption can be deduced:

$$\delta_n = -\frac{e^2\lambda^2 N}{8\pi^2 c^2 \varepsilon_0 n m^*} \tag{13}$$

$$\delta_\alpha = -\frac{e^3\lambda^2 N}{4\pi^2 c^3 \varepsilon_0 n \mu m^{*2}} \tag{14}$$

where

δ_α change in absorption coefficient due to free carriers
δ_n change in refractive index due to free carriers
n background refractive index
λ free space wavelength.

The total free carrier contributions to refractive index and absorption are given by adding the terms for electron and hole contributions.

In order to improve upon this simple description of free carrier effects it is necessary to apply a quantum mechanical treatment, and to consider the various scattering mechanisms individually. An accurate empirical formula has been reported [37], in which the changes in refractive index and absorption coefficient due to free carriers are given as follows

$$\Delta n = -\left(8.8\times10^{-22}N_e + 8.5\times10^{-18}\times N_h^{0.8}\right) \tag{15}$$

$$\Delta\alpha = 8.5\times10^{-18}N_e + 6.0\times10^{-18}\times N_h\,. \tag{16}$$

These relations are plotted in Fig. 17.

6.4 Pin Diode Attenuator

It is clear from figure that a necessary condition for waveguide transparency is that the carrier concentration in the passive waveguide should be low. For

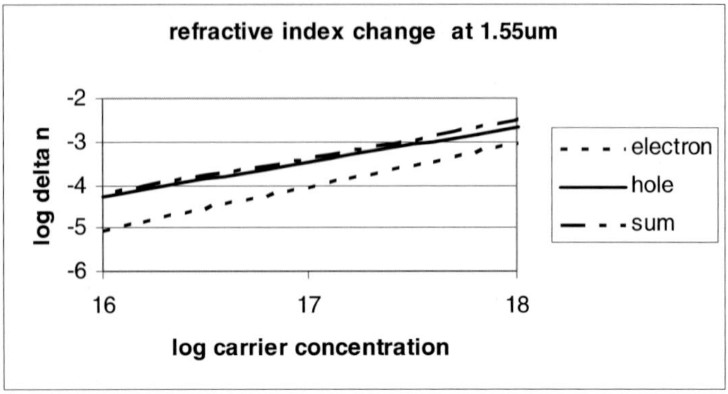

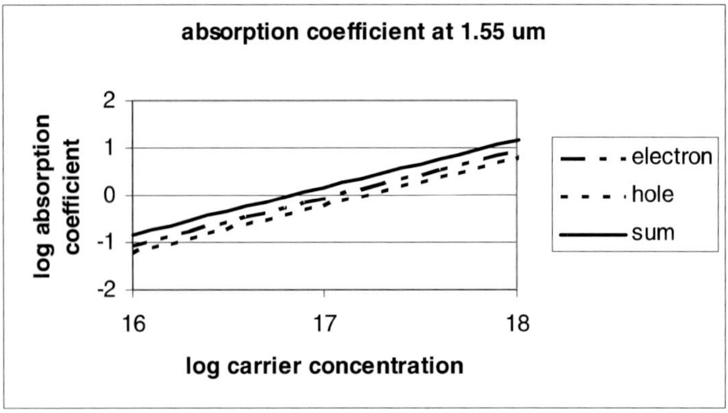

Fig. 17. The dependence of refractive index and absorption length on free carrier concentration

example to achieve absorption below $0.1\,\mathrm{dB\cdot cm^{-1}}$ (an absorption coefficient of less than $0.02\,\mathrm{cm^{-1}}$) a doping level below 10^{-15} is required. Starting from a passive silicon waveguide a sufficiently low carrier concentration, a convenient way in which to vary the carrier concentration within the waveguide is to fabricate a pin diode with the cross-section shown in Fig. 18. This structure allows the carrier concentration to remain low in the inactive state, but to increase on application of a forward bias, whereupon electrons are injected from the n region and holes from the p region. Care must be taken to locate the doped regions sufficiently far from waveguide (in practice several μm) in order to prevent passive loss which would result from exposure of the optical mode to the doped regions.

The pin diode device can be used to control refractive index, and hence produce attenuation via an interferometer, as in the example of a thermooptic interferometric attenuator described above. Alternatively, the absorption due to free carriers can be used directly to produce attenuation in a section of plain waveguide. From the relations of figure it can be deduced that

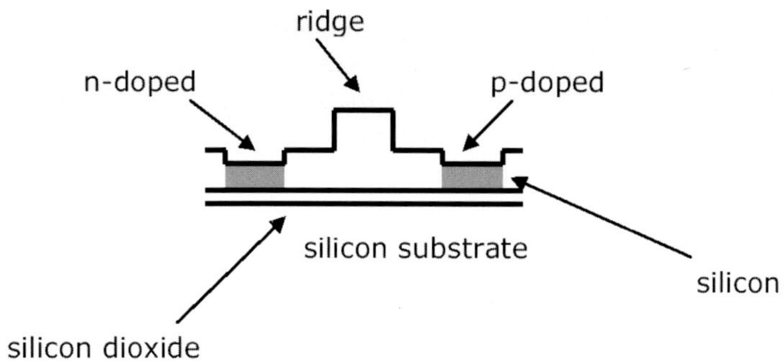

Fig. 18. Cross-section through a waveguide pin diode attenuator

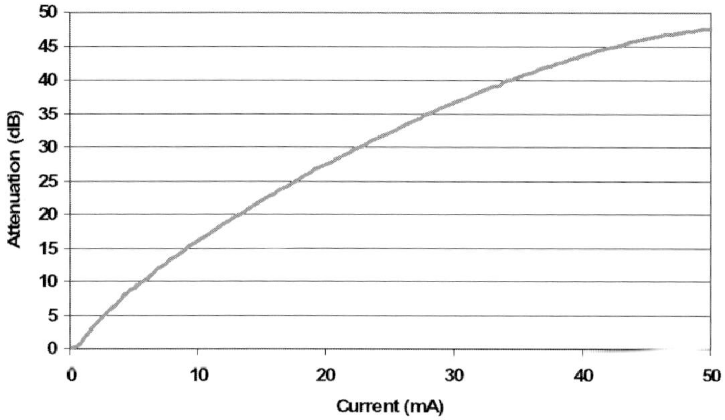

Fig. 19. Attenuation vs current characteristic of four 2.5 mm section pin diode attenuator

a carrier concentration of $10^{18}\,\mathrm{cm}^{-3}$ will produce attenuation of $50\,\mathrm{dB}\cdot\mathrm{cm}^{-1}$ (an absorption coefficient of $10\,\mathrm{cm}^{-1}$) so that a high level of attenuation can be produced in a convenient length of waveguide. The current required to achieve this concentration is of the order of $200\,\mathrm{mA}\cdot\mathrm{cm}^{-1}$ of waveguide length. In practice it is usual to divide the length of the waveguide pin diode into a number of sections and feed the individual diodes in series, thus allowing a convenient voltage/current operating range to be achieved. Figure 19 shows the attenuation vs current characteristics of such a device in which a 10-mm long device is divided into four sections fed in series. Consequently the operating voltage is in the range 3 V to 4 V. Power dissipated at 50 dB attenuation is approximately 200 mW.

An advantage of the absorption attenuator is the high level of extinction possible, and the low polarisation dependence which can be achieved over the entire attenuation range. Both of these parameters are more difficult to control in Mach–Zehnder interferometer based devices, where the photonic properties of the interferometer limit performance.

One of the design aims of the pin diode attenuator is to minimise the current required to provide a given carrier density within the waveguide. This is achieved in practice in the design of figure by etching trenches before introduction of the dopant by ion implantation, in order that the doped regions extend throughout the depth of the SOI layer. This precaution ensures eliminates the 'wasted' current which would otherwise flow if carriers of both polarities were able to escape from the device and diffuse sideways away from the waveguide.

The response time of the pin diode attenuator is governed by the transit time of carriers across the waveguide, and for the device illustrated it is typically in the submicrosecond range. Modulation at frequencies in excess of 100 MHz has been predicted for waveguide pin diodes of cross-sectional dimensions of the order of a few micrometer [38].

References

1. S. V. Kartalopoulos: *Introduction to DWDM Technology* (IEEE Press 2000)
2. D. J. Blumenthal: Integrated devices for wavelength-agile all-optical networks, Integrated photonics research technical digest (2002)
3. K. Maru, T. Chiba, T. Hasegawa, K. Tanaka, H. Nonen, H. Uetsuka: High resolution dynamic gain equalizer using super-high Δ planar lightwave circuit technology, Optical Fiber Communication Conference (OFC) Proceedings (2003), 172–173
4. H. Nishihara, M. Haruna, T. Suhara: *Optical Integrated Circuits* (McGraw-Hill 1997)
5. M. Kawachi: Silica waveguides on silicon and their application to integrated-optic components, Optical and Quantum Electronics **22**, 391–416 (1990)
6. R. Soref: Silicon-Based Optoelectronics, Proceedings of the IEEE **81**, 1687–1706 (1993)
7. S. Suzuki, M. Yanagisawa, Y. Hibino, K. Oda: High-density integrated planar lightwave circuits using SiO_2-GeO_2 waveguides with a high refractive index difference, Journal of Lightwave Technology **12**, 790–796 (1995)
8. R. Soref, J. P. Lorenzo: All-silicon active and passive guided-wave components for $\lambda = 1.3$ E and 1.6 μm, IEEE Journal of Quantum Electronics **22** (1986)
9. J. P. Colinge: *Silicon-On-Insulator Technology: Materials to VLSI* (Kluwer Academic Publishers 1997)
10. S.S. Iyer, A.J. Auberton-Herve (Eds): *Silicon Wafer Bonding Technology for VLSI and MEMs* (EMIS Processing Books 2001)
11. T. M. Benson, M. Burnham, D. E. Davis, N. Mohd Kassim, M. Seifouri: Single-Mode Optical Waveguides in Silicon, IEEE Colloquium Digest 1989/93, Integrated Optics, 11/1–11/4 (1989)
12. R. Soref, J. Schmidtchen, K. Petermann: Large Single-Mode Rib Waveguides in GeSi-Si and Si-on-SiO_2, IEEE Journal of Quantum Electronics **8**, 1971–1974 (1991)
13. A. Rickman, G. Reed, F. Namavar: Silicon-on-insulator optical rib waveguide loss and mode characteristics, Journal of Lightwave Technology **12**, 1771–1776 (1994)

14. K.K. Lee, D.R. Lim, L.C. Kimerling: Fabrication of ultralow-loss Si/SiO_2 waveguides by roughness reduction, Optics Letters **26**, 1888–1890 (2001)
15. I.E. Day, S.W. Roberts, R. O'Carroll, A. Knights, P. Sharp, G.F. Hopper, B.J. Luff, M. Asghari: Single-chip variable optical attenuator and multiplexer subsystem integration, in Optical Fiber Communication Conference (OFC) Proceedings (2002), 72–73
16. M.K. Smit, C.V. Dam: PHASAR-Based WDM-Devices: Principles, Design and Applications, IEEE Journal of Selected Topics in Quantum Electronics **2**, 236–250 (1996)
17. R. Scarmozzino, A. Gopinath,R. Pregla, S. Helfert: Numerical Techniques for Modeling Guided-Wave Photonic Devices, IEEE Journal of Selected Topics in Quantum Electronics **6**, 150–162 (2000)
18. C. Vassallo: *Optical Waveguide Concepts* (Elsevier 1991)
19. R.G. Hunsperger: *Integrated optics: theory and technology* (Springer 1984)
20. M. Kawachi: Silica waveguides on silicon and their application to integrated-optic components, Optical and Quantum Electronics **22**, 391–416 (1990)
21. R.C. Hewson-Browne, P.C. Kendall, D.A. Quinney: Roughness scattering into substrate radiation modes of rib waveguides, IEE Proceedings J. **136**, 281–286 (1986)
22. K.K. Lee, D.R. Lim, L.C. Kimerling: Fabrication of ultralow-loss Si/SiO_2 waveguides by roughness reduction, Optics Letters **26**, 1888–1890 (2001)
23. G.H. Jin, J. Harari, L. Joannes, J.P. Vilcot, D. Decoster: Numerical analysis of the radiation losses due to surface roughness in integrated optics devices, IEEE Photonics Technology Letters **8**, 1202–1204 (1996)
24. F. Ladouceur, J. Love, T. Senden: Effect of side wall roughness in buried channel waveguides, IEE Proceedings J. **141**, 242–248 (1994)
25. F. Payne: private communication (2002)
26. J. Love: Application of a low loss criterion to optical waveguides and devices, IEE Proceedings J **136**, 225–228 (1989)
27. A.W. Snyder, J.D. Love: *Optical Waveguide Theory* (Chapman and Hall, 1993)
28. I. Day, I. Evans, A. Knights, F. Hopper, S. Roberts, J. Johnston, S. Day, J. Luff, H. Tsang, M. Asghari: Tapered silicon waveguides for low insertion loss highly efficient high speed electronic variable optical attenuators, Optical Fiber Communication Conference (OFC) Proceedings (2203), pp. 249–250
29. M.K. Smit: New focussing and dispersive planar component based on an optical phased array, Electronics Letters **34**, 385–386 (1988)
30. C. Dragone: An $N \times N$ optical multiplexer using a planar arrangement of two star couplers, IEEE Photonics Technology Letters **9**, 812–815 (1991)
31. P.D. Trinh, S. Yegnanarayanan, F. Coppinger, B. Jalali: Silicon-on-insulator (SOI) phased-array wavelength multi-demultiplexer with extremely low-polarisation sensitivity, IEEE Photonics Technology Letters **9**, 940–942 (1997)
32. J.C. Chen, C. Dragone: A study of fibre-to-fibre losses in waveguide grating routers, Journal of lightwave technology **15** (1997)
33. C. van Dam: Loss reduction for phased-array demultiplexers using a double etch technique, Integrated Photonics Research Proceedings, 52–55 (1996)
34. C. G. Herben: Low-loss and compact phased array demultiplexers using a double etch process, Proceedings of the 9th Conference on integrated optics (ECIO 1999), pp. 211–214

35. J. H. den Besten: Low-loss, Compact, and polarisation independent PHASAR demultiplexer fabricated by using a double-etch process, IEEE Photonics Technology Letters **14**, 62–64 (2002)
36. Bookham Technology plc patent entitled *Waveguide birefringence control using thermal oxide*
37. R. Soref, B. R. Bennett: IEEE Journal of Quantum Electronics **23** (1987)
38. A. Vonsovici, A. Koster: IEE Journal of Lightwave Technology **17** (1999)

Silicon-Based Photonic Crystals

Torsten Geppert[1], Joerge Schilling[2], Ralf Wehrspohn[3], and Ulrich Gösele[1]

[1] Max Planck Institute of Microstructure Physics,
Weinberg 2, 06120 Halle (Saale), Germany
{tgeppert,goesele}@mpi-halle.de

[2] California Institute of Technology,
M.C.136-93 Pasadena CA 91125-3600, USA
schill@caltech.edu

[3] Nanophotonic Materials Group, Department of Physics, Paderborn University,
Warburger Str. 100, 33098 Paderborn, Germany
wehrspohn@physik.uni-paderborn.de

Abstract. We introduce the concept of silicon-based photonic crystals with the main focus on the macroporous silicon material system. photonic crystals are the optical analog to electronic semiconductors. Due to their periodic modulation of the refractive index they exhibit a band-structure for photons propagating in the Photonic crystal. This photonic band structure and in particular the photonic band-gap provide a number of interesting effects for both fundamental research as well as applied physics. These effects include, e.g., forbidden propagation of photons for certain frequencies or a huge reduction of the group velocity. The macroporous silicon material system allows the fabrication of photonic crystals with high dielectric contrast and a variety of patterns as well as defect incorporation using lithographic prestructuring and subsequent photoelectrochemical etching. The etching process will be discussed and experimental results of two-and three-dimensional photonic crystals based on macroporous silicon will be presented and their optical properties will be explored.

1 Introduction

The concept of photonic crystals was introduced to the scientific community in 1987 independently by *Sajeev* [1] and *Yablonovitch* [2]. John's approach was to study the localization of photons in disordered dielectric superlattices while Yablonovitch investigated the inhibited spontaneous emission of an atom within a photonic crystal.

Photonic crystals are the optical analog of electronic semiconductors. In semiconducting materials the electron waves experience scattering at the periodic electrostatic potential provided by the crystal lattice atoms. Due to the interference of these electron-waves the dispersion relation of free electrons $E^{\text{free}}_{\text{electron}}(\boldsymbol{k}) = \hbar \boldsymbol{k}^2$ is replaced by the electronic band-structure which relates the energy E and the quasi-momentum $\boldsymbol{k}$ of the electron.

The potential which photons experience is the refractive index $n = \sqrt{\epsilon}$ of a dielectric medium with dielectric constant ϵ as depicted in Fig. 1. Analogous

L. Pavesi, D. J. Lockwood (Eds.): Silicon Photonics, Topics Appl. Phys. **94**, 295–322 (2004)

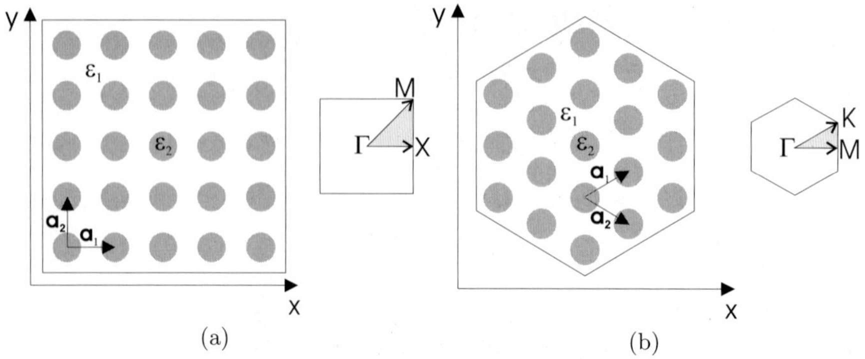

Fig. 1. Arrangement of a material of dielectric constant ϵ_2 within a material of dielectric constant ϵ_1. This could be experimentally realized as, e.g., air pores along the z-direction in bulk Si. (**a**) Square lattice and (**b**) hexagonal lattice. $\vec{a_1}$ and $\vec{a_2}$ represent basis vectors of the underlying photonic crystal lattice. The righthand insets depict the corresponding Brillouin zones with points of high symmetry. The shaded region represents the irreducible Brillouin zone

to the electronic case a periodic modulation of this potential leads to the formation of a photonic band-structure which determines the propagation of light within the photonic crystal and eventually also to the formation of a photonic band-gap. Here the photonic band-structure replaces the dispersion relation of photons $\omega = (c/n)|\boldsymbol{k}|$ in a homogenous dielectric with refractive index n and frequency ω along the direction $\boldsymbol{k}$ where c is the speed of light in vacuum. Photons with an energy that lies within the photonic band-gap cannot propagate through the photonic crystal due to the zero density of photonic states in the photonic band-gap.

The band-structure depends on several parameters of which the most important are the symmetry of the modulation of the refractive index, the difference of the dielectric constants of the materials used to achieve the modulation (the so-called dielectric contrast) and the relative prportions of the different materials within the photonic crystal structure. Figure 2 shows the calculated photonic band-structure for a photonic crystal created by photo-electrochemical etching of air pores into bulk Si. This leads to a hexagonal, periodic arrangement of dielectric structures with a very high dielectric contrast of $\Delta\epsilon = \epsilon_{\mathrm{Si}} - \epsilon_{\mathrm{air}} = 11.6 - 1 = 10.6$. Because this photonic crystal has its periodicity in the x-y-plane and is translationally invariant along the z-axis the photonic crystal modes split up into the two independent polarizations TE (transverse electric) and TM (transverse magnetic) with their $\boldsymbol{H}$ and $\boldsymbol{E}$ along the pore axis, respectively. A complete photonic band-gap, i.e., a photonic band-gap for all directions and all polarizations opens up for normalized frequencies $(\omega a)/(2\pi c)$ between 0.377 and 0.389. The width of the complete photonic band-gap which is defined as the ratio of the difference of the maximum and minimum normalized frequencies of the photonic band-gap divided

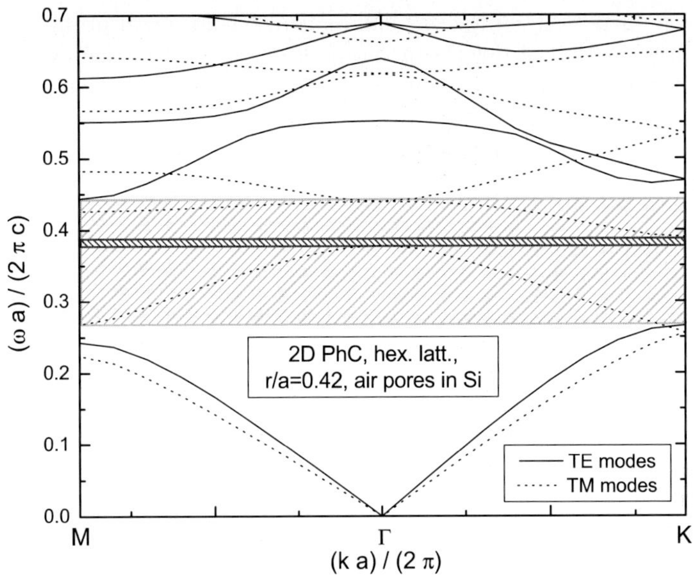

Fig. 2. Calculated photonic band-structure of a hexagonal array of air pores etched into bulk Si. The dielectric constants of the materials are $\epsilon_{\mathrm{Si}} = 11.6$ and $\epsilon_{\mathrm{air}} = 1$. The hatched areas mark the regions of photonic band-gaps for TE (*light grey*) and TM (*dark grey*) polarized modes. There exists a photonic band-gap for both polarizations along both high symmetry directions ΓK and ΓM. But only where these photonic band-gaps overlap a complete photonic band-gap does exist, i.e., a photonic band-gap for all directions and all polarizations. For the calculation the *MIT Photonic Bands* package was used

by the midgap frequency for this structure is about 3%. The TE photonic band-gap is larger than the TM photonic band-gap and the latter one limits the complete photonic band-gap for this structure. The use of the normalized frequency instead of just the conventional angular frequency ω as well as the distinction between TE and TM modes will be explained in Sect. 2 dealing with the theory of photonic crystals.

By changing the above mentioned parameters it is possible to tailor the optical properties of a photonic crystal, e.g., shifting its photonic band-gap to higher or lower frequencies. This possibility to engineer the band-structure offers a large number of opportunities for both fundamental research as well as applications in the field of optoelectronics and photonics.

The lifetime of an excited atom that is placed inside a three-dimensional photonic crystal and which would reach its ground state by the emission of a photon of a frequency that lies within the photonic band-gap should be infinite. This is due to the fact that there are no allowed states for the photon in the photonic band-gap.

For application purposes the photonic band-gap also holds great promises. If, e.g., by the introduction of defects in the photonic crystal lattice, localized states within the photonic band-gap are generated, light with a specific frequency can travel through the photonic crystal without leaking into neighboring modes due to the zero photonic density of states for those neighboring modes. By this method light can be guided not using total internal reflection (TIR) as in conventional optical waveguides but by utilizing a particular property of the photonic crystal structure. Furthermore, based on the photonic band-structure, the group velocity of certain modes travelling in a photonic crystal structure can be strongly reduced which offers a new method for dispersion compensation in photonic circuits.

The above mentioned passive photonic crystal devices can be fabricated reasonably well and their optical properties experimentally verified. But there still remains the challenge to realize active photonic crystal based devices like optical switches or optical transistors which would allow the fabrication of optical processors with very high clock rates and almost no heating problems which limit the performance of today's microelectronic devices.

The practical realization of photonic crystals however is still a challenge for several reasons. The control of light in the photonic crystal is solely based on the periodicity of the photonic crystal's lattice or well controlled deviations thereof, so-called defects. Appropriate means have to be found to guarantee this periodicity during fabrication of a photonic crystal structure. Furthermore, this periodicity has to be ensured on a micrometer scale for the mid-infrared (MIR) down to a nanometer scale for visible (VIS) light.

Approaches to fabricate photonic crystal structures range from self-assembly techniques over interference techniques to the controlled etching of lithographically predefined structures. Self-ordering techniques are relatively inexpensive and easy because no external means are necessary to produce a regular structure. But their major shortcoming consists in the inherent impossibility to incorporate well defined defects such as, e.g., cavities or waveguides. This is however possible using the more expensive and time consuming approaches based on lithography. Table 1 gives a brief overview of some currently used methods for the fabrication of photonic crystals.

Macroporous silicon is a well suited material system for the fabrication of photonic crystals for the near-infrared (NIR) and MIR spectral regions for a number of reasons. Due to its dielectric constant $\epsilon_{\mathrm{Si}} = 11.6$ in the MIR it provides a high dielectric contrast with air and therefore offers the possibility to achieve large photonic band-gaps. Furthermore, Si shows only very little absorption in this spectral region. In addition it is possible to etch highly ordered arrays of macropores into Si with aspect ratios exceeding 100. Using lithographical prestructuring a number of lattice geometries like hexagonal or square lattices can be realized. Furthermore it is also possible to precisely incorporate defects in many shapes using lithographical prestructuring. The lattice constants can be chosen from $a = 500\,\mathrm{nm}$ up to about

Table 1. Overview of different techniques used to fabricate photonic crystals

Material system	Refractive indices n	Dielectric contrast $\Delta\epsilon$	Fabrication principle
air pores in IV semiconductors (Si, Ge)	1, 3.4	10.6	wet and dry etching; self-organiszed or lithographically predefined patterns
air pores in III/V Semiconductors (Ga(in)As, InP ...)	1, 3.4–3.5	10.6–10.9	wet or dry etching after lithographical patterning
air inclusions in polymers (PMMA, photoresist)	1, 1.5	1.25	ordered self assembly of monodisperse spheres after evaporation of solvent; ordered development of photoresist using interference or lithographically defined patterns

$a = 20\,\mu\text{m}$. In addition the variation of the ratio of the pore radius r and the lattice constant a, the so-called r/a-ratio allows the variation of the photonic band-structure by shifting it to higher or lower frequencies. The very small coefficient of thermal expansion ($\alpha = 2.3 \times 10^{-6}\ \text{K}^{-1}$) assures that for moderate temperatures the lattice constant of the photonic crystal remains almost unchanged and therefore the photonic crystal based optical properties of the material are temperature independent. After the etching process the pores can be widened by thermal oxidation of the Si and subsequent etching of the SiO_2 in hydrofluoric acid (HF). Furthermore, due to its dominating role in today's electronics Si is one of the most comprehensively investigated materials.

2 Theory of Photonic Crystals

In this section a brief approach to the theoretical background of photonic crystals is presented. Advanced treatments of photonic crystal theory can, e.g., be found in [3] and [4]. First the relation of Maxwell's equation to photonic crystals is described, followed by brief remarks on how to find solutions for the resulting equations. The subsequent part deals with symmetries, scaling and defects.

2.1 Maxwell's Equations in Periodic Dielectric Media

The propagation of light in dielectric media is governed by the macroscopic Maxwell equations

$$\nabla \times \boldsymbol{D} = \rho \tag{1a}$$

$$\nabla \times \boldsymbol{B} = 0 \tag{1b}$$

$$\nabla \times \boldsymbol{E} = -\frac{\partial \boldsymbol{B}}{\partial t} \tag{1c}$$

$$\nabla \times \boldsymbol{H} = \frac{\partial \boldsymbol{D}}{\partial t} + \boldsymbol{j} \tag{1d}$$

where $\boldsymbol{E}$ and $\boldsymbol{H}$ represent the electric and the magnetic field, $\boldsymbol{D}$ and $\boldsymbol{B}$ the displacement and magnetic induction, $\boldsymbol{\rho}$ and $\boldsymbol{j}$ the free charges and currents and c is the speed of light in vacuum, respectively. Assuming linear, homogenous optical materials we have

$$\boldsymbol{D} = \epsilon_0 \epsilon \boldsymbol{E} \tag{2a}$$

$$\boldsymbol{B} = \mu_0 \mu \boldsymbol{H} \tag{2b}$$

with the magnetic permeability and the dielectric constant μ and ϵ of the material. The corresponding free space values for the magnetic permeability and the dielectric constant are represented by μ_0 and ϵ_0, respectively.

For most of the materials of interest the magnetic permeability μ is close to unity. Furthermore in the following only charge- and current-free media will be treated, i.e., $\boldsymbol{\rho} = \boldsymbol{j} = 0$. Using these simplifications and (2a), (2b) and Maxwell's equations (1c)–(1d) together with the ansatz of harmonic waves

$$\boldsymbol{H}(\boldsymbol{r}, t) = \boldsymbol{H}(\boldsymbol{r}) e^{-i\omega t} \tag{3a}$$

$$\boldsymbol{E}(\boldsymbol{r}, t) = \boldsymbol{E}(\boldsymbol{r}) e^{-i\omega t} \tag{3b}$$

for the magnetic and electric fields, respectively, the time dependence can be separated out and after some algebra the so-called *master equation* for the magnetic field follows:

$$\nabla \times \left(\frac{1}{\epsilon(\boldsymbol{r})} \nabla \times \boldsymbol{H}(\boldsymbol{r}) \right) = \left(\frac{\omega}{c} \right)^2 \boldsymbol{H}(\boldsymbol{r}) \, . \tag{4}$$

Equation (4) determines the distribution of the magnetic field $\boldsymbol{H}(\boldsymbol{r})$ and the corresponding freuqencies ω inside the photonic crystal which is defined by $\epsilon(\boldsymbol{r})$. Furthermore from Maxwell's equations (1a) and (1b) the so-called transversality requirement has to be fulfilled, stating that for a field distribution of the form $\boldsymbol{H}(\boldsymbol{r}) = \boldsymbol{H}_0 e^{i\boldsymbol{k} \times \boldsymbol{r}}$ the relation

$$\boldsymbol{H}_0 \times \boldsymbol{k} = 0 \tag{5}$$

must hold, i.e., the waves are transverse only. From the solution for $\boldsymbol{H}(\boldsymbol{r})$ the electric field distribution can be derived using Maxwell's equations.

2.1.1 Numerical Approaches to Find the Field Distributions

Only for very few $\epsilon(\boldsymbol{r})$ distributions like the very simple case of a homogeneous dielectric medium or the simplest one-dimensional photonic crystal, the Bragg stack or dielectric mirror, it is possible to find analytical solutions for the fields. There are two widely used numerical schemes to calculate the photonic band-structure and field distributions for photonic crystals with more complicated $\epsilon(\boldsymbol{r})$ distributions.

One approach is the Finite-Difference Time-Domain (FDTD) analysis which solves Maxwell's equations in real space in the time domain [5]. FDTD is a general method for numerically solving the time-dependent Maxwell equations in media structured on the order of the wavelength of the light. Therefore it is a well-suited approach to study light propagation in photonic crystals with and without defects. The photonic band-structure can be calculated via a Fourier transformation of the time-dependent FDTD simulation data. The FDTD method can also be used to examine nonlinear and active media. FDTD approaches therefore complement spectral methods used in photonic crystal band-structure calculation described below.

The second widely used method is based on plane wave expansion of the fields $\boldsymbol{H}(\boldsymbol{r})$ and the dielectric distribution $\epsilon(\boldsymbol{r})$. This finally leads to Bloch's theorem, which states that for any kind of periodic potentials the solutions to the master equation are of the form

$$\boldsymbol{H}_{n,\boldsymbol{k}}(\boldsymbol{r}) = e^{i\boldsymbol{k}\times\boldsymbol{r}}\boldsymbol{u}_{n,\boldsymbol{k}}(\boldsymbol{r}) , \tag{6}$$

where n is the band index, $\boldsymbol{k}$ represents the wave vector and $\boldsymbol{u}_{n,\boldsymbol{k}}(\boldsymbol{r})$ is a function with the periodicity of the photonic crystal lattice, i.e., of the $\epsilon(\boldsymbol{r})$ distribution.

The approach based on plane wave expansion using Bloch's theorem is well suited for defect free photonic crystals. However, because this approach is based on perfect periodicity, it can be used for calculations of defect structures only after utilizing the trick of introducing artificial periodicity using supercells similar to what has been used in disordered semiconductors. As photonic crystals are the optical analogs to electronic semiconductors a lot of knowledge from the field of electronic band-structure calculation can be transferred to the photonic crystal field. However, there are two important differences between electronic and photonic band-structure calculations. In optics the potential, i.e., the $\epsilon(\boldsymbol{r})$ distribution is well known while in solid state physics the potentials of the lattice atoms are unknown and self-consistent approaches have to be used. In addition, neglecting the relatively weak contributions of the vectorial nature of the electron spin, the solutions to the Schrödinger equation, i.e., the electron wave functions, are scalar while the solutions to the master equation (4), namely $\boldsymbol{H}(\boldsymbol{r})$ and $\boldsymbol{E}(\boldsymbol{r})$, are vectorial.

2.2 Symmetry, Scaling and Defects

Defining the Hermitian operator Θ

$$\Theta \boldsymbol{H}(\boldsymbol{r}) = \nabla \times \left(\frac{1}{\epsilon(\boldsymbol{r})} \nabla \times \boldsymbol{H}(\boldsymbol{r}) \right) \tag{7}$$

the master equation takes on the form of an eigenvalue problem $\Theta \boldsymbol{H}(\boldsymbol{r}) = (\omega/c)^2 \boldsymbol{H}(\boldsymbol{r})$ where the frequency ω denotes the eigenvalue. Due to the hermiticity of Θ the eigenvalues ω are real and positive and the eigenfunctions, i.e., the field distributions $\boldsymbol{H}(\boldsymbol{r})$, are orthogonal.

Another important property of Θ is its lineatrity from which follows that any linear superposition $a_1 \boldsymbol{H}_1(\boldsymbol{r}) + a_2 \boldsymbol{H}_2(\boldsymbol{r})$ of solutions $\boldsymbol{H}_1(\boldsymbol{r})$ and $\boldsymbol{H}_2(\boldsymbol{r})$ of the master equation (4) with real numbers a_1 and a_2 is also a solution of (4). From the discrete translational symmetry of $\epsilon(\boldsymbol{r})$ follows the form of the solutions as Bloch functions as described in Sect. 2.1.1. Furthermore it can be shown that the photonic band-structure $\omega(\boldsymbol{k})$ posesses the same symmetry as the photonic crystal's Brillouin zone. As a special and important case of this property it follows that for a photonic crystal having mirror symmetry the solutions of the master equation (4) separate into two classes, one for each field polarization. In Fig. 1 the depicted photonic crystal has mirror symmetry with respect to the x-y-plane. Due to this mirror symmetry the modes separate into TE modes having field components $\boldsymbol{E}_z = \boldsymbol{H}_x = \boldsymbol{H}_y = 0$ and TM modes with field components $\boldsymbol{E}_x = \boldsymbol{E}_y = \boldsymbol{H}_z = 0$ using the coordinate system of Fig. 1.

The inversion symmetry of the dispersion relation $\omega_{-\boldsymbol{k}} = \omega_{\boldsymbol{k}}$ follows from time reversal symmetry $t \rightarrow t' = -t$ of the operator Θ.

Another very useful property of photonic crystals is the scaling law. It states that if the photonic crystal structure is rescaled, i.e.,

$$\epsilon'(\boldsymbol{r}) = \epsilon(\boldsymbol{r}/s) \tag{8}$$

the mode profiles and the frequencies are also rescaled according to

$$\boldsymbol{H}'(\boldsymbol{r}) = \boldsymbol{H}(\boldsymbol{r}/s) \tag{9}$$

$$\omega' = \omega/s \tag{10}$$

where s is the scaling factor. This means that there is no fundamental length scale in photonic crystals. For experiments this property brings great relief, because as mentioned before it is quite hard to fabricate photonic crystals on a micro- or nanometer scale. By using the scaling law one can explore a photonic crystal's band-structure using a macroscopic $\epsilon(\boldsymbol{r})$ with a length scale of a few centimeters, knowing that one only has to rescale the field distributions and frequencies according to the above equations (9) and (10). This scale invariance is also the reason for the usage of the normalized frequency $(\omega a)/(2\pi c)$ instead of the conventionally used angular frequency ω because

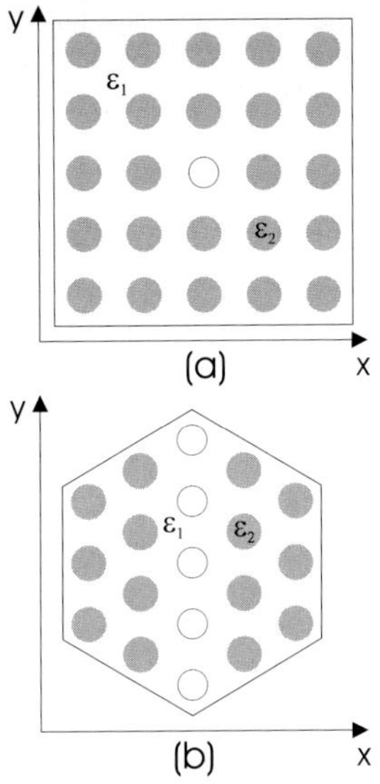

Fig. 3. (**a**) a point defect in a two-dimensional square photonic crystal realized by omitting one air pore and (**b**) a line defect along the ΓK direction in a hexagonal two-dimensional photonic crystal realized by omitting *one* line of pores leading to the name *W*1 defect

the photonic band-structure of, e.g., Fig. 2 holds for any hexagonal photonic crystal with dielectric contrast of $\Delta\epsilon = 10.6$ and $r/a = 0.42$. Rescaling of the photonic crystal's lattice constant a simply shifts the band-structure to higher or lower frequencies. The normalized frequency $(\omega a)/(2\pi c)$ can be rewritten as a/λ from which the wavelength λ can be easily derived from the lattice constant a of the photonic crystal.

As mentioned in the introduction, photonic crystals can be seen as the optical analog to electronic semiconductors. Disturbing the electronic semiconductor host matrix, e.g., by the incorporation of dopant atoms, leads to new electronic states, the so-called acceptor and donor states. Breaking the strict periodicity of the photonic crystal lattice, e.g., by omitting one or several pores as shown in Fig. 3, has as a consequence the formation of new photonic states in the photonic band-structure as depicted in Fig. 4. Introduction of a line defect into the hexagonal photonic crystal leads to several new modes in the photonic band-gap of the bulk crystal.

Figure 3a shows a simple realization of a so-called cavity, while Fig. 3b shows a W1 line defect in a hexagonal lattice along the ΓK direction. Comparison of the TE band-structures of the bulk photonic crystal with undisturbed periodicity and of the photonic crystal with an incorporated W1 waveguide in Fig. 4 clearly reveals some new photonic states in the photonic band-gap which is depicted as the hatched region. Furthermore, the introduction of the W1 waveguide leads to slight shifts of the frequencies of the bulk photonic crystal due to the change in effective refractive index caused by the omitted pores of the waveguide. The frequency of the defect states can also be adjusted by changing, e.g., the radius of the defect pore or by using a W3

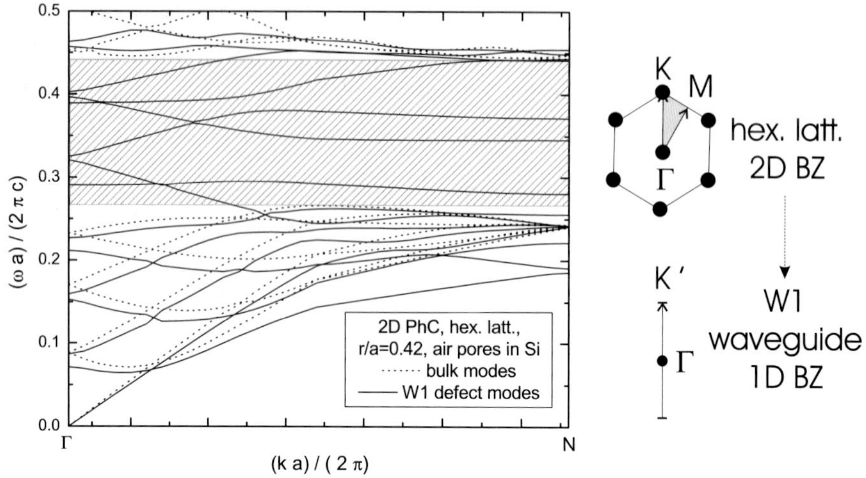

Fig. 4. *Left*: Calculated photonic band-structure for TE modes of a hexagonal array of air pores with $r/a = 0.42$ (*dotted lines*) in bulk Si and of the same structure with an incorporated W1 waveguide along ΓK direction as depicted in Fig. 3b (*full lines*). The hatched region represents the band-gap of the bulk photonic crystal. *Right*: Brillouin zones (BZ) of an undisturbed hexagonal lattice of pores (*top*) and for a W1 waveguide structure (*bottom*)

waveguide with three missing rows of pores instead of only one missing pore row as in the case of the W1 waveguide. For sake of completeness it should be noted that introduction of the waveguide reduces the symmetry from two-dimensional in the bulk photonic crystal to one-dimensional in the photonic crystal with waveguide as shown in the right part of Fig. 4. This has to be taken into account for the photonic band-structure calculation in which the edge of the bulk photonic crystal Brillouin zone K is replaced by the edge of the W1 photonic crystal Brillouin zone K′ [6].

3 Photonic Crystals Based on Silicon

After the short introduction to the theory of photonic crystals in the previous section we will now focus on their experimental realization using Si. In the first part a brief overview of different approaches is given and thereafter we will focus on photoelectrochemically etched macroporous silicon. The etching process will be described. The structure of the photonic crystals is investigated using scanning electron microscopy (SEM). Transmission and reflection measurements are used to characterize the optical properties of the samples. Both, two-dimensional photonic crystals without and with defects will be discussed, followed by consideration of three-dimensional photonic crystals. Regarding poregrowth in Si we will mainly focus on macropore formation in

(a) (b) (c) (d)

Fig. 5. (**a**) Yablonovite structure; (**b**) p-type macroporous Si Kielovite structure [9]; (**c**) Si Lincoln-Log structure [10]; (**d**) [110] and [111] facets of an Si inverse opal structure [11]

n-type Si. More comprehensive reviews of different pore regimes and pore formation mechanisms in p-type and n-type Si can be found in [7] and [8].

3.1 Silicon-Based Photonic Crystals – An Overview

Silicon is the dominating material in today's microelectronics and therefore a lot of experience in microstructuring of Si exists. Due to its high dielectric constant $\epsilon = 11.6$ it provides a high dielectric contrast with air. This makes it a promising candidate for photonic crystal fabrication.

From an intuitive point of view one would expect a complete photonic band-gap for a structure for which the interference occurs along all spatial directions for the same wavelength. This means that the Brillouin zone of such a photonic crystal should be sphere-like. A periodic arrangement of dielectric scatterers in a face-centered-cubic (fcc) lattice was therefore initially favored but theoretical calculations showed that even for arbitrarily high dielectric contrast such a structure would not exhibit a complete photonic band-gap. However, the situation changes when a two-atom basis is used resulting in a diamond structure featuring a complete photonic band-gap [12] because the polarization degeneracy of the one-atomic fcc-structure is lifted. A corresponding structure having diamond symmetry and known as *Yablonovite* [13] was the first experimentally realized photonic crystal showing a complete photonic band-gap although in the microwave range. It consists of holes drilled into a dielectric under certain angles as shown in Fig. 5a. It is possible to fabricate a similar structure, the *Kielovite* [9], by etching pores into (111) oriented p-type silicon. In contrast to the macropores grown perpendicular to the surface of (100) oriented n-type Si focused on later in this article, the macropores in (111) oriented p-type silicon can be etched along the $\langle 113 \rangle$ crystalline direction as shown in Fig. 5b. The difference consists in the angle of only 29.5° off axis orientation for the Kielovite compared to 35° for the Yablonovite. Calculations by *John* [14] suggest that the Kielovite structure should exhibit a complete photonic band-gap for an appropriate pore diameter.

Another approach to fabricate a three-dimensional photonic crystal is the so-called Lincoln-log structure shown in Fig. 5c as fabricated by Lin and coworkers at *Sandia National Laboratories* [10]. It consists of Si bars stacked on top of each other with subsequent layers arranged orthogonally to one another and the bars in each layer are parallel. Every third layer is shifted by half of the distance between two logs in the first layer. Such a Lincoln-log structure can be fabricated by depositing a layer of SiO_2 which is then structured and trenches are etched. The remaining gap is filled with polycrystalline Si. Repetition of this procedure leads to the desired structure after all the remaining SiO_2 is removed in a last step. This procedure is time consuming due to its layer-by-layer fabrication and therefore the vertical extension of these structures is up to now limited to one or two unit cells [15]. However, the layer-by-layer approach allows relatively easy integration of linear defects.

A technique that manages to get along without complex and costly lithography steps is based on self-organization of colloidal particles. Due to their cubic symmetry such structures can not possess a complete photonic band-gap even for the highest dielectric contrast of the constituent materials. However, the inverted structure does have a complete photonic band-gap. Such structures, as the inverted opal in Fig. 5d, can extend over several hundred unit cells. The structure shown in Fig. 5d was fabricated by infiltrating Si into an artificial opal made up by self-organization of silica spheres. The self-organization of the silica spheres solved in aqueous ethylene glycol was achieved by evaporating the solvent. A sintering step is then used to connect neighboring spheres slightly by a tube like SiO_2 cylinder. Now Si is deposited into the space between the connected spheres. Subsequent removal of the silica results in interconnected air spheres within a Si matrix [11]. The first successful realization of such a structure with a midgap frequency at the telecom window at 1.5 μm had a band-gap of about 5% [11].

In principle all of the above structures can also be used as templates for infiltration of high or low refractive index materials, thereby offering further possibilities of tailoring the band-structure of such a photonic crystal.

3.2 Photonic Crystals Based on Macroporous Silicon

After the brief overview of different fabrication techniques for photonic crystals based on Si in general we will now focus on the special case of the macroporous Si material system.

To produce the required periodic variation of the refractive index in order to create a photonic crystal *Grüning* and *Lehmann* [16, 17] as well as *Lau* and *Parker* [18] proposed the macroporous silicon material system. By a photoelectrochemical etching procedure Si is selectively removed from the bulk material resulting in ordered arrays of macropores. Figure 6 schematically shows the relevant aspects during the etching procedure. Before putting

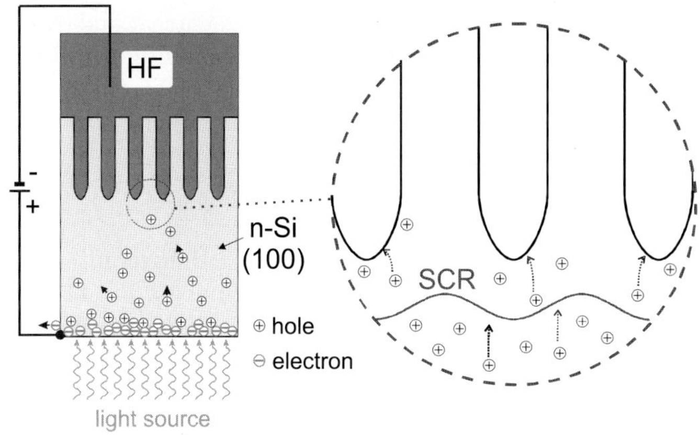

Fig. 6. Schematics of the photoelectrochemical etching process to produce ordered macropores in Si. The backside of the (100) n-Si sample is illuminated to produce holes which diffuse towards the pores where they get focused onto the pore tips by the space charge region (SCR) and promote the dissolution of the Si at the Si/HF interface

the sample into the etching apparatus the surface is lithographically prestructured, e.g., with a square or hexagonal pattern. Using anisotropic KOH etching so-called etch-pits shaped as inverted pyramids are transferred into the Si surface and serve as nucleation sites for the pore growth. Dissolution of Si implies defect electrons, i.e, holes, to be present at the Si/electrolyte interface. Since in the in our experiments used n-type Si holes are minority carriers these holes have to be generated by illuminating the backside of the sample. After photogeneration of the holes they diffuse through the bulk Si towards the pore growth front. The holes are focused onto the pore tips because the space charge region (SCR), resulting from the potential between the Si and the HF electrolyte follows the shape of the pore growth front. If all the parameters like inter pore distance a, potential U and doping of the n-Si are correctly chosen, all holes reaching the pore growth front are focused onto the pore tips where they lead to the dissolution of Si and none of them enter the Si walls separating adjacent pores, thereby passivating the Si walls against dissolution. Increasing or decreasing the intensity of the light source leads to a corresponding change in the number of available holes for the dissolution of Si. By this mechanism it is therefore possible to control the pore diameter. The porosity $p = A_{\text{pore}}/A_{\text{sample}}$ of a macroporous Si photonic crystal is defined as the ratio of the pore area A_{pore} and the total sample area A_{sample}. For a certain current density j_{PS} across the Si/HF interface all of the Si gets dissolved. This is the so called electropolishing regime. According to Lehmann's pore growth model there is a current density $j_{\text{pore}} = j_{\text{PS}}$ across the bottom of each pore during pore growth while there is no current flow across the pore wall/HF interface. Defining an average current density

$j = I/A_{\text{sample}}$ across the sample where I describes the total current and defining the current density across the pores $j_{\text{pore}} = I/A_{\text{pore}}$ and using the above relations the porosity can also be written as $p = A_{\text{pore}}/A_{\text{sample}} = j/j_{\text{PS}}$. The current density j_{PS} can be calculated using the relation

$$j_{\text{PS}} = 3300 \frac{\text{A}}{\text{cm}^2} c_{\text{HF}}^{3/2} e^{\frac{-0.345\,\text{eV}}{k_{\text{B}}T}} \tag{11}$$

where c_{HF} is the concentration of HF in weight percent, k Boltzmann's constant and T the absolute temperature in kelvin. The numbers in (11) are empirically determined parameters. Besides (11) for stable macropore growth the further condition $d_{\text{pore wall}} \approx 2x_{\text{SCR}}$ has to be fulfilled stating that the thickness of the remaining pore walls $d_{\text{pore wall}}$ should be about twice as thick as the extension x_{SCR} of the space charge region. The latter one is determined by

$$x_{\text{SCR}} = \sqrt{\frac{2\epsilon_{\text{Si}}\epsilon_0(V - \phi_0)}{eN_{\text{D}}}} \tag{12}$$

with the dielectric constant of Si being ϵ_{Si}, and the free space permittivity ϵ_0, the fundamental charge $e = 1.69 \times 10^{-19}$ C and the concentration of dopant atoms N_{D}. ϕ_0 describes the curvature of the bands at the Schottky-contact like Si/HF interface in equilibrium and V is the applied external voltage.

Based on these equations conditions can be chosen to allow for the growth of macropores in n-Si. Details on the electrochemical etching of Si and pore formation can be found in [8, 19, 20].

3.3 2-dimensional Photonic Crystals

In this section two-dimensional photonic crystals based on macroporous silicon will be discussed. Using the previously described photoelectrochemical etching procedure it is possible to obtain pores with aspect ratios clearly exceeding 100. Due to this high aspect ratio these photonic crystals can be seen as almost ideal, i.e., infinitely high, two-dimensional photonic crystals. The lattice constant of the photonic crystal can be varied from 500 nm up to about 20 μm. First bulk photonic crystals will be discussed followed by a treatment of defect structures introduced into two-dimensional photonic crystals.

3.3.1 Bulk 2-Dimensional Photonic Crystals

The structural quality of macropores etched into a Si wafer can be judged by cleaving the sample and examining the cleaving edge using SEM. The Moiré pattern in this printed version underlines the almost perfect straightness of the pores. Figure 7a shows the cleaving edge of hexagonally arranged macropores with a lattice constant of $a = 700$ nm. The aspect ratio of these pores

Fig. 7. SEM-image of (**a**) the cleaving edge of a two-dimensional hexagonal lattice of macropores in Si with a lattice constant of a = 700 nm (courtesy of S. Schweizer). The pores are straight and parallel providing the required strict periodicity for a photonic crystal. (**b**) Topview of a hexagonal pore array in Si with a lattice constant of $a = 500$ nm

is about 100 allowing to treat this photonic crystal as two-dimensional. The pores are straight and parallel providing the strict periodicity of the dielectric materials Si and air needed for a photonic crystal. Figure 7b shows the topview of a sample with a hexagonal lattice of air pores in Si with a lattice constant of $a = 500$ nm.

While SEM investigations give an impression of the structural quality of the etched samples they are not sufficient to judge the sample's photonic properties. For this purpose transmission and reflection measurements were performed. In Fig. 8 the calculated reflectivity on the left can be compared to the measured reflectivity in the center diagram and to the photonic band-structure on the right for a photonic crystal consisting of hexagonally arranged air pores in Si with a lattice constant of $a = 500$ nm and $r/a = 0.425$. The upper part shows the results for TE modes and the lower part for TM modes along the ΓM direction. The dashed lines in the photonic band-structure of Fig. 8 represent so-called noncoupling bands. The electric field along the pore axis $\boldsymbol{E}_z$ of a TM polarized plane wave approaching the photonic

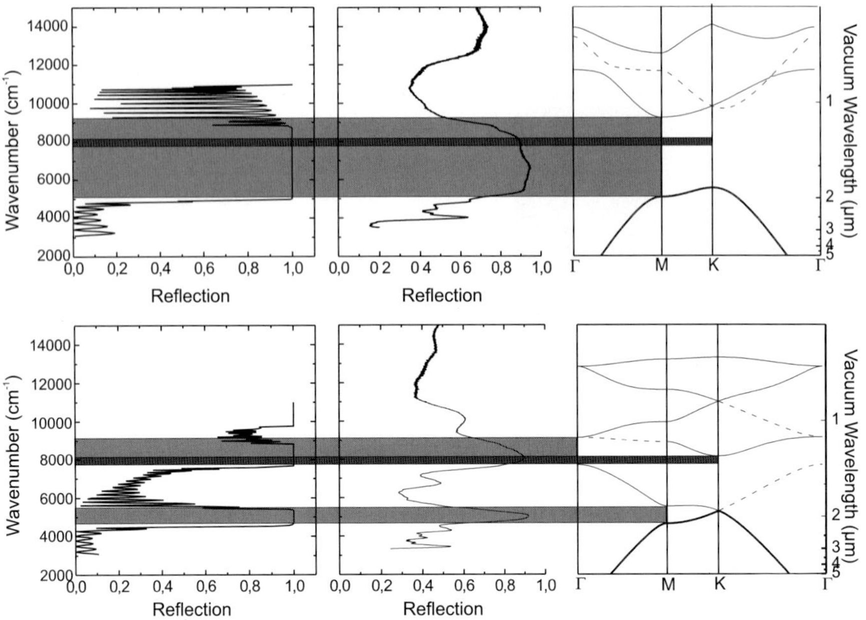

Fig. 8. Reflection of a two-dimensional photonic crystal of hexagonally arranged air-pores with $r/a = 0.425$ in Si along the ΓM direction. *Top*: TE modes, *Bottom*: TM modes. *Left*: Calculation for a finite photonic crystal with 8 pore rows. *Center*: Measurement of a semi-infinite photonic crystal. *Right*: photonic band-structure for comparison. The *dark grey* region represents the complete photonic band-gap

crystal is symmetric with respect to a plane containing the z-axis and the vector of the incoming plane wave. Such a plane wave can only couple to photonic crystal modes if they have the same symmetry.

This is e.g., the case for the second TM band for hexagonally arranged air pores in Si, as can be seen from the upper part of Fig. 9. The lower part of Fig. 9 shows the $\boldsymbol{E}_z$ field intensity for the third TM band which reveals antisymmetry with respect to the previously defined plane. Therefore a plane wave cannot couple to this photonic crystal mode leading to a high reflectivity in the corresponding frequency range although there is no real photonic band-gap. Furthermore, for both polarizations the reflectivity is very high in the photonic band-gap regions. The measurements also agree well with the calculated reflection of a finite photonic crystal structure with 8 pore rows based on a two-dimensional transfer-matrix code using the *Translight* software [21].

Band-gap Tuning and Switching One of the advantages of the macroporous Si/air material system for photonic crystals is that the band-structure can be easily shifted by simply changing the r/a-ratio. For a fixed lattice constant a the porosity p and with it the pore radius r is determined by the

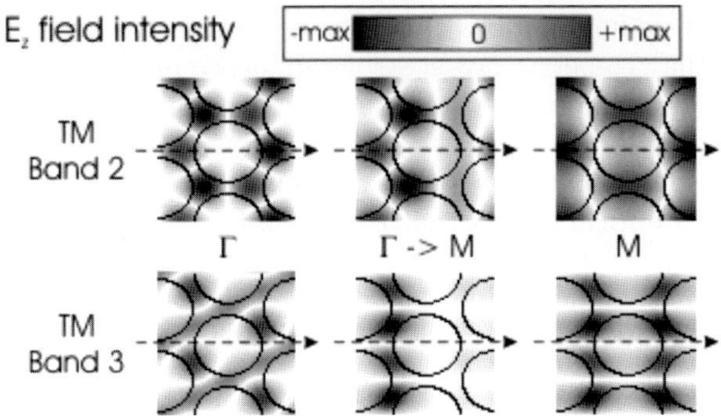

Fig. 9. Field intensity of $\boldsymbol{E}_z$ for the second and third TM bands at the Γ and M points, respectively (*left and right images*) and for $|\boldsymbol{k}| = 1/2\Gamma\text{M}$ (*center image*) along the ΓM direction for a hexagonal photonic crystal of air pores in Si. The dashed arrows mark the ΓM direction

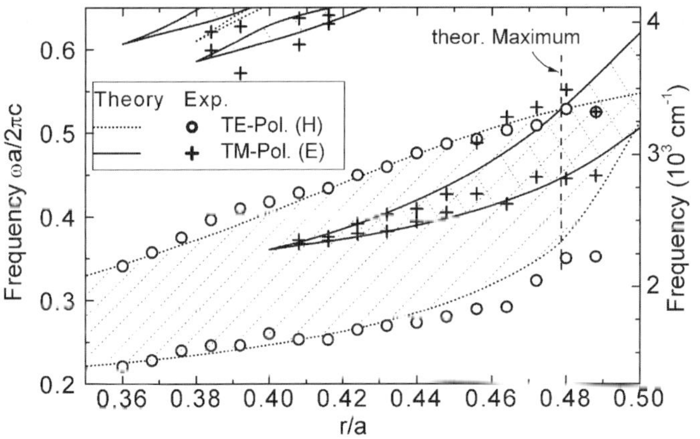

Fig. 10. Theoretical (*lines*) and experimentally verified (*symbols*) position of the TE and TM bandedges depending on the r/a-ratio for a two-dimensional photonic crystal consisting of air pores in Si [23]

current during the etching process. Therefore, it is possible to shift the photonic band-gap by keeping a constant. Figure 10 shows the gapmap, i.e., the bandedges of TE and TM polarized modes for hexagonally arranged air pores in Si depending on a certain parameter as which the r/a-ratio was chosen in this case. The lines represent the results of calculations based on a plane wave expansion method [22] and the symbols are experimentally determined values. Only for the largest r/a-ratios, i.e., the highest porosities, there is a slight discrepancy between theory and experiment. These are due to difficulties in preparing and handling such structures. Hexagonally arranged air pores in Si have a maximum complete photonic band-gap of about 16% for a r/a-ratio of 0.47.

As mentioned in the preceding sections the dielectric contrast as well as the lattice constant and the geometry have a strong influence on the band-structure of a photonic crystal and, consequently, on the position of the photonic band-gap.

A change of one of these parameters will shift the photonic band-gap and allows the realization of an optical switch that blocks light of a certain frequency for one set of parameters while allowing transmission for the changed parameters. A change of the symmetry and the lattice constant could be realized by mechanically stretching or compressing the sample. In the case of a photonic crystal based on macroporous silicon this is not favorable since silicon is a hard and brittle material. Changes of the lattice constant by thermal expansion of the Si are also not practicable due to the very small coefficient of expansion $\alpha = 2.3 \times 10^{-6}\,\mathrm{K}^{-1}$ of Si.

However the dielectric contrast can easily be varied by filling liquid crystals into the macropores. Depending on temperature they are either randomly oriented at 59 °C or higher temperatures or the are in their nematic phase at room temperature, where they are oriented parallel to the pore axis [27]. By this method it was possible to achieve a shift of 70 nm for the low-frequency bandedge of hexagonal macropores with $a = 1.58\,\mu\mathrm{m}$ and a photonic band-gap in the frequency range of 4.4 µm to 6 µm. While this method allows for fine tuning of the photonic band-structure, e.g., to compensate for fabrication tolerances, it is not suitable for high speed switching applications because it takes several microseconds for the liquid crystals to undergo the necessary phase change from their nematic ordering into random orientation.

Another possibility to change the dielectric contrast is to vary the dielectric constant of the Si itself. This can be achieved by the creation of carriers by illumination of the sample leading to a change of the dielectric constant $\epsilon(\omega) = \epsilon_{\mathrm{Si}} - \omega_{\mathrm{P}}/\omega^2$ where ϵ_{Si} is the dielectric constant of Si in the absence of free carriers and ω_{P} represents the plasma frequency in the Drude model $\omega_{\mathrm{P}} = (Ne^2)/(\epsilon_0 m^*)$ with the reduced effective mass $1/m^* = 1/m_e^* + 1/m_h^*$, the number N of generated free carriers and the elementary charge $e = 1.69 \times 10^{-19}$ C. The photonic band-structure and the expected shift of the bandedge depending on plasma frequency is shown in Fig. 11. By this method a 29 nm shift of the bandedge at 1910 nm with free carriers was achieved [28, 29]. The turn on time of about 400 fs of this optically induced bandedge shift is rather short and promising for high speed switching. Since Si is an indirect semiconductor the turn off time is on the order of several nanoseconds limiting the switching speed. Theoretical considerations, however, suggest that it should be possible to reduce the switch off time to about 10 ps using defect induced de-excitation via surface states or impurity atoms incorporated into the Si.

The transmission through an ideal photonic crystal, i.e., an infinitely extended, perfectly periodic structure $\epsilon(\boldsymbol{r})$, is zero for frequencies within its band-gap. Experimentally fabricated photonic crystals however are by na-

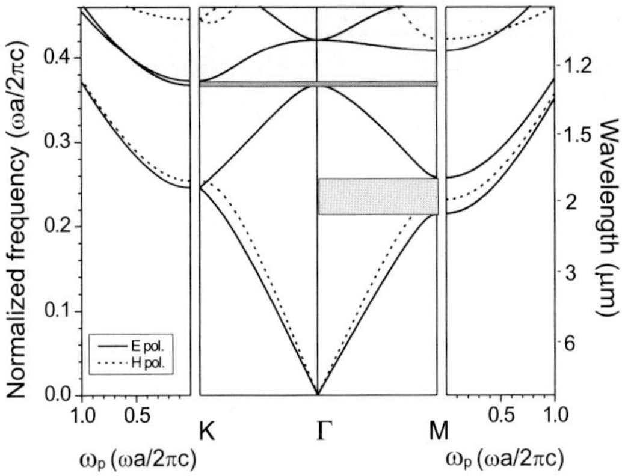

Fig. 11. *Center*: Photonic band-structure of a two-dimensional hexagonal photonic crystal of macroporous Si with $r/a = 0.41$. *Left* and *right*: Changes of the photonic band-gap depending on plasma frequency ω_p at M and K points, respectively [26]

ture finite objects. From this point of view it is desirable to investigate how extended a photonic crystal has to be before light starts to feel the photonic band-gap and cannot travel through the photonic crystal anymore. For this purpose structures consisting of only a few pore rows were prepared. After the standard fabrication of an ordered macropore array by photoelectrochemical etching of Si this photonic crystal was further structured. First the pore walls were passivated by a thermal oxide and a chemical vapor deposited (CVD) nitride. Then an aluminum layer was deposited onto the macropore array and structured by standard photo-lithography. Although the feature size in this step was on the order of 10 μm, the alignment relative to the pores and the precision of these structures was better than 1 μm. In the aluminum mask a window was opened and the passivating oxide and the nitride were removed by chemical etching. The porous Si in the regions without passivating oxide was etched in a subsequent isotropic plasma etching process resulting in structures like the one shown in Fig. 12. Details can be found in [30] and [31]. The resulting structure consisted of only a single pore row along ΓK direction as depicted in the lower right inset in Fig. 12. The lines in Fig. 12 represent the calculated transmission based on an extended plane wave approach [4,31] and the symbols show the results of measurements at structures like the one shown in the lower right inset through 1, 2, 3 and 4 pore rows, respectively. The attenuation of the transmitted signal increases with increasing number of pore rows with a rate of about 10 dB per pore row [24]. Such a signal attenuation by a constant factor per pore row is typical for a evanescent, i.e., exponentially decaying field. These results show that already for rather finite photonic crystals consisting of only a few pore rows practically useful effects are obtained.

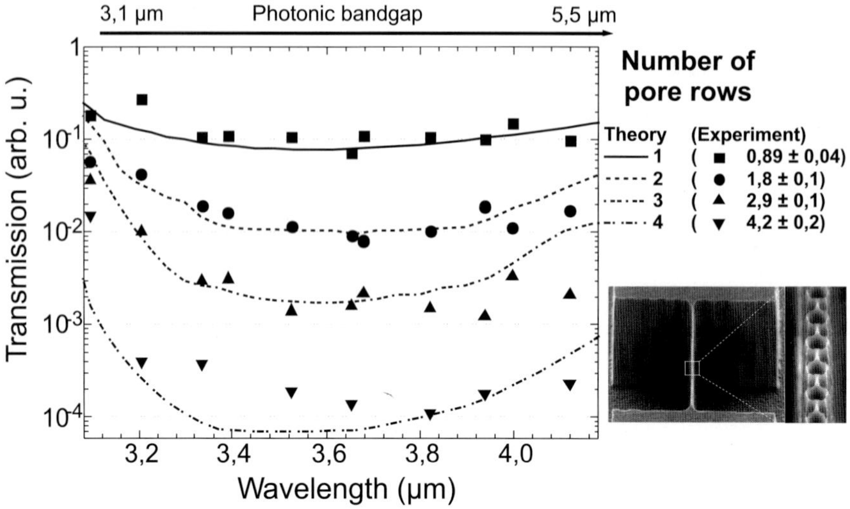

Fig. 12. Calculated (*lines*) and measured (*symbols*) transmission through a few pore rows along ΓK direction. The *lower right inset* shows an SEM picture of the structure [23, 24, 25]

3.3.2 Defects in 2-Dimensional Photonic Crystals

While the bulk photonic crystals treated in the previous section reveal a lot of basic physics it is necessary to implement some functionality for actual applications. This is similar to electronic semiconductors which deploy their full potential and possibilities only after doping which introduces new electronic states in the band-structure. In the case of photonic semiconductors functionality in form of new photonic states is achieved by disturbing the strict periodicity of the underlying crystal lattice by omitting some pores as previously shown in Fig. 3. For macroporous Si defect structures are defined during the lithography step which gives considerable freedom in designing photonic crystals with incorporated defects. The simplest kind of defect is given by a single missing pore as depicted in Fig. 13a. This leads to a new state in the photonic band-gap whose field energy is concentrated in and around this cavity. Omitting subsequent neighboring pores as shown in Fig. 13b–d creates waveguides within the photonic crystal. Although they look similar to conventional waveguides made of bulk Si embedded within a material with lower refractive index and using total internal reflection (TIR) for the guiding the photonic crystal waveguide is based on a fundamentally different waveguiding mechanism. Introducing of a defect leads to the formation of new photonic states within the photonic band-gap. Above and below those defect-induced modes the photonic DOS is zero. As a consequence light travelling in the photonic crystal using the defect modes cannot leak away into neighboring modes and, therefore, ideally the loss is zero. However, the confinement is

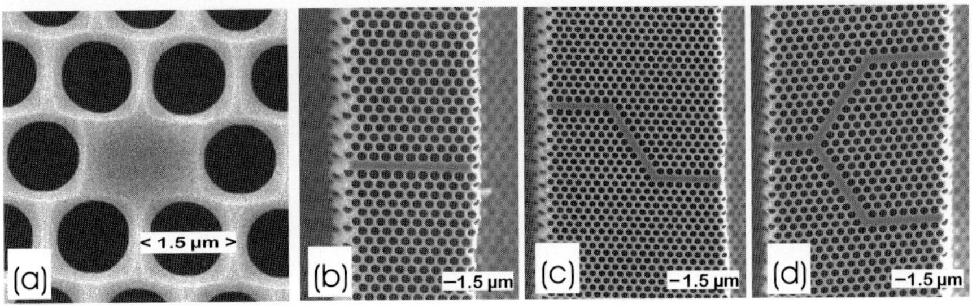

Fig. 13. SEM micrographs of defect structures in a hexagonal lattice of air pores in Si (**a**) a point defect or *cavity* (**b**) and (**c**) straight and bent linear defects and (**d**) Y-branch. Realized by omitting adjacent pores in the otherwise strictly periodic lattice

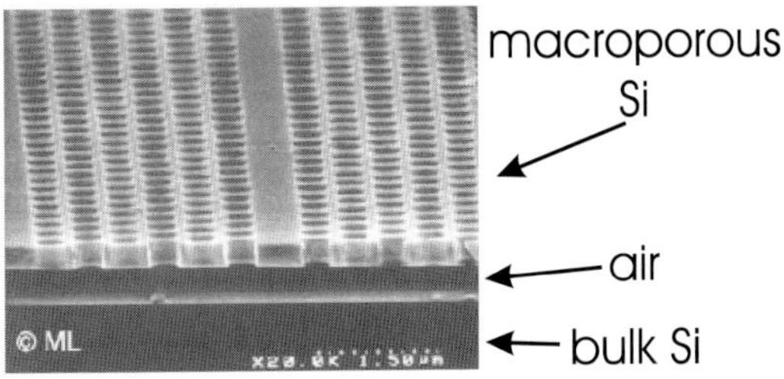

Fig. 14. *Airbridge* structure: macropores etched into Si prepared as a membrane embedded within air [32]

very high only in the plane providing the periodic modulation of the refractive index, i.e., in the x-y-plane, the plane perpendicular to the pore axis. Along the z direction there is no photonic crystal based confinement. By using membrane structures as shown in Fig. 14 consisting of macropore arrays with air or SiO_2 above and below the pores would allow to provide at least partial light confinement along the z direction by TIR.

A closer look onto the SEM micrographs in Fig. 13 reveals that the pores next to defects are somewhat larger than the other pores farther away from the defects because the pores surrounding the defect collect more charge carriers. This effect can be minimized by etching pores with a smaller diameter than the finally desired one and widening of the pores by subsequent thermal oxidation and HF etching of the SiO_2.

After the fabrication of a Si macropore array additional structure can be added. In Fig. 15 a linear defect incorporated into a thin catwalk consisting of 13 pore rows is shown in a zoom series starting from a global overview towards the defect. The high quality of the structure with its high aspect ratio of the catwalk as well as its flat edge are observable. Thinning the catwalk further allows to measure the transmission of single pore rows as shown in Fig. 12.

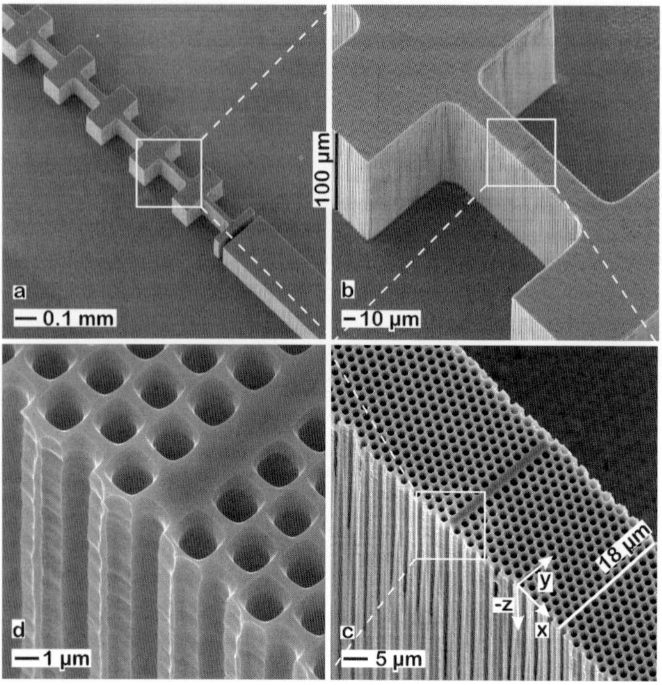

Fig. 15. After photoelectrochemically etching of the Si the macropore array was further structured by subsequent lithographical and wet and dry etching processes. The SEM picture series shows a zoom into the structure finally revealing the high quality line defect incorporated into the thin part of the catwalk [33]

3.4 3-Dimensional Photonic Crystals

Photonic crystal effects only appear in directions where the refractive index changes periodically. This was true for the two-dimensional photonic crystals treated in the previous section in the x-y plane, but not along the z direction. In this section photonic crystals based on macroporous Si will be reviewed whose refractive index is also periodically modulated along the z direction offering the possibility to create three-dimensional photonic crystals.

3.4.1 Bulk 3-Dimensional Photonic Crystals

The diameter of a pore etched into Si depends on the etch current. This current can be controlled by changing the backside illumination intensity which generates the holes necessary for the dissolution of the Si at the pore tip. If the illumination intensity is varied during pore growth the pore diameter will vary correspondingly as is shown in Fig. 16. By this method a nice periodic change in the pore diameter along the growth direction and with it of the refractive index can be achieved. The pore diameter modulation, however, does

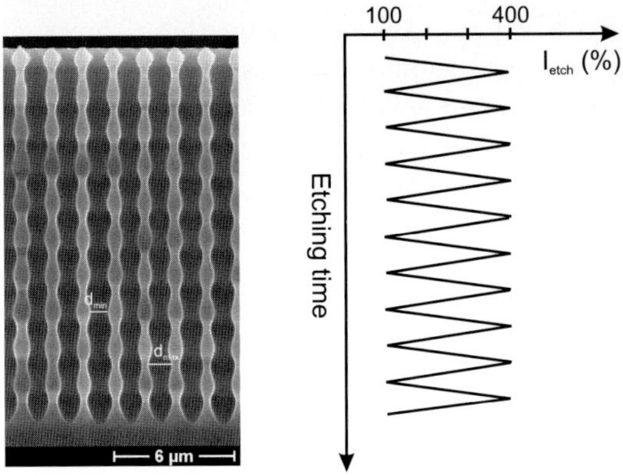

Fig. 16. Variation of the pore diameter along the pore axis is achieved by varying the etch current by changing the illumination intensity. *Left*: Cross-sectional SEM picture of the cleaved sample. *Right*: Schematic diagram of etch current vs. time [29]

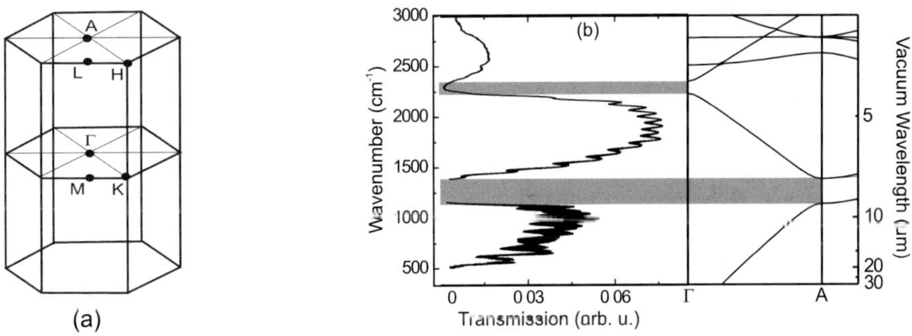

Fig. 17. (**a**) Brillouin zone of a three-dimensional photonic crystal with high symmetry points. (**b**) Transmission along the ΓA direction of a three-dimensional macroporous Si photonic crystal. *Left*: Measured Transmission, *Right*: Photonic band-structure along ΓA. The lattice constant of the hexagonal lattice in the x-y-plane is 1.5 µm. Along the z-axis 15 modulation periods were realized with a length of $l_z = 2$ µm

not exactly follow the etch current modulation [34]. While the current profile exhibits sharp features the corresponding pore profiles show soft edges. The pore profiles are smeared out. This is due to limitations of the theoretical model on which the current profile is based. If the illumination and with it the free carrier concentration is varied too sudden or too strong the etching mechanism at the pore tip changes leading to the observed deviation of the etched pore shape from the designed profile [29]. Detailed studies on how the current profile has to be chosen to produce desired sharp changes of the pore radius are currently underway.

The Brillouin zone of the three-dimensional photonic crystal is depicted in Fig. 17a. In the x-y-plane it maintains the hexagonal symmetry of the two-dimensional photonic crystal and gets extended along the z direction due to the introduced periodicity l_z of the pore diameter modulation. To observe the photonic crystal effects caused by this periodic modulation of the refractive index along the z-axis the transmission along the ΓA direction, i.e., along the pores, was measured. The center panel of Fig. 17 shows the measured transmission spectrum of a photonic crystal consisting of air pores in Si with hexagonal lattice constant of 1.5 μm in the x-y-plane and 15 modulations of the pore diameter with a period length of $l_z = 2\,\mu\text{m}$ along the z-axis. The two lowest band-gaps are clearly observed and agree well with the calculated band-structure given in the right panel of Fig. 17b. While in the two-dimensional photonic crystal *every* plane perpendicular to the pore axis was a mirror plane this is no longer true in the three-dimensional photonic crystal. Therefore the modes do no longer separate into TE and TM polarization.

Reduction of the Group Velocity In the case of electronic semiconductors the slope of the band-structure $\partial E_n(\boldsymbol{k})/\partial \boldsymbol{k}$ can in a semiclassical approximation be interpreted as $\hbar$ times the velocity $\boldsymbol{v}_n(\boldsymbol{k})$ of an electron in a definite Bloch level $\Psi_n(\boldsymbol{k})$ with energy $E_n(\boldsymbol{k})$ [35]. In a photonic crystal the slope of the band-structure represents the group velocity $\boldsymbol{v}_n(\boldsymbol{k})$ of light with frequency ω travelling along the direction $\boldsymbol{k}$ in the n-th band

$$\frac{\partial \omega_n(\boldsymbol{k})}{\partial \boldsymbol{k}} = \boldsymbol{v}_n(\boldsymbol{k}) \,. \tag{13}$$

In a linear approximation the inverse slope $(\partial \omega_n(\boldsymbol{k})/\partial \boldsymbol{k})^{-1} = n_{\text{eff}} = c/\boldsymbol{v}_n(\boldsymbol{k})$ can be interpreted as the effective refractive index n_{eff} for that particular frequency. As a consequence flat bands lead to a low group velocity in a photonic crystal.

Experimentally the group velocity $\boldsymbol{v}_n(\boldsymbol{k})$ can be determined from the frequency difference $\Delta\omega_{\text{max}}$ of the Fabry–Perot transmission maxima and the pore depth d_{pore} together with (13). In $\boldsymbol{k}$-space the difference between adjacent transmission maxima $\Delta\omega_{\text{max}}$ is given by $\Delta\boldsymbol{k}_{\text{max}} = \pi/d_{\text{pore}}$. This leads to

$$\boldsymbol{v}_n(\boldsymbol{k}) = \frac{\partial \omega_n(\boldsymbol{k})}{\partial \boldsymbol{k}} \approx \frac{\Delta\omega_{\text{max}}}{\Delta\boldsymbol{k}_{\text{max}}} \tag{14}$$

from which the group velocity can experimentally be determined using the above difference quotient as approximation for the differential quotient in (14). Using such a three-dimensional photonic crystal based on macroporous silicon a reduction of the maximum group velocity to about 30%–40% of the vacuum speed of light c_0 is achieved for the second band in Fig. 17b as shown in Fig. 18 [36]. The lines represent the calculated group velocity derived from the band-structure of the first two bands in Fig. 17b and polarization of the $\boldsymbol{E}$ field along ΓM direction while the symbols are the experimentally obtained

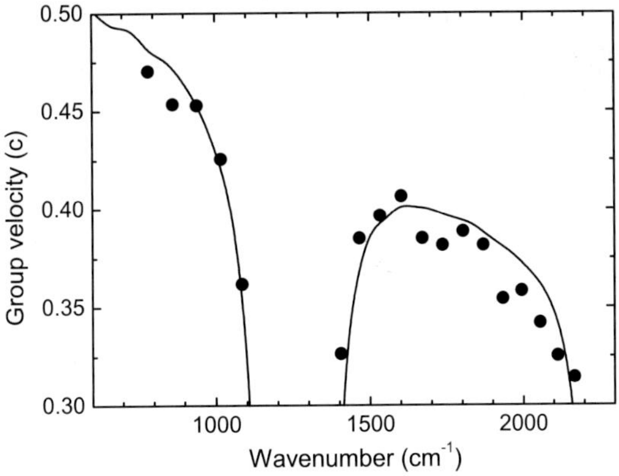

Fig. 18. Group velocity $\boldsymbol{v}_g$ along ΓA direction for the structure in Fig. 17b. The black line is the derivative of the band-structure, the *open symbols* represent the experimentally determined values using the Fabry–Perot resonances in Fig. 17b and (14)

values. It has to be taken into account that simply taking the derivative of, e.g., the second band, in the band-structure in Fig. 17b yields negative values for $\boldsymbol{v}_g$ which makes physically no sense. But this is simply due to the fact that the reduced zone scheme was used in Fig. 17b where the second band seems to have negative slope. In the extended zone scheme the band would have positive slope at every k-point and a physically meaningful positive group velocity would be the result. As mentioned above a flat band corresponds to a high effective refractive index. This in return leads to a high reflection R of incoming light following from the well known equation $R(0°) = \{(n_1 - n_2)/(n_1 + n_2)\}^2$ describing the amount of reflected light R for incidence perpendicular to a dielectric surface in a classical approximation. For this reason it is not possible to experimentally probe the regions of photonic bands with zero slope by transmission.

3.4.2 Defects in 3-Dimensional Photonic Crystals

As shown in Sect. 3.3.2 defects in the x-y-plane in two-dimensional photonic crystals are defined via lithography. As defects are simply well behaved deviations of a photonic crystal's periodicity a defect layer in a three-dimensional macroporous Si photonic crystal can be realized by a current profile like the one shown in Fig. 19a. Leaving the current unchanged for a certain time and thereby keeping the pore radius constant leads to a defect layer within a three-dimensional photonic crystal. An experimental realization of such a structure is shown in Fig. 19b. Experimental results concerning the introduced defect states in the band-structure and their modelling can be found in the thesis of *Schilling* [29] and will be published soon.

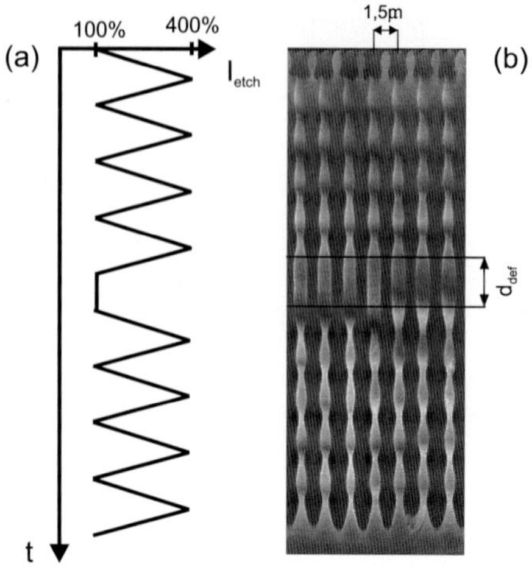

Fig. 19. Defect layer in a three-dimensional photonic crystal based on macroporous Si. (**a**) Schematic representation of the current profile used for the creation of the defect layer within the three-dimensional photonic crystal. (**b**) Cross-sectional SEM picture of the cleaved sample of an etched structure with 5 modulations of the pore diameter above and below the defect layer. The vertical extension of the defect is $d_{\mathrm{def}} = 2.12\,\mu\mathrm{m}$

4 Conclusion and Outlook

The concept of photonic crystals was introduced and a brief introduction to the underlying theory was given. Fabrication methods for Si based photonic crystals including self-organization as well as lithographical approaches were described with special focus to macroporous Si. The latter allows the realization of two-and three-dimensional extended photonic crystals by photoelectrochemical etching of Si in HF. These structures feature ordered pore arrays with high aspect ratios exceeding 100 and provide high dielectric contrast. A variety of macropore patterns like square or hexagonal arrangement can easily be chosen by lithography. Furthermore the lattice constant can be chosen over a wide range from 500 nm up to about 20 µm. This together with the appropriate choice of the r/a ratio allows the fabrication of photonic crystals for the NIR to MIR spectral range. In two-dimensional macroporous Si photonic crystals defects can be incorporated during the lithography process while in three-dimensional structures defects can be incorporated during the photoelectrochemical etching. Macroporous Si pore arrays can also serve as templates which can be filled with, e.g., liquid crystals, to tune the photonic properties such as the position of the photonic band-gap. The basic properties of photonic crystals such as the photonic band-gap, defect states and the reduction of group velocity were experimentally verified with these structures.

Future work includes a better understanding of the photoelectrochemical etching process and application of this knowledge to the fabrication of a three-dimensional photonic crystal with a complete photonic band-gap [37]. For the realization of compact gas sensing devices we plan to use the reduced group velocity to enhance the absorption in two-dimensional photonic crystals [38]. To ensure efficient coupling in spite of the high refractive index and with it the high back reflection of incident light tapered structures will be fabricated and their efficiency evaluated.

Acknowledgements

The authors are gratefull to S. Schweizer, F. Müller, R. Hillebrand, A. Birner, C. Jamois, K. Busch, S. Leonard and H.M. van Driel for theoretical and experimental contributions. Funding by DFG within SPP1113 is gratefully acknowledged.

References

1. S. John: Phys. Rev. Lett. **58**, 2486 (1987)
2. E. Yablonovitch: Phys. Rev. Lett. **58**, 2059 (1987)
3. J. D. Joannopoulos, R. D. Meade, J. N. Winn: *Photonic Crystals: Molding the Flow of Light* (Princeton University Press 1995)
4. K. Sakoda: *Optical Properties of Photonic Crystals* (Springer, Berlin, Heidelberg 2001)
5. O. Hess, C. Hermann, A. Klaedtke: Phys. Stat. Sol. A **197**, 605 (2003)
6. S. G. Johnson, P. R. Villeneuve, S. Fan, J. D. Joannopoulos: Phys. Rev. B **62**, 8212 (2000)
7. P. Allongue: INSPEC Data Series (1997)
8. V. Lehmann: *Electrochemistry of Silicon* (Wiley-VCH, Weinheim 2002)
9. M. Christophersen, J. Carstensen, A. Feuerhake, H. Föll: Mater. Sci. Eng. B. **69**, 194 (2000)
10. S. Y. Lin, J. G. Fleming, D. L. Hetherington, B. K. Smith, R. Biswas, K. M. Ho, M. M. Sigalas, W. Zubytzki, S. R. Kurtz, J. Bur: Nature **394**, 251 (1998)
11. A. Blanco, E. Chomski, S. Grabtchak, M. Ibisate, S. John, S. W. Leonard, C. Lopez, F. Meseguer, H. Miguez, J. P. Mondia, G. A. Ozin, O. Toader, H. M. van Driel: Nature **405**, 437 (2000)
12. K. M. Ho, C. T. Tan, C. M. Soukoulis: Phys. Rev. Lett. **65**, 3152 (1990)
13. E. Yablonovitch, T. J. Gmitter, K. M. Leung: Phys. Rev. Lett. **67**, 2295 (1991)
14. S. John: unpublished results
15. S. Noda, N. Yamamoto, M. Imada, H. Kobayashi, M. Okano: J. Lightwave Technol. **17**, 1984 (1998)
16. U. Grüning, V. Lehmann, C.M. Engelhard: Appl. Phys. Lett. **66**, 3254 (1995)
17. U. Grüning, V. Lehmann, S. Ottow, K. Busch: Appl. Phys. Lett. **68**, 747 (1996)
18. H. W. Lau, G. J. Parker, R. Greef, M. Hölling: Appl. Phys. Lett. **67**, 1877 (1995)
19. V. Lehmann, H. Föll: J. Electrochem. Soc. **137**, 653 (1990)
20. V. Lehmann, H. Föll: J. Electrochem. Soc. **140**, 2836 (1993)
21. P. Bell, J. Pendry, L. Moreno, A. Ward: Comp. Phys. Commun. **85**, 306 (1995)
22. K. Sakoda: Phys. Rev. B **52**, 7982 (1995)
23. J. Schilling, R. B. Wehrspohn, A. Birner, F. Müller, R. Hillebrand, U. Gösele, S. W. Leonard, J. P. Mondia, F. Genereux, H. M. van Driel, P. Kramper, V. Sandoghdar, K. Busch: J. Opt. A: Pure Appl. Opt. **3**, 121 (2001)
24. S. W. Leonard, H. M. van Driel, K. Busch, S. John, A. Birner, A.-P. Li, F. Müller, U. Gösele, V. Lehmann: Appl. Phys. Lett **75**, 3063 (1999)

25. J. Schilling, A. Birner, F.Müller, R. B. Wehrspohn, R. Hillebrand, U. Gösele, K. Busch, S. John, S. W. Leonard, H. M. van Driel: Opt. Mat **17**, 7 (2001)
26. S. W. Leonard, H. van Driel nad J. Schilling, R. B. Wehrspohn: Phys., Rev. B, **66**, 161102(R) (2002)
27. S. W. Leonard, J. Mondia, H. van Driel, O. Toader, S. John, K. Busch, A. Birner, U. Gösele: Phys. Rev. B **61**, R2389 (2000)
28. S. Leonard: Ph.D. Thesis, University of Toronto, Dept. of Physics (2001)
29. J. Schilling: Herstellung und optische Eigenschaften von 2D- und 3D-photonische Kristallen aus makroporösem Silizium: Ph.D. Thesis, Martin-Luther-University Halle-Wittenberg, Halle, Germany (2002)
30. S. Ottow, V. Lehmann, H. Föll: J. Electrochem. Soc., **140**, 385 (1996)
31. A. Birner: Optische Wellenleiter und Mikroresonatoren in zweidimensionalen Photonischen Kristallen aus Makroporösem Silizium: Ph.D. Thesis, Martin-Luther-University Halle-Wittenberg, Halle, Germany (2000)
32. M. Loncar, D. Nedeljkovic, T. Doll, J. Vukovich, A. Scherer, T. Pearsall: Appl. Phys. Lett. **77**, 1937 (2000)
33. A. Birner, A.-P. Li, F. Müller, U. Gösele, P. Kramper, V. Sandoghdar, J. Mlynek, K. Busch, V. Lehmann: Mat. Sci. Semicond. Proc. **3**, 487 (2000)
34. J. Schilling, F. Müller, S. Matthias, R. B. Wehrspohn, U. Gösele, K. Busch: Appl. Phys. Lett. **78**, 1180 (2001)
35. N. W. Ashcroft, N. D. Mermin: *Solid State Physics*, 2nd edn. (Springer, 1976)
36. J. Schilling, F. Müller, R. B. Wehrspohn, U. Gösele, K. Busch: Mat. Res. Soc. Symp. Proc. **722**, L6.8.1 (2002)
37. *Special issue on Photonic Crystals: Optical Materials for the 21st Century*, edited by R. B. Wehrspohn, K. Busch: Phys. Stat. Sol. A **197** (2003)
38. A. Feist, A. Lambrecht, J. Schilling, F. Müller, R. Wehrspohn: *Vorrichtung und Verfahren zur qualitativen und quantitativen Bestimmung von Gasen und Flüssigkeiten mit PBG-Struktur*, Patent DE 100 63 151 a 1

Silicon-Based Waveguide Technology for Wavelength Division Multiplexing

Siegfried Janz

Institute for Microstructural Sciences, National Research Council of Canada, Building M-23A, 1200 Montreal Rd. Ottawa, Ontario Canada K1A 0R6
siegfried.janz@nrc-cnrc.gc.ca

Abstract. This chapter reviews the application of silicon-based planar waveguide components for wavelength division multiplexing (WDM) and demultiplexing. The polarization dependent properties and dominant waveguide loss mechanisms in both silica-on-silicon and silicon-on-insulator (SOI) waveguides are described, and the technologies developed to control loss and eliminate polarization dependence in both material platforms are presented and compared. Although less mature, the development of high index contrast devices in SOI closely parallels the evolution of silica glass waveguide technology. Many of the important design challenges have been resolved, and the performance of SOI-based integrated optic components is rapidly approaching that of the silica waveguide system.

1 Introduction

The multiplexing, demultiplexing, and measurement of wavelength are essential photonic functions, and the need for high performance dispersive elements continues to drive research in planar waveguide optics. Wavelength division multiplexing (WDM) in particular is a flexible technology that provides an elegant solution to the problem of transporting large quantities of data in parallel streams over a single conduit, the optical fibre or perhaps even free space. Therefore the applications of planar waveguide dispersive elements may likely expand to include board-to-board data bus applications, rack-to-rack connections, and finally merge with the telecommunications network at the local area network level. The same wavelength selective element that lies at the heart of any WDM demultiplexer can also be used for spectroscopy, replacing present day bulk optic spectrometers for sensing and chemical analysis wherever small size, stability and portability are essential. These new applications will all require continued improvements in planar waveguide spectrometer performance and reductions in size.

Of the many waveguide materials systems investigated to date, silica glass on silicon has come to dominate the commercial market for WDM planar waveguide devices. Extremely low loss and uniform silica glass waveguides can be deposited on silicon using chemical vapor deposition [1,2,3,4,5,6] and flame hydrolysis [7,8,9,10,11]. Silicon itself has become the most widely used platform for large-scale planar waveguide circuits. Silicon is an inexpensive,

L. Pavesi, D. J. Lockwood (Eds.): Silicon Photonics, Topics Appl. Phys. **94**, 323–360 (2004)

mechanically strong platform that is the natural substrate for both conventional and high index contrast waveguide systems including silica waveguides, silicon oxynitride/silica waveguides, polymers, and more recently the silicon-on-insulator (SOI) system. Single crystal silicon wafers can be obtained with diameters up to eight and twelve inches, as a a result of demand from the microelectronics industry. From another perspective, it is possible that the most viable solution to the integration of active and passive device may be based on hybrid integration schemes. Examples of hybrid systems combining passive glass or silicon waveguide components with III–V based semiconductor lasers and detectors [12, 13, 14] rely on the underlying Silicon optical bench. Here again silicon comes to fore because of its mechanical properties and the existing fabrication and micromachining knowledge that can be applied.

In the future silicon is expected to play an increasing role in photonics as emphasis shifts from conventional waveguide devices to devices where the feature size becomes comparable to or less than the wavelength of light. In this regime the use of sub-μm waveguides and photonic crystal structures may significantly alter the photonic landscape, not only in scale but also in the types of functions and applications that can be implemented. The creation of these new microphotonic devices relies on the use of high-index contrast waveguides, and the commonly used platforms are silicon oxynitride/silica on silicon, and silicon-on-insulator. Although the technology of microphotonics is just beginning to emerge, it is clear that silicon is again the material of choice because of its excellent optical qualities and the scope of the fabrication technologies that can be brought to bear on the technical challenges.

A discussion of the intricate details involved in waveguide spectrometer design is well beyond the scope of this chapter. The objective will be to provide an overview of the interplay between material science and optical engineering that occurs in the development of demultiplexers and spectrometers on the silicon platform. In Sect. 2 of this chapter, the basic design and layout of the AWG and echelle grating demultiplexer will be reviewed. Section 3 will give an overview of the material and fabrication challenges and solutions encountered when making well-established silica waveguide based WDM demultiplexers and spectrometers. A similar discussion of the implementation of planar waveguide demultiplexer technology in the SOI platform is presented in Sect. 4 of this chapter, to highlight the special challenges encountered when working with high index contrast systems. High index contrast silicon oxynitride waveguides are also a topic of much interest. Although a discussion of silicon oxynitride devices has been omitted, the parallel discussions of SOI and silica should have some relevance to this system as well.

2 Planar Waveguide WDM Components

Wavelength division multiplexing technology first attracted serious interest in the integrated optics community more than twenty years ago. It was then not obvious how WDM would be eventually implemented, if ever, and the existing technology for time domain multiplexing (TDM) still had a considerable potential for improvement. Furthermore, prior to the advent of the erbium doped fibre amplifier (EDFA) there was still debate as to the wavelengths that would actually be used in fibre-optic links. The first demultiplexers were essentially miniature versions of grating devices found in bulk optic spectrometers. The basic principles of echelle grating spectrometers have been previously described for both free space and planar waveguide configurations [15, 16, 17]. One early device consisted of a three layer multimode glass waveguide with a diffraction grating mechanically ruled on one facet of the die, and multimode input and output fibres bonded to the other face of the die [18]. Other groups used similar designs but with holographic gratings written in glass [19], or wet etched silicon gratings [20] bonded onto one facet of the waveguide. A somewhat different approach to the same problem was demonstrated by *Chartier* et al. [21], who used ion-exchanged waveguides. The diffraction grating was produced by ion exchange through a mask to define the step grating structure. Because of the low index contrast between the ion-exchanged glass and normal glass, the grating efficiency of this device would be very small. These labor-intensive hybrid approaches illustrate that creating the grating facets was the biggest obstacle in making planar waveguide diffraction grating devices. This is still the key to implement the echelle design, but as will be discussed below, modern fabrication methods and tools can meet and overcome this challenge.

2.1 Arrayed Waveguide Grating Demultiplexer

The most significant development in planar waveguide demultiplexers came with the proposal by *Smit* [22] to replace the diffraction grating by an array of waveguides. The layout of such an arrayed waveguide grating demultiplexer (AWG) is shown in Fig. 1. The light is coupled into a single input waveguide leading to a slab waveguide splitter/combiner where it diffracts in the plane of the waveguide, illuminating an array of many waveguides. Each waveguide in the array section is longer than the adjacent waveguide by a constant length increment ΔL. The array waveguide inputs are distributed evenly on the circular arc of radius R that forms the boundary of the splitter section. In the simplest configuration the input waveguide is at the center of the circle so that the arc defines a line of constant phase for input light coupled to the splitter section. The array waveguides all converge at a second slab coupler which is often identical to the splitter. Because of the length increment between each array waveguide, the corresponding phase difference between light emerging from adjacent waveguides is $\Delta\phi(\lambda) = 2\pi\Delta Ln/\lambda$ where n is the effective index of the fundamental waveguide mode. At the center wavelength λ_0 this phase shift is a multiple of 2π, and the optical path in the output coupler simply mirrors the path through the input coupler, with the light being focused at the central output waveguide. As the wavelength is

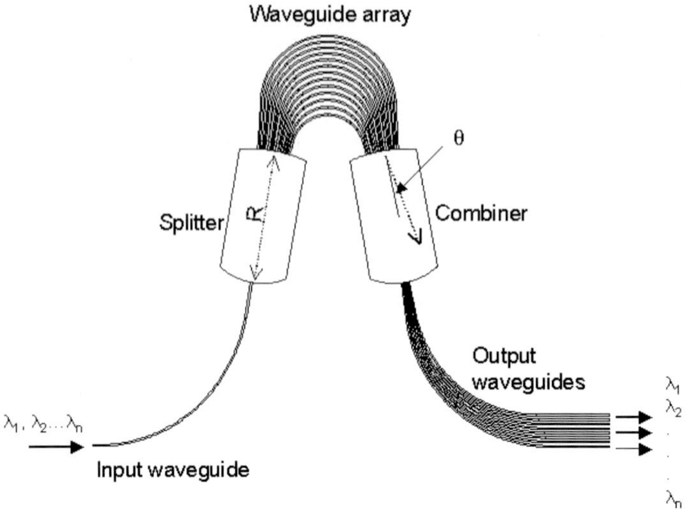

Fig. 1. Schematic layout of an arrayed waveguide grating demultiplexer

changed to $\lambda_0 + \Delta\,\lambda$ however, the phase increment is no longer an exact multiple of 2π, and the beam is directed at an angle θ away from the center waveguide

$$\sin\theta = \frac{\lambda}{n_{\mathrm{c}} d} \times \frac{\Delta\,\phi\,(\lambda) - \Delta\,\phi\,(\lambda_0)}{2\pi}\,. \tag{1}$$

Here d is the separation of the array waveguide outputs, n_{c} is the effective index of the combiner waveguide, and the order M is defined in terms of the center wavelength λ_0 as $M = n_{\mathrm{g}}\Delta\,L/\lambda_0$, where n_{g} is the group effective index. Substituting $\Delta\,\phi\,(\lambda)$ into (1) gives the diffraction angle in terms of the wavelength shift $\Delta\,\lambda$

$$\sin\theta = -\frac{\Delta\,\lambda}{n_{\mathrm{c}} d} \times M\,. \tag{2}$$

This has the same form as the grating equation that describes diffraction from a conventional ruled grating in Littrow configuration, when illuminated at normal incidence. Thus from a theoretical point of view the operation of a conventional grating and AWG are the same. The great advantage of the AWG is that fabrication requires only the ability to make single mode waveguides with low loss, and with sufficient reproducibility and uniformity that the relevant phase relationships between the optical paths through the array section are maintained. The phase relationships between the array waveguide outputs can be arbitrarily specified simply by assigning each array waveguide length. It is therefore possible to design very high order gratings, or gratings with unique phase distributions at the output of the array section – for example to modify the shape of the focused spot on the output waveguide. Since

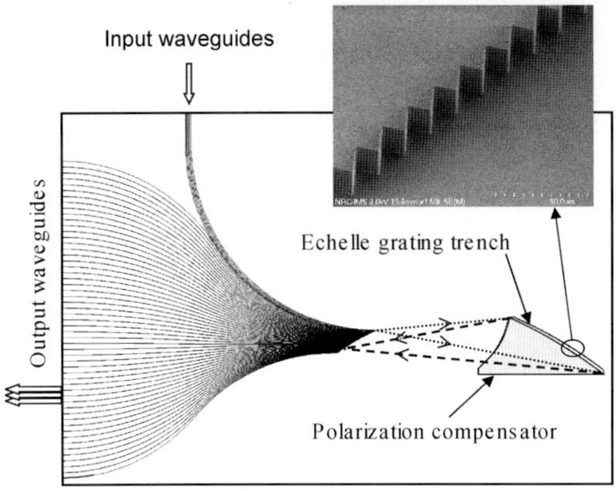

Fig. 2. Schematic layout of 48-channel echelle grating demultiplexer chip. *Inset* photograph shows magnified view of the grating trench

the first proposal by *Smit* [22], AWGs have evolved into a mature technology that is displacing the dielectric filter demultiplexer in the commercial telecommunications market. In the laboratory, AWG demultiplexers with more than 1000 channels have been demonstrated. There is already an extensive literature on the subject of AWG devices, and several recent reviews and articles on this subject can be found in [23].

2.2 Echelle Grating Demultiplexer

Since the development of the AWG demultiplexer, the telecommunications industry has neglected conventional diffraction grating for integrated WDM applications. This occurred because of a perceived difficulty in manufacturing planar waveguide diffraction grating devices. Although the operating principles basic of the diffraction gratings and AWGs are the same, three fundamental challenges in implementing planar waveguide diffraction gratings must be overcome. These are to fabricate vertical grating facets in a waveguide structure, to reduce the polarization dependence of the grating diffraction efficiency, and to eliminate or compensate the birefringence of the device.

Despite the difficulties faced in fabricating the device, the echelle grating (EG) does offer some compelling advantages. The EG configuration shown in Fig. 2 has a folded beam path and does not require the waveguide array, and so will be several times smaller than the equivalent AWG device. An EG device can also easily be scaled up to larger channel counts of more than 160 channels and smaller channel spacing. Similar scaling of AWG devices quickly leads to components of unmanageable size, because of the length increment conditions on the individual array waveguides. Some work has continued on echelle gratings based demultiplexers, but mainly in the III–V semiconductor systems [22, 25, 26, 27]. More recently, the feasibility of planar

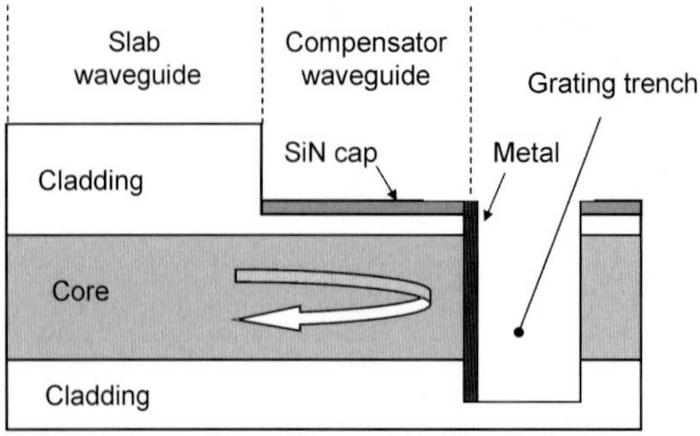

Fig. 3. Cross-sectional schematic view of the echelle grating device picture in Fig. 2, showing the layer structure of the slab waveguide, compensator waveguide, and the grating trench. The Si substrate is not shown

waveguide EG demultiplexers in glass has been demonstrated [28, 29]. The EG layout for a 48-channel glass device is shown in Fig. 2. The input waveguide terminates at the boundary of slab waveguide region where the light diverges and illuminates the echelle grating. Light is diffracted from the concave echelle grating and focused backwards towards the output waveguides. The angular dispersion of the diffracted light is again given by (2) in the case of illumination at normal incidence to the grating facets, where M is the grating order. At any given wavelength, the grating will form an image of the input waveguide at the Rowland circle [16]. The output waveguides are positioned along this arc with spacing chosen to give the required wavelength channel separation. Figure 3 shows the corresponding cross-section view of the waveguide structure including the polarization compensator section and grating.

The prototype echelle waveguide structure consists of a 5 μm thick phosphorous doped silica glass (PSG) core surrounded by upper and lower cladding glass layers, with an index step of approximately 1% between the core and cladding. The layers are grown on a six inch silicon substrate using plasma enhanced chemical vapor deposition (PECVD). As shown in Fig. 3, the diffraction grating in Fig. 2 is formed by etching a 10 μm deep trench through the core layer, using a high density inductively coupled plasma system. The wall of the trench facing the input and output waveguides on the Rowland circle is stepped to form a 20th order echelle grating. After etching the grating facet is metal coated to enhance the reflectivity. The ability to etch grating sidewalls that are flat and vertical to within $\pm 1°$ is the key step to achieving acceptable insertion losses. A grating wall offset of only $1°$ from the vertical will result in a coupling loss of more than 0.5 dB from the incident fundamental mode to the reflected fundamental mode. In addition to the insertion loss penalty, the lost light will be coupled to leaky modes of the slab waveguide that can eventually couple back into the input and output waveguides, and contribute to return loss, cross-talk and ghost peaks in the channel spectra. Using an anisotropic etching process based on C_4F_8 and Ar gases, we have been able to consistently achieve grating walls with a sidewall angle between 89° and 90° as measured by scanning electron microscopy.

The diffraction efficiency of a metallized grating is polarization dependent due to the different electromagnetic boundary conditions for light polarized parallel and perpendicular to the metal–glass interface. By depositing metal only on the reflecting facet of each grating tooth in our planar waveguide grating, we were able to largely eliminate this effect and achieve planar waveguide diffraction gratings with a PDL of less than 0.2 dB across the entire C-band. As will be discussed in Sect. 3, the polarization birefringence effects intrinsic to glass waveguides were compensated using a polarization compensator section shown in Figs. 2 and 3. By adjusting the layer parameters of this compensator structure, TE-TM channel wavelength shifts of $\Delta\lambda = 10$ pm or less are routinely achieved.

Two versions of the final device were fabricated: a 48-channel echelle grating demultiplexer with 100 GHz channel spacing, and a 256-channel spectrometer chip with 25 GHz channel spacing. In both cases the basic fabrication processes were identical. The output channel spectra for the two devices are shown in Figs. 4 and 5. The fibre-to-fibre insertion loss for the 48-channel devices is approximately 4 dB, with a measured adjacent channel cross-talk better than −35 dB. The insertion loss for the 256-channel device is approximately 10 dB, while other performance parameters are similar to the 40-channel devices. In both devices the polarization dependent wavelength shift of the channel wavelengths was reduced to 10 pm or better. This similarity in performance for essentially identical fabrication procedures demonstrates the scalability of echelle gratings. The overall die size was $18 \times 17\,\text{mm}^2$ for the 48-channel device, and of $20 \times 40\,\text{mm}^2$ for the 256-channel device. A comparable 256-channel AWG device has dimensions of $75 \times 55\,\text{mm}^2$ [30], approximately five times the area of the EG chip. In competitive technologies such as AWGs, scaling up to 256 or more channels drastically increases device size, and accumulated phase errors in the array waveguides limit achievable cross-talk. That is not the case here, where the cross-talk is comparable for 40 and 256-channel devices.

We have presented two general approaches to planar waveguide demultiplexers in this section. The AWG is a mature technology that has been commercialized for a number of years, and is based on standard waveguide technology. The echelle grating approach has only been recently demonstrated on glass waveguides, and a number of unique challenges needed to be overcome to successfully fabricate this device. However, it has been established that EG devices with performance comparable to state of the art AWG demultiplexers can be fabricated in the silica-on-silicon system, using standard silicon fabrication tools. There are other approaches to planar waveguide demultiplexing which we cannot include here, such as the use of superprism effects in two-dimensional photonic crystals [31] and planar waveguide transmission

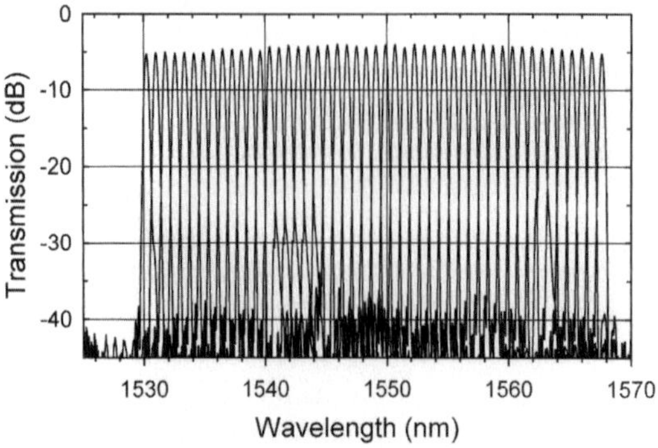

Fig. 4. Measured channel spectra for a 48-channel echelle grating demultiplexer

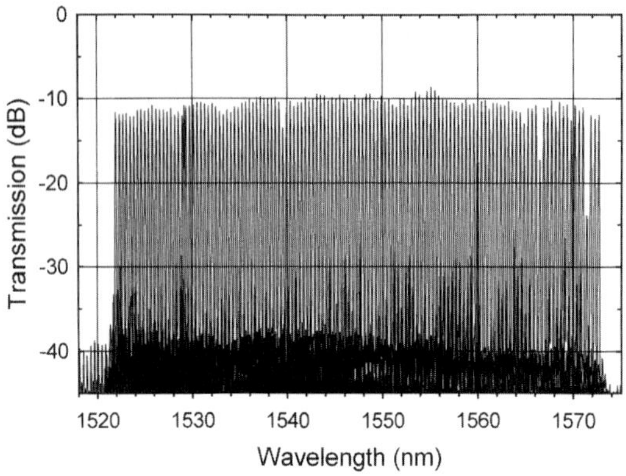

Fig. 5. Measured channel spectra for a 256-channel echelle grating demultiplexer

gratings [32]. These approaches have not yet been as extensively researched as the AWG and EG devices described, but many of the materials and fabrication issues will be similar. In the following sections this chapter will now focus on the specific materials and fabrication challenges in implementing AWG and EG demultiplexers on the silica-on-silicon and silicon-on-insulator platforms. The objective is to review how key issues such as polarization sensitivity and waveguide loss have been resolved in the more mature glass waveguide technology. Then approaches to the similar issues in silicon-on-insulator (SOI) will be explored, with the hope of shedding some light on the path to achieving extremely compact spectrometers and dispersive elements in high index contrast systems.

3 Silica Glass Waveguides on Silicon

Over the last decade, silica glass waveguides on silicon substrates have been the dominant platform for high performance passive waveguide components such as WDM demultiplexers. Losses in glass are intrinsically lower than in most other materials at the wavelengths of interest for telecommunications. In glass optical fibre losses much less than 1 dB per kilometer are achieved. The techniques used to ensure compositional purity and uniformity in optical fibre manufacture cannot yet be reproduced in planar waveguide manufacturing, but waveguide losses of less than 0.1 dB/cm are routinely achieved in planar glass waveguides deposited on silicon [3, 33]. Glass waveguide core dimensions and cladding-core index steps can closely match those of optical fibre. The resulting good overlap between fibre and waveguide mode size eliminates the need for discrete coupling optics or mode matching structures. Although such coupling techniques are well known and used in the packaging of active III–V semiconductor devices, the added complexity and cost become prohibitive for waveguide multiplexers and demultiplexers with up to hundreds of input and output ports. In the case of glass waveguides well matched to optical fibre, coupling of fibre to numerous ports involves only alignment and attachment, and can be done in a single step using fibre ribbons mounted in silicon V-groove arrays.

3.1 Silica Glass Waveguide Growth

One must modify the refractive index of the glass in order to fabricate waveguides. In conventional glass waveguides where index steps of the order of $\Delta n = 0.01$ are used to define the waveguide core, small amount of dopants such as Ti, P, Ge, As, and B are incorporated during growth [7, 9, 11, 34]. In early work Ti was often used, but recently Ge doping has been more commonly employed. Ti doped glass has the disadvantage of having significantly higher losses for a given index step compared with Ge-doped glass. This may arise because Ti doping raises the melting temperature of glass making consolidation and homogenization of the glass by annealing less effective. Increasing the refractive index using P-doping has also been used, but P is a much more reactive element than Ge. As a result the growth chemistry may be more difficult to control, and unless the final glass structure is appropriately capped with an impermeable layer, P-doped silica glass (PSG) can degrade with time due to the absorption of water. Both P and B dopants are more commonly incorporated into glass as a means of lowering the melting temperature and modifying the thermal expansion coefficients, with the objective of reducing birefringence and planarizing the surface of the completed device.

Silica waveguides are commonly deposited using two techniques. Flame hydrolysis deposition (FHD) is capable of growing very thick glass layers [7, 8, 9, 10, 11]. In the case of silica waveguides optimized to match optical fibre waveguide mode sizes, the core is of the order of five to ten μm thick with upper and lower cladding layers up to several tens of μm thick. In FHD, a mixture of gases, including $SiCl_4$ are burned along with hydrogen and oxygen in a torch, depositing a fine soot of silica particles onto a silicon substrate. Rotating the substrates wafer and/or scanning the torch position ensures uniform coverage. This layer of silica particles is then consolidated by heating the wafer to a temperatures of the order of 1200 °C or higher, so that the particles sinter together and a homogeneous silica layer is formed. Doping with Ti, Ge, P, N, B and As is accomplished by including appropriate compounds containing these elements in the burned mixture.

Recently silica deposition by plasma enhanced chemical vapor deposition (PECVD) has become accepted for waveguide manufacture [1, 2, 3, 4, 5, 6]. PECVD can also grow silica layers at rates of several hundred nanometers per minute, with excellent control over the uniformity of refractive index and layer thickness. In PECVD, a plasma is created between two parallel electrodes in the presence of a gas mixture consisting of SiH_4, N_2O and a carrier gas such as He, Ar or N_2. The wafer itself is usually heated to a temperature of the order of 400 °C. These gases react on the surface of the wafer to form a layer of SiO_2. As in the case of FHD, the glass index and material properties are modified through incorporation of dopants during deposition by adding the appropriate precursor gases to the gas mixture. Simply varying the precursor gas ratios can be used to deposit layers of SiO_xN_y with compositions and refractive index ranging from SiO_2 ($n = 1.46$) to Si_3N_4 ($n = 2.0$). Silicon oxynitride in particular has attracted considerable attention as a material for the fabrication of high index contrast waveguides [35]. PECVD is attractive because it is already a standard tool in the microelectronics industry, where PECVD silicon dioxide and silicon nitride are used as dielectrics for various applications. Commercial PECVD reactors intended for the semiconductor industry can easily be adapted to the manufacture of glass waveguides. PECVD deposition can also be carried out on a very large scale, for example some reactors can deposit oxide on tens or even hundreds of wafers at one time.

The quality of as-deposited PECVD glasses tends to be lower than those produced by FHD. The glass network often contains significant concentrations of hydrogen and nitrogen, which can contribute to waveguide absorption. In particular, strong absorption lines centered at $\lambda = 1480$ nm and $\lambda = 1510$ nm associated with overtones of the N-H and Si-H bonds often appear [1, 3, 36]. These lines can be particularly strong in silicon oxynitrides. Since the absorption features are very broad, they may contribute to demultiplexer insertion loss and channel nonuniformity in the C-band ($\lambda = 1525$ nm–1565 nm). The glass network also is not as complete as in a thermal or FHD grown oxide, with defects and voids that contribute to scattering losses and make the material susceptible to changes with time, for example as water is absorbed into the network. For this reason, PECVD grown glass for optical applications must be annealed at high temperatures of the order of $T = 1000$ °C [1, 3]. Annealing releases hydrogen trapped in the glass during growth, and plays a similar role to the high temperature anneal in FHD by consolidating and homogenizing the glass. After annealing, waveguides grown by PECVD have excellent optical properties comparable to those achieved using FHD.

Waveguides fabricated in silica on silicon are usually buried channel waveguides, except where slab waveguides are required as in the combiner section of an AWG. Fabrication usually occurs in a series of steps in which first a lower cladding is deposited using FHD or PECVD, although a lower cladding layer of thermally grown oxide is often used. A doped silica layer with a slightly higher refractive index is then deposited to form the waveguide core. The planar waveguide circuit is then patterned on the core layer using conventional lithography tools, and defined using dry etching. The final step is to deposit an upper cladding layer over the structure.

3.2 Losses in Silica Waveguides

Compared with most optoelectronic materials the transparency of silica glass is excellent. However there are sources of waveguide loss that must eliminated, particularly when building to the specifications of the modern telecommunications industry, where total device insertion losses of less than a few dB are expected. These losses can arise from intrinsic material absorption as discussed previously, and light scattering. Rayleigh scattering is always present to some extent due to small index inhomogeneities and defects in the glass itself. Such intrinsic scattering can usually be minimized to acceptable levels through proper annealing of the glass to temperatures above which the glass becomes plastic. Scattering may also arise from roughness of the sidewall profiles of the ridge waveguides. This roughness is introduced by imperfections in lithography, or as an artifact of the etch process used to define the waveguides. This side-wall scattering is one of the critical issues in high index contrast waveguides systems such as silicon-on-insulator, and will be discussed in more detail in Sect. 4.2. In conventional glass waveguides the side-wall roughness is usually of the order of 10 nm to 50 nm in height, and the resulting loss has not been noted as a serious problem. Since the core-cladding index step is of the order of $\Delta n = 0.01$, scattering is very weak. Calculated losses for typical values of roughness are of the order of 0.1 dB/cm or less for buried glass channel waveguides, as shown in Fig. 6. However, as Fig. 6 illustrates, the induced loss for the same roughness can be much higher for high index contrast waveguides such as SOI.

Another source of scattering is the presence of cracks and defects. Such defects may arise during deposition of cladding layers over the high aspect ratio ridge waveguides, because the high stress occurring at the waveguide corners and variations in growth processes with the local substrate geometry (e.g. for deposition on vertical and horizontal surfaces and near sharp corners). If these defects are close enough to the waveguide core they can interact with the waveguide mode and contribute to the waveguide loss. Conformal overcladding coverage that avoids cracking can be achieved by building up the cladding using a series of depositions of thin layers, with appropriate

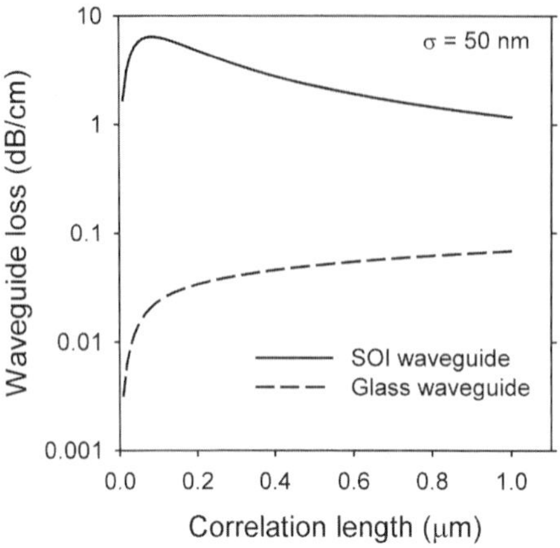

Fig. 6. Variation of the scattering induced loss with roughness correlation length in a buried glass waveguide and an SOI waveguide, both with dimensions 5 μm × 5 μm and an assumed rms sidewall roughness of $\sigma = 50\,\text{nm}$. The *curves* are calculated using the model of Payne and Lacey [37]

anneals between each deposition. Dopants such as P and B are often used to lower the melting point of the cladding glass well below that of the core glass material, so that each layer in the sequence can be made to reflow and conform to the ridge waveguide profile, while the ridge itself retains its shape.

3.3 Polarization Dependent Properties of Silica Glass Waveguides

The polarization dependence of waveguides presents one of the most difficult challenges in the design of planar waveguide components using glass on silicon. Since optical fibers used in communications links are not polarization maintaining, any optical signal arriving at a waveguide chip will be in an arbitrary polarization state. Furthermore, this polarization can fluctuate randomly on time scales of minutes for buried fiber links or less than a second for above ground fibre links. Because of this random time variation, the polarization sensitivity in any component used to process the optical signal will be translated into signal noise.

The two fundamental polarization effects are polarization dependent loss (PDL) and birefringence. Intrinsic PDL is due to waveguide loss by scattering or absorption processes that are polarization dependent. Polarization dependent absorption is unlikely to occur in amorphous materials such as glass. However polarization dependent scattering can arise, for example, from residual roughness in the ridge waveguide side-walls. The boundary conditions for the electric fields normal and parallel to the interface between two dielectric media are different. The parallel component of the electric field E must be continuous, while the normal component of the electric displacement $D = \varepsilon E$ must be continuous across the interface. Clearly this difference will result in polarization dependence of the scattering at a rough interface. Again, since the index contrast between cladding and core found in conventional glass waveguides is small, this source of PDL has not been reported as a significant problem.

Waveguide birefringence refers to difference $\Delta n = n_{\mathrm{TE}} - n_{\mathrm{TM}}$ in effective indices n_{TE} and n_{TM} of the orthogonally polarized TE and TM waveguide modes. The transverse electric (TE) mode refers to the mode with electric field parallel to the plane of the waveguide. However, in the case of a channel waveguide, TE is often used to denote the mode with field parallel to the wafer plane. Similarly the transverse magnetic mode (TM) refers to the mode with the magnetic field parallel to the wafer plane (or electric field approximately normal to the wafer plane). Birefringence is present in any waveguide that is not perfectly symmetric in cross-section. As mentioned above, this geometric or modal birefringence is rooted in the different boundary conditions for electric field tangential and normal to the core-cladding interface. This difference means that the mode shape and effective index for TE and TM polarized modes differ in any waveguide that is not perfectly square or circular in cross-section. In high-index contrast waveguides such as SOI, the birefringence can be very large, unless stringent design rules are followed.

Birefringence changes the phase velocity of orthogonally polarized waveguide modes, so that the TE and TM polarized components of an optical signal become separated in time. This effect is known as polarization mode dispersion (PMD) and is a critical issue in optical fiber transmission. However, since planar waveguide components tend to be a few centimeters in length, the accumulated PMD is small, typically only of the order 100 femtoseconds. In cases where very high bit rates are used, or the signal must traverse many cascaded components, the on-chip PMD can become important. Birefringence also creates a shift in the effective wavelength of orthogonally polarized light, since wavelength inside the waveguide is $\lambda = \lambda_0/n$. Here the vacuum wavelength is λ_0 and waveguide mode effective index is n. For a given waveguide birefringence, Δn, the channel wavelengths in a demultiplexer will have a polarization dependent shift of $\Delta\lambda = \lambda_0 \Delta n / n_{\mathrm{g}}$ between TE and TM polarized light. This is the most important polarization effect in planar waveguide demultiplexers for WDM systems. If this shift is large it can result in significant inter-channel cross-talk and signal attenuation. Even if $\Delta\lambda$ is small relative to the channel pass-band, it can result in unacceptable PDL over much of the pass-band.

In low index contrast glass waveguides the geometrical birefringence can be quite small (e.g. $\Delta n \sim 10^{-5}$). The single most important source of birefringence is stress induced optical anisotropy. The thermal expansion coefficient of silica glass is $\alpha = 0.5\,\mathrm{ppm/^\circ C}$ while for silicon is ten times larger at approximately $\alpha = 5\,\mathrm{ppm/^\circ C}$. Since the glass is deposited at high temperatures, as the silicon substrate and glass system is cooled to room temperature the silicon tends to contract much more than the glass. If the glass layer remains bonded to the substrate and does not develop cracks, it will usually be under strong compressive stress at room temperature. On the other hand

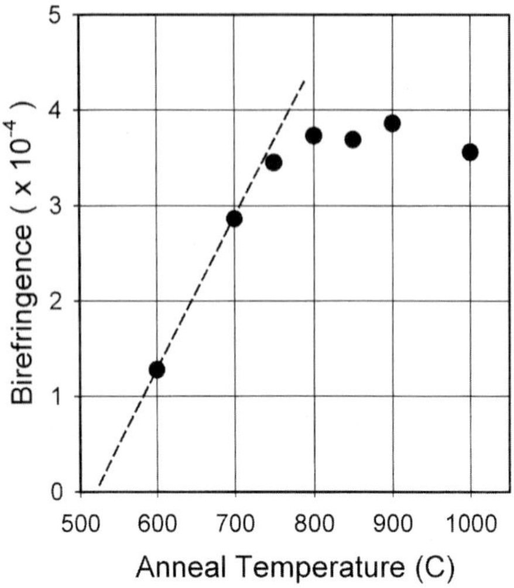

Fig. 7. Birefringence measured by Rayleigh scattering for PSG waveguides that have been annealed at temperatures from 600 °C to 1000 °C. The *dashed line* is the estimated extrapolation to zero birefringence [39]

stresses are also present even in the glass films at the deposition temperature. For example, PECVD glasses in some cases may even be under tensile stress, particularly if the as deposited material has microscopic voids and is non-stoichiometric. For this reason the stress of PECVD films can vary widely, and often displays complex behavior and hysteresis as the temperature is varied [38].

In the case of silica waveguide layers, some form of annealing at high temperatures is a critical step in producing material with acceptable optical properties. Annealing is usually done at temperatures near or above the temperature at which the glass deforms under plastic flow. At such temperatures the glass accommodates itself to the silicon substrate and stress will vanish. Stress is only frozen in when the temperature falls below the point where glass becomes rigid. Figure 7 illustrates that the final thermal stress at room temperature depends mainly on the difference between the plastic flow temperature and room temperature rather than initial deposition conditions. Figure 7 shows the stress induced birefringence measured in PSG glass waveguides for several different anneal temperatures. In this case the stress is frozen in only as the temperature falls below approximately 800 °C. After annealing the properties of the glass depend primarily on chemical composition rather than initial deposition conditions and growth chemistry. As a result, the stresses and corresponding optical birefringence built into glass waveguides of a similar chemical composition are remarkably similar whether FHD or PECVD deposition was employed. The stress in waveguide glass films is usually compressive and in the range of a few hundred MPa.

Stress and the resulting strain modify the optical properties and electronic structure of a material by changing the bond length and crystal symmetry at the microscopic level. Using the language of optically anisotropic crystals, the strained silica behaves as a uniaxial crystal with the optic axis along the surface normal z-direction. The refractive index n_{TE} for TE light (E-field polarized in the x-y or wafer plane) corresponds to the ordinary index while the index for TM polarized light (E-field along the z-direction) is the extraordinary index. The resulting refractive index changes can be cast in the formalism of the photoelastic effect, where the effect of strain e_j on the index ellipsoid is expressed in terms of the photoelastic tensor p_{ij} as

$$\Delta \left(\frac{1}{n^2} \right)_i = p_{ij} e_j \, . \tag{3}$$

Contracted tensor notation is used here (i.e. $1 \rightarrow xx$, $2 \rightarrow yy$, $3 \rightarrow zz$, $4 \rightarrow zy$, $5 \rightarrow zx$, $6 \rightarrow xy$). In isotropic materials this relation can be simplified to give the strain induced change in index $\Delta\, n_i$ for light polarized along the i-axis

$$\Delta\, n_i = -0.5 n^3 p_{ij} e_j \, . \tag{4}$$

In an alternate formulation in terms of the stress σ_i, (2) can be rewritten as

$$\Delta\, n_i = -C_{ij} \sigma_j \, , \tag{5}$$

where C_{ij} is the stress-optic tensor [40]. In the case of isotropic materials only tensor elements p_{11} (or C_{11}) and p_{12} (or C_{12}) and elements related by permutation of the coordinate axes are non-zero. The stress-optic and strain-optic tensor elements are related by the mechanical constants of the materials: Poisson's ratio μ and Young's modulus E,

$$\begin{aligned} C_{11} &= \frac{n^3}{2E} \left[p_{11} - 2\mu p_{12} \right] \, , \\ C_{12} &= \frac{n^3}{2E} \left[(1-\mu) p_{12} - \mu p_{11} \right] \, . \end{aligned} \tag{6}$$

A glass film grown on a silicon substrate will be under biaxial stress in the plane of wafer ($\sigma_1 = \sigma_2$), but is under no stress in the surface normal direction ($\sigma_3 = 0$) since the upper surface is free. The stress induced refractive index changes for light polarized in the wafer plane ($\Delta\, n_1 = \Delta\, n_2$) and normal to the plane ($\Delta\, n_3$) will then have the form

$$\begin{aligned} \Delta\, n_1 &= -\left(C_{11} + C_{12} \right) \sigma_1 \\ \Delta\, n_3 &= -2 C_{12} \sigma_1 \, , \end{aligned} \tag{7}$$

and the corresponding birefringence due to the in plane stress is $\Delta\, n = \Delta\, n_1 - \Delta\, n_3 = (C_{12} - C_{11})\sigma_1$, where in the context of a planar waveguide film n_1 is the index for TE polarized light and n_3 is the index for TM polarized light. Taking the physical parameters for fused silica ($p_{11} = 0.121$, $p_{12} = 0.270$, $E = 70\,\mathrm{GPa}$, $\mu = 0.17$ [41]), and assuming a typical in-plane stress for a silica on silicon layer of $-300\,\text{MPa}$, (1) through (5) predict an optical birefringence of $\Delta n = -0.0013$ and a positive shift of 0.0025 in the overall refractive index (i.e. the average of the TE and TM indices). These shifts are small, but they are two orders of magnitude larger than the geometrical waveguide birefringence. The birefringence will produce an effective wavelength difference between TE and TM polarized light of $\Delta\lambda = 1.3\,\text{nm}$. In a standard WDM system based on the 100 GHz ($\sim 0.8\,\text{nm}$) channel spacing ITU grid, this is almost twice the channel spacing, rendering the device unusable for randomly polarized optical signals. The calculated Δn in this example is somewhat high compared with that actually measured in commercially available planar glass waveguides. However, measurements in our laboratory on waveguides from several vendors yield birefringence values scattered throughout the range from $\Delta n = -0.0004$ to $\Delta n = -0.001$. Without correction, these values will produce polarization induced channel wavelength shifts comparable to a full channel spacing.

The discussion so far has considered the case of a planar slab waveguide only. In silica devices the waveguide is actually in the form of a ridge or buried channel. In cases where the cladding layers have similar thermal expansion coefficients as the core glass layers, the procedure outlined above can still provide a good estimate of the TE-TM waveguide birefringence in buried channel silica waveguides.

From the previous discussion, it is obvious that controlling stress birefringence is one of the fundamental challenges in implementing glass waveguides based components for WDM systems. As WDM technology has evolved over the last decade, a number of approaches have been proposed and demonstrated. The simplest and most conceptually elegant approach is to interchange the polarization states of orthogonally polarized optical signals inside the devices. It was shown by *Takahashi* and *Hibino* [30, 42] that placing a zero-order half-wave plate in the waveguide array such that it intersects each array waveguide at the midpoint, polarization insensitive operation can be achieved. An AWG demultiplexer shown in Fig. 1 is symmetric about this midline, with the optical path length before and after polarization interchange exactly equal. The overall phase accumulation of any optical signal passing through an array waveguide with a wave plate will therefore be independent of initial polarization. This method of polarization compensation does add a number of additional process steps to cut a groove and insert and bond the wave plate. The groove and wave plate should be as thin as possible since the light diverges as it propagates across the wave plate gap, resulting in coupling loss at the opposite facet of the groove. In the original demonstration, the additional losses incurred by introducing the wave plate into 100 μm wide groove are about 5 dB [42]. More recently, by using a thin polyimide wave plate the total groove width has been reduced to only 18 μm, with a corresponding excess insertion loss of only 0.4 dB [30]. Despite these extra losses, the wave plate technique effectively decouples the polarization compensation problem from the material properties, and can be applied in any waveguide material system. The polarization interchange approach is effective in devices such as an AWG where there is a clear symmetry plane in the optical paths, and is now one of the accepted techniques in AWG manufacture. On the other hand, devices such as the echelle grating demultiplexer shown in Fig. 2 have no symmetry plane and other methods of polarization compensation must be used.

In devices based on buried channel waveguides, stress itself can be manipulated to correct for birefringence effects. By including the dopants B and P in silica glass (BPSG), the thermal expansion coefficient of the glass silica can be tailored to closely match that of silicon. One solution is then to use BPSG as the waveguide core material, so that the thermal stress in the core always remains close to zero [43,44]. Unfortunately in practice it has been found that BPSG of the appropriate composition has higher optical loss and may even degrade with time. BPSG can be equally effective though, when used as the waveguide cladding [11, 45] as shown in Fig. 8. In this case, as the wafer is cooled from the annealing temperature, the BPSG cladding applies a stress to the buried silica channel in the surface normal direction that is close to the stress induced by the silicon substrate. The net result is that while the isotropic compressive stress will still shift the index of refraction, the birefringence anisotropy can be largely eliminated. This technique has been referred to as stress balancing, and has been adopted as a manufacturable method for eliminating birefringence effects in glass AWG waveguides. The use of BPSG glass as the upper cladding is also useful because of the low melting point of BPSG glass. Since each cladding layer can be annealed and undergo reflow, good conformal coverage of the cladding is achieved and the final waveguide structure can be fully planarized. The stress balancing method can only be applied to buried channel waveguides, since the stress must be applied along the surface normal direction to balance the in-plane stress applied by the silicon substrate. Therefore, although it may be the most attractive approach for birefringence control in devices based on buried channel waveguide structures, it is not possible to apply stress balancing in devices involving a slab waveguide, such as the echelle grating demultiplexer.

Glass waveguides can be made birefringence free through the use of a birefringence compensating layer (BCL), as shown in Fig. 9 [46]. The BCL concept is an extension of the concept of form birefringence [15]. In an artificial material consisting of alternating layers of high and low index materials, the refractive index will be different for light polarized parallel and perpendicular to the layer plane. This anisotropy, known as form birefringence, has been used to artificially induce a strong birefringence in nonlinear waveguides to allow phase matching of wavelength conversion processes [47]. In the work of *Wörhoff* et al. [46], the form birefringence is used instead to eliminate waveguide birefringence. The BCL is a thin layer of silicon nitride with $n \sim 2$ placed in contact with the waveguide core. If the thickness and index of the

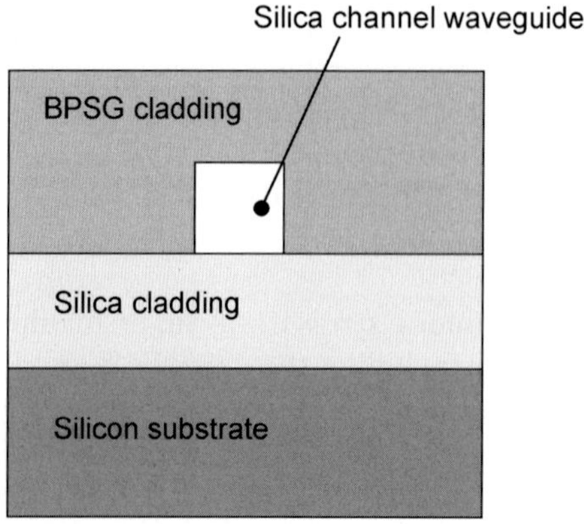

Fig. 8. Buried channel waveguide with a BPSG cladding for birefringence compensation by stress balancing

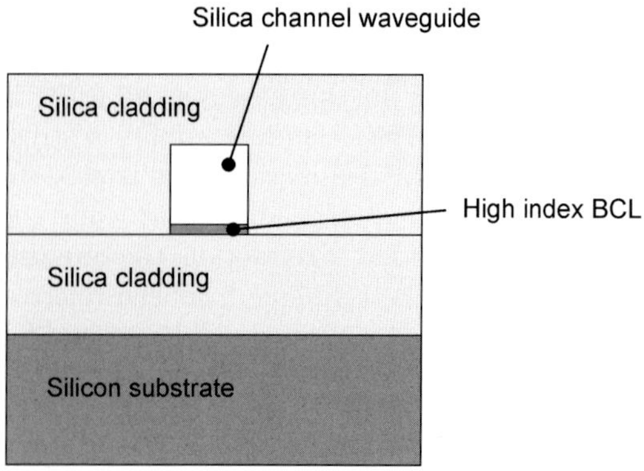

Fig. 9. Cross-section of a silica buried channel waveguide using a high index birefringence compensating layer [46]

BCL are correctly chosen, the resulting form birefringence can be used to cancel out the stress and geometrical components of the birefringence. The thickness of the BCL layer in [46] was approximately 26 nm to eliminate the birefringence of a 3 microm wide silica channel waveguide. An uncertainty of ±10 nm in the deposited BCL thickness can result in potential errors in demultiplexer channel wavelength shift of ±0.2 nm. However, the BCL method can be used to compensate both ridge and slab waveguides, and hence can be used to fabricate polarization independent AWG and echelle grating demultiplexers.

Other solutions to the waveguide birefringence problem involve direct changes to the structure of the waveguide itself. The geometrical contribution to the birefringence depends directly on waveguide aspect ratio, the birefringence being largest for a slab waveguide and zero for a waveguide of square cross section. In glass, manipulation of waveguide aspect ratio is of little use since the stress birefringence completely dominates the waveguide

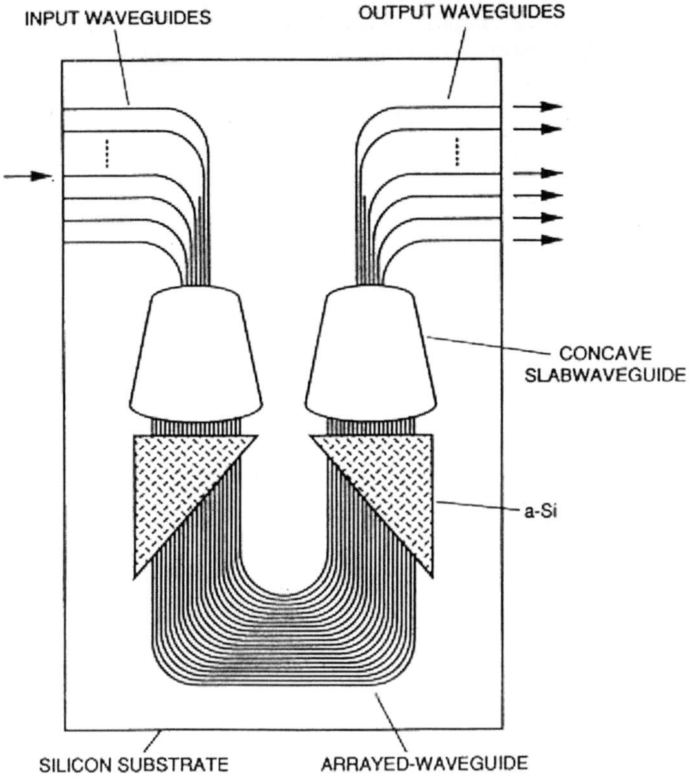

Fig. 10. AWG with two a-Si polarization compensation patches [48]

birefringence. However, aspect ratio can be a very useful tool for controlling birefringence in high index contrast waveguides.

It is not always necessary to eliminate birefringence. If the local waveguide birefringence can be modified it is often possible to design polarization independent devices using polarization compensation techniques. The first such polarization compensator was demonstrated on an AWG device shown in Fig. 10 [48]. The waveguide birefringence in the two triangular patches on the waveguide array is modified using an a-Si overlayer. This overlayer applies an additional stress to the underlying buried channel waveguides, modifying the birefringence. In an AWG with no compensator, the phase of TE polarized light in arriving at the output of array waveguide i is related to the phase of the light at the output of waveguide $i+1$ by a constant $\Delta\,\phi_{\mathrm{TE}} = k_0 n_{\mathrm{TE}}\,(L_{i+1} - L_i)$. Here n_{TE} is the effective index of the TE mode, k_0 is the vacuum wavevector, and $\Delta\,L = (L_{i+1} - L_i)$ is the length increment between array waveguides i and $i+1$. If the waveguide is birefringent, then the phase increment of the TM polarized light arriving at the array output will be $\Delta\,\phi_{\mathrm{TM}} = k_0 n_{\mathrm{TM}}\,(L_{i+1} - L_i)$. The difference between $\Delta\,\phi_{\mathrm{TE}}$ and

$\Delta\phi_{\mathrm{TM}}$ will cause the TM and TE polarized light to be diffracted in slightly different directions in the output combiner, and is the source of $\Delta\lambda$, the polarization dependent shift in demultiplexer channel wavelength. When the polarization compensator sections are added as in Fig. 10, each waveguide will have an unmodified length $L_i - L_i'$ with effective indices n_{TE} and n_{TM}, and a modified length L_i' with indices n_{TE}' and n_{TM}'. The compensator patch dimensions can be chosen such that the phase increment is polarization independent, i.e. $\Delta\phi_{\mathrm{TE}} = \Delta\phi_{\mathrm{TM}}$ for all waveguides in the array. Imposing this condition on the phases defines the lengths L_i' of the modified waveguides in the compensator section

$$L_{i+1}' - L_i' = \frac{\Delta n}{\Delta n - \Delta n'}\left(L_{i+1} - L_i\right) . \tag{8}$$

Here Δn and $\Delta n'$ are the effective index birefringences of the unmodified and modified waveguides, respectively. Compensators of this form have been demonstrated in silica on silicon AWGs [48] and also in InP/InGaAsP AWG devices [48]. Implementing this method requires a manufacturable way of changing the local waveguide birefringence enough that the intrinsic stress birefringence can be compensated. Techniques to achieve this for ridge waveguides in AWG include the deposition of a-Si films on the compensator section, to apply additional stress to the underlying ridge waveguides. Using this technique polarization dependent channel shifts less than 0.01 nm have been achieved in a silica demultiplexer [48].

The important advantage of the compensator approach is that it is the one technique that can easily be adapted to echelle grating demultiplexer devices. It has been shown be *He* et al. [50] that the compensator concept can be extended naturally to the slab waveguide section in an echelle grating spectrometer (see Fig. 2) and also to the combiner/splitter section of an AWG [51]. In both cases the formulation of the compensator shape is identical to (8) except that the lengths L_i and L_i' now refer to the path lengths of two rays traced from the input waveguide through two adjacent array waveguides to the output waveguide, rather than the length of the array waveguides alone. In effect the triangular compensator section has simply been moved from the array section as in Fig. 10 to the slab waveguide sections.

When originally demonstrated in InP/InGaAsP waveguides, the compensator birefringence was modified simply by etching away a thin layer from the top of the waveguide. This same approach can be used to form compensators in glass AWG devices or echelle grating devices. However, because the stress induced birefringence is so large, all of the upper cladding and part of the waveguide core must be removed to achieve full compensation. This in turn causes a significant waveguide mode mismatch at the boundary between the unmodified slab waveguide and the compensator. This mode mismatch contributes to the overall insertion loss of the device, and since the mode mismatch is itself polarization dependent, it also contributes to the polarization dependent loss (PDL). Hence an alternative method for modifying birefringence is necessary to compensate glass echelle grating devices. The solution we have found [28, 29] is similar to the BCL layer technique proposed by *Wörhoff* et al. [46]. Unlike the BCL layer, we use a very thin high index layer to shift the birefringence of the compensator slightly, but not eliminate it. The structure of the compensator is shown in Fig. 3. The upper silica cladding is etched until approximately 1 μm of material is left. This layer of silicon nitride with an index of $n \sim 2$ is deposited on top of the remaining cladding. As the light travels through the compensator the upper tail of the waveguide mode propagates in the silicon nitride layer. This interaction of the mode with the silicon nitride layer is polarization dependent, lowering the TM effective index relative to the TE effective index. The birefringence can be sufficiently enhanced that full polarization compensation can be achieved using a compensator of practical dimensions, as determined by (8). The critical feature of this compensator structure is that there is almost no change in the waveguide mode shape in the unmodified slab waveguide and the compensator waveguide. The addition insertion loss and PDL associated with this compensator structure are negligible. The polarization dependent wavelength shift $\Delta\lambda$ in our echelle grating devices described in Sect. 2 is routinely reduced from about 500 pm to less than 10 pm using this form of compensation.

4 Silicon-On-Insulator Waveguides

Although glass on silicon has evolved rapidly into a commercially viable waveguide platform, there has always been a strong interest in silicon compatible semiconductor waveguides because of the size reduction that could be achieved, and also the hope that silicon based active and passive optoelectronic devices could be integrated on chip. Although silicon-germanium waveguides were an early candidate in this role, the issues of lattice mismatch stress have so far curtailed its application for passive waveguides [52]. Silicon-on-insulator (SOI) wafers consist of a silicon substrate, a thin oxide layer on top of which is a thin single crystal layer of silicon. This top layer of silicon can act as a waveguide with the buried oxide layer providing vertical optical confinement. Photonic circuits can then be defined by etching the silicon to form ridge waveguides. SOI wafers fabricated by implantation of oxygen ions (SIMOX) were initially of poor optical quality [53]. Recently, new fabrication techniques have been developed to produce SOI wafers, and their availability and quality has improved tremendously. This is in large part due to the demand for SOI from the microelectronics industry for a variety of high-speed applications. The SOI waveguide system has now been demonstrated as a commercially viable platform for photonic integrated circuits [14]. Because of its high index contrast, it has also become an attractive platform for research in microphotonics and photonic crystal devices.

The technical issues encountered in creating a viable SOI waveguide technology are nominally similar to those that had to be met in the silica glass on silicon system – minimizing waveguide loss and controlling polarization effects. However, the engineering approaches can be very different. Glass waveguides are intrinsically multicomponent systems and achieving viable optical material requires good understanding and control over chemical composition and thermal processing.

The silicon guiding layer in SOI is a high purity single component material, and there are effectively no chemical or thermal processing degrees of freedom. The Si is also single crystal. Compared with glass technology, the intrinsic material uniformity is one of the advantages of the SOI platform. However, because of the high index contrast between the Si waveguide core and surrounding cladding (e.g. usually an oxide), details of waveguide geometry and dimension play a central role in determining the performance of SOI waveguides. For example in SOI the waveguide losses are largely determined by the roughness of the side-walls of the waveguide. Unlike silica glass on silicon, the waveguide layer in SOI has precisely the same thermal expansion coefficient as the substrate and there should be no stress-induce birefringence. However, because of the high index contrast between core and cladding the boundary conditions for the electric fields tangential and perpendicular to the core-cladding interface create a very strong geometrical birefringence. These effects become stronger as the waveguide size scales down to the microphotonic regime, where the modal electric field becomes quite large at the interface and even outside the waveguide core.

In the following, we again explore the critical issues in designing and fabricating high performance waveguides for WDM and other demanding applications, in particular as they relate to making microphotonic components, an area where SOI will most likely have a true niche. Research in SOI is less mature than in glass waveguides, but many of the design solutions reported to date closely parallel similar developments in glass waveguides.

4.1 Mode Control in SOI Waveguides

The issue of waveguide mode control arises in high index contrast materials systems. In silica waveguides, the core/cladding index step is usually so small that waveguides can easily be designed to be single mode. On the other hand, SOI waveguides larger than a few hundred nanometers in cross-section will support many modes. Each mode has a slightly different effective index, and therefore in a demultiplexer each mode will have a different effective wavelength. Thus when higher order modes are present, the output of each demultiplexer channel will have peaks or side lobes at a series of wavelengths corresponding to the modes in the demultiplexer waveguide array (or the slab waveguide in the case of an echelle grating demultiplexer). The peak corresponding to the fundamental mode is at the design channel wavelength, while the ghost peaks contribute to cross-talk. Side lobes due to higher order modes have been observed in SOI demultiplexer spectra [54]. The amplitude of these side lobes will depend on the details of mode conversion processes occurring in the device.

Higher order slab modes in the combiner section of an AWG will create ghost peaks shifted slightly from the channel center wavelength. The relative shift $\Delta\lambda_{01}$ in peak wavelength λ_m for channel m, between light in the fundamental mode and first higher order mode, will be proportional to the difference in their effective indices and the difference between λ_m and the AWG center wavelength λ_0, i.e. $\Delta\lambda_{01} = (\lambda_m - \lambda_0)(n_0 - n_1)/n_1$. Assuming the usable bandwidth of a typical demultiplexer covers the C-band, this relation predicts that if excited, the second mode could introduce a ghost peak separated by approximately 0.1 nm from the peak wavelength, assuming typical SOI slab waveguide thickness of a few µm. This is much less than the standard 100 GHz (0.8 nm) channel spacing but be enough to cause noticeable distortions in the channel transmission spectrum.

On the other hand, if the fundamental and first modes are also present in the array waveguides, the modal wavelength shift may be estimated using the relation $\Delta\lambda_{01}/\lambda_0 = (n_1 - n_0)/n_0$. In this case modal ghost peaks will occur at a much larger shift of 10 nm or more from the design wavelength, and severely degrade the nonadjacent channel cross-talk of the device. This large wavelength shift would also occur in the case of an echelle grating demultiplexer, where the dispersive section is a slab waveguide. It may be attractive to fabricate an echelle grating demultiplexer in SOI, given the small sizes that can be achieved. However, given the above considerations, it is clear that some form of mode suppression will be a key element in the design of such devices.

Considerable attention has been given to the issue of higher order mode suppression in SOI waveguides [55, 56, 57, 58]. Although slab waveguides will support a number of slab modes, SOI ridge waveguides can be designed to have a width and an etch depth such that only the fundamental mode is supported. It has been shown that ridge waveguide will be single mode if the waveguide width W, height H, and ratio r of lateral cladding thickness h to ridge height H (see Fig. 11) satisfy the relationship

$$\frac{W}{H} \leq 0.3 + \frac{r}{\sqrt{1-r^2}} . \tag{9}$$

Complying with this condition is straightforward for ridge waveguides used in SOI AWG demultiplexers.

As will be discussed in the next section, modifying ridge profile is another method to achieve zero polarization birefringence. As noted by *Pearson* et al. [59], it may not be possible to simultaneously satisfy the single mode condition and the requirement that the waveguide have zero birefringence. When competing requirements of polarization insensitivity and minimizing bend loss make it difficult to satisfy the single mode relationship of (9), it

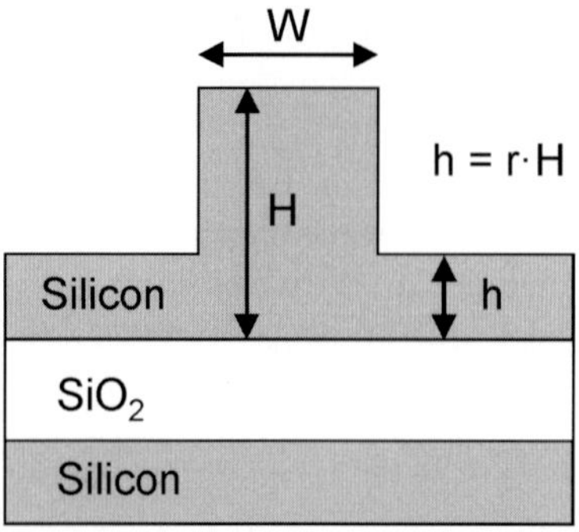

Fig. 11. SOI waveguide cross-section illustrating the critical dimensions for determining the single mode condition

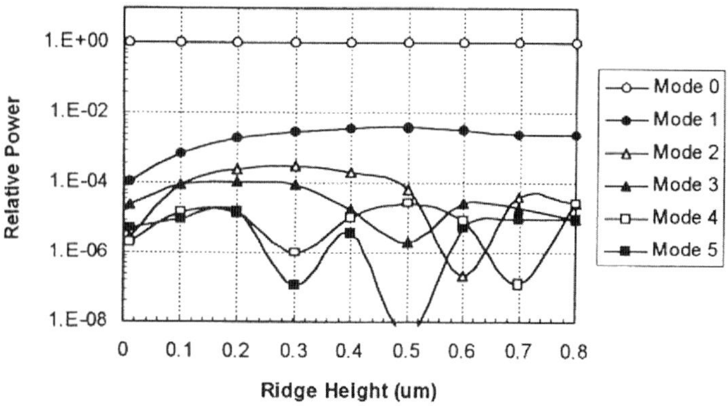

Fig. 12. Coupling efficiency from the fundamental ridge waveguide mode to the fundamental and higher order slab waveguide modes in a 1.5 μm thick SOI structure [59]

is still possible to work with multimode waveguides provided mode filtering is used. *Pearson* et al. [59] showed that in an SOI based AWG, waveguides supporting higher order modes may be used when the waveguide bend radii are small enough that the losses for the fundamental mode are negligible while losses for the second and higher order modes are extremely large. In this case only the fundamental mode will contribute to the AWG channel spectrum.

At present there are no known methods of suppressing higher order modes in a slab waveguide made out of high-index contrast material such as SOI. A multimode slab combiner section of an AWG can be tolerated provided that most of the optical power is coupled into the fundamental mode of the combiner at the point where the ridge waveguide joins the slab section [59]. Figure 11 shows the calculated power coupled to the first few slab modes from a single mode ridge waveguide, in 1.5 μm thick SOI. The ratio of power coupled into the second slab mode to that in the fundamental mode will be less than −20 dB. Since light is coupled to and from the slab by single mode

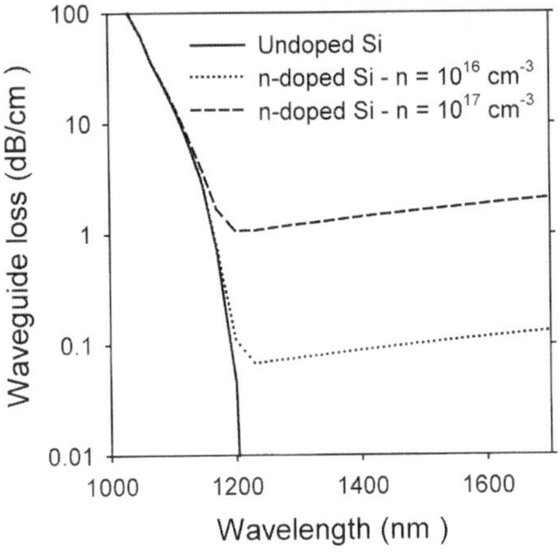

Fig. 13. Optical loss coefficient including interband absorption and free carrier absorption in intrinsic and doped silicon

ridge waveguides, this suggests that in this case the contribution of the second mode to the AWG output spectrum will be suppressed by more than −40 dB.

4.2 Losses in SOI Waveguides

Intrinsic loss in undoped single crystal silicon is very low for wavelengths longer than the absorption edge near 1100 nm, and reported losses in large cross-section processed waveguides are as low as 0.1 dB/cm in the near infrared [60, 61]. In the case of doped silicon waveguides, free carrier induced absorption varies linearly with the density of electrons and holes [62]. Figure 13 shows the calculated absorption due to interband transitions and doping in silicon. The loss due to free carriers does not become significant until carrier densities much higher than 10^{16} are reached. Although doping is not an issue for passive waveguide devices, in cases where active functions such as variable optical attenuators (VOA) [63] are integrated on chip, the overlap of the waveguide mode with doped areas must be minimized.

The more important source of loss in SOI waveguides is scattering from roughness at the waveguide cladding interface. The roughness can be qualitatively characterized by a root-mean square roughness σ, a measure of the typical height of the asperity, and the correlation length l_c, which is a measure of the width of an asperity. *Payne* and *Lacey* [37] derived a relation for the upper bound of the scattering loss coefficient for a waveguide of core index n_1 and width d

$$\alpha \leq \frac{\sigma^2}{k_0 d^4 n_1} \kappa \,. \tag{10}$$

Here κ depends on the details of the waveguide geometry and the statistical distribution of the roughness, but is of order unity for most practical waveguide geometries. This relation predicts that for any given roughness,

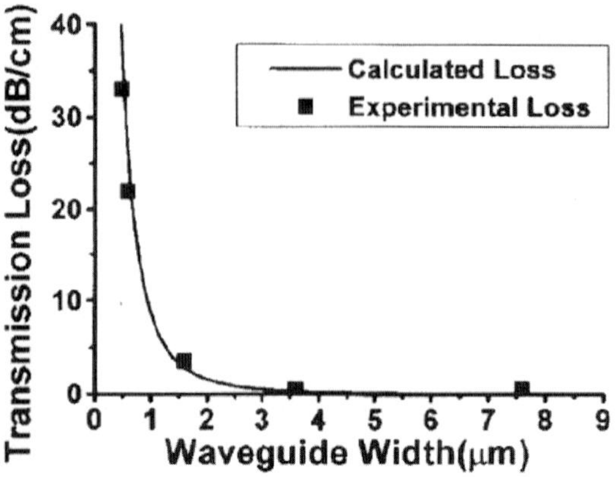

Fig. 14. Calculated and measured variation of waveguide loss with ridge width for silicon-on-insulator waveguides [64]

the scattering loss will increase rapidly as waveguide dimensions are reduced. Reducing scattering loss will be one of the major challenges to overcome in developing high performance microphotonic components. With modern dry etching tools such as reactive ion etching, rms roughness and correlation lengths of the order 100 nm or less may be achieved. Figure 6 shows a comparison of loss for the case of a 5 μm square glass waveguide and a 5 μm square SOI waveguide, and illustrates the importance of scattering loss in SOI. As the waveguide dimensions are scaled down the loss due to roughness scattering becomes very large [37, 64], as illustrated by the calculations of *Lee* et al. shown in Fig. 14 [64]. Unfortunately, it may not be straightforward to improve the side wall roughness that results during conventional lithography and dry etching. Thus if microphotonic waveguide and photonic crystal devices are ever to become practical, other means of creating smooth interfaces must be found.

By far the simplest approach to eliminating loss is to keep the waveguide dimensions so large (e.g. > 5 μm) that the side-wall scattering is negligible. For example, a four channel AWG demultiplexer has been fabricated in the SOI platform using waveguide cross-sections with dimensions of approximately 5 μm [54]. On-chip insertion loss for the 2.7 cm long device is less than 6 dB. More recently commercial SOI based AWGs with much higher channel counts have been developed [63], with performance approaching that of silica glass demultiplexers. However, the reliance on large waveguide size means the devices must also be large – in fact comparable to the size of silica AWGs – and therefore abandons the promise of a silicon-based microphotonic platform for very high levels of integration. Pursuing this promise is one motivation for the ongoing search for improved waveguide fabrication techniques.

Dry etching techniques could also be abandoned altogether in favor of using wet etches for waveguide definition [64]. Lee et al. have reported that wet etching can reduce the rms roughness of a silicon side-wall from 10 nm to 2 nm. Most wet etches are anisotropic. Since the (111) set of planes usually have the slowest etch rate, wet etching often results in the preferential formation of smooth (111) facets. Thus wet etching is highly constrained with respect to the device geometries that can be fabricated. For example, if an anisotropic wet etch process is used to fabricate curved ridge waveguides, the width and profile of the ridge will vary with the angle of the waveguide relative to the crystal planes.

Given that dry etches are necessary to create complex waveguide circuits, roughness may be removed after ridge definition. One approach is to remove the residual roughness using a thermal oxidation [65]. Once the silicon waveguide has been etched using standard processes, the silicon is oxidized at high temperatures in steam. As the oxide forms, silicon is consumed faster in asperities than on flat planes, so that the oxide-silicon interface becomes flatter with increasing oxidation time. Initial results show that this method produces sidewalls with an rms roughness σ of less than 2 nm, comparable to those formed by wet etching. However more recent results also suggest that the degree to which roughness decreases saturates with increasing oxidation time, to σ values near 0.5 nm [66]. This limit may be sufficiently small that for total propagation lengths envisaged in microphotonic components, the scattering losses are acceptable.

To conclude this section, the control of scattering losses at the core-cladding interface is a key step in scaling down high performance optical components base on high index contract systems such as SOI. Preliminary work has demonstrated that methods for eliminating roughness are available, but these have not yet been proven on manufacturable working devices such as compact demultiplexer chips. The polarization dependence of sidewall scattering has not been extensively addressed theoretically or experimentally, and may play a role in determining device PDL in high index contrast waveguides.

4.3 Polarization Dependent Properties of SOI Waveguide Devices

Turning attention once again to dispersive devices specifically for WDM and other high performance applications, as in silica the control of waveguide birefringence is essential. Unlike silica on silicon, there is only minimal built-in stress in the SOI waveguiding layer, since the Si waveguide core has the same mechanical properties as the Si substrate. Hence the primary source of birefringence in SOI waveguides is the geometrical contribution. Figure 15 shows the calculated variation of the geometrical waveguide birefringence of an SOI slab waveguide with thickness. Two curves are shown, one corresponding to the effective index birefringence Δn and the second to the waveguide group index birefringence Δn_{g}. As noted by *Vivien* et al. [58], in the case of SOI

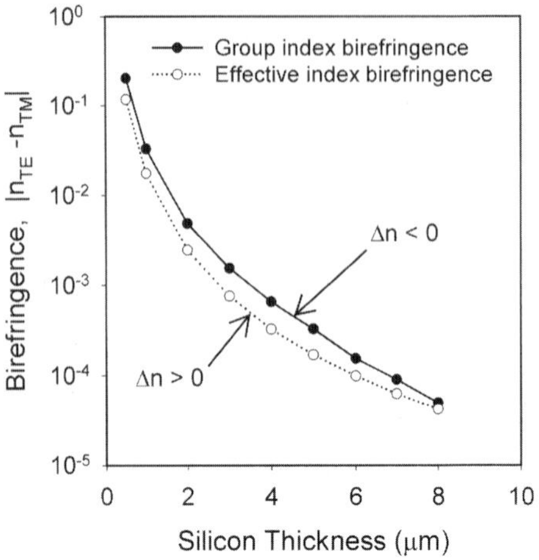

Fig. 15. Variation of the effective index birefringence $\Delta\, n$ and the group index birefringence $\Delta\, n_{\mathrm{g}}$ with silicon thickness for an SOI slab waveguide at $\lambda = 1550\,\mathrm{nm}$

the group index birefringence is opposite in sign and slightly larger than $\Delta\, n$. The polarization dependent wavelength shift $\Delta\, \lambda$ depends on the effective index birefringence, even though the dispersion of a spectrometer will depend on the group index. The effective index birefringence can cause a wavelength shift $\Delta\, \lambda = \lambda_0(\Delta\, n/n_{\mathrm{g}})$, while in the worst case the group index birefringence will cause a shift $\Delta\, \lambda = \lambda_{\mathrm{BW}}(\Delta\, n_{\mathrm{g}}/n_{\mathrm{g}})$, where λ_{BW} is the usable spectrometer bandwidth. This usable wavelength range is typically two orders of magnitude smaller than the wavelength. As a result, the group index birefringence will cause a much smaller change in any channel wavelength than the effective index birefringence. Compensation methods must therefore primarily counteract the polarization dependence of the effective index. Fortunately, the methods that minimize $\Delta\, n$ also suppress $\Delta\, n_{\mathrm{g}}$, although the conditions for zero birefringence will not be exactly the same [58].

Given that the polarization dependent channel wavelength shift in a WDM system demultiplexer should be of the order of $\Delta\, \lambda = 10\,\mathrm{pm}$ or smaller, $\Delta\, n/n_{\mathrm{g}}$ should be less than 10^{-5}. Based on the results in Fig. 15, an SOI slab waveguide must be of the order of 10 μm or thicker to meet this criterium. Since such a large thickness is not practical for most integrated optics applications, some form of polarization compensation is again required.

The high index contrast that is the source of the polarization birefringence in SOI waveguides also provides one method of compensating that birefringence. The simplest way to eliminate the geometrical birefringence is simply to make the waveguide as symmetrical as possible, for a perfectly square buried channel waveguide will have zero birefringence. In the case of SOI ridge waveguides, it has been shown that a ridge width and etch depth can usually be chosen for which either the effective index or group index are polarization independent [58, 59].

Since the dispersive element of an AWG consists of ridge waveguides, ridge shape can be used as a tool for polarization compensation. This has been

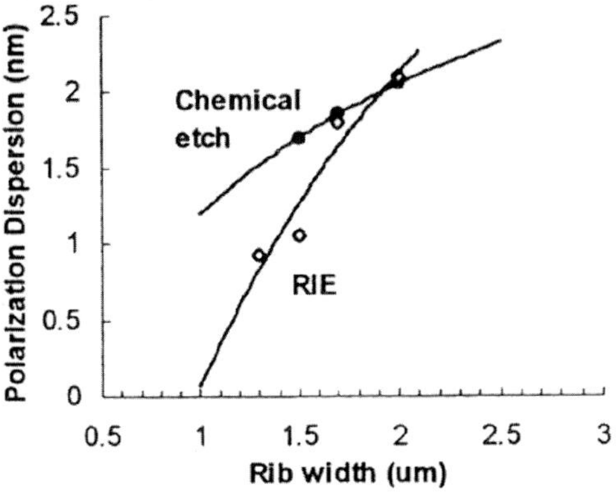

Fig. 16. Measured polarization dispersion (i.e. channel wavelength shift) in SOI AWG demultiplexers for different array waveguide rib widths. The ridge etch depth is 1 μm in all devices. Data is shown for sloped ridge sidewalls (wet etched) and vertical side walls (reactive ion etched) [59]

demonstrated in SOI by *Trinh* et al. in a four channel AWG [54] with a polarization dependent wavelength shift of $\Delta\lambda = 0.04$ nm and no other forms of compensation used. This AWG device was fabricated using 5 μm thick silicon waveguides, for which the birefringence is already small. This represents a factor of four reduction in birefringence compared with the 5 μm thick slab waveguide or a shallow etched ridge. Ridge waveguide compensation was also demonstrated by *Pearson* et al. in SOI microspectrometers based on 1.5 μm thick SOI [59]. Here, the measured channel shift $\Delta\lambda$ was reduced from the 4 nm expected for a slab waveguide, to $\Delta\lambda = 0.8$ nm. In both these examples full compensation was not achieved. One unresolved problem is that deeper etching would cause the array waveguides to become multimode, which would cause unwanted cross-talk. In many cases, single mode waveguides and polarization compensation cannot be achieved simultaneously for a given ridge profile.

Fabrication tolerance is another critical factor in compensating an AWG demultiplexer. Figure 16 shows the $\Delta\lambda$ measured at different ridge etch depths, for the SOI demultiplexer shown in Fig. 17. From the slopes of these curves, uncertainties in the array ridge widths of 100 nm lead to uncertainties in $\Delta\lambda$ of about 0.1 nm (or uncertainties in Δn of 2×10^{-5}). *Vivien* et al. [58] show that for similar 1.5 μm thick waveguides, uncertainties of only ±10 nm in ridge etch depth result in wavelength shift compensation errors of similar magnitude. These results suggest that while ridge profile engineering is effective, it may be difficult to control with sufficient accuracy. As discussed in the previous section on SOI waveguide losses, ridge profile is also a design parameter that determines whether an SOI ridge waveguide is single mode. Since the requirements of mode suppression and birefringence suppression may sometimes conflict, there is a need for other methods of polarization compensation with relaxed fabrication tolerances.

Our research in this area parallels that on glass waveguides. The first approach is to adapt the polarization compensator described in Sect. 3 (see Fig. 2), and previously used in the silica and InP echelle grating demul-

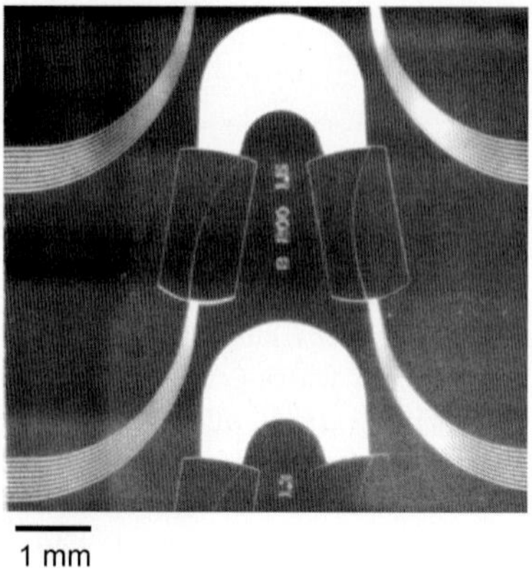

Fig. 17. SOI-based microspectrometer/demultiplexer fabricated using 1.5 µm thick Si waveguides, showing compensation regions in the input and output combiners [67]

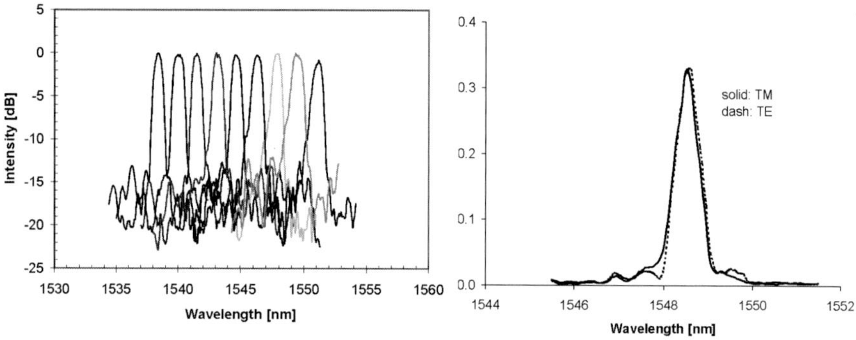

Fig. 18. The measured channel spectra for the demultiplexer shown in Fig. 17, and the superposed TE and TM spectra for one channel [67]

tiplexers. The use of compensators in SOI microspectrometers was demonstrated by *Cheben* et al. [67]. The device is shown in Fig. 17. It is an 8-channel 200 GHz spacing demultiplexer with overall device dimensions of approximately $0.5 \times 0.5\,\mathrm{cm}^2$. The AWG was fabricated using 1.5 µm thick SOI. The device pictured in Fig. 17 takes advantage of the fact that in an AWG complementary pairs of compensators can be placed in the input and output slab waveguide combiners. The birefringence change required in each compensator can then be half that for a single compensator, thereby reducing the mode mismatch loss that occurs at the compensator-slab boundary. The birefringence of the compensator sections was modified by a shallow etch. Etching the compensators did not result in any measurable deterioration in cross talk of the device. Before etching the compensator sections, the polarization dependent wavelength shift in this demultiplexer was $\Delta\lambda = 2.22\,\mathrm{nm}$. Depending on the etch depth used to define the compensator sections, $\Delta\lambda$ could be tuned from $\Delta\lambda = 2.22\,\mathrm{nm}$ to $\Delta\lambda = -3.65\,\mathrm{nm}$. The TE and TM channel spectra are shown in Fig. 18 for a compensated device with $\Delta\lambda = 0.04\,\mathrm{nm}$.

Tunability is a particularly useful feature of any compensator. As waveguide dimensions shrink, the uncompensated birefringence increases. At the same time, the wavelength shift $\Delta\lambda$ becomes more sensitive to the exact dimensions of the waveguide ridge profile and the compensator etch depth. Even with effective compensation methods, the challenge becomes to accurately cancel out the birefringence to give $\Delta\lambda < 10$ pm, given the tolerances of standard lithography and fabrication tools. Therefore it is an attractive option if one can measure the residual polarization dependence and correct it after device fabrication. As demonstrated in the case of the etched AWG compensator of Fig. 17, final correction for birefringence can be made simply by a calibrated wet etch. A change in etch depth of 1 nm corresponds to a shift in $\Delta\lambda$ of approximately 10 pm for the demultiplexer in Fig. 17. Although effective, it may therefore be difficult to reduce $\Delta\lambda$ to less than 10 pm using compensator etch depth due to the small tolerances.

Again in parallel to similar developments in glass waveguide technology, cladding induced stress is another tool that may be used to control polarization sensitivity in SOI demultiplexers. The use of BPSG as a cladding to balance the thermally induced stress in glass ridge waveguides on silicon has already been described. In SOI there is very little intrinsic stress in the waveguide core, but silicon dioxide is often deposited on top of SOI waveguides as a final protective cladding layer. This oxide will be under varying degrees of stress depending on the deposition technique and processing history. In most cases the SiO_2 film will be under compressive stress of the order of a few hundred MPa. In an SOI ridge waveguide, the oxide cladding can therefore create a stress distribution within the ridge, which will in turn create an anisotropic refractive index distribution in the waveguide. Figure 19 shows the calculated distribution of in-plane and out-of-plane stress inside a 2.2 µm high SOI ridge waveguide, with a 1 µm thick compressively stressed oxide layer [68]. The ridge itself is 1.83 µm wide and the etch depth is 1.34 µm. The overlying oxide layer is 1 µm thick. These calculations show that the oxide compresses the Si waveguide in the plane of the wafer, and applies a tensile stress along the surface normal direction.

A simple estimate of the expected stress birefringence can be made using a buried channel waveguide model [69]. We assume an Si channel waveguide embedded in an oxide under compressive stress. In this case the Si will simply be compressed in the in-plane direction and stretched in the surface normal direction. The stress in the Si and corresponding index anisotropy have been evaluated assuming an oxide with an in-plane stress of $\sigma = -380$ MPa. Using the photoelastic constants of silicon, (4) predicts changes in the demultiplexer channel wavelength of $\delta\lambda_{TE} = -0.2$ nm for TE polarized light and $\delta\lambda_{TM} = +2.2$ nm for TM polarized light. Here we make use of the fact that the stress induced channel wavelength shift is $\delta\lambda = \lambda_0(\Delta n/n)$. The oxide induced birefringence change is comparable and of opposite sign to the geometrical birefringence encountered in SOI slab waveguides, and can be used

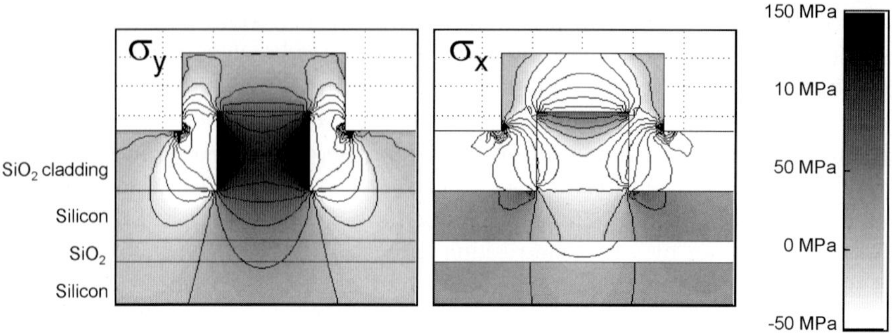

Fig. 19. Out-of plane (σ_y) and in-plane (σ_x) stress distribution in a 2 μm wide SOI ridge waveguide with a 1 μm thick compressively stressed SiO_2 cladding layer (tensile stress is positive, $\sigma_{x,y} > 0$) [68]

as a compensation tool. The effect of stress on real SOI waveguides has been measured using an AWG demultiplexer similar to the 8-channel device pictured in Fig. 17. The effect of oxide induced stress can be measured simply by measuring the change in TE and TM polarized peak channel wavelengths of an AWG as the oxide thickness is varied. The results of these measurements are presented in Fig. 20. For the case of 1 μm oxide upper cladding, the measured stress induced shifts in channel wavelength is $\delta\lambda_{\mathrm{TE}} = -0.4\,\mathrm{nm}$ for TE polarized light and $\delta\lambda_{\mathrm{TM}} = -1.2\,\mathrm{nm}$ for TM polarized light. The corresponding channel wavelength shift will be zero for an oxide thickness near 0.4 μm. These results are only in qualitative agreement with our simplified buried channel model, as would be expected given the complex shape of the realistic stress distribution in a ridge waveguide shown in Fig. 19. However, as in glass channel waveguide devices, these results demonstrate that overcladding induced stress in SOI waveguides can be a very effective tool for birefringence control.

The great advantage of using overcladding stress to compensate birefringence is that it is much easier to make fine adjustments after wafer fabrication, making it possible to correct process variations and hit stringent target polarization specifications at the device level. As the results in Fig. 20 suggest, in the simplest case the stress induced component of the birefringence can be adjusted by adding or removing oxide. The tolerances on the final target thickness of the overclad layer can themselves be increased or decreased by modifying the stress in the deposited oxide cladding. Alternatively the stress applied by the oxide layer may be adjusted using a calibrated thermal anneals. Measurements on PECVD grown oxides demonstrate that the compressive stress and hence the stress induced birefringence can be decreased by almost 50% using short anneals [68]. Corresponding changes in the demultiplexer channel wavelengths have been observed. Although the applications of stress compensation to high index contrast waveguide devices have not been fully explored yet, preliminary measurement and theory suggest that as

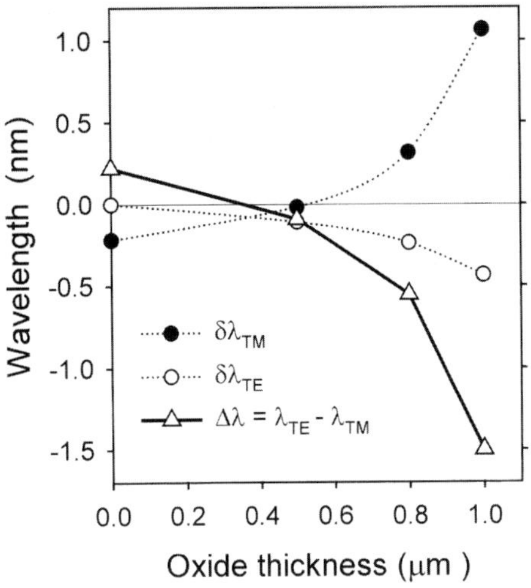

Fig. 20. The variations $\delta\lambda_{\mathrm{TE}}$ and $\delta\lambda_{\mathrm{TM}}$ in the measured AWG demultiplexer channel wavelengths with thickness of the stressed oxide overcladding, for TE and TM polarized light, and the corresponding change in birefringence $\Delta\lambda$

in the case of glass waveguide demultiplexers, cladding induced stress may likely become the preferred tool for polarization compensation is silicon microphotonic waveguide devices.

5 Summary

In this chapter, we have given an overview of some of the important material related design problems and solutions encountered in implementing high performance demultiplexers and spectrometers on silicon substrates. The first section covers the relatively mature subject of silica on silicon waveguide demultiplexers. The basic AWG and echelle grating demultiplexer designs are reviewed, and various approaches used to minimize loss and control the polarization birefringence are presented. In the second part of this chapter, we review progress in demultiplexers implemented in high-index contrast systems, in particular the silicon-on-insulator waveguide platform. Methods of achieving modal control, low waveguide loss, and polarization insensitivity are reviewed, illustrating where there are parallels with previous work in silica glass and where the properties of SOI demand unique solutions. The emphasis is on technical solutions that may enable the scaling down of high performance demultiplexers, spectrometers and other dispersive elements into the microphotonic regime. Such microphotonic spectrometers have applications in compact spectral sensors, optical network monitoring, wavelength channel multiplexing and demultiplexing, and will enable WDM as a low cost solution to parallel data transport inside the computer, between racks, and in smaller local area networks.

Acknowledgements

The author gratefully acknowledges P. Cheben, A. Delâge, D.-X. Xu, and W. N. Ye for the many useful discussions, proof reading, and assistance in the preparation of graphs and results presented in this chapter. The author also acknowledges the National Research Council of Canada for providing the opportunity and facilities to carry out research and development in the field of silicon photonics.

References

1. C. H. Henry, R. F. Kazarinov, H. J. Lee, K. J. Ostrowsky, L. E. Katz: Low loss Si_3N_4-SiO_2 optical waveguides on Si, Appl. Opt. **26**, 2621 (1987)
2. B. H. Verbeek, C. H. Henry, N. A. Olsson, K. J. Orlowsky, R. F. Kazarinow, B. H. Johnson: Integrated four-channel demuliplexer fabricated with phosphorous doped SiO_2 waveguides on Si, J. Lightwave Technol. **6**, 1011 (1988)
3. G. Grand, J. P. Jadot, H. Denis, S. Valette, A. Fournier, A. M. Grouillet: Low loss PECVD silica channel waveguides for optical communications, Electron. Lett. **26**, 2135 (1990)
4. E. S. Bulat, M. Tabasky, B. Tweed, C. Herrick, S. Hankin, N. J. Lewis, D. Oblas, T. Fitzgerald: Fabrication of waveguides using low temperature plasma processing techniques, J. Vac. Sci. Technol. **A114**, 1268 (1993)
5. M. I. Alayo, I. Pereyra, M. N. P. Carreno: Thick SiO_xN_y and SiO_2 films obtained by PECVD technique at low temperatures, Thin Sol. Films **332**, 40 (1998)
6. K. Wörhoff, A. Driessen, P. V. Lambeck, L. T. H. Hilderink, P. W. C. Linders, T. J. A. Popma: Plasma enhanced chemical vapour deposition silicon oxynitride optimized for application in itegrated optics, Sens. Act. A **74**, (1999)
7. M. Kawachi, M. Yasu, T. Edahiro: Fabrication of SiO_2-TiO_2 glass planar optical waveguides by flame hydrolysis deposition, Electron. Lett. **19**, 583 (1983)
8. N. Takato, K. Jinguji, M. Yasu, H. Toba, M. Kawachi: Silica based single mode waveguides on silicon and their application to guided wave optical interferometers, J. Lightwave Technol. **6**, 1003 (1988)
9. T. Kominato, Y. Ohmori, H. Okazaki: Very low loss GeO_2 doped silica waveguides fabricated by flame hydrolysis deposition method, Electron. Lett. **26**, 327 (1990)
10. M. Kawachi: Recent progress in silica-based planar lightwave circuits on silicon, IEE Proc.-Optoelectron. **143**, 257 (1996)
11. A. Kilian, J. Kirchoff, B. Kuhlow, G. Przyrembel, W. Wischmann: Birefringence free planar optical waveguide made by flame hydrolysis deposition (FHD) through tailoring of the overcladding, J. Lightwave Technol. **18**, 193 (2000)
12. C. H. Henry, G. E. Blonder, R. F. Kazarinov: Glass waveguides on silicon for hybrid optical packaging, J. Lightwave Technol. **7**, 1530 (1989)
13. U. Hilleringham, K. Goser: Optoelectronic system integration on silicon: waveguides, photodetectors, VLSI CMOS circuits on one chip, IEEE Trans. Electron. Devices **42**, 841 (1995)
14. R. Morris: Silicon-on-insulator integrated optic tranceivers, Proc. SPIE **3007**, 56 (1997)

15. M. Born, E. Wolf: *Principles of Optics*, edn. 7 (Cambridge University Press 1999)
16. R. März: *Integrated Optics: Design and Modeling* (Artech House, Norwood 1995)
17. E. S. Koteles: Integrated planar waveguide demultiplexers for highdensity WDM applications, in *Wavelength Division Multiplexing: A critical review*, SPIE CR71, (Bellingham, WA, USA 1999), pp. 3–32
18. R. Watanabe, K. Nosu: Slab waveguide demultiplexer for multimode optical transmission in the 1.0–1.4 μm wavelength region, Appl. Opt. **19**, 3588 (1980)
19. H. W. Yen, H. R. Friedrich, R. J. Morrison, G. L. Tangonan: Planar Rowland Spectrograph for fibre optic demultiplexing, Opt. Lett. **6**, 639 (1981)
20. Y. Fujii, J. Minowa: Optical demultiplexer using a silicon concave diffraction grating, Appl. Opt. **22**, 974 (1983)
21. G. H. Chartier, J. L. Coutaz, A. Girod, F. Jaussaud, O. Parriaux: Optical waveguides made by ion exchange in glass, J. Non-Cryst. Solids **47**, 259 (1982)
22. M. K. Smit: New focussing and dispersive planar waveguide component based on optical phased array, Electron. Lett. **24**, 385 (1988)
23. J. Capmany, C. Doerr, K. Okamoto, M. K. Smit: Introduction to the special issue on arrayed waveguide grating routers/WDM mux demuxs and related applications/uses, IEEE J. Sel. Top. Quant. Electron. **8**, 6, 1087–1089 (2002)
24. J. B. D. Soole, A. Scherer, H. P. Leblanc, N. C. Andreakis, R. Bhat, M. A. Koza: Monolithic InP/InGaAsP/InP grating spectrometer for the 1.48–1.56 μm wavelength range, Appl. Phys. Lett. **58**, 1949 (1991)
25. C. Kremer, G. Ebbinghaus, G. Heise, R. Muller-Nawrath, M. Schienle, L. Stoll: Grating spectrograph in InGaAsP/InP for dense wavelength division multiplexing, Appl. Phys. Lett. **59**, 627 (1991)
26. M. Fallahi, K. A. McGreer, A. Delage, I. M. Templeton, F. Chatenooud, R. Barber: Grating demultiplexer integrated with MSM detector array in InGaAs/AlGaAs/GaAs for WDM, IEEE Photon. Technol. Lett. **5**, 794 (1993)
27. J.-J. He, B. Lamontagne, A. Delage, L. Erickson, M. Davies, E. S. Koteles: Monolithic Integrated wavelength demultiplexer based on waveguide Rowland circle grating in InGaAsP/InP, J. Lightwave Technol. **16**, 631 (1998)
28. S. Janz, M. Pearson, B. Lamontagne, L. Erickson, A. Delâge, P. Cheben, D.-X. Xu, M. Gao, A. Balakrishnan, J. Miller, S. Charbonneau: Planar waveguide echelle gratings: an embeddable diffractive element for photonic integrated circuits, Tech. Dig. Optical Fiber Communication Conference (OFC 2002), OSA Trends in Optics and Photonics **70**, (Optical Society of America, Washington DC, 2002) pp. 69–70
29. S. Janz, M. Pearson, B. Lamontagne, L. Erickson, A. Delage, P. Cheben, D.-X. Xu, M. Gao, A. Balakrishnan, J. Miller, S. Charbonneau: The scalable planar waveguide component technology: 40 and 256 channel echelle grating demultiplexers, Integrated Photonics Research, Postconference Edition, IFE1.1, OSA Trends in Optics Photonics **78** (Optical Society of America, Washington DC 2002)
30. Y. Hibino: Recent Advances in high density and large scale AWG multi-demultiplexers with higher index contrast silica based PLCs, IEEE J. Sel. Top. Quant. Electron. **8**, 1090 (2002)
31. H. Kosaka, T. Kawashima, A. Tomita, M. Notomi, T. Tamamura, T. Sato, S. Kawakimi: Superprism Phenomena in Photonic Crystals: Toward Microscale Lightwave Circuits, J. Lightwave Technol. **17**, 2032 (1999)

32. C. N. Morgan, I. H. White, R. V. Penty: Improved performance of compact silica 2DIO wavelength filters, OSA Trends in Optics and Photonics **78**, Integrated Photonics Research, Postconference Edition, IFE4.1 (Optical Society of America, Washington DC, 2002)
33. K. Okamoto: Recent progress of integrated optics planar lightwave circuits, Opt. Quant. Electron. **31**, 107 (1999)
34. H. J. Lee, C. H. Henry, K. J. Orlowsky, R. F. Karazinov, T. Y. Kometani: Refractive-index dispersion of phosphosilicate glass, thermal oxide, silicon nitride films on silicon, Appl. Opt. **27**, 4104 (1988)
35. K. Wörhoff, P. V. Lambeck, A. Driessen: Design, tolerance analysis and fabrication of silicon oxynitride based planar optical waveguides for communication devices, J. Lightwave Technol. **17**, 1401 (1999)
36. M. V. Bazylenko, M. Gross, P. M. Allen, P. L. Chu: Fabrication of low-temperature PECVD channel waveguides with significantly improved loss in the 1.50–1.55 µm wavelength range, IEEE Phot. Technol. Lett. **7**, 774 (1995)
37. F. P. Payne and J. P. R. Lacey: A theoretical analysis of scattering loss from planar optical waveguides, Opt. Quant. Electron. **26**, 977 (1994)
38. J. K. Cramer and S. P. Murarka: Stress-temperature behaviour of electron cyclotron resonance oxides and their correlation to hydrogenous species concentration, J. Appl. Phys. **77**, 3048 (1995)
39. S. Janz, P. Cheben, H. Dayan, R. Deakos: Measurement of birefringence in thin film waveguides by Rayleigh scattering, Opt. Lett. **28** (2003) in press
40. M. Huang: Stress effects on the performance of optical waveguides, Int. J. Solids Structures **40**, 1615 (2003)
41. CRC handbook on laser science and technology (CRC, Boca Raton 1971)
42. H. Takahashi, Y. Hibino, I. Nishi: Polarization-insensitive arrayed waveguide grating wavelength demultiplexer on silicon, Opt. Lett. **17**, 499 (1992)
43. Y. Y. Chun, Y. T. Lee, H. J. Lee, S. J. Chung: Birefringence reduction in a high core boron-doped silica-on-silicon planar optical waveguide, J. Korean Phys. Soc. **29**, 140 (1996)
44. S. Suzuki, S. Sumida, Y. Inoue, M. Ishii, Y. Ohmori: Polarisation insensitive arrayed waveguide gratings using dopant-rich silica-based glass with thermal expansion adjusted to Si substrate, Electron. Lett. **33**, 1173 (1997)
45. S. M. Ojha, C. Cureton, T. Bricheno, S. Day, D. Moule, A. J. Bell, J. Taylor: Simple method of fabricating polarization insensitive and very low crosstalk AWG grating devices, Electron. Lett. **34**, 78 (1998)
46. K. Wörhoff, B. J. Offrein, P. V. Lambeck, G. L. Bona, A. Driessen: Birefringence compensation applying double core waveguiding structures, IEEE Phot. Technol. Lett. **11**, 206 (1999)
47. A. Fiore, S. Janz, L. Delobel, P. van der Meer, P. Bravetti, V. Berger, E. Rosencher, J. Nagle: Second-harmonic generation at $\lambda = 1.6\,\mu\text{m}$ in $AlGaAs/Al_2O_3$ waveguides using birefringence phase matching, Appl. Phys. Lett. **72**, 2942 (1998)
48. H. Takahashi, Y. Hibino, Y. Ohmori, M, Kawachi: Polarizatio-insensitive arrayed-waveguide wavelength multiplexer with birefringence compensating film, IEEE Phot. Technol. Lett. **5**, 707 (1993)
49. C. M. G. Vreeburg, C. G. P. Herben, X. J. M. Leitjens, M. K. Smit, F. H. Groen, G. M. van der Tol, P. Demeester: A low loss 16-channel polarization dispoersion-compensated PHASAR demultiplexer, IEEE Phot. Technol. Lett. **10**, 382 (1998)

50. J.-J. He, E. S. Koteles, B. Lamontagne, L. Erickson, A. Delâge, M. Davies: Integrated polarization compensator for WDM waveguide demultiplexers, IEEE Phot. Technol. Lett. **11**, 224 (1999)
51. J.-J. He, B. Lamontagne, E. S. Koteles: Polarisation compensated AWG demultiplexer fabricated in single shallow etching step, Electron. Lett. **35**, 737 (1999)
52. S. Janz, J.-M. Baribeau, A. Delâge, H. Lafontaine, S. Mailhot, R. L. Williams, D.-X. Xu, D. M. Bruce, P. E. Jessop, M. Robillard: Optical properties of pseudomorphic SiGe for Si based waveguides at the $\lambda = 1300$ and 1550 nm telecommunications wavelengths bands, IEEE J. Sel. Top. Quant. Electron. **4**, 990 (1998)
53. J.-P. Colinge: *Silicon-on-insulator technology* 2nd edn. (Kluwer Academic, Norwell 1997)
54. P. D. Trinh, S. Yegnanarayanan, F. Coppinger, B. Jalali: Silicon-on-insulator (SOI) phased array wavelength multi/demultiplexer with extremely low polarization sensitivity, IEEE Phot. Technol. Lett. **9**, 940 (1997)
55. R. A. Soref, J. Schmidtchen, K. Petermann: Large single mode ridge waveguides in GeSi-Si and Si-on-SiO_2, IEEE J. Quantum Electron. **27**, 1971 (1991)
56. S. P. Pogossian, L. Vescan, A. Vonsovici: The single mode condition for semiconductor rib waveguides with large cross sections, J. Lightwave Technol. **16**, 1851 (1998)
57. O. Powell: Single-mode condition for silicon waveguides, J. Lightwave Technol. **20**, 1851 (2002)
58. L. Vivien, S. Laval, B. Dumont, S. Lardenois, A. Koster, E. Cassan: Polarization-indpendent single-mode rib waveguides on silicon on insulator for telecommunications wavelengths, Opt. Commun. **210**, 43 (2002)
59. M. R. T. Pearson, A. Bezinger, A. Delage, J. W. Fraser, S. Janz, P. E. Jessop, D.-X. Xu: Arrayed waveguide grating demultiplexers in silicon on insulator, in *Silicon-based Optoelectronics II*, Proc. SPIE **3953**, 11 (2000)
60. A. G. Rickman, G. T. Reed, F. Namavar: Silicon-on-insulator optical rib waveguide loss and mode characteristics, IEEE J. of Lightwave Technol. **12**, 1771 (1994)
61. U. Fisher, T. Zinke, J. R. Knopf, F. Arndt, K. Petermann: 0.1 dB/cm waveguide losses in single mode SOI rib waveguides, IEEE Phot. Technol. Lett. **8**, 647 (1996)
62. R. A. Soref: Silicon-based optoelectronics, Proc. IEEE **81**, 1687 (1993)
63. I. E. Day, S. W. Roberts, R.O'Carroll, A. Knights, P. Sharp, G. F. Hopper, B. J. Luff, M. Asghari: Single chip variable optical attenuator and multiplexer subsystem integration, Tech. Dig. Optical Fiber Communications Conference (OFC 2002), OSA Trends in Optics and Photonics **70** (Optical Society of America, Washington DC 2002), p. 72
64. K. K. Lee, D. R. Lim, H.-C. Luan, A. Agarwal, J. Foresi, L. C. Kimmerling: Effect of size and roughness on light transmission in a Si/SiO_2 waveguide: Experiments and model, Appl. Phys. Lett. **77**, 1617 (2000)
65. K. K. Lee, D. R. Lim, L. C. Kimmerling, J. Shin, F. Cerrina: Fabrication of ultra-low loss Si/SiO_2 waveguides by roughness reduction, Opt. Lett. **26**, 1888 (2001)
66. D. K. Sparacin, K. Wada, L. C. Kimmerling: Oxidation kinetics of waveguide roughness minimization in silicon photonics, in *Integrated Photonics Research*, OSA Technical Digest **129** (Optical Society of America, Washington DC 2003)

67. P. Cheben, A. Bezinger, A. Delage, L. Erickson, S. Janz, D.-X. Xu: Polarization compensation in silicon-on-insulator arrayed waveguide grating devices, in *Silicon based Hybrid Optoelectronics III*, Proc. SPIE **4293**, 15 (2001)
68. D.-X. Xu, P. Cheben, D. Dalacu, S. Janz, M.-J, Picard, N. G. Tarr, W. N. Ye: Control and compensation of birefringence in SOI waveguides, Proc. 16th Ann. Meet. IEEE Lasers Electro-Optics Soc. (IEEE/LEOS 2003), p. 590–591
69. D.-X. Xu, P. Cheben, S. Janz, D. Dalacu: Influence of cladding material properties on SOI waveguide birefringence, Proc. CLEO-Pacific RIM 2003, paper TU3I-(8)-5

Light Modulation with Silicon Devices

Andrea Irace[1], Giovanni Breglio[1], Mario Iodice[2], and Antonello Cutolo[3]

[1] University of Naples "*Federico II*",
Department of Electronic and Telecommunication Engineerging,
Via Claudio, 21, 80125 Naples, Italy
{irace,breglio}@diesun.die.unina.it

[2] Istituto per la Microelettronica e Microsistemi – Sezione di Napoli – IMM-CNR,
Via Pietro Castellino, Napoli, Italy
iodice@irece.na.cnr.it

[3] University of Sannio, Corso Garibaldi, Benevento, Italy
cutolo@unisannio.it

Abstract. In this chapter we deal with different approaches to obtain fast and low power consumption amplitude modulation with silicon optoelettronic devices. After a description of the various physical effects that allow the interaction of material modification with travellling light we describe the optical configurations and different driving schemes that, so far, have been presented to the scientific community.

1 Introduction

Among optoelectronic materials, silicon is acquiring more and more importance for the fabrication of photonic integrated circuits. From its initial use as a mere substrate for the realization of silica-based devices, or as a motherboard for the assembly of lasers, photodetectors, and switches, the technologically best-known semiconductor has rapidly attained to the role of an active material. For example, silicon-on-silicon or silicon-on-insulator waveguides have been successfully fabricated, while a wide range of light modulators and switches, exploiting electro-optical, optomechanical, or thermo-optical working principles have been proposed.

Moreover, in the last ten years we are looking at new and very promising *weddings* between standard electronic and different aspects of others fields. In particular, the simultaneous use of electrical and optical signals on the same chip gives us the possibility to realize the so called optoelectronic integrated circuit (OEIC), characterized by high speed, high efficiency and electromagnetic immunity. Since Silicon is the basic material for electronics, in recent times it has been also tested in optoelectronic applications.

Near-infrared ($\lambda > 1.2\,\mu$m) silicon photonic integrated circuits (Si-PICs) have received great interest in the past few years. Even if III–V compounds are the most suitable candidates for fast switching purposes, the wide knowledge of silicon processing technologies makes silicon-based devices very attractive [1].

L. Pavesi, D. J. Lockwood (Eds.): Silicon Photonics, Topics Appl. Phys. **94**, 361–391 (2004)

It is, then, reasonable to foresee that the performance of Si-PICs will eventually surpass that of III–IV PICs in several areas. The principal areas of application of Si-based optoelectronics are wherever a simple low-cost optical device is required and when it is instrumental to realize the optical circuit and the electronic circuit onto the same substrate.

Even though silicon physical properties, such as indirect band-gap and low carrier mobility, may pose some obstacles to the success of optoelectronics circuits and systems, there are several market niches where silicon-based optoelectronics can play a crucial role. The motivation for this new vision have to be found in the need of cheap systems which can be integrated in consumer electronic apparatus, for communication and sensing purposes. In particular, for modulation applications, different configurations of two or three terminal [2] devices based on free carrier absorption [3] or interferometric configurations, such as Fabry–Perot [4] or Mach–Zehnder modulators, have been realized.

With several exceptions therefore most practical modulators are based on a shift in the real or imaginary refractive index of the material in which the optical field propagates. The induced index changes then often lead to variations in macroscopically measurable quantities, usually intensity or direction. We therefore start this chapter from the description of the physical effects that can cause these variations and after we move onto the description of the optical modulation mechanisms and finally the analysis of the most convenient electrical driving schemes.

2 Physical Effects

Since any optical modulator presented in literature is based on charge/field or temperature induced variation of the optical properties of silicon, before entering the details of the optical modulation schemes and the electrical driving options, some attention has to be paid to the physical mechanisms which underly the variation of refractive index and absorption coeffcient in this material.

Even if the totality of the optical modulators are based on carrier-induced or thermo-optical effect, also field-induced variation of refractive index and absorption coefficent will be dealt with since they have been theoretically predicted and experimentally observed.

2.1 Field and Charge-Carrier Effect

The complex refractive index of any material can be written as $n + ik$ where n is the refractive index and the imaginary part k is the optical extinction coefficient. k is related to α, the linear absorption coefficient, by the relation $k = \alpha\lambda/4\pi$ where λ is the optical wavelength. It is known that n and k are related by the Kramers–Kronig dispersion relations and the same relations hold for Δn and Δk as discussed below.

In Silicon the linear electro-optical effect is zero due to crystal symmetry while it has been known that the optical absorption spectrum of silicon is modified by external electric fields (the Franz–Keldysh effect) or by changes in the material's charge-carrier density. Starting with an experimental knowledge of the modified spectrum $\Delta\alpha(\omega, E)$ or $\Delta\alpha(\omega, \Delta N)$, *Soref* and *Bennett* have computed [5] the change in the index Δn.

The Kramers–Kronig coupling between Δn and $\Delta\alpha$ has been specified in several textbooks and journal articles, as follows:

$$\Delta n(\omega) = \frac{c}{\pi} P \int \frac{\Delta\alpha(\omega')\mathrm{d}\omega'}{\omega'^2 - \omega^2}, \qquad (1)$$

where P denotes the principal value of the integral. By injecting free carriers, absorption may be modified:

$$\Delta\alpha(\omega, \Delta N) = \alpha(\omega, \Delta N) - \alpha(\omega, 0) \qquad (2)$$

or α may be changed by an applied electric field:

$$\Delta\alpha(\omega, E) = \alpha(\omega, E) - \alpha(\omega, 0) \qquad (3)$$

2.1.1 Electric Field Effects

The Franz–Keldysh effect, which alters the α spectrum of c-Si, is field-induced tunneling between valence and conduction band states. In recent years, the generic term electroabsorption has been adopted for $\Delta\alpha$ versus E effects.

The electroabsorption spectrum at the indirect edge has been measured in detail by *Wendlund* and *Chester* [6]. Their experimental data are given in Fig. 1.

Soref and *Bennett* [5] have performed numerical caluclation to investigate electrorefraction end the results are reported in Fig. 2. Δn was then expressed as a function of optical wavelength from 1.00 to 1.60 µm. It is found that Δn is positive for $\lambda > 1.05$ µm, and that Δn is a strong function of wavelength. Starting at 1.3 µm, as λ is decreased towards the gap wavelength, Δn rises rapidly and reaches a maximum at 1.07 µm, a wavelength slightly below the nominal λ_g. Then, as λ decreases further, Δn decreases and becomes negative. They find at 1.07 µm that $\Delta n = 1.3 \times 10^{-5}$ for and external field of 100 kV/cm.

The change in index as a function of applied E field is plotted in Fig. 3 at the optimum 1.07 µm wavelength and at the nearby 1.09 µm wavelength.

Another "pure field effect", the Kerr effect, is present in Si. *Soref* [5] has estimated the strength of the Kerr effect in c-Si using the anharmonic oscillator model of *Moss* [7]:

$$\Delta n = -3e^2(n^2 - 1)\frac{E^2}{2nM^2\omega_0^4 x^2} \qquad (4)$$

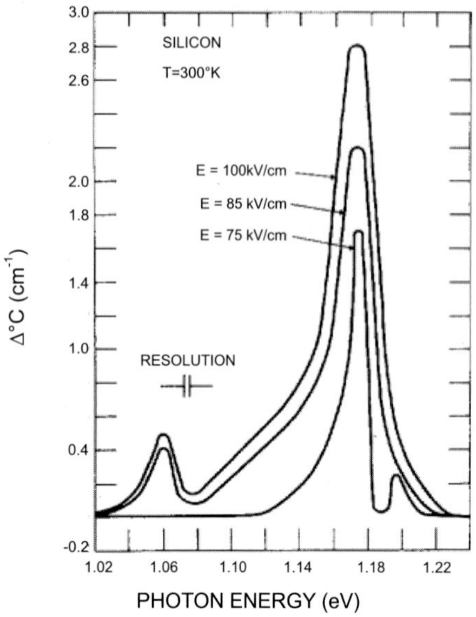

Fig. 1. Experimental electroabsorption spectrum of c-Si after *Wendland* and *Chester* [6]

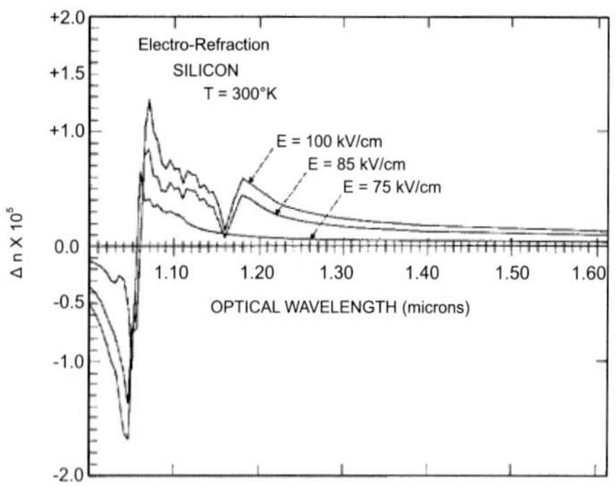

Fig. 2. Electrorefraction spectrum of c-Si [5]

here, e is the electronic charge, n is the unperturbed refractive index, M is the effective mass, ω_0 is the oscillator resonance frequency, and x is the average oscillator displacement.

2.1.2 Charge-Carrier Effect

The optical properties of silicon are strongly affected by injection of charge carriers into an undoped sample (ΔN) or by the removal of free carriers from a doped sample ($-\Delta N$). Optically, it does not make much difference whether carriers come from impurity ionization or from injection. Thus, an equivalence is assumed here.

Three carrier effects are important: 1. traditional free-carrier absorption, 2. Burstein–Moss bandfilling that shifts the α spectrum to shorter wavelengths, and 3. Coulombic interaction of carriers with impurities, an effect

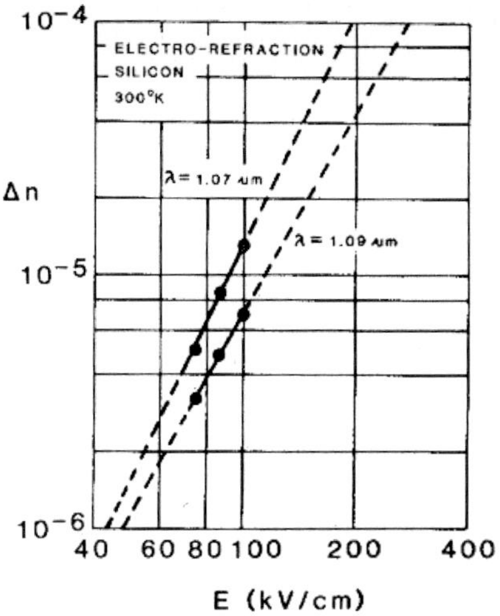

Fig. 3. Field dependence of electrorefraction [5]

that shifts the spectrum to longer wavelengths. These act simultaneously. What is experimentally observed is a redshift; thus, Coulombic effects are stronger than bandfilling in c-Si (see [8] and [9]).

For impurity concentrations in the $1 \times 10^{18}\,\mathrm{cm}^{-3}$ to $1 \times 10^{20}\,\mathrm{cm}^{-3}$ range, Schmid [8] has given detailed absorption spectra for n type and p-type silicon in both the transparent ($h\nu < E_g$) and opaque ($h\nu > E_g$) regions covering $0.6\,\mathrm{eV} < h\nu < 1.5\,\mathrm{eV}$. Spitzer and Fan [10] also present data for n-type material over a wide infrared range from 0.025 eV to 1.1 eV. They find a weak absorption band from 0.25 eV to 0.82 eV.

Using the above data, Fig. 4 and Fig. 5 are a composite drawing of optical absorption versus $h\nu$ for various concentrations ΔN of free electrons and holes. For the undoped material in Fig. 4, the data of *Dash* and *Newman* [11], *Spitzer* and *Fan* [10], and *Schmid* [8] has been used. Over the 0.25 eV to 0.82 eV range, *Schmid*'s doping curves have been extrapolated to follow the spectral shape of *Spitzer* and *Fan*'s data.

In n-type Si, for wavelengths longer than the 5 μm absorption band edge, the experimental results of [10] show that free-carrier absorption follows a λ^2 law out to $\lambda = 50\,\mu\mathrm{m}$ for samples with $8 \times 10^{16}\,\mathrm{cm}^{-3}$ doping. In addition, *Balkanski* and *Besson* [12] observe that a in n-type samples saturates or "levels off" over the 50 μm to 87 μm range. *Randal* and *Rawcliffe* [13] find that the $\alpha(\lambda)$ spectrum is flat from $\lambda = 100$ to 500 μm. For these reasons, Fig. 4 curves should include a leveling off of α in the far infrared, as well as the α-λ^2 middle-infrared behavior mentioned above.

Schumann et al. [14] have presented a curve of α versus ΔN in n-type Si at the 87 μm wavelength, a curve reproduced here as Fig. 6.

On this curve, the experimental result of [12] have been reported. Also plotted in Fig. 6 is a theoretical curve for p-type Si, a curve drawn through

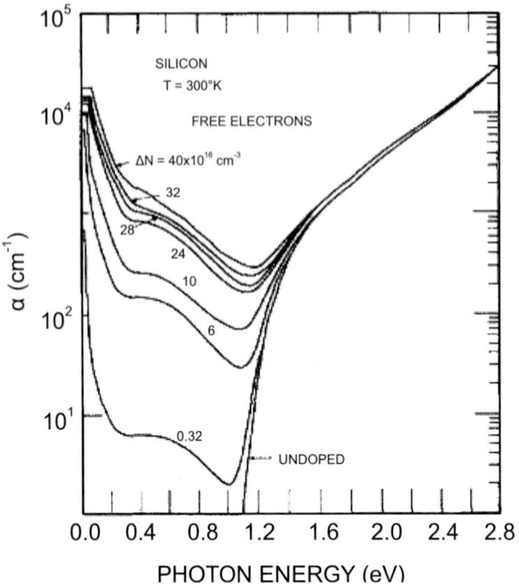

Fig. 4. Free electrons density dependence of absorption as a function of the photon energy after *Soref* and *Bennett*

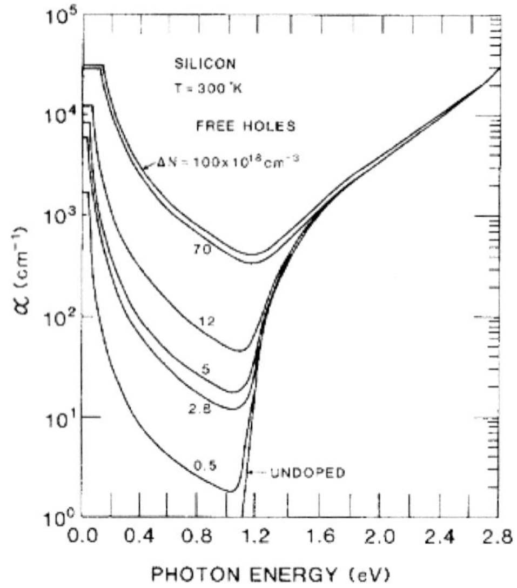

Fig. 5. Free holes density dependence of absorption as a function of the photon energy after *Soref* and *Bennett*

the two experimental data points of *Walles* and *Boija* [15,16]. The saturation of α at low frequencies is also consistent with the reflectivity measurements of *Schumann* and *Phillips* [17]. Above each plateau, an ω^{-2} extrapolation is used in Fig. 4. In the higher frequency portion of Fig. 4, ($h\nu > 1.2\,\mathrm{eV}$), the various curves are assumed to merge smoothly into the undoped curve as shown. The merger is complete above 2.8 eV. In Fig. 4, the range of integration used for the Kramers–Kronig inversion was 0.001 eV to 2.8 eV.

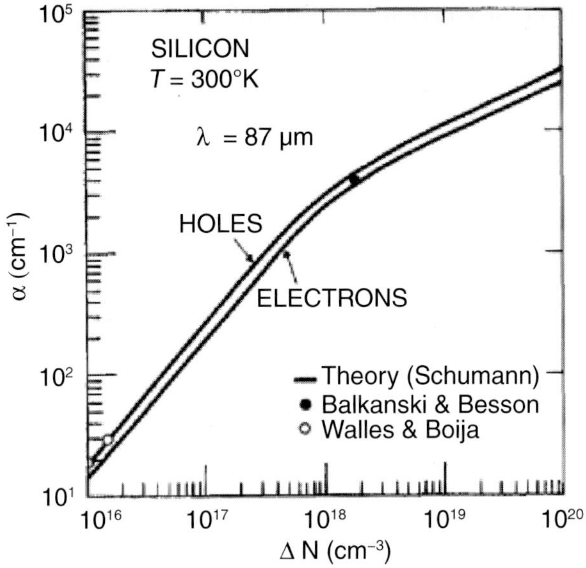

Fig. 6. Field dependenca of electrorefraction after *Soref* and *Bennett*

2.1.3 Drude's Model

The most used way to predict the change in refractive index and absorbtion coefficient is a simple free-carrier or Drude model of c-Si. The well-known formulas for refraction and absorption due to free electrons and free holes are as follows:

$$\Delta n = -\frac{e^2\lambda^2}{8\pi^2 c^2 \epsilon_0 n} \times \left(\frac{\Delta N_e}{m^*_{ce}} + \frac{\Delta N_h}{m^*_{ch}}\right) \tag{5a}$$

$$\Delta \alpha = \frac{e^3\lambda^2}{4\pi^2 c^3 \epsilon_0 n} \times \left(\frac{\Delta N_e}{m^*_{ce}\mu_e} + \frac{\Delta N_h}{m^*_{ch}\mu_h}\right) \tag{5b}$$

where e is the electronic charge, ϵ_0 is the permittivity of free space, n is the refractive index of unperturbed c-Si, m^*_{ce} is the conductivity effective mass of electrons, m^*_{ch} is the conductivity effective mass of holes, μ_e is the electron mobility, and μ_h is the hole mobility. An empirical expression for both Δn and $\Delta \alpha$ evaluation at $\lambda = 1.55\,\mu m$ has been reported by *Soref* and *Lorenzo* in [33] ans is recalled in the followng equations:

$$\Delta n = \Delta n_e + \Delta n_h = -\left[8.8 \times 10^{-22} \times \Delta N_e + 8.5 \times 10^{-18} \times (\Delta N_h)^{0.8}\right] \tag{6a}$$

$$\Delta \alpha = \Delta \alpha_e + \Delta \alpha_h = 8.5 \times 10^{-18} \times \Delta N_e + 6.0 \times 10^{-18} \times \Delta N_h\,. \tag{6b}$$

2.2 Thermo-optic Effect in Crystalline Silicon

The refractive index of the optical materials is not a constant parameter over the temperature range in which the materials, such as crystals, semiconductors, and glasses are used in different optical devices or systems. The variation of the refractive index with the temperature at a constant pressure is called the thermo-optic coefficient. It is denoted as $\mathrm{d}n/\mathrm{d}T$, where n and T are the refractive index and the temperature, respectively. Its unit is per degree centigrade or Kelvin. Normally, the value of $\mathrm{d}n/\mathrm{d}T$ is very small, of the order of $10^{-3}\,\mathrm{K}^{-1}$ to $10^{-6}\,\mathrm{K}^{-1}$. Although the value is quite small, this phenomenon can determine the incorrect functioning of devices where the wavelength dependence scales with the refractive index, like distributed feedback lasers, couplers, interferometers, and waveguides. It can be also exploited for the design of thermally activated devices, like switches, modulators, tunable lasers, wavelength demultiplexers, and filters.

The increasing attention to the optoelectronic use of silicon has clearly solicited further optical characterization efforts. Among the optical properties of this material, the temperature dependence of the absorption and the refractive index have obviously received wide attention, both in crystalline and amorphous silicon, at photon energies around and below the energy gap.

The direct extraction of $\mathrm{d}n/\mathrm{d}T$ from the measurement of the refractive index usually poses problems arising from the weakness of the effect. For this reason, some standard techniques, which may seem simple in principle, like those requiring the use of prism-shaped specimens, are often ineffective, or require in fact complicated experimental setups to give reliable results. Recently proposed techniques, on the other hand, allowing the indirect determination of the thermo-optic coefficient from the analysis of the temperature-dependent operation of diffraction-grating-based photonic devices, present the drawback of the ad hoc realization of complex integrated test structures.

Lautenschlager et al., for instance, have measured the complex dielectric function of Si [18] in the energy range 1.7 eV to 5.7 eV, and from 20 K to 850 K, by using an ellipsometric technique. In [19] the optical absorption for both crystalline and amorphous silicon was measured at $\lambda = 1.06\,\mu\mathrm{m}$ from room temperature to about 350 °C. Recommended values of n measured by many authors at various wavelengths and temperatures were collected in [20]. A simple linear dependence of n from T in the range between 25 °C and 750 °C was reported in [21] for crystalline silicon at photon energies above 0.8 eV. *Bertolotti* et al. [22] reported on the measurement of the refractive index in several semiconductors, including silicon, for $\lambda = 0.8\,\mu\mathrm{m}$ to $12\,\mu\mathrm{m}$. Unfortunately, these data were taken in the limited temperature range between 15 °C and 35 °C. A first physically meaningful model of $n(\lambda, T)$, whose validity extends to deep infrared radiation, was proposed in [23]. Indirect measurements of crystalline silicon thermo-optic coefficient at $\lambda = 1.55\,\mu\mathrm{m}$, and at room temperature, are reported in [24], returning a value of $2.0 \times 10^{-4}\,\mathrm{K}^{-1}$.

It is obvious that for more than twenty years researches are carried out for measuring thermo-optic coefficient, but only in recent times it is possible to find in literature direct characterization of the thermo-optic effect at typical telecommunication wavelengths of 1.3 μm and 1.55 μm, and theoretical models useful to explain it. In fact, a simple technique exists, based on the multiple-beam interference principle, which has demonstrated to give reliable results [25]. This technique is direct, and is based on the measurement of temperature variation necessary to cause a complete optical detuning in a Fabry–Perot filter. In its simplest form, this test structure is a one dimensional device, actually an etalon obtained by optical polishing on both sides of a chip made of the material under testing. A first direct measurement of this important parameter, based on this method, was carried out in [25]. The interferometric technique permitted also taking into account the etalon thermal expansion, producing a more accurate evaluation $\mathrm{d}n/\mathrm{d}T = (1.86 \pm 0.08) \times 10^{-4}\,\mathrm{K}^{-1}$. The authors note that thermal expansion effect contributes very little to the étalon detuning. In fact, the ratio between the phase changes induce by thermo-optic effect and by thermal expansion is about 20. In [26] authors intend to contribute to the available data pool with a large set of direct measurements of the thermo-optic coefficient $\mathrm{d}n/\mathrm{d}T$ in crystalline silicon from 300 K to 550 K. The reported values have been obtained at $\lambda = 1.55$ μm, one of the important wavelengths in fiber-optic communication applications.

Measurements in [26] were carried out on c-Si samples with different crystal plane orientations and thicknesses. In particular, the samples were taken from several (111), (100), or (110) wafers, with resistivities in the range of $10\,\Omega \cdot \mathrm{cm}$ to $100\,\Omega \cdot \mathrm{cm}$, indifferently p- or n-type. The wafers were previously lapped for thinning and then polished on both sides to an optical degree with a colloidal silica polishing suspension. The uncertainty in the measurement of the sample thickness was 3 μm. The values of $\mathrm{d}n/\mathrm{d}T$ are reported in Fig. 7 from room temperature to 550 K, distinguished for sample type.

The authors estimate that the plotted data can be affected by an 8% error due to uncertainties in the sample thickness, experimental setup alignment procedure, and in the actual sample temperature at each measurement step. It is worthwhile noting that, given this experimental error, no difference seems to hold for $\mathrm{d}n/\mathrm{d}T$ among the considered sample types. Moreover, in [26] it is also provided a polynomial interpolation of experimental data, which gives a weak quadratic dependence of $\mathrm{d}n/\mathrm{d}T$ from T expressed in kelvin:

$$\frac{\mathrm{d}n}{\mathrm{d}T} = 9.48 \times 10^{-5} + 3.47 \times 10^{-7} \times T - 1.49 \times 10^{-10} \times T^2 \quad \left(\mathrm{K}^{-1}\right) \tag{7}$$

A recent and comprehensive theoretical paper on physical modeling of thermo-optic effect is [27], where the analysis of the temperature dependence of the thermo-optic effect in silicon is reported, and the results obtained are compared with experimental data. The analysis is carried out in the technologically exploitable temperature range 300 K to 600 K, and centered at

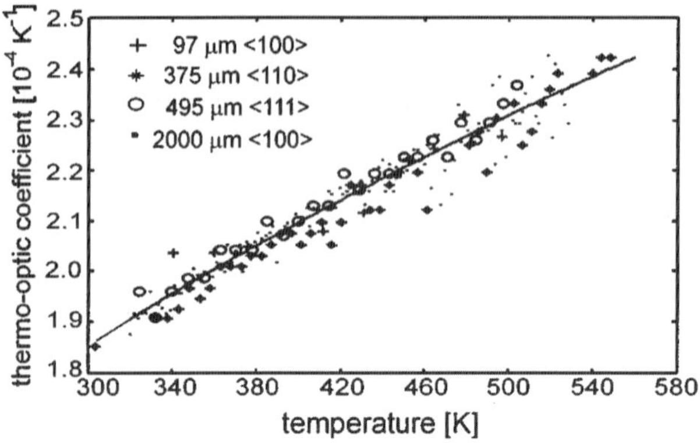

Fig. 7. Measured thermo-optic coefficient in various silicon samples. Interpolation of data as a second-order polynomial is shown as a *solid line* [25]

the principal optical communication wavelength of 1.55 μm. In particular, single and double oscillator approximations, previously proposed to model the refractive index and the thermo-optic coefficient frequency dispersions, have been extended to explain its temperature dependence in the cited temperature range. The reported approach highlights the role of the parameters determining the temperature dependence of the interband transition energies at the critical points of the combined density of states, whose values are also estimated by best fitting procedures. Such values are in good agreement with those reported in other works.

Using a single oscillator approximation, authors obtain:

$$\frac{\mathrm{d}n}{\mathrm{d}T} = \frac{n^2-1}{2n}\left[-3k_{\mathrm{ex}} - \frac{2}{E_g}\frac{\mathrm{d}E_g}{\mathrm{d}T}\frac{1}{1-(E/E_g)^2}\right] \tag{8}$$

where n and k_{ex} are both dependent on the temperature. Experimental data (in Fig. 8) are fitted by means of (8), expressing $\mathrm{d}E_g/\mathrm{d}T$ as the sum of two terms, at constant volume and at constant temperature respectively:

$$\frac{\mathrm{d}E_g}{\mathrm{d}T} = \left(\frac{\partial E_g}{\partial T}\right)_V + 3k_{\mathrm{ex}}V\left(\frac{\partial E_g}{\partial V}\right)_T \tag{9}$$

where, for semiconductors, the partial derivative with respect to the volume can be neglected.

Concerning the band-gap temperature dependence, we adopt the simple and widely quoted empirical relation given by:

$$E_g(T) = E_g(0) - \frac{\alpha_g T^2}{(T+\beta_g)} \tag{10}$$

where $E_g(0)$ is the average band-gap energy at 0 K, the constant α_g is related to the electron–phonon interaction, and β_g is related to the Debye temperature of the semiconductor. The thermal expansion coefficient k_{ex} of silicon is

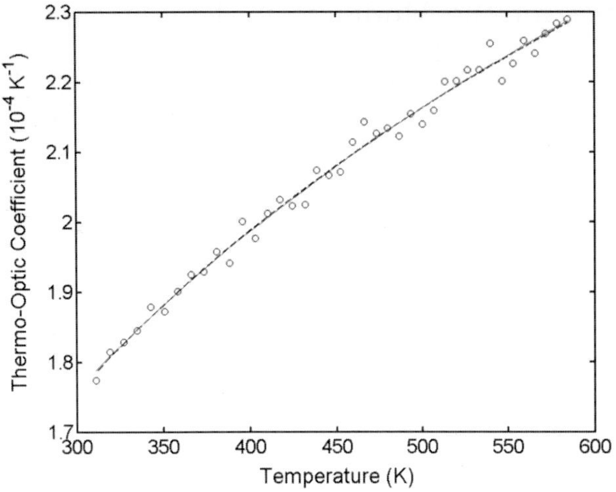

Fig. 8. Experimental results of the thermo-optic coefficient in silicon, and best-fit curves obtained with single (*dashed line*) and double (*solid line*) oscillator approximations. [26]

a well known parameter. The single oscillator fitting procedure involves the three floating parameters $E_g(0)$, α_g and β_g (see (8), (9), and (10)). Table 1 summarizes the parameter values resulting from the fitting procedure with the corresponding uncertainties. In the same paper [27] is discussed another theoretical model based on a semi empirical method for the calculation of the refractive index of semiconductors with diamond and zinc-blende structures at photon energies smaller than the band-gap energy [28]. Using this approach the authors have demonstrated that the thermo-optic coefficient of an indirect gap semiconductor can be written as a function of two elementary oscillators and, once again, it is possible to fit experimental data with a function of six floating parameters. In Fig. 8 the resulting best fitting curve is reported, which is not so different from the single oscillator model curve. The discussion of this model is beyond our purpose, and for that reason we refer to [27].

Table 1. Parameters involved in the single oscillator fitting procedure. The 95% confidence limit is given in parentheses

Single Oscillator	$E_g(0)$ (eV)	α_g $(1 \times 10^{-4}\,\mathrm{eV \cdot K^{-1}})$	β_g (K)
	4.03 (0.02)	3.41 (0.03)	439 (13)

3 Optical Schemes for the Modulation of Light

A modulator is a device with one input port and one output port, contrasted with the N optical inputs and M optical outputs of a switch. As light travels through the modulator, or is reflected off it, the optical amplitude and phase change in response to the electrical information signal. Modulation requires an active effect that modifies the complex refractive index of silicon by an amount $\Delta n + i\Delta k$, where Δn is the real index shift and Δk is the change in optical extinction coefficient. Typically, these changes are induced by plasma dispersion effect or thermo-optic effect as it has been discussed in detail in the previous sections. It is obvious that a modulator is characterized by the so called chirp factor $\Delta n/\Delta k$, which indicates the efficiency of the phase modulation compared to the pure amplitude modulation. The chirp factor for plasma dispersion effect in silicon decreases when the wavelength increases. With the exception of early works, where the authors have demonstrated the feasibility of bulk devices [29], in the last years researches were carried out exclusively on active components integrated in optical waveguides. Experimentally, three basic types of silicon host waveguides were utilized: lightly doped silicon on heavily doped silicon, silicon on insulator (SOI) and epitaxial $Si_{1-x}Ge_x$ on Si. We will present devices which are realized or proposed in the first two types of waveguide, excluding the last, because it is not a pure silicon-based technology. Together with the host waveguide classification, it can be useful to distinguish the modulator on the basis of the modulation behavior, that is if they mainly modify the phase or the amplitude of the optical signal, or, in other words, if the chirp factor is much higher than 1 or about equal to 1.

3.1 Free-Carrier Absorption Based Modulators

In this paragraph we address to the operating principle of silicon optoelectronic modulators based on the pure absorption induced by Plasma Dispersion Effect, that is devices where the chirp factor is about 1. This effect has been treated in the beginning of the chapter, where we have reported the losses variation vs. plasma density relation:

$$\Delta \alpha = \frac{e^3 \lambda^2}{4\pi^2 c^3 \epsilon_0 n} \times \left(\frac{\Delta N_{\mathrm{e}}}{m^*_{\mathrm{ce}}\mu_{\mathrm{e}}} + \frac{\Delta N_{\mathrm{h}}}{m^*_{\mathrm{ch}}\mu_{\mathrm{h}}}\right) \tag{11}$$

where α is the absorption coefficient and $\Delta N_{\mathrm{e,h}}$ is the variation of free carrier density.

For an integrated optoelectronic device, there are two ways to induce a variation of free carrier concentration into the active volume of the waveguide modulator, in order to modify the optical absorption of it: the injection in an undoped layer or the depletion from a highly doped one. The electronic device that can be integrated into the waveguide can be a p–i–n diode or a three terminal active device (either unipolar or bipolar).

The active electrical devices can be integrated in a low-loss Silicon-On-Insulator (SOI) waveguide (typically for p–i–n diode), or in a Silicon-On-Silicon waveguide (for the three terminal device). In what follows we will also discuss about different approaches in the electrical driving schemes.

In the late eighties, research groups have explored silicon as a material for optical interconnects [30]. All-silicon waveguides, for hosting the active device, were realized and characterized, but only recently these structures have reached performances useful for practical applications. In fact, in standard all-silicon waveguide, the high doping difference between n^{++} substrate and guiding layer causes the refractive index difference, which ensures vertical confinement for the infrared radiation propagating into the film. The lateral confinement can be provided by the etching of the top layer, so realizing the so-called *ridge* waveguide. The propagation losses for these first waveguides were in the range of 10 dB/cm to 15 dB/cm. An optimized version of the standard ridge waveguide can be designed [31] to show a single-mode operation, supporting only the first order modes for each polarization. The propagation losses of such a structure can be further reduced implanting a lower highly-doped cladding, reaching a value of about 1.2 dB/cm [32]. The second approach for the fabrication of silicon-based waveguides is characterized by very low optical losses (< 0.2 dB/cm), due to the good vertical confinement ensured by the presence of the non-absorbing SiO_2 bottom layer.

Pioneering papers in silicon optoelectronics were published by *Soref* et al. [5, 3, 33]. In these papers the authors have described the actual possibilities for silicon in the waveguiding and control of infrared optical radiation. After these preliminary proposals, several different solutions were reported in literature. Among the early works, it is useful to cite the paper by *Lorenzo* et al. [34], where a first practical application of the plasma dispersion effect is reported. In this proposal, an all-silicon waveguide hosts a X-switch, driven by a vertical p–i–n diode, working at current density of 1.26 kA/cm^2 and producing a contrast of about 3 dB. Of course, it is unrealistic to expect this unoptimized multimode device to have acceptable performances. The importance of such initial proposal, even if the subject is not a modulator in strictly sense, is in the novelty it suggests, explicitly affirming that the silicon, together with its microelectronic technology, can be used to transmit and manipulate optical signals.

In 1991 *Treyz* et al. [35] have proposed a simple absorption modulator based on the integration of a vertical p–i–n diode in an all-silicon waveguide. The performances of this device are strongly limited by the high propagation losses of the waveguide (> 20 dB/cm). The achieved modulation depth is 76% (-6.2 dB) for a driving current density of 3 kA/cm^2. Switching time is about 50 ns. In Fig. 9 is reported the transverse cross section of the modulator active region.

A similar proposal was published in 1994 by *Leistiko* et al. [36]. The modulator, realized by means of standard microelectronic techniques, showed high

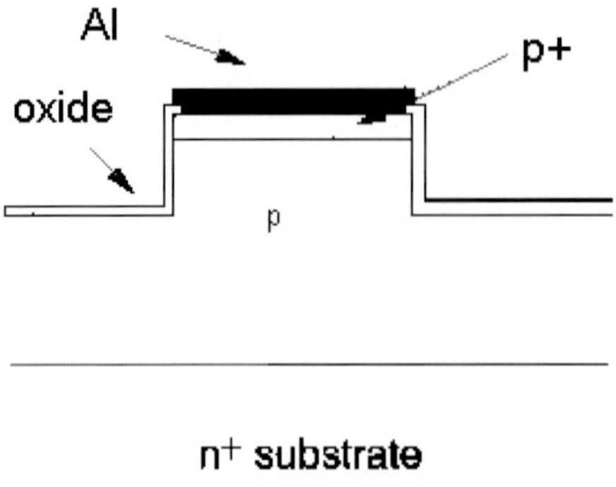

Fig. 9. Schematic diagram of the cross section of the active region for the modulator proposed in [35]

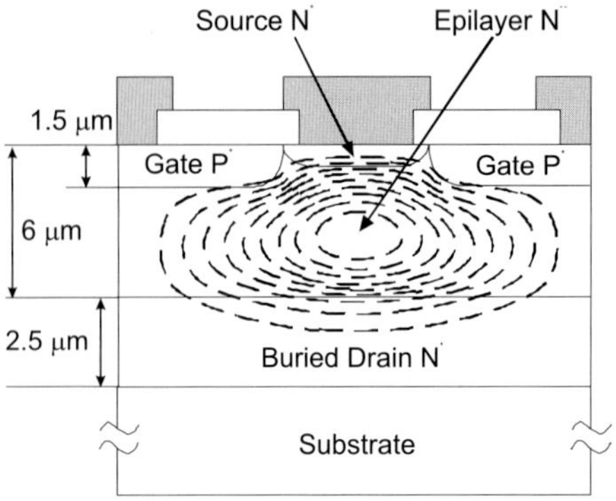

Fig. 10. Cross section of a BMFET optical modulator device

insertion losses, in the range 10 dB to 15 dB, on a total length of about 9 mm. The integrated p–i–n was formed by the n^{++} heavily doped substrate and two p^{+} diffused regions, forming also the optical channel. The measured modulation depth was nearly 100% (−30 dB) and the characteristic turn-on and turn-off time constant are approximately 20 μs.

A more efficient solution was proposed by *Cutolo* et al. [37]. A three terminal device is used to reach a high value of modulation depth with low power consumption, by exploiting phase shift and concurrent increasing optical absorption. Another advantage of such configuration is the lower value for the switching time constants. The cross section of the structure is sketched in Fig. 10.

This structure is capable, at the same time, to generate a complex refractive index variation, in order to modulate the infrared beam, and guarantee a good confinement of the optical radiation. The modulator can be considered as a 2-D optical channel waveguide, although not optimized for this purpose. From the electrical point of view, this device, to be a modulator, should be able to produce an electron–hole plasma into the central waveguiding region. Both surface highly doped regions, together with the n^{++} buried layer, carry out this function. The described structure is the elementary cell of a bipolar mode field effect transistor (BMFET). The two lateral p-type regions act as

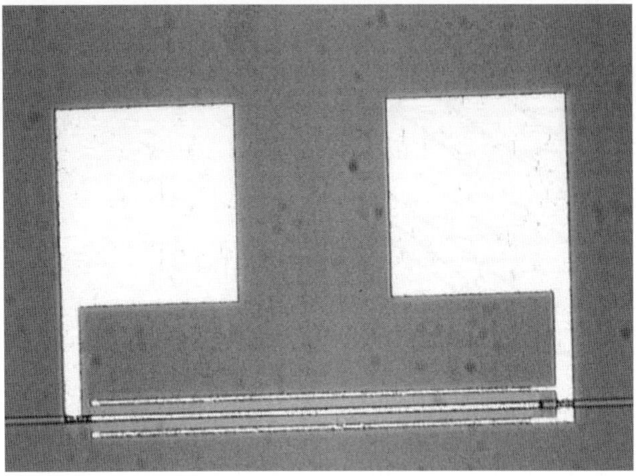

Fig. 11. Top view of a 500 μm long BMFET modulator

the gate, the central n-type region as the source, and the n^{++} buried layer plays the role of the drain.

In this kind of modulator the variation in the charge density, and consequently in the refractive index, can be much larger than carrier density resulting from injection of single polarity, like occurs in a standard FET. Moreover this structure, characterized by the presence of a control terminal connected to the n^{++} buried layer, is preferred to the simple p–i–n diode because of the use of the third electrode which allows a more accurate spatial control of the injected plasma and higher switching frequencies, thanks to the possibility to quickly control the plasma distribution into the channel by using an appropriate bias, as will be described afterwards. The predicted performances are, for a 1000 μm long amplitude modulator, M = 20% (limited to a useful value, even if in principle it is possible to reach 100%), switching power of 27.8 mW, and a switching time of about 5.6 ns.

The top view of a realized BMFET modulator is reported in Fig.11.

It is important to notice that the integration of a three terminal active device in a silicon-based waveguide force the usage of an all-silicon host structure, because of the need to realize a buried junction to control the optical characteristics of the channel, despite the higher propagating losses than a SOI waveguide. Moreover, an all-silicon waveguide is very cheap and easy to make, if compared to a SOI structure, and its thermal conductivity is very high. This fact allows a safe operation, because permits to better dissipate the heat generated during the operation, and reduces also the undesired refractive index variation produced by thermo-optic effect. In addiction, the homogeneous transverse cross section of an all-silicon modulator ensures an uniform temperature distribution in it, so avoiding undesired spatial refractive index variations, and consequently optical mode distortion, due to localized heating.

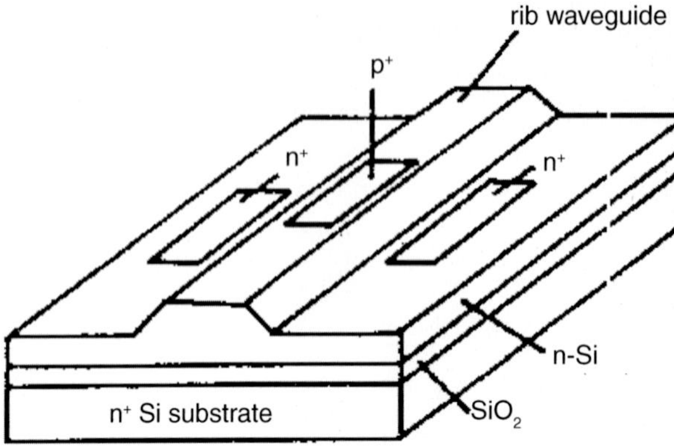

Fig. 12. Schematic of the modulator proposed in [39]

Preliminary experimental measurements carried out on this class of modulator are reported in [38], where authors show that a modulation depth of 75% can be achieved by injecting a constant hole current in the channel region from the gate ($I_{\mathrm{G}} = 10$ mA), thus generating a plasma of carriers, and biasing drain-source junction from 0 V to 16 V.

With the intention to reduce the propagation losses, *Zhao* et al. [39] proposed and realized the integration of a lateral p–i–n diode into a SOI ridge waveguide (see Fig. 12).

The modulator operation is based on the waveguide-vanishing effect, induced by free carrier injection together with usual free carrier absorption. The total insertion losses of the device are measured to be 3.65 dB, the modulation depth is 96% at an injection current of 45 mA. Response time is about 160 ns. The waveguide-vanishing effect is induced by the reduction of the refractive index in the waveguide core. This causes a lot of guided-mode energy to be lost and absorbed in the ridge waveguide, which will cause the guide itself to cut-off. Consequently, there is a concurrent effect of the plasma induces absorption and refraction.

3.1.1 Electrical Driving Schemes

Almost all the modulators presented so far rely on the electrical mechanism of free carriers injection using a p–i–n diode. This two-terminal device, from the electrical point of view, acts as a diffusion capacitor which has to be filled with carriers through a suitable current source.

On the other hand, when dealing with three terminal devices some discussion can be made about the optimal driving scheme to be exploited in order to guaranteed that good optical switching is not limited by slow electrical mechanisms. In order to shed some light on the latter assumption we refer to the paper of *Cutolo* et al. [37] where a three-terminal optical modulator has been analyzed in detail with respect to the electrical driving conditions.

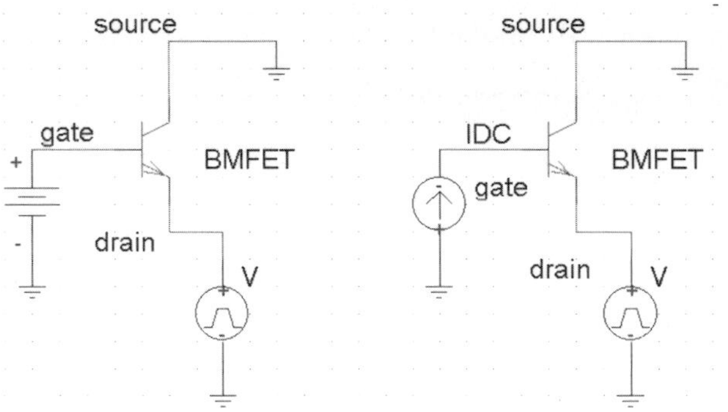

Fig. 13. Electrical driving circuit to switch optical BMFET device

They compare how it is possible to use a three terminal device as a p–i–n diode simply switching the current on and off on the GATE terminal. This way, though easy, results in very low switching characteristics because the epilayer has to be filled and depleted and little can be done about the storage of electrons and holes inside the epilayer after the removal of the GATE current [2, 37].

From steady-state simulations performed on a 1000-μm-long device, *Cutolo* et al. [37] have evaluated that, in order to reach the desired modulation depth, the epilayer must be filled with a concentration of electrons and holes of $5 \times 10^{16}\,\text{cm}^{-3}$. This, on its turn, is obtained by biasing the GATE terminal with a 0.1 mA/μm current. This information can be exploited in different ways in order to switch the optical signal off.

One possiblity to obtain faster performances is by fixing the DRAIN bias with a positive voltage source and then switching the gate control current on and off. They also presented the so-called *Plasma-Drift* driving scheme which has been also used by *Sciuto* et al. [38] to obtain their preliminary experimental results.

These are only two of the possible driving schemes and the first is the one which mostly resembles the switching of a bipolar transistor. In the *Plasma-Drift* configuration, the driving voltage is applied simultaneously to the GATE and to the DRAIN terminal (see Fig. 13). By doing so total charge present in the epilayer is fixed by the GATE current to the amount required to have the required modulation depth while the positive voltage pulse applied to the DRAIN depletes the central region of the waveguide. This results in a faster switching speed since it is not necessary to deplete the epilayer but it is just needed to remove the positive voltage from the DRAIN allowing the electron–hole plasma to redistribute uniformly across the optical channel. In addition to that the recombination process do no affect the switching speed in a relevant way.

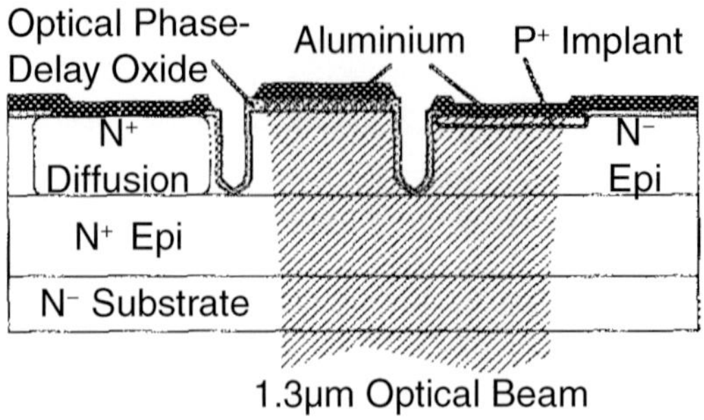

Fig. 14. Schematic cross section of the modulator proposed by *Hemenway* et al. [40]

3.2 Interferometric Structures for Optical Modulation in Silicon

In this paragraph we will discuss on pure phase modulators, $(\Delta n/\Delta k \gg 1)$, and modulators where the phase modulation is converted in amplitude modulation using an interferometric optical structure, external or internal to the modulator itself, working in the II and III window for fiber optic communications ($\lambda = 1.3\,\mu m$ and $1.55\,\mu m$). The physical effects exploited for the generation of a phase change are typically the plasma dispersion effect and the thermo-optic effect, discussed in a previous paragraph. Starting from early proposals, several authors have published papers on pure phase modulators for infrared light beam. The compatibility with microelectronic technologies is another attractive characteristic common to most of the proposed or realized devices.

3.2.1 All-Silicon Modulators

The first proposals of phase modulator integrated in all-silicon waveguides are contemporaneous to the early amplitude modulators we discussed in the previous paragraph. *Hemenway* et al. [40] in 1989 have reported an all-silicon reflection-type optical amplitude modulator, which uses free-carrier dispersion effect to spatially modulate the phase of an optical beam. Subsequent phase-modulation to amplitude-modulation conversion occurs by mode selection in the coupled single-mode fiber. Inside the device (see Fig. 14) one half of the beam is phase delayed relative to the other half. The phase delay consists of two parts, a static delay of $\lambda/4$ and a separate, modulated delay due to free-carrier injection. This device allows reaching a modulation depth of about 35%, at a driving peak current is 26 mA. At low and intermediate drive levels, the frequency response is flat, showing a -3 dB bandwidth of 200 MHz.

Treyz et al. [41] proposed a more canonical interferometric configuration in 1991. Device operation is based on the plasma dispersion effect and is implemented in the Mach–Zehnder interferometer configuration. As shown

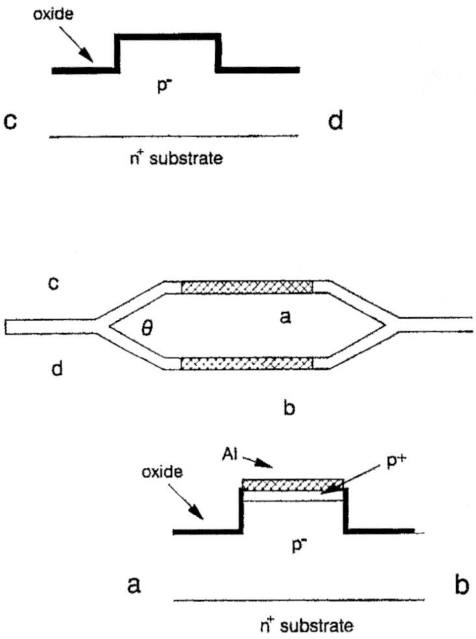

Fig. 15. Schematic diagram of the interferometer proposed by *Treyz* et al. [41]

in Fig. 15 the interferometer consists of passive all-silicon waveguide and p–i–n diode. For a current density of $1.6\,\mathrm{kA/cm^2}$, the injected carrier is about $6.5 \times 10^{17}\,\mathrm{cm^{-3}}$, and the achieved modulation depth is 4.9 dB. Response time of less than 50 ns has been obtained. Despite the simplicity of this modulator, its drawback is the poor quality of the passive host waveguide, whose propagation losses are greater than 20 dB/cm.

In the same year *Xiao* et al. [42] proposed a non-integrated modulator which consists essentially of a high-finesse Fabry–Perot resonant cavity formed by two silicon dioxide mirrors, and a p–i–n diode optical phase modulator placed inside the cavity. Although the novelty of the proposal, the performances are rather low; the modulation depth of 10% is achieved with a current density of $6.0\,\mathrm{kA/cm^2}$. The -3 dB bandwidth is about 40 MHz. In 1994 the same research group realized an optimized version of a similar structure [43], where the carrier effect was produced by optical generation and not by an electronic device integrated in the optical structure. The 100-fs-long pulsed pump beam, with a wavelength of 850 nm, was focused into an asymmetric Fabry–Perot cavity, which is sketched in Fig. 16. The resulting real refractive index variation $\Delta n = 5.0 \times 10^{-3}$, corresponding to a generated free carriers density is about $2.0 \times 10^{18}\,\mathrm{cm^{-3}}$, is converted in an amplitude modulation of 10 %, with a typical switching time of about 2 ns.

Another version of partially integrated Fabry–Perot optical modulator was proposed by *Cocorullo* et al. [44] in 1994. The device is a simple silicon chip, whose opposite transverse facets were polished in order to act as a cavity

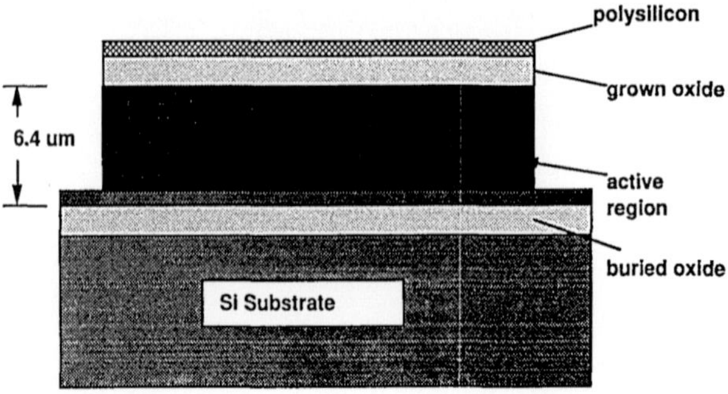

Fig. 16. Cross section of the asymmetric Si Fabry–Perot étalon proposed by *Liu* et al. [43]

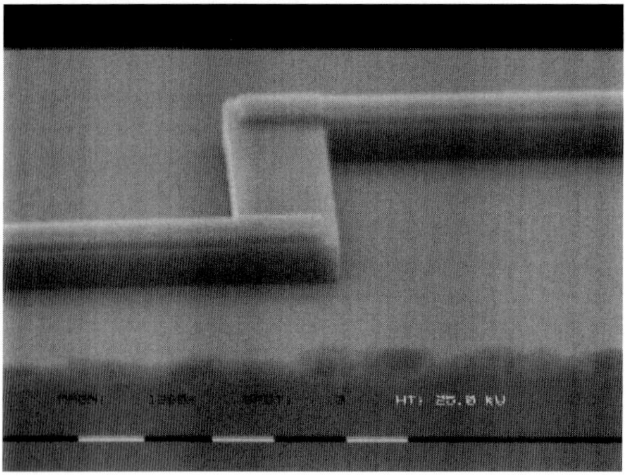

Fig. 17. SEM picture of the Si Fabry–Perot modulator realized by *Cocorullo* et al. [45]

mirrors. The structure can be considered a large slab waveguide, heated by Joule effect, thanks to the current flowing in an aluminum film deposited on the top. A thin n^+ layer under the metal heater avoids large optical field attenuation. In this way it is possible to induce a refractive index variation by thermo-optic effect. Maximum achieved modulation depth is 55% and typical switching time constants are greater of 200 μs. The above described work was preparatory for an integrated version of a similar device. In effect, the first proposal of a Fabry–Perot modulator integrated into an all-silicon optical waveguide was reported by *Cocorullo* et al. [45] in 1997. In this device, whose SEM picture is shown in Fig. 17, an unoptimized all-silicon multimode waveguide hosts a thermo-optic Fabry–Perot modulator.

The cavity mirrors are realized by wet etching and the heating element is made by a doped polysilicon film. The whole structure is fully compatible

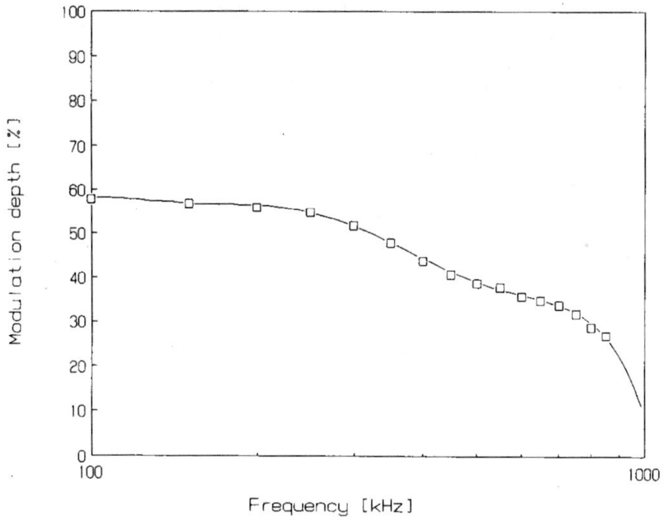

Fig. 18. Frequency response of the Si Fabry–Perot integrated modulator [45]

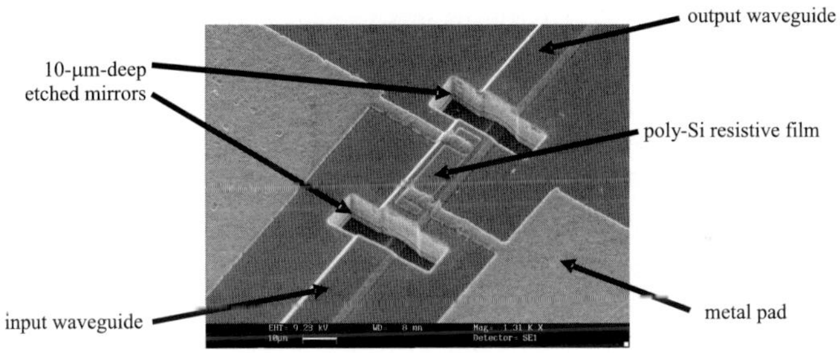

Fig. 19. SEM picture of the Si Fabry–Perot integrated modulator realized by *Iodice* et al. [46]

with microelectronic technologies. The performances are, nowadays the best reported in literature for thermo-optic silicon modulator. The modulation depth is about 60%, for a driving energy of 0.2 μJ. The −3 dB bandwidth is 700 kHz (see Fig. 18) and operation up to 1.2 MHz, with a reduced modulation depth, was observed.

A further optimization of this integrated modulator was proposed by *Iodice* et al. [46], where the device is integrated in a low-loss single-mode all-silicon waveguide (propagation losses $\simeq$ 1.0 dB/cm) and the cavity is defined by means of anisotropic reactive ion etching (see Fig. 19).

In this paper it is also discussed an optimization of the driving signal which permits to achieve a transmission rate of 2.2 Mb/s, at maximum modulation depth of 70% and energy expenses per cycle of about 0.9 μJ.

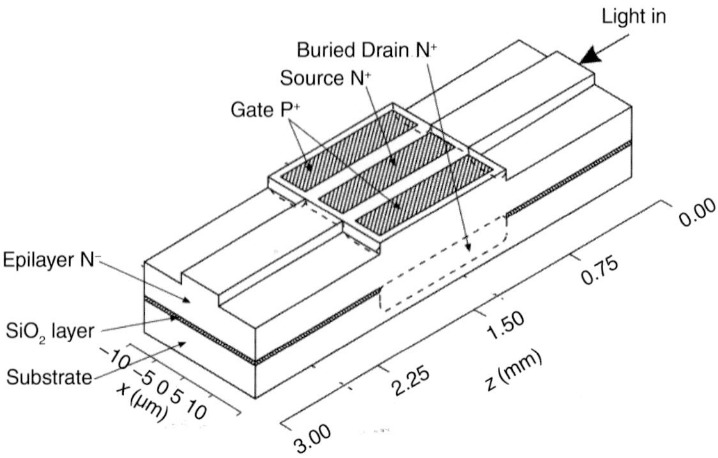

Fig. 20. BMFET optical modulator proposed by *Cutolo* et al. [37]

Beside the experimental papers above discussed, several theoretical proposals are reported in literature. Actually, in most cases these proposals are limited to numerical or analytical models with difficult or even impossible realization, but some papers suggest interesting operation principles or optical configurations. *Friedman* et al. [47] proposed and analyzed a singular integration between a new dual-injection field effect transistor in a standard all-silicon waveguide for light modulation.

For a driving source-drain voltage of 320 mV, the effective refractive index of the waveguide mode changes by $\Delta n = 1 \times 10^{-3}$ when the gate voltage is 12 V. The figure of merit $\Delta\phi/(V \times L)$ is 23°/(V · mm) and the chirp factor is $\Delta n/\Delta k = 16$. Switching characteristics of the gate-controlled modulator are very fast and it is governed by the sweep-out of the mobile charges.

The authors estimated a −3 dB bandwidth of about 3.2 GHz, limited by the RC time constant of the device. Another original proposal published by *Cutolo* et al. [37] describes a bipolar-mode field effect transistor integrated in a silicon waveguide acting as pure amplitude or pure phase modulator. In Fig. 20 is sketched the structure of such a device.

When driven at low injection level ($\Delta N = 2.0 \times 10^{17}\,\mathrm{cm}^{-3}$, that is with $\Delta n/\Delta k = 21.6$, the figure of merit for pure phase modulation behavior attains the record value of 215°/(V · mm). The operation power is also very low, about 43 mW. Accepting the price of rather high switching power, forcing the injected plasma by using the third control terminal (the so-called plasma-drift operation mode), it was calculated that the typical time constant can reach a value smaller than 3.5 ns. Realization in standard microelectronic technologies and preliminary characterization are reported in [2].

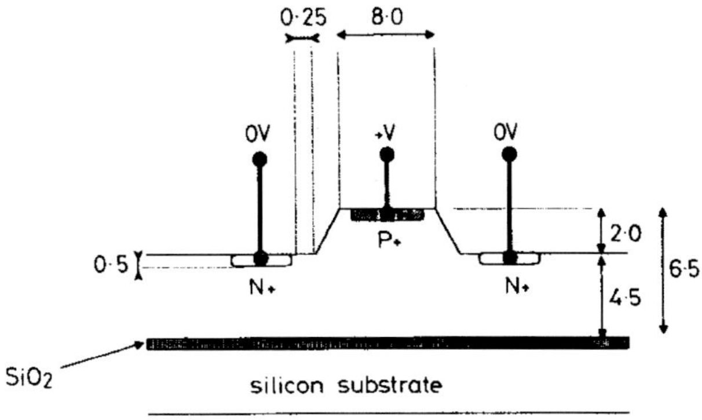

Fig. 21. Transverse cross section of the modulator realized by *Tang* et al. [49]

3.2.2 Silicon-On-Insulator Modulators

The need for a low-loss host waveguide for the integration of active devices pushed during last ten years several research groups in investigating the so-called SOI (Silicon-On-Insulator) structures. With this name is generally indicated an optical waveguide, slab or ridge, where the lower cladding is a silicon dioxide buried layer and the guiding film is in crystalline silicon. Several fabrication techniques can be exploited for realizing such structures: oxygen ions implantation (SIMOX) and bonded and etch back (BESOI) are practically useful for commercial production. The first proposals and realizations of optoelectronic modulator begun to appear in literature when commercial SOI wafers with good characteristics became available at low cost. In 1991 *Treyz* et al. [48] realized the first Mach–Zehnder modulator integrated in a SOI waveguide. The modulation mechanism was based in the thermally induced refractive index variation in the crystalline silicon guiding film. A driving power of 30 mW, dissipated in the NiCr heater realized on the top of one interferometer arm, allows reaching a modulation depth of 40% with time constant of about 50 µs. The first realization of an electrically controlled phase modulator, which integrated a lateral p–i–n diode into a SOI waveguide, was published by *Tang* et al. [49]. In Fig. 21 is reported the transverse cross section of the realized device.

The host waveguide exhibited low propagation losses (< 1.0 dB/cm) and the modulator, characterized in an external Mach–Zehnder interferometer, had a high figure of merit of about $200°/(\mathrm{V} \cdot \mathrm{mm})$. The electrical power needed for a complete detuning of the interferometer was about 12 mW, the injection level $\Delta N = 2.5 \times 10^{17}\,\mathrm{cm}^{-3}$, and the refractive index variation $\Delta n = 1.5 \times 10^{-3}$. This device was not characterized dynamically. An integrated version of the Mach–Zehnder interferometer in a SOI waveguide, with a lateral p–i–n diode as driving element, was published by *Zhao* et al. [50] in 1995. The structure layout and its cross section are reported in Fig. 22.

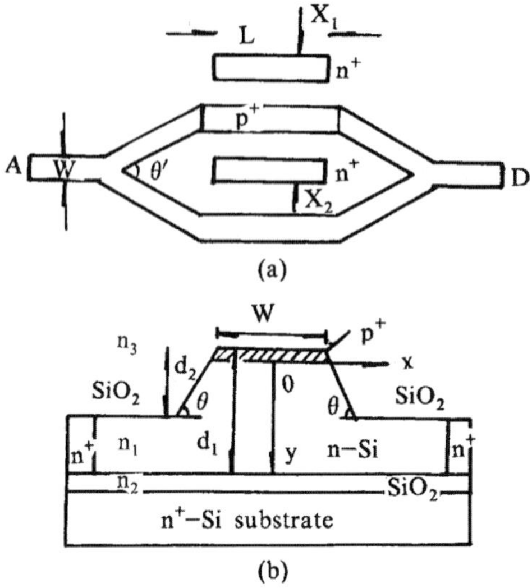

Fig. 22. Schematic structure and transverse cross section of the modulator realized by *Zhao* et al. [50]

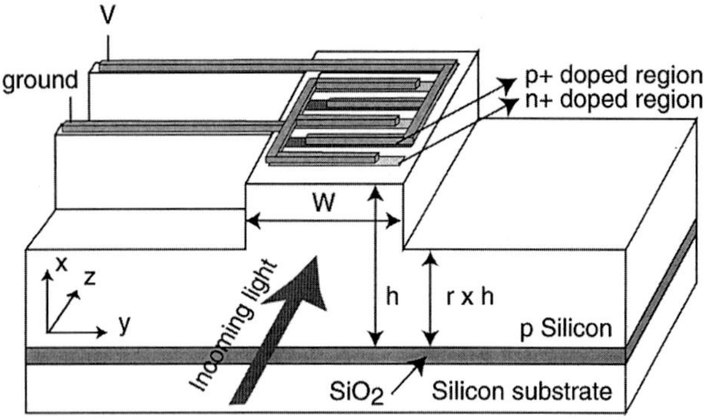

Fig. 23. Schematic view of the phase modulator realized by *Dainesi* et al. [51]

The insertion loss and modulation depth are measured to be 4.81 dB and 98%, respectively, at the wavelength of 1.3 μm when the forward bias voltage applied to the p–i–n diode is 0.95 V. The switching time constant of the device was measured to be less than 0.2 μs. To improve response speed of the modulator, electron irradiation is performed at electron energy of 14 MeV. Another CMOS-compatible fully-integrated Mach–Zehnder interferometer, realized in a SOI waveguide, and utilizing the plasma dispersion effect, was proposed by *Dainesi* et al. [51] in 2000. In this modulator a series of 66 p–i–n diodes is realized longitudinally along the top of the waveguide (see Fig. 23) at both interferometer arms.

Once again the injection level was $\Delta N = 2.5 \times 10^{17}\,\mathrm{cm}^{-3}$, which is needed to obtain a π phase shift over 1 mm long active region. A total driving current of 136 mA produces an ideal 100% modulation depth. As an injection

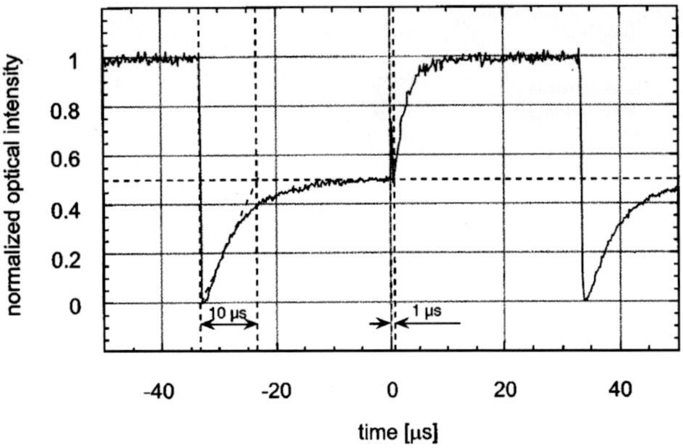

Fig. 24. Experimental optical output showing thermal and electrical phenomena [51]

of charge is always associated to power dissipation, a phase modulator based only on plasma dispersion effect is not feasible and thermo-optic effect will always be present. Because free-carrier and thermo-optic effects are opposite in sign, the resulting net modulation depth is about 50%, but the two phenomena are characterized by very different time constant and it is very easy in the time-domain output pattern to distinguish between the two effects using their respective rise times (see Fig. 24). Authors measured a bandwidth of 100 kHz for thermally induced modulation and of 10 MHz for plasma dispersion induced operation.

An abounding literature is available on theoretical proposals of SOI waveguide integrated phase modulator. Amongst the early papers, it is worthwhile to cite the proposal made by *Giguere* et al. [52]. In this paper it is depicted the integration of a MOSFET in a SOI waveguide. The plasma dispersion effect is exploited for producing an index variation of about 6.0×10^{-4}. At driving current density of $400\,\mathrm{A/cm^2}$ to $800\,\mathrm{A/cm^2}$, the resulting figure-of-merit is $40°/(\mathrm{V} \cdot \mathrm{mm})$. *Huang* et al. [53] in 1993 proposed a novel electronic device configuration, based on the impact ionization mechanism for carrier generation. The simulated turn-on and turn-off time was less than 1 ns, but the driving current density is greater than $100\,\mathrm{kA/cm^2}$. A more conventional approach, that is a lateral p–i–n diode integrated in a rib waveguide, was followed by *Tang* et al. [54] in their paper published in 1994. The publication is particularly interesting for the accurate geometrical definition and optimization of the optoelectronic structure. The resulting figure-of-merit attains the value of $200°/(\mathrm{V} \cdot \mathrm{mm})$ for a driving current of only 10 mA. The theoretical estimated switching time is about 100 ns. A more complicated and more performing approach was proposed by *Wang* et al. [55], in a paper presenting an electrically induced Bragg reflector in a SOI waveguide. The device is sketched in Fig. 25.

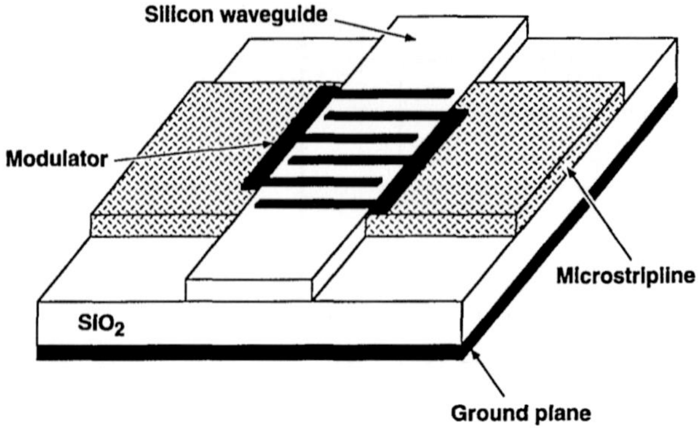

Fig. 25. Schematic view of the Bragg induced modulator proposed by *Wang* et al. [55]

The interdigitated metal fingers are alternately biased and form forward and reverse biased Schottky junctions with the heavily doped silicon film. The resulting longitudinal periodic refractive index structure, whose variation Δn is about 5.0×10^{-3}, acts as a Bragg mirror. For a 500-µm-long active region the predicted achievable modulation depth is 70%. The speed of such modulator seems to be limited by the RC time constant of the structure itself, indicating a -3 dB bandwidth in the range of 50 GHz. The main drawbacks are related to the high propagating losses of the very thin and high-doped core waveguide and to the need of electron beam lithography for defining the Bragg control structure. Further improvement of the optical structure are reported by *Liu* et al. [56], where authors discuss about a Fabry–Perot interferometer integrated in a SOI waveguide, whose mirrors are realized by means of a couple of Bragg reflectors. In this paper the realization and characterization of Bragg mirrors is presented, but the overall modulator performances are only theoretically discussed. They affirm that with an injection level $\Delta N = 10^{18}\,\mathrm{cm}^{-3}$ it is possible to completely detune the F–P cavity, reaching a 100% of modulation depth. A different approach in Bragg mirrors exploitation was suggested by *Cutolo* et al. [57]. The Bragg reflector was placed in the middle of a lateral p–i–n diode (see Fig. 26) and the amplitude modulation was achieved changing, by injection of free carriers, the Bragg resonance wavelength. Numerical simulations show a modulation depth of 50% achieved in about 12 ns with a power dissipation of 4.0 mW and an insertion loss of 1.0 dB.

Another proposal of Fabry–Perot interferometer based modulator was presented by *Vonsovici* et al. [58], where waveguide discontinuities, realized by RIE etching, play the role of the two mirrors of the interferometer. The electronic device driving the modulator is a standard lateral p–i–n diode and the reported maximum modulation depth, for a $\Delta n = 5.0 \times 10^{-4}$, is 60% at current density of 500 $\mathrm{A/cm^2}$, and with -3 dB bandwidth of about 100 MHz.

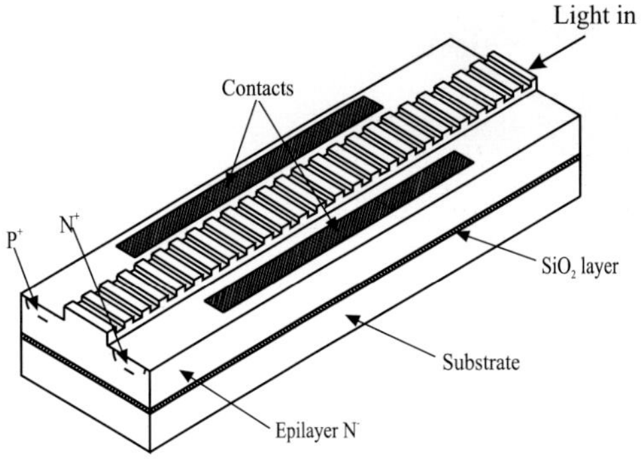

Fig. 26. Schematic view of the Bragg modulator proposed by *Cutolo* et al. [57]

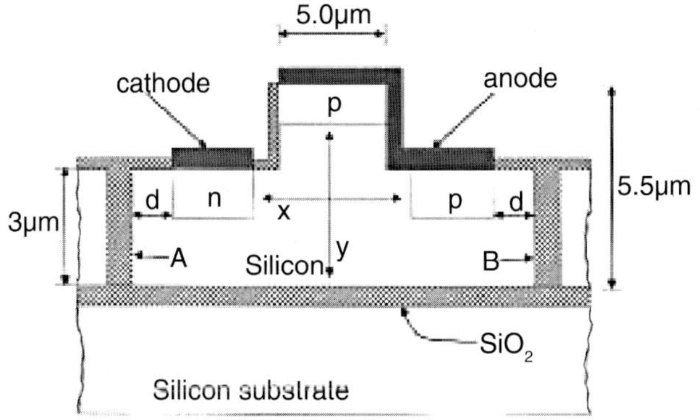

Fig. 27. Cross section of the active region of the modulator proposed by *Hewitt* et al. [61]

In the last years an optimization of the Bragg modulator presented in [57] was proposed by *Coppola* et al. [59], where are expected 4.0 ns of rise time and 0.3 mW Of Dissipated Power. A final optimization of the Bragg modulator presented in [57] has been recently presented (see [60]) where, after careful design of the host waveguide, together with the shaping of the driving signal which allows better injection/depletion of carriers, over 1.4 GHz maxiumum operation frequency has been reached.

Another solution for the enhancement of the modulation efficiency was discussed by *Hewitt* et al. [61], whose paper describes a simple p–i–n diode integrated in a SOI waveguide. The novelty of such a proposal is the presence of vertical isolation trenches laterally to the diode. The modeling indicates that increased DC and transient performances results from these trenches defined adjacent to the outer edge of the contact region (see Fig. 27).

With driving current $I = 4\,\mathrm{mA}$ it is possible to produce a index variation of $\Delta n = 5.0 \times 10^{-3}$, which for an active region length of 500 µm allows to generate a π phase shift. Switching characteristics are limited by the rise

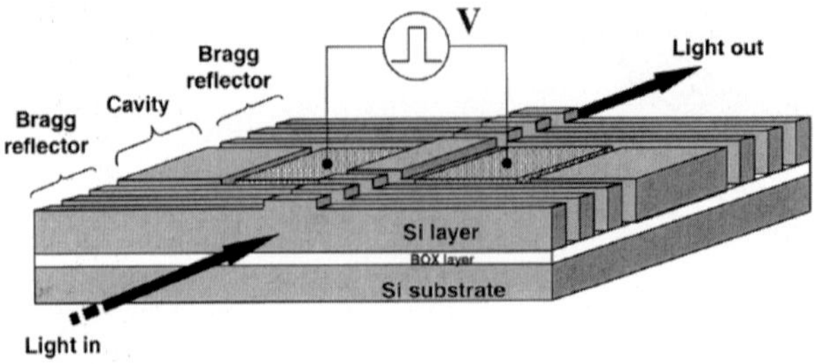

Fig. 28. Sketch of the modulator proposed by *Barrios* et al. [62]

time, which is about 30 ns long. Finally, it is interesting to mention a recent paper by *Barrios* et al. [62], where the authors perform an accurate theoretical analysis of a Fabry–Perot modulator with deep Si/SiO_2 Bragg reflectors (see Fig. 28).

Carriers are also laterally confined in the cavity region by employing deep-etched trenches. The refractive index of the cavity is varied by using the free-carrier dispersion effect produced by a p–i–n diode. The device has been designed and analyzed using electrical and optical simulations. Their calculations predict, for a 20-μm-long device, a modulation depth of around 80% and a transmittance of 86% at an operating wavelength of 1.55 μm by using an electrical power under dc conditions on the order of 25 μW. The switching speed of this device is calculated to be about 16 ns.

References

1. R. A. Soref: Silicon-based optoelectronics, Proc. IEEE **81**, 1687–1706 (1993)
2. G. Breglio, A. Cutolo, A. Irace, P. Spirito, L. Zeni, M. Iodice, P. Sarro: Two silicon optical modulators realizable with a fully compatible bipolar process, IEEE J. Sel. Top. Quantum Electron. **4**, 1003–1010 (1998)
3. R. A. Soref, J. P. Lorenzo: Single-crystal silicon-A new material for 1.3 and 1.6 μm integrated-optical components, Electron. Lett. **21**, 953–955 1985
4. G. Cocorullo, F. D. Corte, I. Rendina, A. Cutolo: New possibilities for efficient silicon integrated electro-optical modulators, Opt. Commun. **86**, 228–235 (1991)
5. R. A. Soref, B. R. Bennett: Electrooptical effect in Silicon, IEEE J. Quantum Electron. **QE-23**, 123–129 1987
6. P. H. Wendland, M. Chester: Electric field effects on indirectoptical transitions in silicon, Phys. Rev. **140**, 1384 (1965)
7. T. S. Moss, G. J. Burrell, B. Ellis: *Semiconductor Opto-Electronics*, (Butterworths, London 1975)
8. P. E. Schmid: Optical absorption in heavily doped silicon, Phys. Rev. B **23**, 5531 (1981)
9. A. A. Volfson, V. K. Subashiev: Fundamental absorption edgeof silicon heavily doped with donor or acceptor impurities, Sov. Phys. Semicond. **1**, 327–332 (1967)

10. W. Spitzer, H.Y. Fan: Infrared absorption in n-type silicon, Phys. Rev. **108**, 268 (1957)
11. W.C. Dash, R. Newman: Intrinsic optical absorption in single-crystal germanium and silicon, Phys. Rev. **99**, 1151–1155 (1955)
12. M. Balkanski, J.M. Besson: Optical properties of degenerate silicon, Tech. Note 2, Contract AF61(052)-789, Defense Tech. In-form. Cen., AD 619581 (1965)
13. C.M. Randall, R.D. Rawcliffe: Refractive indices of germanium, silicon, and fused quartz in the far infrared, Appl. Optics, **6**, 1889 (1967)
14. P.A. Schumann Jr., W.A. Keenan, A.H. Tong, H.H. Gegenwarth, C.P. Schneider: Optical constants of silicon in the wavelength range 2.5 to 40 μm, IBM Tech. Rep. TR-22.1008, East Fishkill,NY , May 20 (1970)
15. S. Walles: Transmission of silicon between 40 and 100 μm, ArkivFysik, **25**, 33 (1964)
16. S. Walles, S. Boija: Transmittance of doped silicon between 40and 100 μm, J. Opt. Soc. Amer. **54**, 133–134 (1964)
17. P.A. Schumann, R.P. Phillips: Comparison of classical approximation to free carrier absorption in semiconductors, Solid State Electron. **10**, 943 (1967)
18. P. Lautenschlager, M. Garriga, L.Viña, M. Cardona: Temperature dependence of the dielectric function and interband critical points in silicon, Phys. Rev. B, **36**, 4821 (1987)
19. M.R. Siregar, M. von Allmen, W. Luthy: Temperature Dependence of Optical Absorption at 1.06 μm in Amorphous and Crystalline Silicon, Rapport de la session d'automne de la Société Suisse de Physique **52**, 45 (1979)
20. H.H. Li: Refractive Index of Silicon and Germanium and Its Wavelength and Temperature Derivatives, J. Phys. Chem. Ref. Data **9**, 561 (1980)
21. G.E. Jellison, Jr., H.H. Burke: The temperature dependence of the refractive index of silicon at elevated temperatures at several laser wavelengths, J. Appl. Phys. **60**, 841 (1986)
22. M. Bertolotti, V. Bogdanov, A. Ferrari, A. Jascow, N. Nazarova, A. Pikhtin, L. Schirone: Temperature dependence of the refractive index in semiconductors, J. Opt. Soc. Am. B, **7**, 918 (1990)
23. G. Ghosh: Temperature dispersion of refractive indices in crystalline and amorphous silicon, Appl. Phys. Lett. **66**, 3570 (1995)
24. N.A. Nazarova, G.I. Romanova, A.D. Yaskov: Refractometric characteristics of silicon, Sov. J. Opt. Technol. **55**, 220 (1988)
25. G. Cocorullo, I. Rendina: Thermo-optical modulation at 1.5 μm in silicon etalon, Electron. Lett. **28**, 83 (1992)
26. G. Cocorullo, F.G. Della Corte, I. Rendina: Temperature dependence of the thermo-optic coefficient in crystalline silicon between room temperature and 550 K at the wavelength of 1523 nm, Appl. Phys. Lett. **74**, 3338 (1999)
27. F.G. Della Corte, M. Esposito Montefusco, L. Moretti, I. Rendina, G. Cocorullo: Temperature Dependence Analysis of the Thermo-Optic Effect in Silicon by Single and Double Oscillator Models, J. Appl. Phys. **88**, 7115 (2000)
28. A.N. Pikhtin, A.D. Yaskov: Dispersion of refractive index of semiconductor with diamond and zinc-blende structures, Sov. Phys. Semicond. **12**, 622 (1978)
29. I.M. Grodnenskiy, I.N. Dyuzhikov, I.O. Ogloblin, Y.F. Sokolov, K.V. Starostin: The use of silicon p–i–n structures for controlling radiation with a wavelength of 1.15 μm, Radio Eng. Electron. Phys. **26**, 147 (1981)

30. D. F. Clark, D. Uttamchandani, I. Andonovic, B. Culshaw: Optical waveguides in silicon for optical interconnects, Optical Interconnects, IEE Colloquium (1988), pp. 12/1–12/61
31. S. P. Pogossian, L. Vescan, A. Vonsovici: The single-mode condition for semiconductor rib waveguides with large cross section, J. Lightwave Technol. **16**, 1851–1853 (1998)
32. G. Cocorullo, F. G. Della Corte, M. Iodice, T. Polichetti, I. Rendina, P. M. Sarro: Low Loss All-Silicon Single Mode Optical Waveguide with Small Cross-Section, Fiber and Integrated Optics, **20**, 207–219 (2001)
33. R. A. Soref, J. P. Lorenzo: All-silicon active and passive guided-wave components for $\lambda = 1.3$ and 1.6 μm, IEEE J. Quantum Electron. **QE-22**, 873–879 (1986)
34. J. P. Lorenzo, R. A. Soref: 1.3 μm electro-optic silicon switch, Appl. Phys. Lett. **51**, 6 (1987)
35. G. V. Treyz, P. G. May, J. M. Halbout: Silicon optical modulators at 1.3 μm based on free-carrier absorption, Electron. Dev. Lett. **12**, 276–278 (1991)
36. O. Leistiko, H. Bak: Planar silicon optical waveguide light modulators, 24th European Solid State Device Research Conference, Edinburgh, Scotland (1994), pp. 645–648
37. A. Cutolo, M. Iodice, P. Spirito, L. Zeni: Silicon Electro-Optic Modulator Based on a Three Terminal Device Integrated in a Low-Loss Single-Mode SOI waveguide, J. Lightwave Technol. **15**, 505 (1997)
38. A. Sciuto, S. Libertino, A. Alessandria, S. Coffa, G. Coppola: Design, fabrication, and testing of an integrated Si-based light modulator, J. Lightwave Technol. **21**, 228–235 (2003)
39. C. Z. Zhao, E. K. Liu, G. Z. Li, L. Gu: Silicon-on-insulator optical intensity modulator based on waveguide-vanishing effect, Electron. Lett. **32**, 1667–1668 (1996)
40. B. R. Hemenway, O. Solgaard, D. M. Bloom: All-silicon integrated optical modulator for 1.3 μm fiber-optic interconnects, Appl. Phys. Lett. **55**, 349 (1989)
41. G. V. Treyz, P. G. May, J. M. Halbout: Silicon Mach–Zehnder waveguide interferometers based on the plasma dispersion effect, Appl. Phys. Lett. **59**, 771 (1991)
42. X. Xiao, J. C. Sturm, K. K. Goel, P. V. Schwartz: Fabry–Perot optical intensity modulator at 1.3 μm in silicon, Photon. Technol. Lett. **3**, 230 (1991)
43. Y. M. Liu, X. Xiao, P. R. Prucnal, J. C. Sturm: All-optical switching in an asymmetric Fabry-Perot étalon based on the free-carrier plasma effect, Appl. Opt. **33**, 3871 (1994)
44. G. Cocorullo, M. Iodice, I. Rendina: All-silicon Fabry-Perot modulator based on the thermo-optic effect, Opt. Lett. **19**, 420 (1994)
45. G. Cocorullo, M. Iodice, I. Rendina, P. M. Sarro: Silicon thermooptical micromodulator with 700-kHz −3-dB bandwidth, Photon. Technol. Lett. **7**, 363 (1995)
46. M. Iodice, F. G. Della Corte, I. Rendina, P. M. Sarro, M. Bellucci: Transient analysis of a high-speed thermo-optic modulator integrated in an all-silicon waveguide, Opt. Eng. **42**, 169 (2003)
47. L. Friedman, R. A. Soref, J. P. Lorenzo: Silicon double-injection electro-optic modulator with junction gate control, J. Appl. Phys. **63**, 1831 (1988)
48. G. V. Treyz: Silicon Mach–Zehnder waveguide interferometers operating at 1.3 μm, Electron. Lett. **27**, 118 (1991)

49. C. K. Tang, G. T. Reed: Highly efficient optical phase modulator in SOI waveguides, Electron. Lett. **31**, 451 (1995)
50. C. Z. Zhao, G. Z. Li, E. K. Liu, Y. Gao, X. D. Liu: Silicon on insulator Mach–Zehnder waveguide interferometers operating at 1.3 μm, Appl. Phys. Lett. **67**, 2448 (1995)
51. P. Dainesi, A.Küng, M. Chabloz, A. Lagos, Ph. Flückiger, A. Ionescu, P. Fazan, M. Declerq, Ph. Renaud, Ph. Robert: CMOS compatible fully integrated Mach–Zehnder interferometer in SOI technology, Photon. Technol. Lett. **12**, 660 (2000)
52. S. R. Giguere, L. Friedman, R. A. Soref, J. P. Lorenzo: Simulation studies of silicon electro-optic waveguide devices, J. Appl. Phys. **68**, 4964 (1990)
53. H. C. Huang, T. C. Lo: Simulation and analysis of silicon electro-optic modulators utilizing the carrier-dispersion effect and impact-ionization mechanism, J. Appl. Phys. **74**, 1521 (1993)
54. C. K. Tang, G. T. Reed, A. J. Walton, A. G. Rickman: Low-loss, single-mode, optical phase modulator in SIMOX material, J. Lightwave Technol. **12**, 1394–1400 (1994)
55. C. C. Wang, M. Currie, S. Alexandrou, T. Y. Hsiang: Ultrafast, all-silicon light modulator, Optics Lett. **19**, 1453 (1994)
56. M. Y. Liu, S. Y. Chou: High-modulation-depth and short-cavity-length silicon Fabry–Perot modulator with two grating Bragg reflectors, Appl. Phys. Lett. **68**, 170 (1996)
57. A. Cutolo, M. Iodice, A. Irace, P. Spirito, L. Zeni: An electrically controlled Bragg reflector integrated in a rib silicon on insulator waveguide, Appl. Phys. Lett. **71**, 199 (1997)
58. A. Vonsovici, R. Orobtchouk, A. Koster: Numerical simulation of a silicon-on-insulator waveguide Fabry–Perot interferometer for intensity light modulators at 1.3 μm, J. Lightwave Technol. **15**, 2124 (1997)
59. G. Coppola, A. Irace, M. Iodice, A. Cutolo: Simulation and analysis of a high-efficiency silicon optoelectronic modulator based on a Bragg mirror, Opt. Eng. **40**, 1076 (2001)
60. A. Irace, G. Breglio, A. Cutolo: All-silicon optoelectronic modulator with 1 GHz switching capability, Electron. Lett. **39**, 232 (2003)
61. P. D. Hewitt, G. T. Reed: Improved modulation performance of a silicon p–i–n device by trench isolation, J. Lightwave Technol. **19**, 387 (2001)
62. C. A. Barrios, V. Rosa de Almeida, M. Lipson: Low-power-consumption short-length and high-modulation-depth silicon electrooptic modulator, J. Lightwave Technol. **21**, 1089–1098 (2003)

Index

Topics in Applied Physics

78 **High-Power Diode Lasers**
By R. Diehl (Ed.) 2000, 260 Figs. XIV, 416 pages

79 **Frequency Measurement and Control**
Advanced Techniques and Future Trends
By A. N. Luiten (Ed.) 2001, 169 Figs. XIV, 394 pages

80 **Carbon Nanotubes**
Synthesis, Structure, Properties, and Applications
By M. S. Dresselhaus, G. Dresselhaus, Ph. Avouris (Eds.) 2001, 235 Figs. XVI, 448 pages

81 **Near-Field Optics and Surface Plasmon Polaritons**
By S. Kawata (Ed.) 2001, 136 Figs. X, 210 pages

82 **Optical Properties of Nanostructured Random Media**
By Vladimir M. Shalaev (Ed.) 2002, 185 Figs. XIV, 450 pages

83 **Spin Dynamics in Confined Magnetic Structures I**
By B. Hillebrands and K. Ounadjela (Eds.) 2002, 166 Figs. XVI, 336 pages

84 **Imaing of Complex Media with Acoustic and Seismic Waves**
By M. Fink, W. A. Kuperman, J.-P. Montagner, A. Tourin (Eds.) 2002, 162 Figs. XII, 336 pages

85 **Solid–Liquid Interfaces**
Macroscopic Phenomena – Microscopic Understanding
By K. Wandelt and S. Thurgate (Eds.) 2003, 228 Figs. XVIII, 444 pages

86 **Infrared Holography for Optical Communications**
Techniques, Materials, and Devices
By P. Boffi, D. Piccinin, M. C. Ubaldi (Eds.) 2003, 90 Figs. XII, 182 pages

87 **Spin Dynamics in Confined Magnetic Structures II**
By B. Hillebrands and K. Ounadjela (Eds.) 2003, 179 Figs. XVI, 321 pages

88 **Optical Nanotechnologies**
The Manipulation of Surface and Local Plasmons
By J. Tominaga and D. P. Tsai (Eds.) 2003, 168 Figs. XII, 212 pages

89 **Solid-State Mid-Infrared Laser Sources**
By I. T. Sorokina and K. L. Vodopyanov (Eds.) 2003, 263 Figs. XVI, 557 pages

90 **Single Quantum Dots**
Fundamentals, Applications, and New Concepts
By P. Michler (Ed.) 2003, 181 Figs. XII, 352 pages

91 **Vortex Electronis and SQUIDs**
By T. Kobayashi, H. Hayakawa, M. Tonouchi (Eds.) 2003, 259 Figs. XII, 302 pages

92 **Ultrafast Dynamical Processes in Semiconductors**
By K.-T. Tsen (Ed.) 2004, 190 Figs. XI, 400 pages

93 **Ferroelectric Random Access Memories**
Fundamentals and Applications
By H. Ishiwara, M. Okuyama, Y. Arimoto (Eds.) 2004, Approx. 200 Figs. XIV, 278 pages

94 **Silicon Photonics**
By L. Pavesi, D.J. Lockwsood (Eds.) 2004, 100 Figs. XIV, 397 pages

The New Springer Global Website
Be the first to know
Benefit from new practice-driven features.
Search all books and journals – now faster and easier than ever before.
Enjoy big savings through online sales.
springeronline.com – the innovative website with you in focus.
springeronline.com
The interactive website for all Springer books and journals
1842
Springer
010048x

Printing: Saladruck, Berlin
Binding: Stein+Lehmann, Berlin